Interpretation of Classical Electromagnetism

Fundamental Theories of Physics

An International Book Series on The Fundamental Theories of Physics: Their Clarification, Development and Application

Volume 78

Interpretation of Classical Electromagnetism

by

W. Geraint V. Rosser

formerly of the University of Exeter,
Exeter, U.K.

KLUWER ACADEMIC PUBLISHERS

DORDRECHT / BOSTON / LONDON

Library of Congress Cataloging-in-Publication Data

Rosser, W. G. V. (William Geraint Vaughan)
Interpretation of classical electromagnetism / by W. Geraint V. Rosser.
p. cm. -- (Fundamental theories of physics ; v. 78)
Includes index.
ISBN 0-7923-4187-2 (acid free paper)
1. Electromagnetism. I. Title. II. Series.
QC760.R588 1996
537--dc20 96-9778

ISBN 0-7923-4187-2

Published by Kluwer Academic Publishers,
P.O. Box 17, 3300 AA Dordrecht, The Netherlands.

Sold and distributed in the U.S.A. and Canada
by Kluwer Academic Publishers,
101 Philip Drive, Norwell, MA 02061, U.S.A.

In all other countries, sold and distributed
by Kluwer Academic Publishers Group,
P.O. Box 322, 3300 AH Dordrecht, The Netherlands.

Printed on acid-free paper

Printed in the Netherlands

There's always a hole in theories somewhere, if you look close enough.

Mark Twain
(Tom Sawyer Abroad)

Table of Contents

Preface

The aim of this book is to interpret all the laws of classical electromagnetism in a modern coherent way. In a typical undergraduate course using vector analysis, the students finally end up with Maxwell's equations, when they are often exhausted after a very long course, in which full discussions are properly given of the full range of applications of individual laws, each of which is important in its own right. As a result, many students do not appreciate how limited is the experimental evidence on the basis of which Maxwell's equations are normally developed and they do not always appreciate the underlying unity of classical electromagnetism, before they go on to graduate courses in which Maxwell's equations are taken as axiomatic. This book is designed to be used between such an undergraduate course and graduate courses. It is written by an experimental physicist and is intended to be used by physicists, electrical engineers and applied mathematicians. The main aim is to interpret Maxwell's equations and the laws of classical electromagnetism starting from the expressions for the electric and magnetic fields due to an accelerating classical point charge. It is also hoped that the book will be useful for long standing graduates, who missed out on such an interpretation. To help students in their formative years and readers in their rusty maturity, the reader is taken slowly through what at times are subtle arguments with summaries on the way and generally at the end. In many cases, topics are approached in different complementary ways. It is hoped that this overall approach will make the book suitable for individual study to complement traditional courses.

In Chapter 1, an account is given of a typical development of Maxwell's equations, that is devoid of applications so that the reader can see clearly what assumptions are made. Particular emphasis is given to the interpretations of what makes a conduction current flow and the role of the vacuum displacement current. Then in Chapter 2, the equations for the potentials ϕ and $\mathbf{A}$ are developed from Maxwell's equations, leading up to the retarded potentials, which are then applied to determine the electric and magnetic fields due to an oscillating electric dipole. As an introduction to the approach we shall develop in Chapter 4, it is illustrated in Chapter 2 how Maxwell's equations apply to the fields of the oscillating electric dipole. The Liénard-Wiechert potentials are derived in Chapter 3 and then applied to derive the standard expressions for the electric and magnetic fields due to an accelerating classical point charge, which will then be used as the bases of our interpretation of classical electromagnetism in Chapters 4–7. In Chapter 4 a modern interpretation of Maxwell's equations is developed using the expressions for the electric and magnetic fields due to an accelerating classical point charge. In particular, we shall strip the vacuum displacement current of the last vestiges of the nineteenth century aether theories. In Chapter 5, a new approach is developed for the interpretation of the origin of the induction and radiation electric fields due to varying charge and current distributions. A similar approach is used in Chapter 6 to interpret the origin of the magnetic field due to a varying current distribution. Then in Chapter 7 we shall interpret the behaviour of AC circuits in terms of the electric and magnetic fields due to the conduction electrons responsible for the conduction current flow. Until Chapter 8 we shall generally avoid the use of methods based on energy, since the underlying physical principles are not always apparent in those methods. However, to complete the picture, we shall give a review of energy methods in Chapter 8, presented in a way that is consistent with the approach developed in earlier chapters. This will lead up to a discussion of the conservation laws for a system of spatially separated moving charges and to a critique of the Poynting vector hypothesis. Since the theory of the electric and magnetic fields due to dielectrics and magnetic materials is treated comprehensively in many excellent text books, we shall only give a brief review in Chapter 9 of the extension of Maxwell's equations to field points inside dielectrics and magnetic materials, which will be presented and interpreted in a way that is consistent with the approach developed in earlier chapters. We shall avoid the use of special relativity until Chapter 10, so that readers do not get the impression that the new ideas presented in this book are exotic relativistic effects. Since special relativity and its applications are covered in detail in many text books, we shall concentrate in Chapter 10 on topics that illustrate the essential unity of classical electromagnetism and special relativity.

The author would like to thank Mrs Val Barnes and Mrs Eileen Satterly for typing the manuscript.

CHAPTER 1

A typical conventional development of Maxwell's equation

1.1. Introduction

It will be assumed from the outset that the reader has already done an introductory course on classical electromagnetism leading up to Maxwell's equations, and that the reader is fully familiar with vector analysis. A summary of the relevant formulae of vector analysis is given in Appendix A1. In this chapter, a brief review is given of the way Maxwell's equations can be developed in an introductory course, so as to illustrate how limited is the experimental evidence used to develop Maxwell's equations in introductory courses. Maxwell's equations will then be discussed in greater detail in later chapters and interpreted in a way consistent with the retarded potentials (Lorentz gauge). A reader, interested in the historical background to the development of classical electromagnetism, is referred to books such as Whittaker [1] Tricker [2] Schaffner [3] etc. The discussions in this book will be restricted to classical electromagnetism, and all quantum effects will be generally be neglected.

The electric field intensity **E** and the magnetic induction, or magnetic flux density **B** at a field point in empty space will be defined in terms of the Lorentz force law

$$\mathbf{F} = \frac{\mathrm{d}}{\mathrm{d}t}\left(\frac{m_0\mathbf{u}}{(1 - u^2/c^2)^{1/2}}\right) = q\mathbf{E} + q\mathbf{u} \times \mathbf{B} \tag{1.1}$$

acting on a test charge of magnitude q and rest mass m_0, that is moving with velocity **u** at the field point. For the sake of brevity, we shall generally use the abbreviation 'electric field **E**' instead of the phrase 'electric field of intensity **E**'. Similarly we shall use the abbreviation 'magnetic field **B**' instead of the phrase 'magnetic field of magnetic induction (or magnetic flux density) **B**'.

Maxwell's equations for the fields due to macroscopic charge and current distributions made up of moving and accelerating atomic charges, such as electrons and positive ions will be developed in three stages.

Stage 1. In this chapter, Maxwell's equations will be developed first for continuous charge and current distributions in otherwise empty space. This will lead up to equations (1.115), (1.116), (1.117) and (1.118).

Stage 2. Following Lorentz, charged atomic particles will be treated as continuous charge distributions of exceedingly small but finite dimensions. It will be assumed that equations (1.115), (1.116), (1.117) and (1.118) hold at field points inside such classical point charges. This will lead up to equations (1.137), (1.138), (1.139) and (1.140) for the microscopic fields **e** and **b**. These equations will be called the *Maxwell-Lorentz equations.*

Stage 3. Macroscopic charge and current distributions are made up of moving charged atomic particles such as electrons and positive ions. In atomic physics one is sometimes interested in the microscopic fields **e** and **b**, near and inside atoms and molecules. However, the scale of many electromagnetic phenomena is so large on the atomic scale that one only needs to know the macroscopic fields **E** and **B**, which are defined as the average values of the microscopic fields **e** and **b**, averaged over volumes large on the atomic scale, but kept small on the laboratory scale. The development of Maxwell's equations for the macroscopic fields **E** and **B**, for the case of charge and current distributions in empty space, in which case the relative permittivity ε_r and the relative permeability μ_r are both equal to unity everywhere, will be outlined in Section 1.11 of this chapter. A brief discussion of Maxwell's equations for the macroscopic fields **E** and **B** inside stationary dielectrics and stationary magnetic materials will be given in Chapter 9.

1.2. Electrostatics and the equation $\nabla \cdot \mathbf{E} = \rho/\epsilon_0$

1.2.1. *Coulomb's law*

The equation $\nabla \cdot \mathbf{E} = \rho/\varepsilon_0$ is generally developed from Coulomb's law of electrostatics, according to which, if there are two stationary point electric charges of magnitudes q and q_1 at positions $\mathbf{r}$ and $\mathbf{r}_1$ respectively, as shown in Figure 1.1, then the force $\mathbf{F}_1$ on the charge q at $\mathbf{r}$, due to the charge q_1 at $\mathbf{r}_1$, is given by:

$$\mathbf{F}_1 = \frac{qq_1(\mathbf{r} - \mathbf{r}_1)}{4\pi\varepsilon_0|\mathbf{r} - \mathbf{r}_1|^3} = \frac{qq_1\mathbf{R}_1}{4\pi\varepsilon_0 R_1^3} \tag{1.2}$$

where $\mathbf{R}_1 = (\mathbf{r} - \mathbf{r}_1)$ is a vector from the position of the charge q_1 at $\mathbf{r}_1$ to the position of the charge q at $\mathbf{r}$ and $R_1 = |\mathbf{r} - \mathbf{r}_1|$ is the magnitude of the vector $\mathbf{R}_1 = (\mathbf{r} - \mathbf{r}_1)$. If the charges q and q_1 are both positive, or both negative, the force given by equation (1.2) is a force of repulsion. If q_1 and q have opposite signs, the force given by equation (1.2) is a force of attraction. According to equation (1.2) the force $\mathbf{F}_1$ is proportional to $1/R_1^2$. Priestly was probably the first to develop the inverse square law. Priestly started from the observation

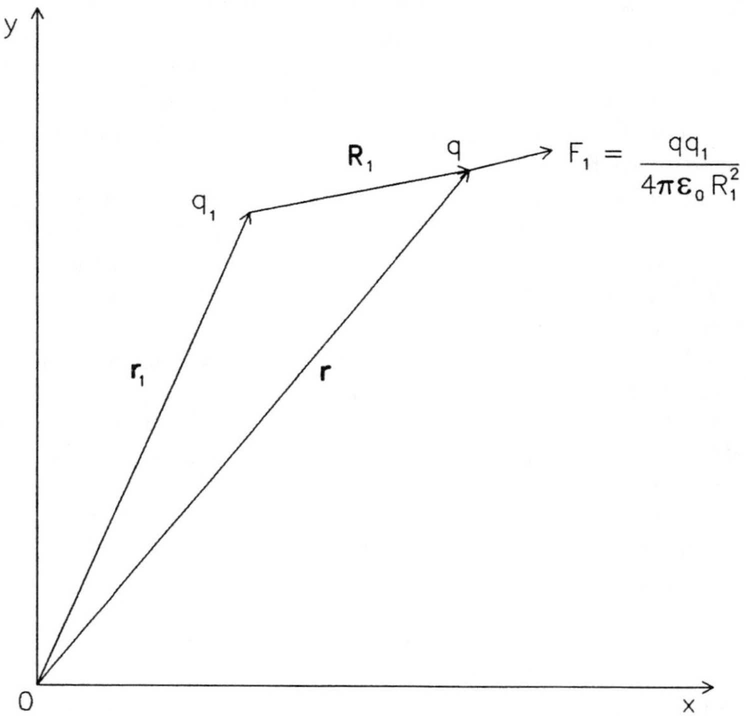

Figure 1.1. The electrostatic force on the stationary charge q due to the stationary charge q_1.

that there is no electrostatic force on a stationary charge inside a charged, hollow conductor. This approach was extended by Cavendish, Maxwell, Plimpton and Lawton and more recently in 1971 by Williams, Faller and Hill who showed that, if the force between two stationary point electric charges is proportional to $1/R_1^n$, where R_1 is the separation of the two charges, then experimentally $n = 2$ to one part in 10^{15}. The inverse square law was checked directly by Coulomb in 1785.

Equation (1.2) can be used to compare the magnitudes of two charges. The ratio q_1/q_2 of the magnitudes of two charges is given by the ratio of the electrostatic forces the stationary charges q_1 and q_2 would give on a stationary test charge of magnitude q, when the experiment is repeated with first q_1 and then with q_2 at the same distance from the test charge q. If q_2 is known, the magnitude of q_1 can then be determined.

1.2.2. *SI units*

The SI (or MKSA) system of units will be used throughout the text. In this system, the unit of mass is the kilogramme (denoted kg), the unit of length is the metre (denoted by m) and the unit of time is the second (denoted by s). The unit of electric current is the ampere (denoted A). The ampere will be defined in Section 1.4.1. The unit of charge is the coulomb (denoted by C) and is equal to the total charge passing any cross section of a circuit per second, when the steady current in the circuit is 1A. The unit of force is the newton (denoted by N).

If in equation (1.2), $\mathbf{F}_1$ is expressed in newtons, q and q_1 in coulombs and R_1 in metres, then ε_0 has the numerical value of 8.854 187 817 . . . × 10^{-12} F m^{-1} where F m^{-1} stands for farad per metre. The constant ε_0 is generally called the absolute permittivity of free space, though we shall prefer to call it the electric constant.

Using [M], [L], [T] and [Q] to represent the dimensions of mass, length, time and electric charge respectively, it follows from equation (1.2) that the dimensions of the electric constant ε_0 are $[\mathrm{M}^{-1}\ \mathrm{L}^{-3}\ \mathrm{T}^2\ \mathrm{Q}^2]$. Some readers, who are more familiar with cgs units, may think it strange to find that the electric constant ε_0 has dimensions. This is not very different to the case of the gravitational constant G in Newton's law of universal gravitation. The gravitational constant G has the dimensions $[\mathrm{M}^{-1}\ \mathrm{L}^3\ \mathrm{T}^{-2}]$ and the experimental value of $6.672\ 59 \times 10^{-11}$ N m^2 kg^{-2}.

1.2.3. *The principle of superposition*

Consider a system of N stationary point charges of magnitudes $q_1, q_2, \ldots,$ q_N at positions $\mathbf{r}_1, \mathbf{r}_2, \ldots, \mathbf{r}_N$ respectively in empty space. According to the principle of superposition the force on a test charge of magnitude q at $\mathbf{r}$ due to the charge q_1 are $\mathbf{r}_1$ is unaffected by the presence of the other charges $q_2,$ $q_3, \ldots, q_N$ and is still given by equation (1.2). The resultant force on the test charge q due to all the other N charges is given by the vector sum of the forces $\mathbf{F}_1, \mathbf{F}_2, \ldots, \mathbf{F}_N$, on q due to $q_1, q_2, \ldots q_N$ respectively, that is

$$\mathbf{F} = \mathbf{F}_1 + \mathbf{F}_2 + \ldots + \mathbf{F}_N = \sum_{i=1}^{N} \frac{q q_i(\mathbf{r} - \mathbf{r}_i)}{4\pi\varepsilon_0|\mathbf{r} - \mathbf{r}_i|^3} . \tag{1.3}$$

A reader interested in a full discussion of Coulomb's law is referred to the article: "The teaching of Electricity and Magnetism at College Level", *American Journal of Physics*. Vol. 18, page 1, 1950.

1.2.4. *The electric field*

Equation (1.2) is often rewritten in the form:

$$\mathbf{F}_1 = q\mathbf{E}_1 \tag{1.4}$$

with

$$\mathbf{E}_1(\mathbf{r}) = \frac{q_1(\mathbf{r} - \mathbf{r}_1)}{4\pi\varepsilon_0|\mathbf{r} - \mathbf{r}_1|^3} = \frac{q_1\mathbf{R}_1}{4\pi\varepsilon_0 R_1^3} \tag{1.5}$$

where $\mathbf{R}_1 = (\mathbf{r} - \mathbf{r}_1)$. It is then said that the charge q_1 at $\mathbf{r}_1$ in Figure 1.1 gives rise to an electric field, of intensity $\mathbf{E}_1$ given by equation (1.5), at the position of the test charge q at $\mathbf{r}$. If there are N stationary point charges of magnitudes $q_1, q_2, \ldots, q_N$ at $\mathbf{r}_1, \mathbf{r}_2, \ldots, \mathbf{r}_N$ respectively, then, according to equation (1.3) which was derived using the principle of superposition, the resultant force on the test charge of magnitude q at $\mathbf{r}$ is

$$\mathbf{F} = q\mathbf{E}_1 + q\mathbf{E}_2 + \ldots + q\mathbf{E}_N = q\mathbf{E}(\mathbf{r}) \tag{1.6}$$

where $\mathbf{E}(\mathbf{r})$ is the intensity of the resultant electric field at $\mathbf{r}$. We have

$$\mathbf{E}(\mathbf{r}) = \mathbf{E}_1 + \mathbf{E}_2 + \ldots + \mathbf{E}_N = \sum_{i=1}^{N} \frac{q_i(\mathbf{r} - \mathbf{r}_i)}{4\pi\varepsilon_0|\mathbf{r} - \mathbf{r}_i|^3} . \tag{1.7}$$

The contributions of the individual charges to the resultant electric field intensity $\mathbf{E}$ must be added vectorially. Instead of using the phrase electric field of electric intensity $\mathbf{E}$ we shall generally use the shorter phrase 'the electric field $\mathbf{E}$', where the symbol $\mathbf{E}$ stands for the electric field intensity.

Consider the continuous charge distribution shown in Figure 1.2. The total charge inside the volume element $dV_s = dx_s\, dy_s\, dz_s$, at the position $\mathbf{r}_s$ in Figure 1.2, will be treated as a point charge of magnitude $\rho(\mathbf{r}_s)\, dV_s$, where $\rho(\mathbf{r}_s)$ is the charge density, that is the charge per cubic metre, at $\mathbf{r}_s$. It follows from

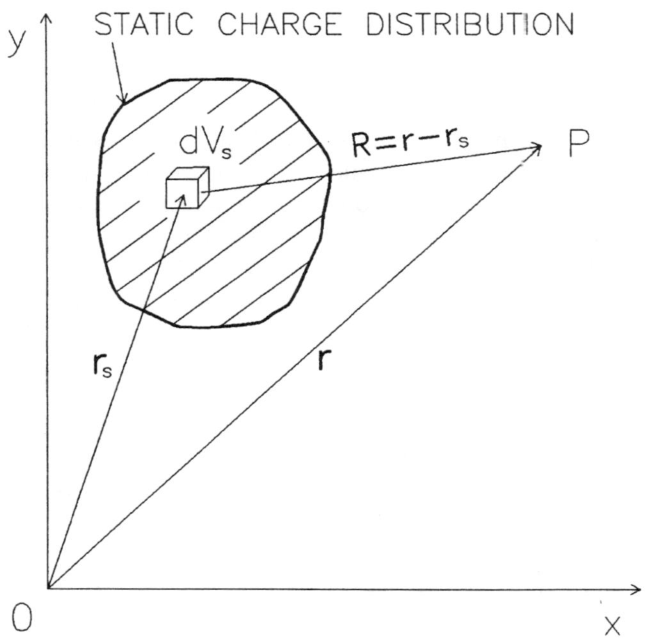

Figure 1.2. Determination of the electric field due to a continuous, electrostatic charge distribution.

equation (1.7) that the resultant electric field $\mathbf{E}(\mathbf{r})$ at the field point at $\mathbf{r}$ is given by

$$\mathbf{E}(\mathbf{r}) = \frac{1}{4\pi\varepsilon_0}\int \frac{\rho(\mathbf{r}_s)\,(\mathbf{r} - \mathbf{r}_s)\mathrm{d}V_s}{|\mathbf{r} - \mathbf{r}_s|^3}\,. \tag{1.8}$$

Notice that we are using the position vector $\mathbf{r}_s$ to denote the position of the volume element $\mathrm{d}V_s$ of the charge distribution at the position (x_s, y_s, z_s), which we shall call the source point. The position vector $\mathbf{r}$ is used to denote the position of the field point at (x, y, z). This convention will be used throughout the book.

If the test charge q placed at the field point at $\mathbf{r}$ were as large as one coulomb, the test charge q would give rise to enormous charge distributions on nearby conductors, which would change the value of the electric field at the field point at $\mathbf{r}$. The force on the stationary test charge q would then be equal to $q\,\mathbf{E}'$, where $\mathbf{E}'$ would be the new value of the electric field resulting from both the original charge distribution and the new extra charge distributions. In order to measure the value the electric field $\mathbf{E}$ had before the test charge was introduced, we would have to make the magnitude of the test charge as small as possible, so that, in the general case, the electric field $\mathbf{E}$ at a field point in empty space is defined in terms of the force $\mathbf{F}_{\mathrm{elec}}$ acting on a *stationary* test charge of magnitude q placed at the field point using the relation:

$$\mathbf{E} = \text{Limit}\left(\frac{\mathbf{F}_{\mathrm{elec}}}{q}\right) \tag{1.9}$$

in the limit when the magnitude of the test charge q tends to zero. Notice $\mathbf{E}$ is parallel to $\mathbf{F}_{\mathrm{elec}}$, so that the electric field $\mathbf{E}$ at a point is in the direction a stationary positive point charge would start to move, if it were placed at that point. Equation (1.9) is also used to define the electric field $\mathbf{E}$ in the general case when the charge distributions are varying with time.

It is assumed in classical electromagnetism that, if the charge q is moving and accelerating in an electric field, the electric force on the moving charge is still given by

$$\mathbf{F}_{\mathrm{elec}} = q\mathbf{E}. \tag{1.10}$$

For example, if an electron of charge q is moving in the electric field $\mathbf{E}$ between the plates of a charged parallel plate capacitor, it is assumed in classical electromagnetism that the electric force on the electron is given by $q\mathbf{E}$, whatever the position, velocity, acceleration and direction of motion of the electron.

When equation (1.10) is applied to moving charges, it is assumed in classical electromagnetism that the magnitude q of the total charge on the charged particle is independent of the velocity of the particle. There is now direct experimental evidence in favour of this assumption. If the total charge on a particle did vary with the velocity $\mathbf{u}$ of the particle, for example, if $q = q_0(1 - u^2/c^2)^{1/2}$, then, since on average the electrons move faster than the protons inside a hydrogen molecule, the hydrogen molecule would have a resultant electric charge and would be deflected by electric fields. In 1960 King [4] showed that the charges on the electrons and protons inside hydrogen molecules were equal in magnitude, but opposite in sign, to within one part in 10^{20}. We therefore conclude that the total charge on a particle is independent of its velocity.

It follows from equation (1.10) that in the SI (or MKSA) system of units, if the force **F** is measured in newtons and the charge q is measured in coulombs, then the electric field **E** can be expressed in newtons per coulomb (denoted N C^{-1}). It will be shown in Section 1.2.9 that **E** is related to the electrostatic potential ϕ by the equation $\mathbf{E} = -\nabla\phi$. Since the unit of ϕ is the volt (denoted by V), the electric field **E** can also be expressed in volts per metre (denoted V m^{-1}).

The rate $\mathrm{d}W/\mathrm{d}t$ at which the electric force, given by equation (1.10), is doing work on the charge q, when it is moving with velocity **u**, is

$$\frac{\mathrm{d}W}{\mathrm{d}t} = \mathbf{F}_{\text{elec}} \cdot \mathbf{u} = q\mathbf{E} \cdot \mathbf{u}. \tag{1.11}$$

1.2.5. *Electric field lines*

It is convenient to represent the electric field at a given instant of time on a diagram using imaginary electric field lines, drawn such that the direction of the tangent to the electric field line at a point is in the same direction as the electric field vector **E** at that point. The number of electric field lines is generally limited such that, on field line diagrams, the number of electric field lines per square metre crossing a surface, that is at right angles to the direction of the field line, is equal to E, the magnitude of the electric field at that point. The electric field lines are closest together where the magnitude of the electric field is highest. The electric field lines due to a stationary positive point charge and a stationary negative point charge are shown in Figures 1.3(a) and 1.3(b) respectively. The electric field lines diverge from the positive charge and converge on the negative charge. In practice, depending on the magnitude of **E**, one often gets better electric field line diagrams by making the number of electric field lines proportional to, not equal to the magnitude of **E**. The concept of using diagrams of electric field lines to represent both the direction and magnitude of the electric field will be used extensively throughout the text, for example when we interpret Gauss' flux law of electrostatics in the next section.

1.2.6. *Gauss' flux law of electrostatics*

Consider the isolated, stationary, positive, point charge of magnitude q shown in Figure 1.4(a). Consider the arbitrary shaped surface S_0 that surrounds the charge, as shown in Figure 1.4(a). Such a surface drawn for the application of Gauss' flux law is often called a Gaussian surface. Consider the infinitesimal element of area $\mathrm{d}\mathbf{S}$ of the surface S_0 that is at a distance **r** from the

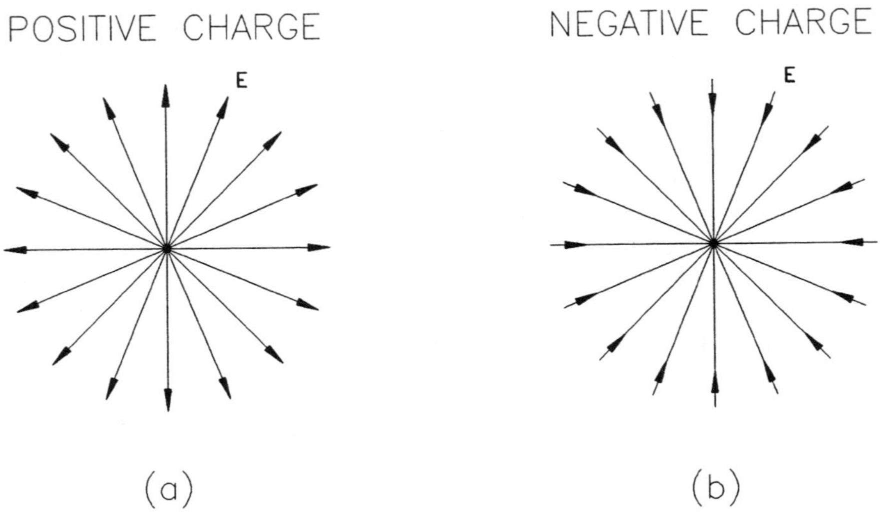

Figure 1.3. The electric fields due to (a) a stationary, positive, point hcarge (b) stationary negative point charge.

charge q, as shown in Figure 1.4(a). The magnitude of the vector **dS** is equal to the magnitude dS of the element of the surface, and the direction of **dS** is the direction of the normal (perpendicular) to the element of surface pointing outwards from the surface S_0. According to Coulomb's law, the electric field **E** at a distance **r** from the charge q is $q\mathbf{r}/4\pi\,\varepsilon_0\,r^3$. The scalar product $\mathbf{E}\cdot\mathrm{d}\mathbf{S}$ is given by

$$\mathbf{E}\cdot\mathrm{d}\mathbf{S} = \frac{q\mathbf{r}\cdot\mathrm{d}\mathbf{S}}{4\pi\varepsilon_0 r^3} = \frac{q\mathrm{d}S\cos\theta}{4\pi\varepsilon_0 r^2} \tag{1.12}$$

where θ is the angle between **r** and **dS**. Now dS cos θ is equal to dS_n, the projection of the area **dS** on to a surface perpendicular to **r**. Since $\mathrm{d}S_n/r^2$ is equal to the solid angle dΩ subtended by **dS** at the position of the charge q, we have

$$\mathbf{E}\cdot\mathrm{d}\mathbf{S} = \left(\frac{q}{4\pi\varepsilon_0}\right)\mathrm{d}\Omega.$$

Integrating over the area of the arbitrary shaped surface S_0 in Figure 1.4(a) and remembering that $\int\mathrm{d}\Omega = 4\pi$, we have

$$\int\mathbf{E}\cdot\mathrm{d}\mathbf{S} = \frac{q}{4\pi\varepsilon_0}\int\mathrm{d}\Omega = \frac{q}{\varepsilon_0}. \tag{1.13}$$

Consider now the case shown in Figure 1.4(b), where the charge q is outside the arbitrary shaped surface S_0. Consider the two elements of area $\mathrm{d}\mathbf{S}_1$ and $\mathrm{d}\mathbf{S}_2$ shown in Figure 1.4(b). Since the electric field lines point in a direction directly away from the positive, point charge q, then $\mathbf{E}\cdot\mathrm{d}\mathbf{S}_1$ is negative whereas $\mathbf{E}\cdot\mathrm{d}\mathbf{S}_2$ is positive. Since the magnitudes of $\mathbf{E}\cdot\mathrm{d}\mathbf{S}_1$ and $\mathbf{E}\cdot\mathrm{d}\mathbf{S}_2$ are both

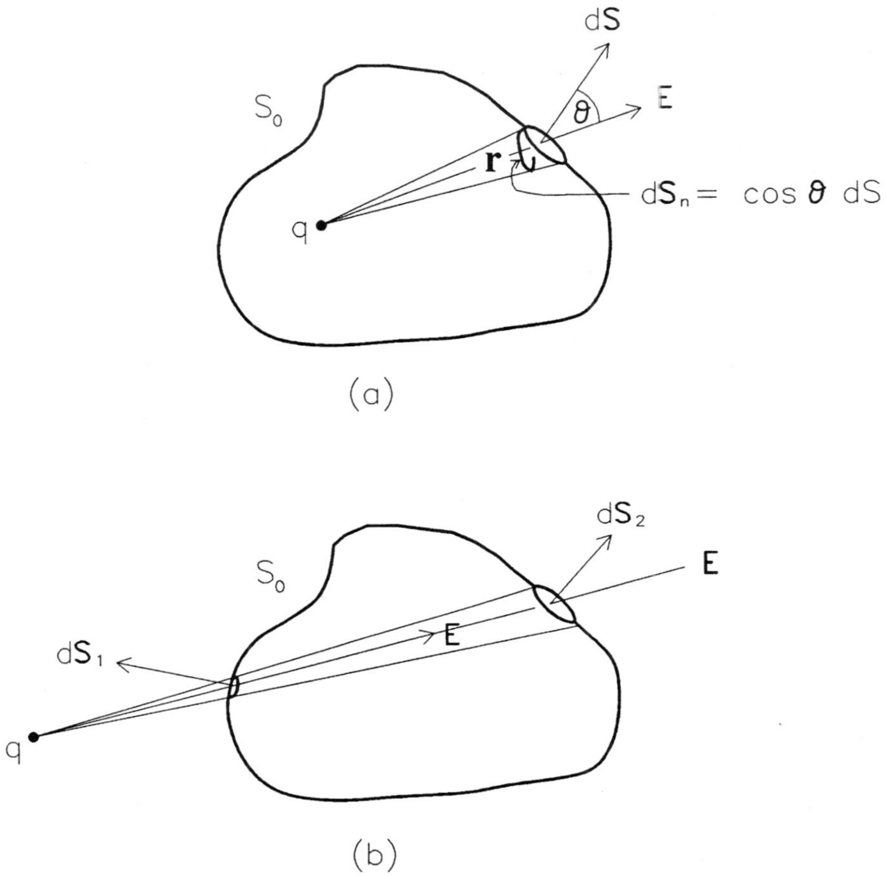

Figure 1.4. Derviation of Gauss' flux law. (a) The charge is inside the Gaussian surface. (b) The charge is outside the Gaussian surface.

equal $(q/4\pi\varepsilon_0)\mathrm{d}\Omega$ their contributions to $\int \mathbf{E}\cdot \mathrm{d}\mathbf{S}$, evaluated over the surface S_0 in Figure 1.4(b), cancel each other. The other elements of area can be treated in pairs in a similar way, so that $\int \mathbf{E}\cdot \mathrm{d}\mathbf{S}$, evaluated over the surface S_0 is zero, when the charge q is outside the surface S_0 in Figure 1.4(b). Summarizing:

$$\int \mathbf{E}\cdot \mathrm{d}\mathbf{S} = \frac{q}{\varepsilon_0} \qquad [q \text{ inside the surface } S_0] \tag{1.13}$$

$$\int \mathbf{E}\cdot \mathrm{d}\mathbf{S} = 0 \qquad [q \text{ outside the surface } S_0] \tag{1.14}$$

This is Gauss' flux law of electrostatics.

In the case of an isolated, stationary, point charge, Gauss' flux law can be illustrated using the concept of electric field lines. It will be assumed that the number of electric field lines per square metre crossing an area perpendicular to **E** is equal to the magnitude of **E**. Assume that a spherical surface of radius r is drawn around the point charge in Figure 1.3(a), with the charge $+q$ at the centre of the spherical surface. The total number of electric field lines crossing the spherical area $4\pi r^2$ is given by the product of this area and the magnitude of the electric field **E**. This product is equal to q/ε_0. Hence the total number of electric field lines crossing the spherical surface is independent of its radius r, so that in the special case of an isolated, stationary, point charge, whose electric field obeys Coulomb's inverse square law, the electric field can be represented by straight electric field lines going all the way from the isolated, stationary, point charge to infinity.

Now consider the flux $\int \mathbf{E} \cdot d\mathbf{S}$ crossing the arbitrary shaped Gaussian surface S_0 in Figure 1.4(a), when the charge q is inside the Gaussian surface. If the number of electric field lines crossing each square metre of an area perpendicular to **E** is equal to the magnitude of **E**, then the scalar product $\mathbf{E} \cdot d\mathbf{S} = E\, dS_n$ is equal to the number of electric field lines crossing $d\mathbf{S}$. Hence the integral $\int \mathbf{E} \cdot d\mathbf{S}$ is equal to the total number of electric field lines crossing the surface S_0 which, since the electric field lines carry on in straight lines to infinity is equal to q/ε_0 whatever the shape of the Gaussian surface. This illustrates Gauss' flux law of electrostatics. If the stationary charge q is outside the Gaussian surface, its electric field lines just cross the surface in straight lines. As many electric field lines leave the surface as enter it, so that in this case $\int \mathbf{E} \cdot d\mathbf{S}$ is zero.

1.2.7. *Gauss' flux law for a continuous charge distribution*

Consider first a system of N stationary point charges of magnitudes $q_1, q_2, \ldots, q_N$. Applying equations (1.13) and (1.14) as appropriate to each of these charges and adding, we find that for any arbitrary Gaussian surface,

$$\int \mathbf{E}_1 \cdot d\mathbf{S} + \int \mathbf{E}_2 \cdot d\mathbf{S} + \cdots + \int \mathbf{E}_N \cdot d\mathbf{S} = \Sigma' \frac{q_i}{\varepsilon_0}. \tag{1.15}$$

The summation Σ' on the right hand side of equation (1.15) is only over those charges that are inside the Gaussian surface. Using the distributive rule, equation (A1.2) of Appendix A1.1, we have

$$\begin{aligned}\mathbf{E}_1 \cdot d\mathbf{S} + \mathbf{E}_2 \cdot d\mathbf{S} + \ldots \mathbf{E}_N \cdot d\mathbf{S} &= (\mathbf{E}_1 + \mathbf{E}_2 + \ldots + \mathbf{E}_N) \cdot d\mathbf{S} \\ &= \mathbf{E} \cdot d\mathbf{S}\end{aligned}$$

where **E** is the resultant electric field at a point on the Gaussian surface due to all the N charges in the system. Substituting in equation (1.15) and using Gauss' integral theorem of vector analysis, which is equation (A1.30) of Appendix A1.7, we have

$$\int \mathbf{E} \cdot d\mathbf{S} = \int \nabla \cdot \mathbf{E}\, dV = \Sigma' \frac{q_i}{\varepsilon_0}. \tag{1.16}$$

For a continuous charge distribution, of charge density ρ, $\Sigma' q_i$ is equal to

$\int \rho \, dV$ evaluated over the volume V_0 enclosed by the Gaussian surface. Hence for a continuous charge distribution

$$\int \mathbf{E} \cdot d\mathbf{S} = \int \nabla \cdot \mathbf{E} \, dV = \int \frac{\rho dV}{\varepsilon_0}. \tag{1.17}$$

If the volume V_0 enclosed by the Gaussian surface is inside the continuous charge distribution and if V_0 is made small enough for the variations of $\nabla \cdot \mathbf{E}$ and ρ inside V_0 to be negligible, then after cancelling V_0, equation (1.17) reduces to

$$\nabla \cdot \mathbf{E}(\mathbf{r}) = \frac{\rho(\mathbf{r})}{\varepsilon_0}. \tag{1.18}$$

In equation (1.18), $\mathbf{E}(\mathbf{r})$ is the total electric field at the field point at $\mathbf{r}$ due to all the charge distributions in the system, and $\rho(\mathbf{r})$ is the value of the charge density at the field point at $\mathbf{r}$ where equation (1.18) is applied. Equation (1.18) is one of Maxwell's equations.

1.2.8. *Generalization of the equation $\nabla \cdot \mathbf{E} = \rho/\varepsilon_0$ to moving charge distributions*

Equation (1.18) was developed for a stationary, continuous charge distribution, that is for electrostatics. It is assumed in classical electromagnetism that equation (1.18) applies at a field point inside a moving and accelerating continuous charge distribution. Following Lorentz we shall use a simplified classical model for individual charged atomic particles such as protons and electrons, each of which will be treated as a continuous charge distribution of finite, but exceedingly small dimensions. Lorentz assumed that equation (1.18) held inside and outside such an accelerating classical point charge. Draw a Gaussian surface of finite extent to enclose such a moving and accelerating classical point charge. Integrating equation (1.18) at a fixed time and applying Gauss' theorem of vector analysis, which is equation (A1.30) of Appendix A1.7, since $\int \rho \, dV$ evaluated at a fixed time is equal to q, we have

$$\int \nabla \cdot \mathbf{E} \, dV = \int \mathbf{E} \cdot d\mathbf{S} = \frac{1}{\varepsilon_0} \int \rho dV = \frac{q}{\varepsilon_0}. \tag{1.19}$$

It will be shown in Section 3.4 of Chapter 3 that the electric field lines due to an accelerating point charge are curved and more complicated than in the case of the stationary point charge shown in Figure 1.3(a). However, if equation (1.19) is valid in the general case of an accelerating classical point charge of magnitude q then the total flux of $\mathbf{E}$ crossing the Gaussian surface surrounding the charge, which is also equal to the total number of electric field lines coming from the accelerating charge, is still equal to q/ε_0 whatever the speed and acceleration of the charge. This is an extremely important result, which will appear from time to time throughout our discussions of Maxwell's equations.

1.2.9. *The electrostatic scalar potential*

In many cases it is easier to determine electrostatic fields by introducing the scalar electrostatic potential ϕ. Consider again the two stationary point charges of magnitudes q_1 and q that are at positions $\mathbf{r}_1$ and $\mathbf{r}$ respectively in Figure 1.1. It is straightforward for the reader to show using cartesian coordinates that

$$\nabla\left(\frac{1}{|\mathbf{r}-\mathbf{r}_1|}\right) = \nabla\left(\frac{1}{[(x-x_1)^2+(y-y_1)^2+(z-z_1)^2]^{1/2}}\right)$$
$$= -\frac{(\mathbf{r}-\mathbf{r}_1)}{|\mathbf{r}-\mathbf{r}_1|^3} \tag{1.20}$$

where the operator ∇ is given by

$$\nabla = \hat{\mathbf{i}}\frac{\partial}{\partial x} + \hat{\mathbf{j}}\frac{\partial}{\partial y} + \hat{\mathbf{k}}\frac{\partial}{\partial z}$$

and $\hat{\mathbf{i}}$, $\hat{\mathbf{j}}$ and $\hat{\mathbf{k}}$ are unit vectors pointing in directions parallel to the x, y and z axes respectively. Hence the expression for the electric field $\mathbf{E}_1$ due to the charge q_1 in Figure 1.1, which is given by equation (1.5), can be rewritten in the form

$$\mathbf{E}_1 = -\nabla\phi_1 \tag{1.21}$$

where

$$\phi_1 = \frac{q_1}{4\pi\varepsilon_0|\mathbf{r}-\mathbf{r}_1|} + C_1. \tag{1.22}$$

The zero of the electrostatic potential ϕ_1 is generally specified by assuming that ϕ_1 is zero at an infinite distance $|\mathbf{r} - \mathbf{r}_1|$ from the charge q_1 in which case the constant C_1 in equation (1.22) is zero.

It follows from equation (1.21) that, for a system of N charges q_1, q_2, . . . , q_N at positions $\mathbf{r}_1$, $\mathbf{r}_2$, . . . , $\mathbf{r}_N$ respectively, the resultant electrostatic field at the field point at $\mathbf{r}$ is given by

$$\mathbf{E} = \mathbf{E}_1 + \mathbf{E}_2 + \ldots + \mathbf{E}_N = -\nabla\phi_1 - \nabla\phi_2 - \ldots - \nabla\phi_N$$
$$= -\nabla\phi \tag{1.23}$$

where according to equation (1.22)

$$\phi(\mathbf{r}) = \Sigma\phi_i = \Sigma\frac{q_i}{4\pi\varepsilon_0|\mathbf{r}-\mathbf{r}_i|}\,. \tag{1.24}$$

Notice that the electrostatic potential ϕ can be determined by adding the numerical values of the contributions ϕ_i due to the individual charges, whereas the resultant electric field $\mathbf{E}$ is the vector sum of the contributions $\mathbf{E}_i$ due to the individual charges. The electrostatic potential ϕ is a scalar quantity.

Consider the continuous volume charge distribution shown in Figure 1.2. By treating the charge inside the infinitesimal volume element dV_s at $\mathbf{r}_s$ in Figure 1.2 as a point charge of magnitude $\rho\, dV_s$ where ρ is the charge density at $\mathbf{r}_s$, it follows from equation (1.24) that the total electrostatic potential $\phi(\mathbf{r})$ at a field point at $\mathbf{r}$, due to the continuous charge distribution shown in Figure 1.2, is

$$\phi(\mathbf{r}) = \frac{1}{4\pi\varepsilon_0} \int \frac{\rho(\mathbf{r}_s)\mathrm{d}V_s}{|\mathbf{r} - \mathbf{r}_s|}. \tag{1.25}$$

If there is a surface charge distribution of magnitude σ coulombs per square metre, the charge $\sigma(\mathbf{r}_s)\mathrm{d}S_s$ on an infinitesimal element of area $\mathrm{d}S_s$ at $\mathbf{r}_s$ can be treated as a point charge. This leads to the expression

$$\phi(\mathbf{r}) = \frac{1}{4\pi\varepsilon_0} \int \frac{\sigma(\mathbf{r}_s)\mathrm{d}S_s}{|\mathbf{r} - \mathbf{r}_s|}. \tag{1.26}$$

for the contribution of the surface charge distribution to the electrostatic potential.

Since according to equation (A1.26) of Appendix A1.6 the curl of the gradient of any scalar function of position is zero, it follows by taking the curl of both sides of equation (1.23) that

$$-\nabla \times \nabla\phi = \nabla \times \mathbf{E} = 0. \tag{1.27}$$

Integrating equation (1.27) over a finite surface and applying Stokes' theorem of vector analysis, which is equation (A1.34) of Appendix A1.8, we have for the electrostatic field

$$\int \nabla \times \mathbf{E} \cdot \mathrm{d}\mathbf{S} = \oint \mathbf{E} \cdot \mathrm{d}\mathbf{l} = 0. \tag{1.28}$$

The condition $\nabla \times \mathbf{E} = 0$ (or its integral form $\oint \mathbf{E} \cdot \mathrm{d}\mathbf{l} = 0$) is a condition that the electrostatic field $\mathbf{E}$ must satisfy if it is a conservative field. In the more general case of varying charge distributions $\nabla \times \mathbf{E}$ is not zero but is equal to $-\partial\mathbf{B}/\partial t$, where $\mathbf{B}$ is the magnetic field. When we come to discuss varying charge and current distributions in Section 2.2 of Chapter 2 we shall find that equation (1.23) must be replaced by equation (2.7) of Chapter 2.

Putting $\mathbf{E} = -\nabla\phi$ in equation (1.18) and using the relation $\nabla \cdot (\nabla\phi) = \nabla^2\phi$, we obtain for a system of continuous charge distributions in empty space

$$\nabla^2\phi = -\frac{\rho}{\varepsilon_0}. \tag{1.29}$$

This is Poisson's equation of electrostatics. The solution of Poisson's equation is given by equation (1.25). At a field point in empty space, $\rho = 0$ and Poisson's equation reduces to Laplace's equation

$$\nabla^2\phi = 0. \tag{1.30}$$

Laplace's equation can be solved using the methods of potential theory. No new physical principles are involved. The interested reader is referred to the standard text books such as Jeans [5], Smythe [6] and Jackson [7].

1.2.10. *The potential energy of a charge in an electrostatic field*

So far our development of the properties of the scalar electrostatic potential ϕ has been entirely mathematical. It is possible in the context of electrostatics to give ϕ an operational interpretation in terms of the potential energy of a test charge of magnitude q placed at the field point. To move a test charge q at zero speed in an electrostatic field, we must apply a force

$$\mathbf{F}_{\text{app}} = -q\mathbf{E}$$

equal and opposite to the electric force $q\mathbf{E}$ acting on the test charge. The difference $\mathrm{d}U$ in the potential energy U at two points a distance $\mathbf{dl}$ apart can be defined as the work done by the applied force $\mathbf{F}_{\text{app}}$ in an infinitesimal displacement $\mathbf{dl}$ of the test charge q against the electric force $q\mathbf{E}$ acting on the test charge, that is

$$\mathrm{d}U = \mathbf{F}_{\text{app}} \cdot \mathbf{dl} = -q\mathbf{E} \cdot \mathbf{dl}. \tag{1.31}$$

Since for electrostatics $\mathbf{E} = -\boldsymbol{\nabla}\phi$, equation (1.31) can be rewritten in the form

$$\mathrm{d}U = q\boldsymbol{\nabla}\phi \cdot \mathbf{dl} = -q\mathbf{E} \cdot \mathbf{dl}. \tag{1.32}$$

Using equation (A1.9) of Appendix A1.2, we have

$$\mathrm{d}U = q\mathrm{d}\phi = -q\mathbf{E} \cdot \mathbf{dl} \tag{1.33}$$

where $\mathrm{d}\phi$ is the total change in ϕ in the infinitesimal displacement $\mathbf{dl}$. Using equation (1.33), we find that the total work done in moving the test charge q at zero speed from infinity, where by definition $\phi = 0$ and $U = 0$, to the position $\mathbf{r}$, where the electrostatic potential is $\phi(\mathbf{r})$ and the potential energy is $U(\mathbf{r})$, is given by

$$U(\mathbf{r}) = -q\int_{\infty}^{r} \mathbf{E} \cdot \mathbf{dl} = q\int_{\infty}^{r} \mathrm{d}\phi = q\phi(\mathbf{r}). \tag{1.34}$$

It follows from equation (1.28) that $\int \mathbf{E} \cdot \mathbf{dl}$ and hence U are independent of the path taken from infinity. Hence, the electrostatic scalar potential ϕ can be defined in terms of the potential energy of a stationary test charge q by the relation

$$\phi = \text{Limit}\left(\frac{U}{q}\right) \tag{1.35}$$

as q tends to zero. According to equation (1.34), if the test charge q is at a field point where the electrostatic potential is $\phi(\mathbf{r})$, then the potential energy of the charge is $q\phi(\mathbf{r})$.

It follows from equation (1.33) that

$$\mathrm{d}\phi = -\mathbf{E} \cdot \mathbf{dl}.$$

Integrating from position 1 where the electrostatic potential is ϕ_1, to position 2 where the electrostatic potential is ϕ_2, we find that

$$\phi_2 - \phi_1 = -\int_1^2 \mathbf{E} \cdot \mathrm{d}\mathbf{l}. \tag{1.36}$$

The quantity $(\phi_2 - \phi_1)$ is called the potential difference between the positions 2 and 1. The idea of an electrostatic potential difference, given by equation (1.36), will be used extensively in our discussions of Ohm's law and conduction current flow in Section 1.3 and in our discussions of quasi-stationary phenomena in Chapter 7.

Since the electric force on a test charge of magnitude q, which is in an applied electric field $\mathbf{E}$, is equal to $q\mathbf{E}$, the work done on the test charge by the applied electric field in a displacement $\mathrm{d}\mathbf{l}$ of the test charge is $q\mathbf{E} \cdot \mathrm{d}\mathbf{l}$. If all this work goes into increasing the kinetic energy T of the test charge by $\mathrm{d}T$, then

$$\mathrm{d}T = q\mathbf{E} \cdot \mathrm{d}\mathbf{l}.$$

Integrating we find that, if the test charge q is released from rest at a point in empty space at the position $\mathbf{r}$, where the scalar potential is $\phi(\mathbf{r})$, then it will reach infinity, where its potential energy is zero, with a kinetic energy T given by

$$T = q\int_r^\infty \mathbf{E} \cdot \mathrm{d}\mathbf{l} = -q\int_\infty^r \mathbf{E} \cdot \mathrm{d}\mathbf{l}.$$

But according to equation (1.34)

$$-q\int_\infty^r \mathbf{E} \cdot \mathrm{d}\mathbf{l} = q\phi(\mathbf{r}) = U(\mathbf{r}).$$

Hence the test charge reaches infinity with a kinetic energy T given by

$$T = q\phi(\mathbf{r}) = U(\mathbf{r}). \tag{1.37}$$

Using the relativistic expression for the kinetic energy T of a charged particle of (rest) mass m_0 and velocity $\mathbf{u}$ we can rewrite equation (1.37) in the form

$$m_0c^2\left(\frac{1}{(1 - u^2/c^2)^{1/2}} - 1\right) = q\phi. \tag{1.38}$$

In the zero velocity limit, equation (1.38) reduces to

$$\frac{1}{2}m_0u^2 = q\phi.$$

If the test charge q is accelerated from position 1 to position 2 in the electrostatic field, the reader can show, by using different limits of integration, that

$$T_2 - T_1 = \int_1^2 q\mathbf{E} \cdot \mathrm{d}\mathbf{l} = -q\int_2^1 \mathbf{E} \cdot \mathrm{d}\mathbf{l} = q(\phi_1 - \phi_2)$$

where T_1 and T_2 are the kinetic energies of the test charge at positions 1 and 2 respectively, where the electrostatic potentials are ϕ_1 and ϕ_2. Rearranging we have

$$T_1 + q\phi_1 = T_2 + q\phi_2. \tag{1.39}$$

Now $q\phi_1$ and $q\phi_2$ are the potential energies U_1 and U_2 of the test charge at positions 1 and 2 respectively in the electrostatic field. Hence equation (1.39) can be rewritten in the form:

$$T_1 + U_1 = T_2 + U_2 \tag{1.40}$$

which is the law of conservation of energy for a charge moving in an electrostatic field.

1.2.11. *Summary of electrostatics*

The basic equations of electrostatics, which we have derived from Coulomb's law for continuous charge distributions, can be summarized as follows

$$\nabla \cdot \mathbf{E} = \frac{\rho}{\varepsilon_0}. \tag{1.18}$$

$$\nabla \times \mathbf{E} = 0. \tag{1.27}$$

The solution of these equations is given in terms of the scalar electrostatic potential ϕ by

$$\mathbf{E} = -\nabla\phi \tag{1.23}$$

where

$$\phi(\mathbf{r}) = \frac{1}{4\pi\varepsilon_0}\int \frac{\rho(\mathbf{r}_s)\mathrm{d}V_s}{|\mathbf{r} - \mathbf{r}_s|}. \tag{1.25}$$

The differential equation for ϕ, which is generally called Poisson's equation is

$$\nabla^2\phi = -\frac{\rho}{\varepsilon_0}. \tag{1.29}$$

In empty space, where ρ is zero, equation (1.29) reduces to Laplace's equation

$$\nabla^2\phi = 0. \tag{1.30}$$

1.3. Conduction current flow in stationary conductors

1.3.1. *Introduction*

So far in this chapter, we have only considered charges at rest (electrostatics). Before going on to discuss the electric and magnetic fields due to moving charges, which will lead up to Maxwell's equations, we shall make a few comments about conduction current flow in a stationary electrical conductor. A fuller account of this important, but often neglected topic, is given in Appendix B. The reader should be thoroughly familiar with the contents of Appendix B, which should be read in conjunction with this section.

1.3.2. *The equation* $\mathbf{J} = \sigma\mathbf{E}$ *and Ohm's law*

The conduction current I flowing in a conductor is equal to the rate at which charge is crossing any cross section of the conductor. The conduction current density $\mathbf{J}$ is defined as the current per square metre crossing an area perpendicular to the direction of current flow, so that for uniform current flow $J = I/A$, where A is the area of cross section perpendicular to the direction of current flow.

It is found experimentally that, when a source of emf maintains a steady electrostatic potential difference across a stationary metallic conductor, that forms part of a complete electrical circuit, a steady conduction current flows in the conductor. The ratio of the electrostatic potential difference ϕ across the conductor to the current I flowing in the circuit is called the resistance R of the conductor. We have

$$R = \frac{\phi}{I}. \tag{1.41}$$

Consider the uniform conductor of length l and of uniform cross sectional area A shown in Figure 1.5. The conductor forms part of a complete electrical circuit carrying a steady current I. The emf maintains an electrostatic potential ϕ across the conductor. According to the equation $\mathbf{E} = -\nabla\phi$, there is a steady electric field of magnitude E equal to ϕ/l inside the conductor. Putting $I = JA$ and $\phi = El$ in equation (1.41) and rearranging, in vector form we have

$$\mathbf{J} = \sigma\mathbf{E} \tag{1.42}$$

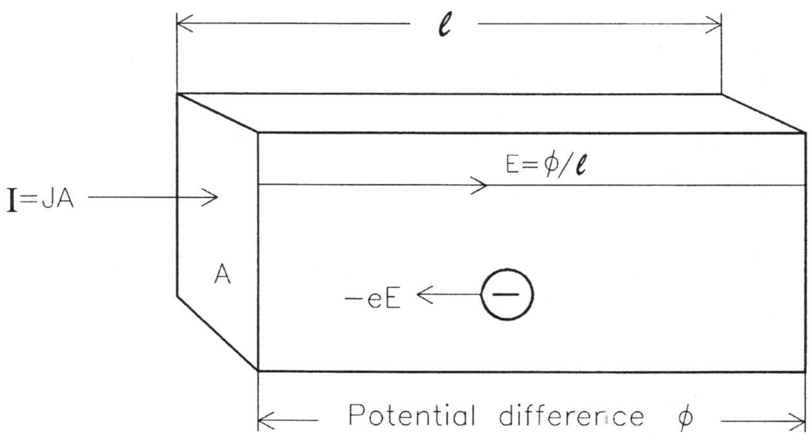

Figure 1.5. An example to illustrate the relation $\mathbf{J} = \sigma\mathbf{E}$.

where

$$\sigma = \frac{l}{RA}. \tag{1.43}$$

The electrical conductivity σ can be defined as the ratio of the current density **J** at a point inside the conductor to the value of the total electric field **E** at that point. The resistivity of the conductor is equal to the reciprocal of the electrical conductivity. A conduction current, given by equation (1.42), flows in a stationary conductor, whenever there is an electric field **E** inside the conductor.

When Ohm's law is valid, the electrical conductivity σ and the resistance R of a circuit are constants, which are independent of the potential difference across the conductor, provided the external conditions, such as temperature, are kept constant. Ohm's law is not a universal law of nature, though it does hold for many metallic conductors over a wide range of values of the applied potential difference. In general, particularly for semiconductors, the electrical conductivity $\sigma = J/E$ is a function of the electric field E inside the conductor, as well as of the temperature of the conductor. For single crystals σ may be a tensor. When a conductor is made up of a large number of such anisotropic crystals orientated at random, the average conductivity is generally isotropic.

The equation $\mathbf{J} = \sigma\mathbf{E}$ is not one of Maxwell's equations. It is called a constitutive equation. The value of σ depends on the properties of the conductor and on the experimental conditions.

1.3.3. *Classical model of conduction current flow*

There are in a metal a large number of electrons, generally called free or conduction electrons, that are able to move about inside the metal under the influence of an applied electric field. In a p-type semiconductor, most of the charge carriers are positive holes. Consider a conductor of cross sectional area A, and which is carrying a steady conduction current I. Let the number of moving charges per cubic metre, each of charge q, be equal to n. Since I is equal to the total charge passing any cross section of the conductor per second, we have

$$I = qnAv \tag{1.44}$$

where v is the mean drift velocity of the moving charges. The current density J is given by

$$J = \frac{I}{A} = qnv. \tag{1.45}$$

When the conduction current is due to the drift of electrons, q is negative and v is in the direction opposite to the direction of the electric field **E** inside the conductor. For the conduction electrons in copper at room temperature

$n = 8.5 \times 10^{28}$ m^{-3} and $q = -1.602 \times 10^{-19}$ C. Consider a current of 1A flowing in a copper wire of cross sectional area 1 mm^2. Substituting in equation (1.44), we find that $v = 7.3 \times 10^{-5}$ m s^{-1}. At this velocity it takes an electron 3.8 hours to drift one metre. The conduction electrons in copper have kinetic energies of about 7eV. The corresponding velocity is about 1.6×10^6 m s^{-1}, or approximately $c/200$, where c is the speed of light in empty space. Thus, in the presence of the electric field inside a copper wire, the conduction electrons acquire a mean drift velocity of only about 10^{-4} m s^{-1} superimposed on their thermal velocities, which are in all directions and have magnitudes of about 1.6×10^6 m s^{-1}. The conduction electrons gain momentum and kinetic energy when they are accelerated by the electric field inside the conductor, but they lose this extra momentum and extra kinetic energy in the collisions they make with impurities, lattice defects and phonons, which reduce their mean drift velocity to the low value of about 10^{-4} m s^{-1}. According to the classical theory of electrical conductivity the electrical conductivity σ is given by

$$\sigma = \frac{nq^2}{mf} \,. \tag{1.46}$$

where f is the collision frequency. For copper at room temperatures $\sigma = 5.8 \times 10^7$ S m^{-1}. Putting $n = 8.5 \times 10^{28}$ m^{-3}, $q = -1.602 \times 10^{-19}$ C and $m = 9.108 \times 10^{-31}$ kg in equation (1.46), we find that the collision frequency f is 4.10×10^{13}, so that the mean time between collisions is about 2.4×10^{-14} s. In this time interval, at a speed of $c/200$, a conduction electron will travel a distance of 3.6×10^{-8} m, which is approximately 150 atomic diameters. This shows that, on average, a conduction electron must pass through or pass very close to many atoms before making a collision. This illustrates how the use of quantum mechanics and the idea of quantum mechanical tunnelling through potential barriers is essential in a full theory of electrical conduction in metals. An introduction to the theory of electrical conduction is given by Weisskopf [8].

1.3.4. *Joule heating*

According to equation (1.11), the instantaneous rate at which the electric field $\mathbf{E}$ inside a stationary conductor is doing work on a charge q that is moving with velocity $\mathbf{u}$ inside the conductor is $q\mathbf{E} \cdot \mathbf{u}$. The average rate at which the steady electric field $\mathbf{E}$ is doing work on a conduction electron is

$$\langle q\mathbf{E} \cdot \mathbf{u} \rangle = q\mathbf{E} \cdot \langle \mathbf{u} \rangle = q\mathbf{E} \cdot \mathbf{v}$$

where $\mathbf{v}$ is the mean drift velocity. A conduction electron gains kinetic energy at this average rate, but loses this extra kinetic energy in collisions, leading to Joule heating. If there are n conduction electrons per cubic metre, the rate of generation of Joule heat per cubic metre inside a conductor is

$$n(q\mathbf{E} \cdot \mathbf{v}) = \mathbf{E} \cdot (qn\mathbf{v}) = \mathbf{E} \cdot \mathbf{J}$$

where, according to equation (1.45), $\mathbf{J} = qn\mathbf{v}$ is the conduction current density. If the conductor is of length l and of cross sectional area A, the rate of generation of Joule heat in the conductor is $lA(EJ)$. Since $E = J/\sigma = I/A\sigma$, the rate of generation of Joule heat is equal to I^2R, where $R = l/\sigma A$ is the resistance of the conductor and I is the current flowing through the conductor.

1.3.5. *Origin of the electric fields inside current carrying conductors*

In electrostatics, when an electric field is applied to an isolated conductor a transient electric current flows in the conductor until a surface charge distribution of such a magnitude is built up that the electric field inside the isolated conductor is zero and the conductor becomes an equipotential. However, according to equation (1.42), there is steady electric field inside a conductor that forms part of a stationary electrical circuit, when a steady conduction current flows in the circuit. A full account of how a source of emf can maintain such an electric field inside a stationary conductor is given in Appendix B1. The illustrative example of a source of emf given in Appendix B1 is that of a Van de Graaff generator, whose terminals are joined by a long conducting wire. When a state of dynamic equilibrium has been reached, the charges, that are moved mechanically by the belt of the Van de Graaff from one terminal to the other, replace the charges removed from the terminals of the Van de Graaff by the current flow into and from the wire connecting the terminals. In this way the Van de Graaff can maintain an electric field inside the connecting wire. It is shown in Appendix B1 that during the transient state, before the current in the connecting wire is constant, electric charge distributions are built up on the surfaces of the connecting wire and at boundaries where conductors of different conductivities are joined, that are of such magnitudes that the resultant electric field $\mathbf{E}$ inside the connecting wire is parallel to the wire and is of such a magnitude that it gives the same value of current in all parts of the connecting wire. The resultant electric field $\mathbf{E}$ inside the wire is equal to $(\mathbf{E}_0 + \mathbf{E}_s)$, where $\mathbf{E}_0$ is the electric field due to the charges on the terminals of the Van de Graaff and $\mathbf{E}_s$ is the electric field due to the surface and boundary charge distributions on the connecting wires. Well away from the Van de Graaff generator, it is these surface and boundary charge distributions that give the main contribution to the electric fields inside the conductors and the potential differences across the resistors in the distant parts of the circuit. It is shown in Appendix B3 that the magnitudes of these surface and boundary charge distributions are exceedingly small.

An interesting query, often raised by students of geophysics is that, when we carry out a resistivity survey by putting two metallic probes into the ground and join them to a battery, such that a current flows in the ground from one probe to the other, how does the current that enters the ground through one probe know where the other probe is and how does the current get there? The answer is that, though individual conduction electrons have velocities of about $c/200$ in all directions, on average they drift in the direction opposite to the direction of the local electric field. When the battery is first connected across the probes, there is an electric field associated with the potential difference between the probes. Initially this electric field is present in both the ground and in the air above ground. We can ignore the current flow in the air above ground, but the electric field will give a current flow in the conducting ground, which will build up charge distributions on the surface of the ground and in regions where the conductivity σ and the relative permittivity ε_r vary with position. These charge distributions are of such magnitudes that they give resultant electric field lines which join the two probes. It does not matter to a conduction electron how this local electric field is produced or where it comes from. The conduction electron just responds to the local resultant electric field **E** and drifts in the direction opposite to **E**. In this way current flows from one probe to the other along the resultant electric field lines due to the applied potential difference and the surface and boundary charge distributions.

1.3.6. *A moving conductor*

If a conductor (or plasma) is moving with velocity **u** in the laboratory reference frame in an electric field **E** and a magnetic field **B**, the force, measured in the laboratory reference frame, on a charge of magnitude q that is at rest relative to the conductor, but which is moving with velocity **u** in the laboratory reference frame, is given by the Lorentz force $q\mathbf{E} + q\mathbf{u} \times \mathbf{B}$ and not by $q\mathbf{E}$. This suggests that for a conductor moving with velocity **u**, the constitutive equation $\mathbf{J} = \sigma\mathbf{E}$ must be changed to

$$\mathbf{J} = \sigma(\mathbf{E} + \mathbf{u} \times \mathbf{B}). \tag{1.47}$$

Equation (1.47) can be derived more rigourously using relativistic methods. Reference Rosser [9].

1.3.7. *The continuity equation for a varying charge and current distribution*

Consider a closed surface S_0 that encloses part of a charge and current distribution. It is found experimentally that the total charge of a complete system is always conserved. The charge crossing the surface S_0 per second, due to the electric current flow, leads to a corresponding decrease in the total charge left inside the surface S_0, so that

$$\int \mathbf{J} \cdot \mathrm{d}\mathbf{S} = -\frac{\partial}{\partial t}\int \rho \mathrm{d}V.$$

Applying Gauss' theorem of vector analysis, which is equation (A1.30) of Appendix A1.7, to the left hand side and rearranging we obtain

$$\int \left(\nabla \cdot \mathbf{J} + \frac{\partial \rho}{\partial t}\right) \mathrm{d}V = 0. \tag{1.48}$$

Equation (1.48) must hold whatever the value of the volume enclosed by the surface S_0. If the volume enclosed by the surface S_0 is increased by an infinitesimal amount dV_0 then $(\nabla \cdot \mathbf{J} + \partial\rho/\partial t)dV_0$ must be zero if equation (1.48) is to be valid for the new total volume. Hence, we must have

$$\nabla \cdot \mathbf{J} + \frac{\partial \rho}{\partial t} = 0. \tag{1.49}$$

This is the continuity equation for a charge and current distribution.

When the current in a circuit is steady, the surface and boundary charge distributions, that give the appropriate value of electric field inside the conductors making up the circuit to give the same value of current in all parts of the circuit, do not vary with time, so that $\partial\rho/\partial t$ is zero for steady currents in complete circuits and equation (1.49) reduces to

$$\nabla \cdot \mathbf{J} = 0 \qquad \text{(steady currents).} \tag{1.50}$$

1.4. Magnetic fields due to steady current distributions (magnetostatics)

1.4.1. *Introduction*

Previously in Section 1.2, we only considered the forces between stationary electric charges. We shall now go on to consider the forces between moving and accelerating charges. No experiments have been carried out with moving atomic charged paricles, such as an electron, to determine the precise expression for the magnetic field due to a moving and accelerating classical point charge. (We shall return in Chapter 3 to derive the appropriate expressions after developing Maxwell's equations). Neither can we isolate part of an electrical circuit that is carrying a steady conduction current, to determine experimentally the precise expression for the magnetic field due to a current element that forms part of the circuit. The most accurate experiments to start from are those using steady currents in rigid current balances of different geometrical configurations. It will be assumed throughout this section that there are no magnetic materials, such as iron, present in the system, so that the relative permeability $\mu_r = 1$ everywhere.

Consider two complete, rigid, stationary electrical circuits in empty space, which are carrying steady conduction currents I_1 and I_2 respectively, as shown in Figure 1.6. Experiments have confirmed, to an accuracy of about 1 part in 10^7, that the total magnetic force $\mathbf{F}_2$ on circuit 2 due to the current I_1 flowing in circuit 1 can be calculated using the formula:

$$\mathbf{F}_2 = \frac{\mu_0 I_1 I_2}{4\pi} \oint_2 \oint_1 \frac{d\mathbf{l}_2 \times (d\mathbf{l}_1 \times \mathbf{r}_{12})}{r_{12}^3}. \tag{1.51}$$

In equation (1.51) $d\mathbf{l}_1$ is an element of length dl_1 of circuit 1 pointing in the direction that the current I_1 is flowing in circuit 1, $d\mathbf{l}_2$ is an element of length

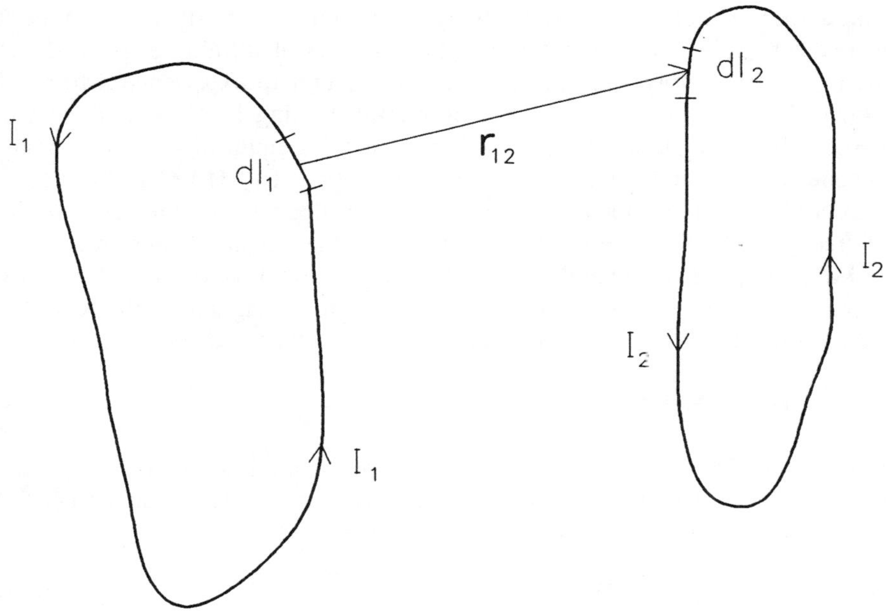

Figure 1.6. The forces between two stationary, rigid circuits carrying steady currents I_1 and I_2.

dl_2 of circuit 2 in the direction of I_2 and $\mathbf{r}_{12}$ is a vector from the position of $\mathbf{dl}_1$ to the position of $\mathbf{dl}_2$, as shown in Figure 1.6. Equation (1.51) is sometimes called Grassmann's equation. There are other formulae, such as Ampère's original formula

$$\mathbf{F}_2 = -\frac{\mu_0 I_1 I_2}{4\pi} \oint_2 \oint_1 \mathbf{r}_{12} \left\{ \frac{2}{r_{12}^3} (\mathbf{dl}_1 \cdot \mathbf{dl}_2) - \frac{3}{r_{12}^5} (\mathbf{dl}_1 \cdot \mathbf{r}_{12}) (\mathbf{dl}_2 \cdot \mathbf{r}_{12}) \right\} \tag{1.52}$$

which give the same value for $\mathbf{F}_2$ for steady conduction currents in complete circuits. For a general discussion see Whittaker [10]. Equation (1.51) is however the traditional starting point in classical electromagnetism since it is the most convenient formula for the development of the concept of the magnetic field due to the current in a circuit. We shall return to discuss the general applicability of equation (1.51) in Chapter 6, after deriving the expression for the magnetic field due to an accelerating charge in Chapter 3.

It is shown in elementary text books on electromagnetism that, according to equation (1.51), if there are two thin, infinitely long, straight, parallel wires at a distance r apart in empty space and carrying steady conduction currents I_1 and I_2 amperes, then the magnetic force f per metre length on either of the two conductors is given by

$$f = \frac{\mu_0 I_1 I_2}{2\pi r} . \tag{1.53}$$

The ampere is defined as 'that unvarying current which, if present in each of two infinitely thin, straight, parallel conductors of infinite length and one metre apart in empty space, causes each conductor to experience a force of exactly 2×10^{-7} newtons per metre of length'. Putting $I_1 = I_2 = 1$, $r = 1$ and $f = 2 \times 10^{-7}$ in equation (1.53), we find that μ_0 is equal to $4\pi \times 10^{-7}$ henry per metre. This shows that the value of $\mu_0 = 4\pi \times 10^{-7}$ H m^{-1} follows from the definition of the ampere. The constant μ_0 is called the absolute permeability of free space, though we shall prefer to call it the magnetic constant.

Equation (1.51) is generally divided into two parts. It is said that the steady current I_1 in circuit 1 in Figure 1.6 gives rise to a magnetic field $\mathbf{B}_1$ at the position of the current element I_2 d$\mathbf{l}_2$ of circuit 2, where $\mathbf{B}_1$ is given by

$$\mathbf{B}_1 = \left(\frac{\mu_0 I_1}{4\pi}\right) \oint_1 \frac{\mathrm{d}\mathbf{l}_1 \times \mathbf{r}_{12}}{r_{12}^3}. \tag{1.54}$$

Equation (1.54) is generally called the Biot-Savart law, in honour of the contributions of Biot and of Savart to electromagnetism. Using equation (1.54) we can now rewrite equation (1.51) in the form

$$\mathbf{F}_2 = I_2 \oint_2 \mathrm{d}\mathbf{l}_2 \times \mathbf{B}_1. \tag{1.55}$$

Equation (1.55) will be developed from the Lorentz force law in Section 1.4.3.

1.4.2. *Definition of the magnetic induction (or magnetic flux density)* B

According to the Lorentz force law, equation (1.1), if a test charge of magnitude q is at rest at a field point where there is an electric field of intensity **E** and a magnetic field of magnetic induction (or magnetic flux density) **B** the only force on the stationary test charge is the electric force

$$\mathbf{F}_{\mathrm{elec}} = q\mathbf{E}. \tag{1.56}$$

Equation (1.56) was used to define the electric field intensity **E** in Section 1.2.4. If the test charge q is moving with velocity **u**, according to the Lorentz force law, equation (1.1), there is an extra contribution to the total force on q, over and above the electric force $\mathbf{F}_{\mathrm{elec}}$, namely the magnetic force which is given by

$$\mathbf{F}_{\mathrm{mag}} = q\mathbf{u} \times \mathbf{B}. \tag{1.57}$$

The magnitude of the magnetic force $\mathbf{F}_{\mathrm{mag}}$ depends on the magnitude and direction of **u** the velocity of the test charge. The magnetic force is zero when **u** is parallel to **B**. It is a maximum when **u** is perpendicular to **B**. The strength of the magnetic field **B** can be defined in terms of the maximum measured magnetic force using the relation

$$B = \text{Limit} \frac{(F_{\mathrm{mag}})_{\mathrm{max}}}{qu}. \tag{1.58}$$

in the limit as the magnitude q of the test charge tends to zero. The direction of the magnetic field $\mathbf{B}$ can be defined as the direction in which the test charge q would be moving if it experienced no magnetic force. The sense of $\mathbf{B}$ is defined such that $\mathbf{F}_{\text{mag}}$, $\mathbf{B}$ and $\mathbf{u}$ obey the left hand rule, according to which, if the thumb of the left hand points in the direction of $\mathbf{F}_{\text{mag}}$ and the second finger points in the direction of $\mathbf{u}$, then for a positive test charge the first finger of the left hand points in the direction of $\mathbf{B}$. This direction for $\mathbf{B}$ is in agreement with the direction in which the north pole of a compass needle would point in empty space. In the SI (or MKSA) system of units, in which $\mathbf{F}_{\text{mag}}$ is measured in newtons, q in coulombs and $\mathbf{u}$ in metres per second, the value of $\mathbf{B}$ given by equation (1.58) is expressed in tesla (denoted T), or alternatively in webers per square metre (denoted Wb m^{-2}). The weber is the SI unit of magnetic flux. In some text books the magnetic vector $\mathbf{B}$ is called the magnetic induction, whereas in other text books it is called the magnetic flux density. To avoid the continued use of the long phrase: a magnetic field of magnetic induction (or magnetic flux density) $\mathbf{B}$, we shall generally us the abbreviation 'a magnetic field $\mathbf{B}$', where the symbol $\mathbf{B}$ stands for the magnetic induction (or magnetic flux density).

It is convenient to represent magnetic fields on diagrams using imaginary magnetic field lines, drawn such that the direction of the tangent to the magnetic field line at a field point is in the direction of the magnetic field $\mathbf{B}$ at that point. The number of magnetic field lines is generally limited by spacing the lines, such that on the diagrams the number of lines per square metre crossing a surface perpendicular to the field line is equal to (or, if more convenient, is proportional to) the magnitude of the magnetic field $\mathbf{B}$ at that point.

1.4.3. *The force on a current element in an external magnetic field*

It will now be shown that the expression

$$\mathbf{F} = I\, \mathrm{d}\mathbf{l} \times \mathbf{B}$$

for the magnetic force on a current element of length $\mathrm{d}\mathbf{l}$, that forms part of a stationary, rigid circuit which carries a steady conduction current I and which is in an external magnetic field $\mathbf{B}$, can be derived using the Lorentz force law, equation (1.1). Let the area of cross section of the current element be equal to A and let the number of moving charges, each of charge q, be equal to n per cubic metre. Since the volume of the current element is $A\, \mathrm{d}l$, the total number of moving charges in the current element is $N = nA\, \mathrm{d}l$. According to equation (1.1), the magnetic force on a charge moving with velocity $\mathbf{u}_i$ in a magnetic field $\mathbf{B}$ is equal to $q\mathbf{u}_i \times \mathbf{B}$. Summing over all the moving charges in the current element, we find that the total magnetic force on the $N = nA\, \mathrm{d}l$ moving charges due to the external magnetic field $\mathbf{B}$ is

$$\mathrm{d}\mathbf{F}_{\text{mag}} = \Sigma q\mathbf{u}_i \times \mathbf{B} = q(\Sigma \mathbf{u}_i) \times \mathbf{B}.$$

Since

$$\Sigma \mathbf{u}_i = N\mathbf{v} = nA\mathbf{v}\ \mathrm{d}l$$

where $\mathbf{v}$ is the mean drift velocity of the moving charges, and since according to equation (1.44)

$$I = qnAv$$

we find that the total magnetic force on the moving charges inside I $\mathbf{dl}$ is

$$\mathrm{d}\mathbf{F}_{\text{mag}} = qnA\ \mathrm{d}l\ \mathbf{v} \times \mathbf{B} = I\ \mathbf{dl} \times \mathbf{B}. \tag{1.59}$$

Integrating equation (1.59) around the complete circuit, we find that the total magnetic force on the moving charges in the complete circuit is

$$\mathrm{d}\mathbf{F}_{\text{mag}} = I \oint \mathbf{dl} \times \mathbf{B}. \tag{1.60}$$

This in agreement with equation (1.55). At this point it is generally assumed that equation (1.55) gives the resultant force on an electrically neutral current element in the external magnetic field $\mathbf{B}$. A brief discussion will now be given of how the magnetic force on the moving charges is transmitted to the current element.

Consider the stationary conductor of length l and of cross sectional area A, that lies parallel to the z axis in Figure 1.7(a). A simplified model will be used in which there are n free electrons per cubic metre, each of charge $-e$ and all moving in the $-z$ direction with the same velocity v in a uniform potential well. There is a uniform external magnetic field $\mathbf{B}$ in the $+y$ direction in Figure 1.7(a). According to equation (1.59), the magnetic force on the conduction electrons inside the conductor in Figure 1.7(a) is

$$F_{\text{mag}} = BIl \tag{1.61}$$

in the $-x$ direction. The magnetic force on one of the free conduction electrons is equal to evB in the $-x$ direction. This force tries to move the conduction electrons in the $-x$ direction. Initially, just after the current is switched on, there will be a drift of conduction electrons in the $-x$ direction until negative and positive charge distributions are built up on the side surfaces of the conductor, as shown in Figure 1.7(b). This is the Hall effect. These surface charge distributions are of such magnitudes that they give a Hall electric field $\mathbf{E}_H$ in the $-x$ direction of such a magnitude that the electric force $-e\mathbf{E}_H$ on a moving conduction electron is in the $+x$ direction and balances the magnetic force $-\ e\mathbf{v} \times \mathbf{B}$ in the $-x$ direction. Hence, after the transient state is over

$$E_H = vB. \tag{1.62}$$

Since the magnetic force on a moving conduction electron is balanced by the electric force on the conduction electron due the the Hall electric field, there is no resultant force on a moving conduction electron so that on our

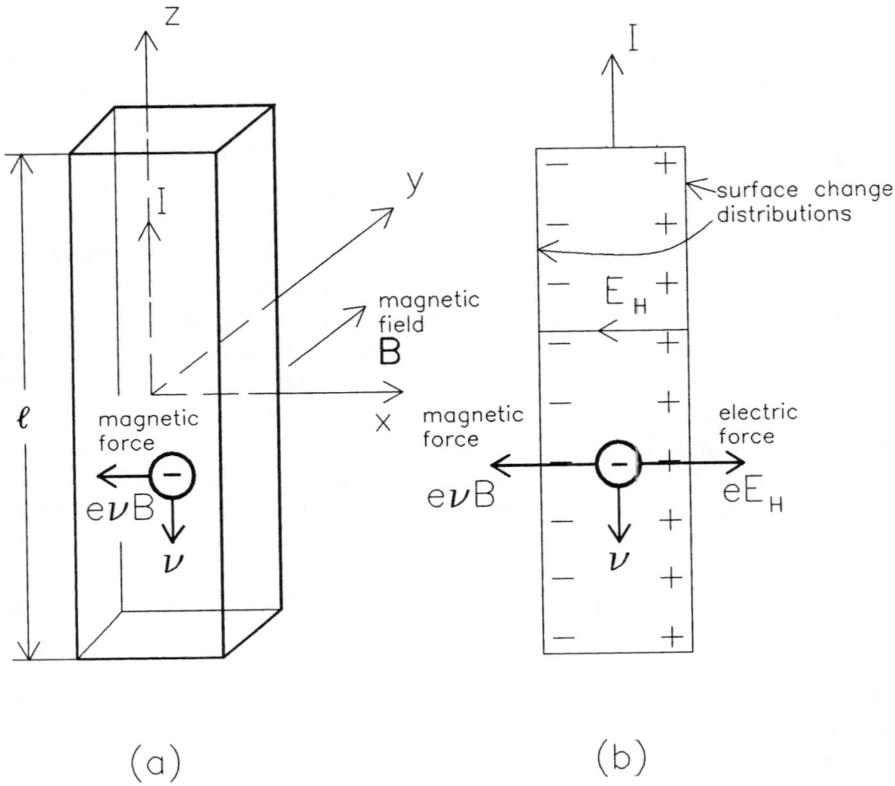

Figure 1.7. Derviation of the magnetic force BIl on a conductor of length l which is carrying a conduction current I in a direction perpendicular to a magnetic field **B**. (a) The magnetic force on one of the conduction electrons. (b) The electric force on the conduction electron due to the Hall electric field.

simplified model the resultant forces on the conduction electrons cannot give rise to the experimentally observed resultant force BIl on the stationary conductor in Figure 1.7(a). It will now be assumed that there are n stationary positive ions per cubic metre each of charge $+e$. There is no magnetic force on the stationary positive ions. The positive and negative electrostatic charge distributions, that give rise to the Hall electric field $\mathbf{E}_H$, attract each other with equal and opposite electrostatic forces which add up to zero. In our low velocity limit the electric forces on the surface charge distributions due to the positive ions and the moving conduction electrons compensate each other. In our simplified model the forces between the positive ions and the conduction electrons give the cohesive forces holding the metallic conductor together, and the resultant of these forces is zero. The force we have not included so far is the force on the stationary positive ions due to the Hall electric field $\mathbf{E}_H$. Since $\mathbf{E}_H$ is in the negative x direction in Figure 1.7(a), the electric force on each positive ion is of magnitude $eE_H = evB$ in the $-x$ direction. Since there are $N = nAl$ stationary positive ions in a length l of the conductor, the total force on the positive ions is $nAl\ evB = BIl$ in the $-x$ direction in Figure 1.7(a). Hence on our simplified model, the resultant force BIl on the stationary current carrying conductor in Figure 1.7(a) comes from the unbalanced force on the stationary positive ions due to the Hall electric field $\mathbf{E}_H$.

The model used so far is very much over simplified. First the mean drift velocity $\mathbf{v}$ of a conduction electrons in a metal, which is only of the order of 10^{-4} m s^{-1}, is superimposed on a velocity distribution, in which individual electrons have speeds up to about $c/200$. It is straightforward for the reader to show that equation (1.59) represents the average magnetic force on the conduction electrons. Secondly, the conduction electrons do not move in a uniform potential well, but in a periodic potential which leads to a band structure, which in turn affects the equation of motion of a conduction electron, leading to the introduction of the concept of effective mass. A reader interested in the application of this more refined model is referred to McKinnon, McAlister and Hurd [11], who show that the resultant force on the current carrying conductor in Figure 1.7(a) is still equal to BIl.

1.4.4. *The Biot-Savart law for the magnetic field due to a steady current in a complete circuit*

According to equation (1.54), the magnetic field $\mathbf{B}$ at a field point at a position $\mathbf{r}$ due to steady current I in a stationary, rigid circuit can be calculated using the formula

$$\mathbf{B}(\mathbf{r}) = \left(\frac{\mu_0 I}{4\pi}\right) \oint \frac{d\mathbf{l}_s(\mathbf{r}_s) \times (\mathbf{r} - \mathbf{r}_s)}{|\mathbf{r} - \mathbf{r}_s|^3} . \qquad (1.63)$$

where $d\mathbf{l}_s(\mathbf{r}_s)$ is an element of the circuit at the source point at $\mathbf{r}_s$ and $(\mathbf{r} - \mathbf{r}_s)$ is a vector from $d\mathbf{l}_s$ to the field point at $\mathbf{r}$. Equation (1.63) is known as the Biot-Savart law.

Equation (1.63) has been checked in its integral form, to an accuracy of about 1 part in 10^7, by experiments with rigid current balances of different geometrical configurations. Since it is not possible to carry out experiments with isolated current elements carrying steady currents, equation (1.63) has not been confirmed directly by experiment in the differential form:

$$d\mathbf{B}(\mathbf{r}) = \left(\frac{\mu_0 I}{4\pi}\right) \frac{d\mathbf{l}_s(\mathbf{r}_s) \times (\mathbf{r} - \mathbf{r}_s)}{|\mathbf{r} - \mathbf{r}_s|^3} . \qquad (1.64)$$

It is not possible to go from the integral form of the Biot-Savart law to the differential form given by equation (1.64), since we can add a function, such as the gradient of a scalar function of position, to the right hand side of equation (1.64) and still obtain equation (1.63) when we integrate around the complete circuit, provided the contribution of the extra function to the magnetic field is zero when it is integrated around the complete circuit. Reference: Whittaker [10]. We shall return in Chapter 6, after developing the full theory of electromagnetism, to discuss the validity of the Biot-Savart law in both the integral form given by equation (1.63) and the differential form given by equation (1.64). It will be sufficient for our initial development of the theory of the magnetic fields due to steady currents to assume that the Biot-Savart law is valid in the integral form given by equation (1.63).

To generalize equation (1.63) to the case of the steady, continuous current distribution shown in Figure 1.8, divide the current distibution into a number of filiamentary current loops. Equation (1.63) can then be applied to each of these. Let the current density at the position $\mathbf{r}_s$, having coordinates x_s, y_s and z_s, be $\mathbf{J}(\mathbf{r}_s)$. The product $\mathbf{J}(\mathbf{r}_s)\,dV_s$ can be treated as a current element. It then follows from the Biot-Savart law, equation (1.63), that the magnetic field $\mathbf{B}(\mathbf{r})$ at a field point at $\mathbf{r}$ is given by

$$\mathbf{B}(\mathbf{r}) = \left(\frac{\mu_0}{4\pi}\right) \int \frac{\mathbf{J}(\mathbf{r}_s) \times (\mathbf{r} - \mathbf{r}_s)\, dV_s}{|\mathbf{r} - \mathbf{r}_s|^3} \,. \tag{1.65}$$

It is assumed in introductory courses that equation (1.65) holds both at a field point outside the current distribution and at a field point, such as P in Figure 1.8, which is inside the steady continuous current distribution.

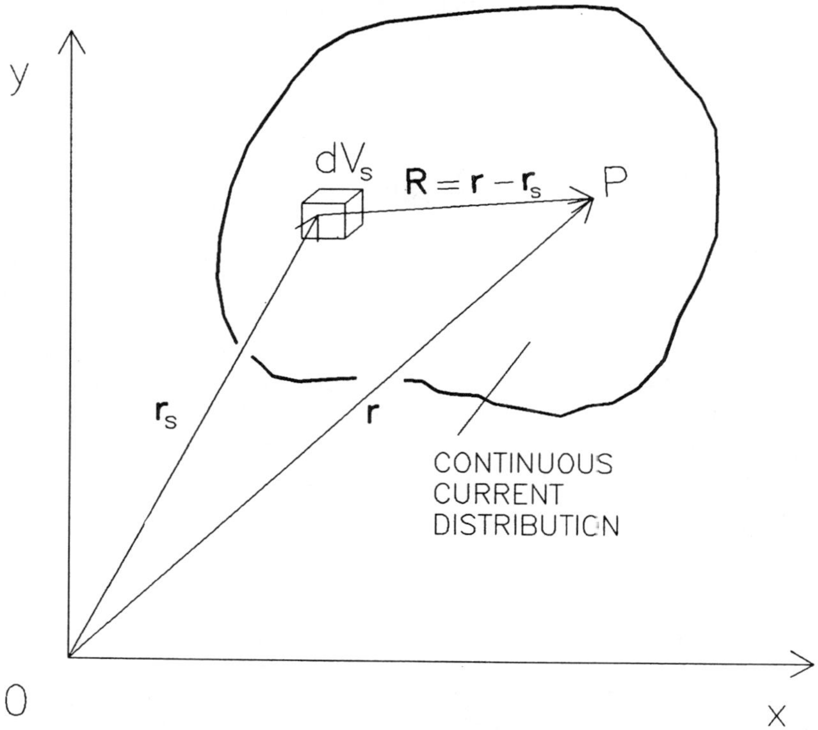

Figure 1.8. Derivation of the equation $\nabla \times \mathbf{B} = \mu_0 \mathbf{J}$ at a field point inside a steady, continuous current distribution.

1.4.5. *The equation* $\nabla \cdot \mathbf{B} = 0$

Consider the magnetic field **B** at a field point that can be either inside or outside the steady current distribution shown in Figure 1.8. According to equation (1.20)

$$\frac{(\mathbf{r} - \mathbf{r}_s)}{|\mathbf{r} - \mathbf{r}_s|^3} = -\nabla\left(\frac{1}{|\mathbf{r} - \mathbf{r}_s|}\right) \tag{1.66}$$

where the operator

$$\nabla = \hat{\mathbf{i}}\frac{\partial}{\partial x} + \hat{\mathbf{j}}\frac{\partial}{\partial y} + \hat{\mathbf{k}}\frac{\partial}{\partial z} \tag{1.67}$$

involves partial differentiation with respect to the coordinates x, y, z of the field point, keeping the position $\mathbf{r}_s$ of the source point fixed. Equation (1.65) can now be rewritten in the form

$$\mathbf{B}(\mathbf{r}) = \left(\frac{\mu_0}{4\pi}\right)\int \nabla\left(\frac{1}{|\mathbf{r} - \mathbf{r}_s|}\right) \times \mathbf{J}(\mathbf{r}_s)\, \mathrm{d}V_s. \tag{1.68}$$

According to equation (A1.23) of Appendix A1.6, for any scalar function ψ of position and any vector function **C** of position

$$\nabla\psi \times \mathbf{C} = \nabla \times \psi\mathbf{C} - \psi(\nabla \times \mathbf{C}).$$

Putting $\psi = 1/|\mathbf{r} - \mathbf{r}_s|$ and $\mathbf{C} = \mathbf{J}(\mathbf{r}_s)$ and then substituting in equation (1.68), we obtain

$$\mathbf{B} = \left(\frac{\mu_0}{4\pi}\right)\int \nabla \times \left(\frac{\mathbf{J}(\mathbf{r}_s)}{|\mathbf{r} - \mathbf{r}_s|}\right) \mathrm{d}V_s - \left(\frac{\mu_0}{4\pi}\right)\int \frac{\nabla \times \mathbf{J}(\mathbf{r}_s)}{|\mathbf{r} - \mathbf{r}_s|}\, \mathrm{d}V_s. \tag{1.69}$$

Now $\nabla \times \mathbf{J}(\mathbf{r}_s)$ involves partial differentiation with respect to the coordinates x, y, z of the field point at **r**. The steady current density $\mathbf{J}(\mathbf{r}_s)$ at the fixed source point at $\mathbf{r}_s$ does not change, if the coordinates x, y, z of the field point at **r** are changed. Hence $\nabla \times \mathbf{J}(\mathbf{r}_s)$ is zero, so that the second integral on the right hand side of equation (1.69) is zero. Since integrating with respect to $\mathrm{d}V_s = \mathrm{d}x_s\, \mathrm{d}y_s\, \mathrm{d}z_s$ and taking the curl by varying the coordinates x, y, z of the field point are independent linear operations, the order in which the operations are carried out can be reversed. Hence, for steady currents, equation (1.69) can be rewritten in the form

$$\mathbf{B}(\mathbf{r}) = \nabla \times \left(\frac{\mu_0}{4\pi}\right)\int \frac{\mathbf{J}(\mathbf{r}_s)}{|\mathbf{r} - \mathbf{r}_s|}\, \mathrm{d}V_s. \tag{1.70}$$

Since, according to equation (A1.25) of Appendix A1.6, the divergence of the curl of any vector is zero, it follows by taking the divergence of both sides of equation (1.70) that at field points both inside and outside the steady current distribution in Figure 1.8

$$\nabla \cdot \mathbf{B} = 0. \tag{1.71}$$

Integrating equation (1.71) over a finite volume and applying Gauss' theorem of vector analysis, which is equation (A1.30) of Appendix A1.7, we obtain

$$\int \nabla \cdot \mathbf{B} \, dV = \int \mathbf{B} \cdot d\mathbf{S} = 0. \tag{1.72}$$

Since according to equation (1.72) $\int \mathbf{B} \cdot d\mathbf{S}$ is zero, there is no net flux of the magnetic field **B**, due to a steady stationary current distribution, from any arbitrary closed surface. It is assumed in classical electromagnetism that equation (1.71) also holds for the magnetic field **B** due to a varying current distribution.

It has been suggested, mainly on theoretical grounds, that magnetic monopoles might exist, and that magnetic field lines would diverge from such magnetic monopoles. The divergence of **B** would not be zero for a system of such magnetic monopoles. However, even if it were proved that magnetic monopoles do exist, they would be so rare that they would play no significant role in classical electromagnetism and equation (1.71) would almost invariably be valid in practice.

1.4.6. *The vector potential* A *due to a steady current distribution*

Consider again the steady, continuous current distribution shown in Figure 1.8. Equation (1.70) can be rewritten in the form

$$\mathbf{B} = \nabla \times \mathbf{A} \tag{1.73}$$

where the value of the vector **A** at the field point at a position **r** is given by

$$\mathbf{A}(\mathbf{r}) = \left(\frac{\mu_0}{4\pi}\right) \int \frac{\mathbf{J}(\mathbf{r}_s) \, dV_s}{|\mathbf{r} - \mathbf{r}_s|} . \tag{1.74}$$

In equation (1.74), $\mathbf{J}(\mathbf{r}_s)$ is the current dentsity at the source point at $\mathbf{r}_s$ having coordinates x_s, y_s, z_s and $dV_s = dx_s \, dy_s \, dz_s$ is a volume element at $\mathbf{r}_s$. The vector **A** is called the vector potential.

Taking the divergence of both sides of equation (1.74) at the field point at **r** having coordinates x, y, z, we have

$$\begin{aligned} \nabla \cdot \mathbf{A}(\mathbf{r}) &= \nabla \cdot \left[\left(\frac{\mu_0}{4\pi}\right) \int \frac{\mathbf{J}(\mathbf{r}_s) \, dV_s}{|\mathbf{r} - \mathbf{r}_s|}\right] \\ &= \left(\frac{\mu_0}{4\pi}\right) \int \nabla \cdot \left(\frac{\mathbf{J}(\mathbf{r}_s)}{|\mathbf{r} - \mathbf{r}_s|}\right) dV_s \end{aligned} \tag{1.75}$$

where the operator ∇ is given by equation (1.67). Putting $\mathbf{C} = \mathbf{J}(\mathbf{r}_s)$ and $\psi = 1/|\mathbf{r} - \mathbf{r}_s|$ in equation (A1.20) of Appendix A1.6, which is

$$\nabla \cdot \psi\mathbf{C} = \psi\nabla \cdot \mathbf{C} + \mathbf{C} \cdot \nabla\psi \tag{1.76}$$

and then substituting in equation (1.75), we obtain

$$\nabla \cdot \mathbf{A}(\mathbf{r}) = \left(\frac{\mu_0}{4\pi}\right) \int \frac{\nabla \cdot \mathbf{J}(\mathbf{r}_s)\, dV_s}{|\mathbf{r} - \mathbf{r}_s|} + \left(\frac{\mu_0}{4\pi}\right) \int \mathbf{J}(\mathbf{r}_s) \cdot \nabla \left(\frac{1}{|\mathbf{r} - \mathbf{r}_s|}\right) dV_s. \tag{1.77}$$

Since, for a steady current distribution the value of the steady current density $\mathbf{J}(\mathbf{r}_s)$ at the fixed source point at $\mathbf{r}_s$ does not vary when the coordinates x, y and z at the field point are changed, then

$$\nabla \cdot \mathbf{J}(\mathbf{r}_s) = 0. \tag{1.78}$$

Hence, the first integral on the right hand side of equation (1.77) is zero. The reader can check by carrying out the partial differentiations that

$$\nabla \left(\frac{1}{|\mathbf{r} - \mathbf{r}_s|}\right) = \nabla \left(\frac{1}{\{(x - x_s)^2 + (y - y_s)^2 + (z - z_s)^2\}^{1/2}}\right)$$

$$= -\nabla_s \left(\frac{1}{|\mathbf{r} - \mathbf{r}_s|}\right) \tag{1.79}$$

where the new operator

$$\nabla_s = \hat{\mathbf{i}} \frac{\partial}{\partial x_s} + \hat{\mathbf{j}} \frac{\partial}{\partial y_s} + \hat{\mathbf{k}} \frac{\partial}{\partial z_s} \tag{1.80}$$

is evaluated by keeping the coordinates x, y, z of the field point fixed and varying the coordinates x_s, y_s and z_s of the source point. Using equations (1.78) and (1.79), equation (1.77) becomes

$$\nabla \cdot \mathbf{A}(\mathbf{r}) = -\left(\frac{\mu_0}{4\pi}\right) \int \mathbf{J}(\mathbf{r}_s) \cdot \nabla_s \left(\frac{1}{|\mathbf{r} - \mathbf{r}_s|}\right) dV_s. \tag{1.81}$$

Putting $\mathbf{C} = \mathbf{J}(\mathbf{r}_s)$ and $\psi = 1/|\mathbf{r} - \mathbf{r}_s|$ in equation (1.76) and remembering that, according to equation (1.50), $\nabla_s \cdot \mathbf{J}(\mathbf{r}_s)$ is zero for steady currents, we have

$$\mathbf{J}(\mathbf{r}_s) \cdot \nabla_s \left(\frac{1}{|\mathbf{r} - \mathbf{r}_s|}\right) = \nabla_s \cdot \left(\frac{\mathbf{J}(\mathbf{r}_s)}{|\mathbf{r} - \mathbf{r}_s|}\right).$$

Substituting in equation (1.81) and applying Gauss' theorem of vector analysis, which is equation (A1.30) of Appendix A1.7, we obtain

$$\nabla \cdot \mathbf{A}(\mathbf{r}) = -\left(\frac{\mu_0}{4\pi}\right) \int \nabla_s \cdot \left(\frac{\mathbf{J}(\mathbf{r}_s)}{|\mathbf{r} - \mathbf{r}_s|}\right) dV_s = -\left(\frac{\mu_0}{4\pi}\right) \int \frac{\mathbf{J}(\mathbf{r}_s) \cdot d\mathbf{S}_s}{|\mathbf{r} - \mathbf{r}_s|}. \tag{1.82}$$

Provided the current distribution is not of infinite extent, which is true in all practical cases, we can always find a surface that is completely outside the current distribution such that $\mathbf{J}(\mathbf{r}_s)$ is zero at all points on that surface, in which case the right hand side of equation (1.82) is zero so that

$$\nabla \cdot \mathbf{A} = 0. \tag{1.83}$$

This shows that the divergence of the vector potential $\mathbf{A}$, that is given by equation (1.74), is zero. The expression for the vector potential **A**, given by equation (1.74), was derived from the integral form of the Biot-Savart law for a steady current distribution.

1.4.7. *The equation* $\nabla \times \mathbf{B} = \mu_0 \mathbf{J}$ *and Ampère's circuital theorem*

Ampère's circuital theorem can be developed from the integral form of the Biot-Savart law, equation (1.63), using the magnetic scalar potential. Reference: Scott [12]. The alternative method we shall use will give the reader some useful practice using vector analysis.

Consider again the stationary, steady, continuous current distribution shown in Figure 1.8. The field point P is inside the continuous current distribution, as shown in Figure 1.8. Taking the curl of both sides of equation (1.73) and using equation (A1.27) of Appendix A1.6, we have

$$\nabla \times \mathbf{B} = \nabla \times (\nabla \times \mathbf{A}) = \nabla(\nabla \cdot \mathbf{A}) - \nabla^2 \mathbf{A}.$$

According to equation (1.83), $\nabla \cdot \mathbf{A}$ is zero when **A** is given by equation (1.74). Substituting for **A** from equation (1.74), we have

$$\nabla \times \mathbf{B}(\mathbf{r}) = -\nabla^2 \left[\left(\frac{\mu_0}{4\pi} \right) \int \frac{\mathbf{J}(\mathbf{r}_s)\, \mathrm{d}V_s}{|\mathbf{r} - \mathbf{r}_s|} \right]. \tag{1.84}$$

Since integrating with respect to $\mathrm{d}V_s = \mathrm{d}x_s\, \mathrm{d}y_s\, \mathrm{d}z_s$ and the application of the Laplacian

$$\nabla^2 = \frac{\partial^2}{\partial x^2} + \frac{\partial^2}{\partial y^2} + \frac{\partial^2}{\partial z^2} \tag{1.85}$$

are independent linear operations, the order in which the operations are carried out can be reversed, so that equation (1.84) can be rewritten in the form

$$\nabla \times \mathbf{B}(\mathbf{r}) = -\left(\frac{\mu_0}{4\pi} \right) \int \nabla^2 \left(\frac{\mathbf{J}(\mathbf{r}_s)}{|\mathbf{r} - \mathbf{r}_s|} \right) \mathrm{d}V_s. \tag{1.86}$$

Since for a steady current distribution the value of $\mathbf{J}(\mathbf{r}_s)$ at the fixed source point at $\mathbf{r}_s$ is independent of changes in the coordinates x, y, z of the field point when the operator ∇^2 is applied, equation (1.86) can be rewritten in the form

$$\nabla \times \mathbf{B}(\mathbf{r}) = -\left(\frac{\mu_0}{4\pi} \right) \int \mathbf{J}(\mathbf{r}_s) \nabla^2 \left(\frac{1}{|\mathbf{r} - \mathbf{r}_s|} \right) \mathrm{d}V_s. \tag{1.87}$$

The reader can check by differentiating that, when **r** is not equal to $\mathbf{r}_s$

$$\nabla^2 \left(\frac{1}{|\mathbf{r} - \mathbf{r}_s|} \right) = \nabla^2 \left(\frac{1}{\{(x - x_s)^2 + (y - y_s)^2 + (z - z_s)^2\}^{1/2}} \right) = 0.$$

Hence the integral on the right hand side of equation (1.87) is zero except in the near vicinity of the point $\mathbf{r} = \mathbf{r}_s$. Consider a very small spherical volume with its centre at the field point at **r**. The variation of $\mathbf{J}(\mathbf{r}_s)$ inside this very small volume can be neglected, so that very close to $\mathbf{r}$ we have $\mathbf{J}(\mathbf{r}_s) = \mathbf{J}(\mathbf{r})$. Hence, when equation (1.87) is applied to only the small spherical volume around the point $\mathbf{r}$, equation (1.87) can be rewritten in the form

$$\begin{aligned}\nabla \times \mathbf{B}(\mathbf{r}) &= -\frac{\mu_0 \mathbf{J}(\mathbf{r})}{4\pi} \int \nabla \cdot \left[\nabla\left(\frac{1}{|\mathbf{r} - \mathbf{r}_s|}\right)\right] \mathrm{d}V_s \\ &= +\frac{\mu_0 \mathbf{J}(\mathbf{r})}{4\pi} \int \nabla \cdot \left(\frac{\mathbf{R}}{R^3}\right) \mathrm{d}V_s \end{aligned} \tag{1.88}$$

where we have rewritten the Laplacian ∇^2 as $\nabla \cdot \nabla$ then used equation (1.20) and put $\mathbf{r} - \mathbf{r}_s = \mathbf{R}$. The integration in equation (1.88) is only over the volume of the small spherical volume with centre at $\mathbf{r}$. The integrand is zero elsewhere. Since $\partial R/\partial x = -\partial R/\partial x_s$, we can rewrite the first terms in the expressions for $\nabla \cdot (\mathbf{R}/R^3)$ and $\nabla_s \cdot (\mathbf{R}/R^3)$ as follows:

$$\begin{aligned}\frac{\partial}{\partial x}\left(\frac{x - x_s}{R^3}\right) &= (x - x_s)\left(-\frac{3}{R^4}\right)\left(\frac{\partial R}{\partial x}\right) + \frac{1}{R^3} \\ &= (x - x_s)\left(-\frac{3}{R^4}\right)\left(-\frac{\partial R}{\partial x_s}\right) + \frac{1}{R^3}\end{aligned}$$

$$\frac{\partial}{\partial x_s}\left(\frac{x - x_s}{R^3}\right) = (x - x_s)\left(-\frac{3}{R^4}\right)\left(\frac{\partial R}{\partial x_s}\right) - \frac{1}{R^3} = -\frac{\partial}{\partial x}\left(\frac{x - x_s}{R^3}\right)$$

with similar expressions for the other components. Hence

$$\nabla \cdot \left(\frac{\mathbf{R}}{R^3}\right) = -\nabla_s \cdot \left(\frac{\mathbf{R}}{R^3}\right).$$

Using this relation in equation (1.88) and applying Gauss' theorem of vector analysis, which is equation (A1.30) of Appendix A1.17, we find that

$$\begin{aligned}\nabla \times \mathbf{B} = -\frac{\mu_0 \mathbf{J}}{4\pi} \int \nabla_s \cdot \left(\frac{\mathbf{R}}{R^3}\right) \mathrm{d}V_s &= -\frac{\mu_0 \mathbf{J}}{4\pi} \int \left(\frac{(\mathbf{r} - \mathbf{r}_s)}{|\mathbf{r} - \mathbf{r}_s|^3}\right) \cdot \mathrm{d}\mathbf{S}_s \\ &= \frac{\mu_0 \mathbf{J}}{4\pi} \int \mathrm{d}\Omega\end{aligned}$$

where

$$\mathrm{d}\Omega = \frac{(\mathbf{r}_s - \mathbf{r}) \cdot \mathrm{d}\mathbf{S}_s}{|\mathbf{r}_s - \mathbf{r}|^3} = -\frac{(\mathbf{r} - \mathbf{r}_s) \cdot \mathrm{d}\mathbf{S}_s}{|\mathbf{r} - \mathbf{r}_s|^3}$$

is the solid angle subtended at the field point at $\mathbf{r}$ by the element of area $\mathrm{d}\mathbf{S}_s$ at $\mathbf{r}_s$ of the small sphere with centre at $\mathbf{r}$. Since $\int \mathrm{d}\Omega$ is equal to 4π, we finally find that at the field point P at the position $\mathbf{r}$ inside the steady, stationary, continuous current distribution in Figure 1.8,

$$\nabla \times \mathbf{B}(\mathbf{r}) = \mu_0 \mathbf{J}(\mathbf{r}). \tag{1.89}$$

Equation (1.89) relates the curl of the total magnetic field $\mathbf{B}(\mathbf{r})$ at the field point P at the position $\mathbf{r}$ in Figure 1.8 due to all the currents in the system, to the local value of the current density $\mathbf{J}(\mathbf{r})$ at the field point. Equation

(1.89) was derived from the integral form of the Biot-Savart law, which, for a steady, continuous current distribution, is given by equation (1.65).

Integrating equation (1.89) over any arbitrary surface, we have

$$\int \nabla \times \mathbf{B} \cdot d\mathbf{S} = \mu_0 \int \mathbf{J} \cdot d\mathbf{S}.$$

Applying Stokes' theorem of vector analysis, which is equation (A1.34) of Appendix A1.8, we obtain

$$\oint \mathbf{B} \cdot d\mathbf{l} = \mu_0 I \tag{1.90}$$

where $I = \int \mathbf{J} \cdot d\mathbf{S}$ is the total current crossing any surface bounded by the line integral on the left hand side of equation (1.90). Equation (1.90) is **Ampère's circuital theorem.**

1.4.8. *The differential equation for the vector potential due to a stationary, steady continuous current distribution*

Putting $\mathbf{B} = \nabla \times \mathbf{A}$ in equation (1.89) and using equation (A1.27) of Appendix A1.6, we have

$$\nabla \times (\nabla \times \mathbf{A}) = \nabla(\nabla \cdot \mathbf{A}) - \nabla^2\mathbf{A} = \mu_0\mathbf{J}. \tag{1.91}$$

Since according to equation (1.83), $\nabla \cdot \mathbf{A}$ is zero for the time independent vector potential $\mathbf{A}$ given by equation (1.74), it follows from equation (1.91) that, for steady, continuous current distributions,

$$\nabla^2\mathbf{A} = -\mu_0\mathbf{J} \tag{1.92}$$

in the gauge in which

$$\nabla \cdot \mathbf{A} = 0. \tag{1.93}$$

Each cartesian component of $\mathbf{A}$ satisfies a differential equation of the form

$$\nabla^2 A_x = -\mu_0 J_x.$$

This is similar to Poisson's equation of electrostatics, which is equation (1.29) and the solution of which is given by equation (1.25). Hence

$$A_x(\mathbf{r}) = \left(\frac{\mu_0}{4\pi}\right) \int \frac{J_x(\mathbf{r}_s)\, dV_s}{|\mathbf{r} - \mathbf{r}_s|}.$$

Combining the solutions of A_x, A_y and A_z leads to equation (1.74). Equations (1.92) and (1.74) will be extended to the case of a varying current distribution in Chapter 2.

1.5. Magnetic forces as a second order effect

In electrostatics we had

$$\nabla \cdot \mathbf{E} = \frac{\rho}{\varepsilon_0}; \qquad \nabla \times \mathbf{E} = 0.$$

For the magnetic fields due to steady currents (magnetostatics) we had

$$\nabla \cdot \mathbf{B} = 0; \qquad \nabla \times \mathbf{B} = \mu_0 \mathbf{J}.$$

The above equations of electrostatics and of magnetostatics are completely independent of each other, so that initially these two branches of electromagnetism can be developed independently. This often leaves students without any idea of the relative magnitudes of the electric and magnetic forces between moving charges. In this section we shall consider a simple example that shows that the magnetic forces between moving charges is of the order of (v^2/c^2) times the electric forces between the charges, where v is the velocity of the charges.

Consider two infinitely long, straight, thin, uniformly charged, non-conducting wires a distance r apart in empty space, as shown in Figure 1.9. The

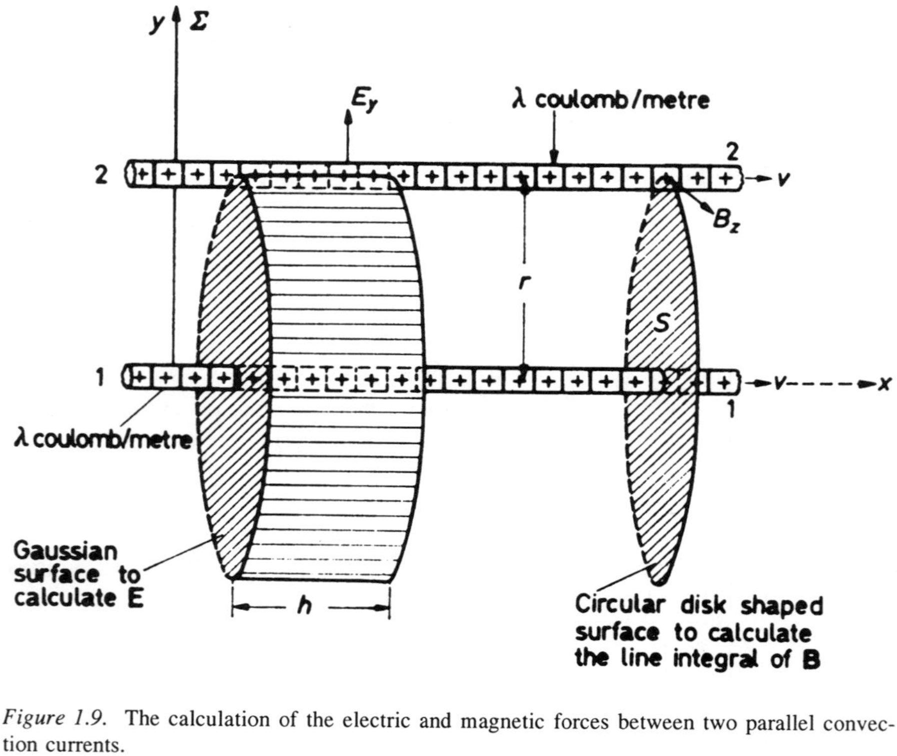

Figure 1.9. The calculation of the electric and magnetic forces between two parallel convection currents.

wires are in the xy plane of a cartesian coordinate system fixed in the laboratory frame, which will be referred to as the inertial reference frame Σ. Both wires are moving in the positive x direction with uniform velocity v in the laboratory system Σ, as shown in Figure 1.9. Let the electric charge on each wire be λ coulombs per metre, measured in the laboratory frame Σ. Since the wires are moving with uniform velocity v, both charge distributions give electric convection currents of magnitudes $I = \lambda v$ in Σ. If it is assumed that the equation $\nabla \cdot \mathbf{E} = \rho/\varepsilon_0$ holds for moving charge distributions, Gauss' flux law equation (1.17), can be used to determine the electric fields in Σ. By symmetry, the electric field due to the charge distribution on wire 1 must diverge radially from wire 1. Applying Gauss' flux law, equation (1.17), to the cylindrical Gaussian surface of radius r and height h, shown in Figure 1.9, we find that

$$\int \mathbf{E} \cdot \mathrm{d}\mathbf{S} = 2\pi r h E_r = \frac{\lambda h}{\varepsilon_0}.$$

Hence the electric field $E_y = E_r$ at the position of wire 2 due to the electric charge on wire 1 is

$$E_y = E_r = \frac{\lambda}{2\pi\varepsilon_0 r}. \tag{1.94}$$

According to the equation (1.10) this electric field gives rise to an electric force of repulsion on wire 2, which is the $+y$ direction and is of magnitude

$$F_{\text{elec}} = \frac{\lambda^2}{2\pi\varepsilon_0 r} \quad \text{newtons per metre length.} \tag{1.95}$$

Consider now the circular disk-shaped surface S of radius r that has wire 1 at its centre as shown in Figure 1.9. Applying Ampère's circuital theorem, equation (1.90), we have

$$\oint \mathbf{B} \cdot \mathrm{d}\mathbf{l} = \mu_0 I = \mu_0 \lambda v.$$

By symmetry $\mathbf{B}$ has the same value at all points on the circumference of the surface S, so that

$$2\pi r B = \mu_0 \lambda v.$$

Hence the magnetic field at the position of wire 2 in Figure 1.9 due to the convection current λv due to the motion of wire 1 is

$$B_z = \frac{\mu_0 \lambda v}{2\pi r}.$$

Comparing with equation (1.94) we see that in the present case

$$\frac{B_z}{E_y} = v\varepsilon_0\mu_0 = \frac{v}{c^2}. \tag{1.96}$$

According to equation (1.59), the magnetic force on wire 2 is $\int I_2 \mathrm{d}\mathbf{l}_2 \times \mathbf{B}_1$, where $I_2 = \lambda v$. Hence the magnetic force on wire 2 is an attractive force given by

$$F_{\text{mag}} = -\frac{\mu_0 \lambda^2 v^2}{2\pi r} \qquad \text{newtons per metre length.} \tag{1.97}$$

Adding questions (1.95) and (1.97), we find that the total force $\mathbf{F} = \mathbf{F}_{\text{elec}} + \mathbf{F}_{\text{mag}}$ on one metre length of wire 2, measured in the laboratory frame Σ, is

$$F_y = \frac{\lambda^2}{2\pi\varepsilon_0 r}(1 - \mu_0\varepsilon_0 v^2).$$

Numerical substitution of $\varepsilon_0 = 8.85 \times 10^{-12}$ F m^{-1} and $\mu_0 = 4\pi \times 10^{-7}$ H m^{-1} shows that $\mu_0\varepsilon_0 = 1/c^2$, where $c = 3 \times 10^8$ m s^{-1} is the speed of light in empty space. Hence

$$F_y = \frac{\lambda^2}{2\pi\varepsilon_0 r}\left(1 - \frac{v^2}{c^2}\right) \qquad \text{newtons per metre length.} \tag{1.98}$$

In this simple example, the ratio of the magnetic force of attraction to the electric force of repulsion between the moving, charged wires in Figure 1.9 is v^2/c^2, where v is the velocity of the charged wires in the laboratory frame Σ. This example illustrates how the magnetic forces between electric charges moving with velocity v are only of the order of v^2/c^2 times the electric forces between the moving charges. Furthermore it can be shown that equation (1.98) can be derived from Coulomb's law of electrostatics and the transformations of the theory of special relativity, illustrating how the magnetic forces between moving charges are second order relativistic effects compared with the electric forces between the charges. References: Section 10.6 of Chapter 10 and Rosser [13]. This relativistic approach illustrates the essential unity of electrostatics and magnetostatics in a vivid way.

It was shown in Section 1.3.3 that when a current of 1A flows in a copper wire of cross sectional area 1 mm^2 (10^{-6} m^2), the mean drift velocity of the conduction electrons is only $v = 7.3 \times 10^{-5}$ m s^{-1}. If the charges on the wires in Figure 1.9 moved with this speed, according to equation (1.94) the ratio of the magnetic force to the electric force on wire 2 would be about 5×10^{-26}. Why then are the magnetic forces between electric circuits so important? In practice, there is no resultant volume charge density inside a metallic conductor such as copper. The negative charges on the conduction electrons are compensated by the charges on the positive ions which are at rest in a stationary conductor as illustrated in Figure 1.10. (The exceedingly small surface and boundary charge distributions associated with conduction current flow, discussed in Section 1.3.5 and Appendix B, are being neglected). Let the total positive charge on the positive ions be equal to $+\lambda$ coulombs per metre length and let the total charge on the conduction electrons be equal to $-\lambda$ coulombs per metre length.

According to equation (1.98), the force on the moving electrons in con-

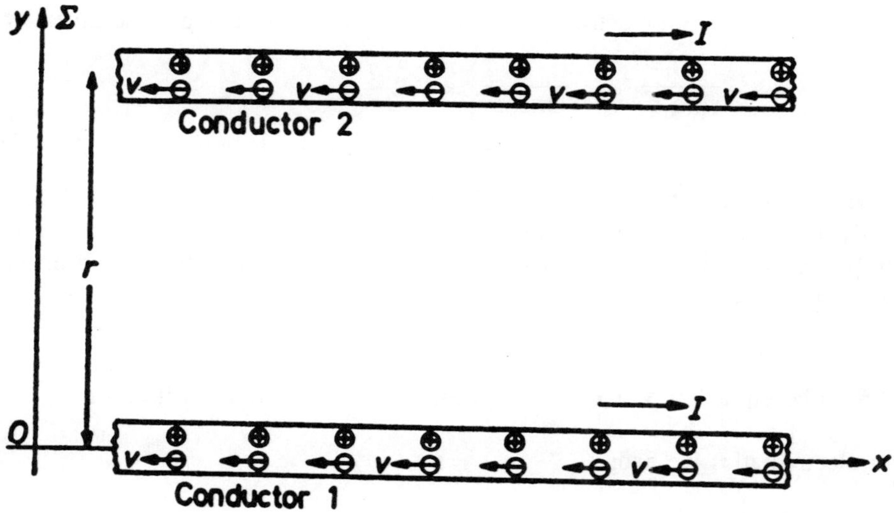

Figure 1.10. The forces between two conduction currents I. In the simplified model used, the positive ions are at rest and the negative electrons all move with the same uniform velocity v. The positive charge per unit length is $+\lambda$ and the negative charge per unit length $-\lambda$. The electric forces cancel leaving only the second order magnetic forces between the charges.

ductor 2 in Figure 1.10 due to the moving electrons in conductor 1 is a repulsive force (magnetic plus electric) given by

$$F_{--} = \frac{\lambda^2}{2\pi\varepsilon_0 r}\left(1 - \frac{v^2}{c^2}\right) \quad \text{newtons per metre length.} \tag{1.99}$$

Since the positive ions in conductor 2 are at rest, there is no magnetic force on them, so that the total force on the stationary positive ions in conductor 2 due to the moving electrons in conductor 1 is an attractive electric force given by

$$F_{-+} = -\frac{\lambda^2}{2\pi\varepsilon_0 r} \quad \text{newtons per metre length.} \tag{1.100}$$

The force on the moving electrons in conductor 2 due to the stationary positive ions in conductor 1 is an attractive force given by

$$F_{+-} = -\frac{\lambda^2}{2\pi\varepsilon_0 r} \quad \text{newtons per metre length.} \tag{1.101}$$

The force on the positive ions in conductor 2 due to the stationary positive ions in conductor 1 is a repulsive electric force given by

$$F_{++} = \frac{\lambda^2}{2\pi\varepsilon_0 r} \quad \text{newtons per metre length.} \tag{1.102}$$

Adding equations (1.99), (1.100), (1,101) and (1.102), we find that the resultant force per metre length on conductor 2 due to all the charges (positive ions and conduction electrons) in conductor 1 is an attractive force given by

$$F = -\frac{\lambda^2 v^2}{2\pi\varepsilon_0 c^2 r} = -\frac{I^2}{2\pi\varepsilon_0 c^2 r} = -\frac{\mu_0 I^2}{2\pi r}$$

newtons per metre length. (1.103)

where we have put $1/\varepsilon_0 c^2 = \mu_0$. Thus the electric forces between the electrically neutral current carrying conductors in Figure 1.10 add up to zero, leaving only the second order attractive magnetic force $\mu_0 I^2/2\pi r$ that is observed experimentally.

1.6. The equation $\nabla \times \mathbf{E} = -\partial\mathbf{B}/\partial t$ and electromagnetic induction

In electrostatics we had:

$$\nabla \cdot \mathbf{E} = \frac{\rho}{\varepsilon_0}; \qquad \nabla \times \mathbf{E} = 0.$$

For the magnetic fields due to steady currents (magnetostatics), we had:

$$\nabla \cdot \mathbf{B} = 0; \qquad \nabla \times \mathbf{B} = \mu_0 \mathbf{J}.$$

It can be seen that, when the charge density ρ and the current density $\mathbf{J}$ are both constant, that is for electrostatics and magnetostatics, the equations for $\mathbf{E}$ and the equations for $\mathbf{B}$ are independent of each other. Experiments will now be outlined which show that, when the charge and current distributions are varying with time, the vectors $\mathbf{E}$ and $\mathbf{B}$ at a field point are related. We shall start by considering electromagnetic induction in this section and then go on to introduce the displacement current in Section 1.7.

It is important to separate two distinct phenomena that can both contribute to what is generally called the induced emf in a closed circuit.

(a) *Motional induced emf (or dynamo emf)*

A motional induced emf is generated when a conductor moves in a magnetic field as, for example, in a dynamo. The origin of motional emfs will be treated in detail in Section 7.7 of Chapter 7, where it will be shown that motional induced emfs arise from the magnetic forces acting on the conduction electrons that are moving with a conductor that is moving in a magnetic field.

(b) *Transformer induced emf*

Consider two stationary coils that are in a vacuum, as shown in Figure 1.11(a). Since both coils are at rest, there are no motional induced emfs. It is found experimentally that, when the current in the primary coil is varying with time, for example when the key K in Figure 1.11(a) is closed or opened, a current flows in the secondary coil. This current flow in the secondary coil

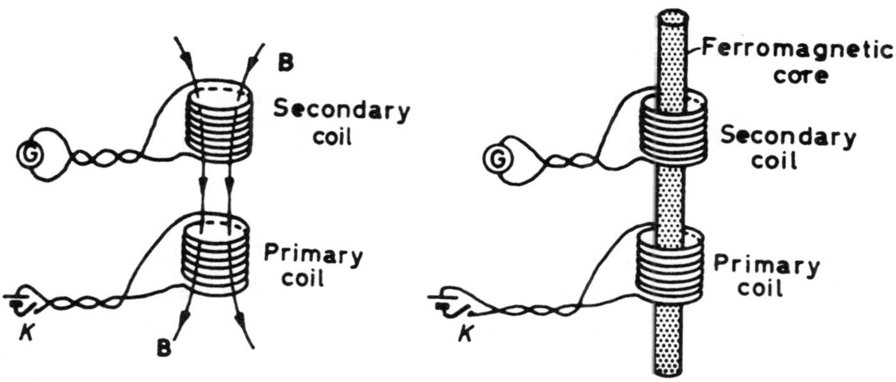

Figure 1.11. (a) A stationary air-cored transformer; (b) a stationary iron-cored transformer.

is only present when the current in the primary coil is varying. The current in the secondary coil depends on the resistance of the secondary, so that it is an emf that is induced in the secondary coil and not a fixed current. It is found experimentally that the emf, denoted ε_{sec}, induced in the secondary coil is proportional to the rate of change of the magnetic flux Φ_{sec} passing through the secondary coil due to the current flowing in the primary coil, that is

$$\varepsilon_{sec} = -\frac{\partial \Phi}{\partial t} = -\frac{\partial}{\partial t}\int \mathbf{B} \cdot d\mathbf{S} \tag{1.104}$$

where **B** is the magnetic field due to the current flowing in the primary coil and the integral is evaluated over a surface bounded by the secondary coil. This induced emf is an example of a transformer induced emf. Equation (1.104) is generally called **Faraday's law of electromagnetic induction**. If an alternating current flows in the primary coil, an alternating current flows in the secondary coil. This is the principle of the air-cored transformer. Since there are no moving parts in the air-cored transformer in Figure 1.11(a), there are no motional induced emfs. According to the constitutive equation $\mathbf{J} = \sigma\mathbf{E}$, the conduction current in the secondary coil in Figure 1.11(a) should be due to an electric field **E** inside the wire making up the secondary coil. The emf in the stationary secondary coil is equal to $\oint \mathbf{E} \cdot d\mathbf{l}$, which is the line integral of the electric field **E** taken around the secondary coil, so that using equation (1.104) we have

$$\varepsilon_{sec} = \oint_{sec} \mathbf{E} \cdot d\mathbf{l} = -\frac{\partial}{\partial t}\int_{sec} \mathbf{B} \cdot d\mathbf{S} \tag{1.105}$$

Using Stokes' theorem of vector analysis, which is equation (A1.34) of Appendix A1.8, equation (1.105) becomes

$$\oint \mathbf{E} \cdot d\mathbf{l} = \int \nabla \times \mathbf{E} \cdot d\mathbf{S} = \int \left(-\frac{\partial \mathbf{B}}{\partial t}\right) \cdot d\mathbf{S}. \tag{1.106}$$

If the area $\Delta\mathbf{S}$ of the secondary coil is small enough for the variations of $\nabla \times \mathbf{E}$ and $\partial\mathbf{B}/\partial t$ over $\Delta\mathbf{S}$ to be negligible, then equation (1.106) becomes

$$\nabla \times \mathbf{E} \cdot \Delta\mathbf{S} = \left(-\frac{\partial\mathbf{B}}{\partial t}\right) \cdot \Delta\mathbf{S}$$

so that

$$\nabla \times \mathbf{E} = -\frac{\partial\mathbf{B}}{\partial t}. \tag{1.107}$$

It is assumed in classical electromagnetism that, when the magnetic field due to the current in the primary coil is varying, there is an induction electric field, whose curl is given by equation (1.107), present at the position of the secondary coil in Figure 1.11(a) and in other parts of empty space, whether the secondary coil is present or not. Equation (1.107) is a relation between the field vectors **E** and **B**, which is valid at any field point. The full properties of induction electric fields will be developed later in Chapters 5 and 7.

Even though equation (1.107) is normally developed from experiments on transformers carried out at mains frequency, it is assumed in classical electromagnetism that equation (1.107) holds at all frequencies, for example for the radiation fields due to high frequency radio transmitters.

If there is a ferromagnetic material passing though the primary and secondary coils, as shown in Figure 1.11(b), the transformer induced emf in the secondary coil is increased. The ferromagnetic core increases both the total magnetic flux passing through the secondary coil and the emf induced in the secondary coil. It is found experimentally that equations (1.104) and (1.107) are still valid.

1.7. The equation $\nabla \times \mathbf{B} = \mu_0(\mathbf{J} + \epsilon_0\dot{\mathbf{E}})$ and the displacement current

In Section 1.4.7 we developed equation (1.89), which is

$$\nabla \times \mathbf{B} = \mu_0\mathbf{J} \tag{1.108}$$

from the Biot-Savart law, equation (1.65), for the magnetic field due to a steady charge and current distribution. Since, according to equation (A1.25) of Appendix A1.6, the divergence of the curl of any vector is zero we must always have

$$\nabla \cdot (\nabla \times \mathbf{B}) = 0. \tag{1.109}$$

It follows by taking the divergence of both sides of equation (1.108) and using equation (1.109) that, when equation (1.108) is applicable, we must have

$$\nabla \cdot \mathbf{J} = 0. \tag{1.110}$$

Comparing equation (1.110) with equation (1.50) we see that equation (1.108) can only be valid for steady charge and current distributions. According to the continuity equation (1.49), in the case of a varying charge and current distribution, instead of equation (1.110) we have

$$\nabla \cdot \mathbf{J} + \frac{\partial \rho}{\partial t} = 0. \tag{1.111}$$

According to equation (1.18) ρ is equal to $\varepsilon_0 \nabla \cdot \mathbf{E}$. Substituting for ρ in equation (1.111) we obtain

$$\nabla \cdot \mathbf{J} + \frac{\partial}{\partial t}(\varepsilon_0 \nabla \cdot \mathbf{E}) = \nabla \cdot (\mathbf{J} + \varepsilon_0 \dot{\mathbf{E}}) = 0. \tag{1.112}$$

A dot over a vector denotes partial differentiation with respect to time. The correct modified form of equation (1.108) must lead to equation (1.112) in the general case of a varying charge and current distribution. If instead of equation (1.108), we had

$$\nabla \times \mathbf{B} = \mu_0(\mathbf{J} + \varepsilon_0 \dot{\mathbf{E}}) \tag{1.113}$$

then, since the divergence of the curl of any vector is zero, we would have

$$\nabla \cdot (\nabla \times \mathbf{B}) = \mu_0 \nabla \cdot (\mathbf{J} + \varepsilon_0 \dot{\mathbf{E}}) = 0$$

which would be in agreement with equation (1.112). It is assumed in classical electromagnetism that equation (1.113) holds in the general case of a varying charge and current distribution in empty space whatever the frequency of the variations in electric current.

At a field point in empty space where **J** is zero, equation (1.113) reduces to

$$\nabla \times \mathbf{B} = \mu_0 \varepsilon_0 \dot{\mathbf{E}}. \tag{1.114}$$

Equation (1.114) relates the curl of the resultant magnetic field **B** at a field point in empty space due to all the current distributions in the system to the rate of change of the resultant electric field **E** at the same field point in empty space due to the same charge and current distributions. The $\varepsilon_0 \dot{\mathbf{E}}$ term in equations (1.113) and (1.114) is generally called the vacuum displacement current. Since it has the same dimensions of ampere per square metre as the current density **J**, the $\varepsilon_0 \dot{\mathbf{E}}$ term should strictly be called the vacuum displacement current density.

Equation (1.113) was developed from equation (1.108) by intelligent guesswork by seeing how equation (1.108) must be modified such that, for varying charge and current distributions, it becomes consistent with the continuity equation (1.49). Another approach used in introductory courses is to develop equation (1.113) by seeing how Ampère's circuital theorem must be extended, when it is applied to the magnetic field between the plates of a parallel plate capacitor that forms part of an AC circuit. Maxwell's original development of the vacuum displacement current density was based on a very complicated mechanical model of the aether. A reader interested in the historical approach is referred to Maxwell [14] or Tricker [15].

In the approach presented in this chapter, the evidence in favour of the $\varepsilon_0\dot{\mathbf{E}}$ term in equations (1.113) and (1.114) is that predictions based on equations (1.113) and (1.114) are in agreement with the experimental results. For example, we shall shown in Section 1.9.2 that the prediction, based on Maxwell's equations, that there are electromagnetic waves, that travel at a speed $c = 1/(\mu_0\varepsilon_0)^{1/2}$ in empty space, depends on the presence of the vacuum displacement current density term $\varepsilon_0\dot{\mathbf{E}}$ in Maxwell's equations.

If we take Maxwell's equations as axiomatic, then by taking the curl of equation (1.114) and using equation (1.18) we obtain the equation of continuity.

1.8. Summary of Maxwell's equations for continuous charge and current distributions in empty space

The experimental evidence used in our development of Maxwell's equations for continuous charge and current distributions in this chapter can be summarized as follows:

1. Coulomb's law of electrostatics leads to the equation (1.18), which is

$$\nabla \cdot \mathbf{E} = \frac{\rho}{\varepsilon_0}. \qquad \text{(M1)} \qquad (1.115)$$

2. The Biot-Savart law for a steady current distribution leads to the equation (1.71) which is

$$\nabla \cdot \mathbf{B} = 0. \qquad \text{(M2)} \qquad (1.116)$$

3. Faraday's law of electromagnetic induction, which is generally developed on the basis of experiments carried out at mains frequency, leads to the equation (1.107), which is

$$\nabla \times \mathbf{E} = -\dot{\mathbf{B}}. \qquad \text{(M3)} \qquad (1.117)$$

4. The Biot-Savart law for steady currents leads to the equation $\nabla \times \mathbf{B} = \mu_0\mathbf{J}$. This equation was then extended by adding the vacuum displacement current density term $\varepsilon_0\dot{\mathbf{E}}$ to the conduction current density $\mathbf{J}$ to give equation (1.113), which is then consistent with the continuity equation (1.49). According to equation (1.113)

$$\nabla \times \mathbf{B} = \mu_0(\mathbf{J} + \varepsilon_0\dot{\mathbf{E}}). \qquad \text{(M4)} \qquad (1.118)$$

The field vectors **E** and **B** at a field point in empty space can be related to experiments by the Lorentz force law, equation (1.1), according to which

$$\mathbf{F} = q\mathbf{E} + q\mathbf{u} \times \mathbf{B}. \qquad (1.119)$$

The Lorentz force law gives the magnitude of the force on a charge of magnitude q that is moving with a velocity **u** at a field point in empty space, where the electric field is **E** and the magnetic field is **B**. Equation (1.119) is only valid provided the charge q is not emitting electromagnetic radiation.

Maxwell's equations divide into two pairs. The first pair, namely equations (1.115) and (1.116), which are sometimes referred to as M1 and M2 respectively are expressions for the divergences of the field vectors **E** and **B**. The second pair of equations, namely equations (1.117) and (1.118) which are sometimes referred to as M3 and M4 respectively, are expressions for the curls of the field vectors **E** and **B**. It is equations (1.115) and (1.118), that is equations M1 and M4, which bring in the sources ρ and **J** of the electromagnetic field.

Maxwell's equations, equations (1.115), (1.116), (1.117) and (1.118), were developed in this chapter on the basis of very limited experimental evidence for very special cases only. It is assumed in classical electromagnetism that Maxwell's equations can be applied in a far wider context than this very limited experimental evidence. For example, it is assumed that Maxwell's equations are valid for the general case of rapidly varying charge and current distributions and for accelerating charged particles moving at relativistic speeds. In our approach, the validity of Maxwell's equations does not depend on whether or not each individual equation has been verified by experiments for all possible experimental situations, but rather on whether or not the predictions of the theory taken as a whole are in agreement with the experimental results. It is found experimentally that, provided all quantum effects can be neglected, Maxwell's equations do make predictions that are in agreement with the experimental results. We have so far avoided a full discussion of the historical development of Maxwell's equations, so that we could go on directly to develop a modern interpretation of classical electromagnetism, not influenced by obsolete historical models such as the mechanical theories of the aether. We shall defer discussion of these obsolete historical models until Section 4.13 of Chapter 4.

1.9. The differential equations for the fields E and B

1.9.1. *Introduction*

According to the Maxwell equation (1.118)

$$\nabla \times \mathbf{B} = \mu_0(\mathbf{J} + \varepsilon_0\dot{\mathbf{E}}). \tag{1.120}$$

Taking the curl of both sides of equation (1.120), we have

$$\nabla \times (\nabla \times \mathbf{B}) = \mu_0 \nabla \times \mathbf{J} + \mu_0\varepsilon_0 \nabla \times \frac{\partial \mathbf{E}}{\partial t}.$$

Using equation (A1.27) of Appendix A1.6 to expand $\nabla \times (\nabla \times \mathbf{B})$ and rearranging the right hand side, we obtain

$$\nabla(\nabla \cdot \mathbf{B}) - \nabla^2\mathbf{B} = \mu_0 \nabla \times \mathbf{J} + \mu_0\varepsilon_0 \frac{\partial}{\partial t}(\nabla \times \mathbf{E}). \tag{1.121}$$

According to the Maxwell's equations (1.116) and (1.117), $\nabla \cdot \mathbf{B}$ is zero and $\nabla \times \mathbf{E} = -\dot{\mathbf{B}}$. Hence equation (1.121) reduces to

$$\nabla^2\mathbf{B} - \mu_0\varepsilon_0 \frac{\partial^2\mathbf{B}}{\partial t^2} = -\mu_0\nabla \times \mathbf{J}. \tag{1.122}$$

It can be seen from equation (1.122) that the magnetic field $\mathbf{B}$ depends only on the current density $\mathbf{J}$ via the $\nabla \times \mathbf{J}$ term in equation (1.122). The $\mu_0\varepsilon_0\ddot{\mathbf{B}}$ term in equation (1.122) comes from the $\varepsilon_0\dot{\mathbf{E}}$ term in equation (1.120). If Maxwell had not introduced the vacuum displacement current term $\varepsilon_0\dot{\mathbf{E}}$, then instead of equation (1.122), we would have obtained

$$\nabla^2\mathbf{B} = -\mu_0\nabla \times \mathbf{J} \tag{1.123}$$

which is only valid for steady currents (magnetostatics). To obtain equation (1.123) take the curl of both sides of equation (1.89). The reader can show that, if the $\varepsilon_0\dot{\mathbf{E}}$ term were present in equation (1.120) but the $\dot{\mathbf{B}}$ term were absent from the Maxwell equation

$$\nabla \times \mathbf{E} = -\dot{\mathbf{B}} \tag{1.124}$$

we would again obtain equation (1.123). Both the $\varepsilon_0\dot{\mathbf{E}}$ term in equation (1.120) and the $-\dot{\mathbf{B}}$ term in equation (1.124) are necessary to give the $\mu_0\varepsilon_0\ddot{\mathbf{B}}$ term in equation (1.122). The operator

$$\left(\nabla^2 - \mu_0\varepsilon_0 \frac{\partial^2}{\partial t^2}\right) = \left(\nabla^2 - \frac{1}{c^2}\frac{\partial^2}{\partial t^2}\right)$$

is called the D'Alembertian. The operator ∇^2 on its own is called the Lapacian. Since the $-\dot{\mathbf{B}}$ term in the Maxwell equation (1.124) was known before Maxwell introduced the vacuum displacement current term into equation (1.120), historically, it was the addition of the vacuum displacement current term to Maxwell's equations that converted the Laplacian in equation (1.123) into the D'Alembertian in equation (1.122). It was this addition of the displacement current term to Maxwell's equations that led to the prediction that the electromagnetic interaction was propagated at a finite speed, namely the speed of light in empty space.

Taking the curl of both sides of the Maxwell equation (1.124), then using equation (A1.27) of Appendix A1.6 to expand $\nabla \times (\nabla \times \mathbf{E})$ and finally substituting for $\nabla \times \mathbf{B}$ using equation (1.120), we obtain

$$\nabla(\nabla \cdot \mathbf{E}) - \nabla^2\mathbf{E} = -\frac{\partial}{\partial t}(\nabla \times \mathbf{B}) = -\frac{\partial}{\partial t}\mu_0(\mathbf{J} + \varepsilon_0\dot{\mathbf{E}}).$$

Substituting ρ/ε_0 for $(\nabla \cdot \mathbf{E})$ from equation (1.115), we finally obtain

$$\nabla^2\mathbf{E} - \mu_0\varepsilon_0 \frac{\partial^2\mathbf{E}}{\partial t^2} = \nabla\left(\frac{\rho}{\varepsilon_0}\right) + \mu_0\frac{\partial\mathbf{J}}{\partial t}. \tag{1.125}$$

It can be seen from equation (1.125) that the electric field $\mathbf{E}$ depends on both the charge density ρ and the current density $\mathbf{J}$ via the $\nabla\rho$ and $\dot{\mathbf{J}}$ terms respectively in equation (1.125).

The reader can show that we need both the displacement current density term $\varepsilon_0\dot{\mathbf{E}}$ in equation(1.120) and the $-\dot{\mathbf{B}}$ term in equation (1.124) to obtain the $\mu_0\varepsilon_0\ddot{\mathbf{E}}$ term in equation (1.125). The reader can show that if the $-\dot{\mathbf{B}}$ term were absent from equation (1.124), but the vacuum displacement current term $\varepsilon_0\dot{\mathbf{E}}$ were still present in equation (1.120) then, instead of equation (1.125), we would obtain

$$\nabla^2\mathbf{E} = \frac{1}{\varepsilon_0}\nabla\rho \tag{1.126}$$

which is only valid for electrostatics.

If the $-\dot{\mathbf{B}}$ term were present in equation (1.124) but the vacuum displacement current term $\varepsilon_0\dot{\mathbf{E}}$ were absent from equation (1.120), which was the historical situation before Maxwell introduced the vacuum displacement current term, then, instead of equation (1.125), we would obtain

$$\nabla^2\mathbf{E} = \nabla\left(\frac{\rho}{\varepsilon_0}\right) + \mu_0\dot{\mathbf{J}}. \tag{1.127}$$

In an electrical conductor where ρ is zero and $\mathbf{J} = \sigma\mathbf{E}$, where σ is the constant electrical conductivity, equation (1.127) reduces to

$$\nabla^2\mathbf{E} - \mu_0\sigma\frac{\partial\mathbf{E}}{\partial t} = 0. \tag{1.128}$$

Equation (1.128) is used extensively in the quasi-stationary limit, that is at low frequencies, when the contribution of the $\mu_0\varepsilon_0\ddot{\mathbf{E}}$ term can be neglected. At high frequencies the $\mu_0\varepsilon_0\ddot{\mathbf{E}}$ term must be included. Equation (1.128) is the same as the equation of diffusion. It is shown in text books on electromagnetism that in a conducting medium of infinite extent there are plane wave solutions of equation (1.128) propagation in the $+x$ direction of the type

$$\mathbf{E}_y = \mathbf{E}_0 \exp\left(-\frac{x}{\delta}\right)\cos\omega\left(t - \frac{x}{c}\right) \tag{1.129}$$

where $\delta = (2/\omega\sigma\mu_0)^{1/2}$ is the skin depth. The waves described by equation (1.129) are attenuated. Such waves cannot propagate in empty space, where $\sigma = 0$.

1.9.2. *Electromagnetic waves in empty space*

In empty space where $\sigma = 0$ and $\mathbf{J} = 0$, equations (1.122) and (1.125) reduce to

$$\nabla^2\mathbf{B} - \mu_0\varepsilon_0\frac{\partial^2\mathbf{B}}{\partial t^2} = 0, \tag{1.130}$$

$$\nabla^2\mathbf{E} - \mu_0\varepsilon_0 \frac{\partial^2\mathbf{E}}{\partial t^2} = 0. \tag{1.131}$$

Full solutions of equations (1.130) and (1.131) are given in the standard test books on classical electromagnetism. Both equations (1.130) and (1.131) are wave equations. In both cases the velocity of the waves is

$$c = \frac{1}{(\mu_0\varepsilon_0)^{1/2}} . \tag{1.132}$$

The electromagnetic waves in empty space are identified with light waves, so that the velocity c in equation (1.132) is the velocity of light in empty space. If Maxwell had not introduced the vacuum displacement current, then at a field point in empty space equations (1.122) and (1.125) would reduce to

$$\nabla^2\mathbf{B} = 0; \qquad \nabla^2\mathbf{E} = 0.$$

These equations do not have propagating wave solutions in empty space. The existence of electromagnetic waves, whose properties can be predicted using Maxwell's equations, is strong evidence in favour of the vacuum displacement current term $\varepsilon_0\dot{\mathbf{E}}$ in equation (1.120).

1.9.3. *Solution of the differential equations for* **E** *and* **B**

The fields **E** and **B** can be determined independently using the differential equations (1.125) and (1.122) respectively, provided the values of ρ and **J** and their spatial and temporal variations are given. The methods of solution and their interpretations are similar to the case of the retarded potentials, which will be described in detail in Section 2.3 of Chapter 2. Hence at this stage we shall only give a brief discussion. The reader should return to this section after reading Chapter 2. By analogy with the retarded potentials a solution of equation (1.122) giving the magnetic field **B** at a field point at a position **r** at the time of observation t is

$$\mathbf{B}(\mathbf{r}, t) = \frac{1}{4\pi\varepsilon_0 c^2} \int \frac{[\nabla_s \times \mathbf{J}(\mathbf{r}_s)]}{R} \, dV_s \tag{1.133}$$

where $\mathbf{R} = (\mathbf{r} - \mathbf{r}_s)$ is a vector from a source point at $\mathbf{r}_s$, where the current density is $\mathbf{J}(\mathbf{r}_s)$, to the field point at **r**. Quantities measured at the retarded time $t^* = (t - R/c)$ are placed inside square brackets. Jefimenko [16] showed that equation (1.133) can be rewritten in the form

$$\mathbf{B}(\mathbf{r}, t) = \frac{1}{4\pi\varepsilon_0 c^2} \int \left(\frac{[\mathbf{J}(\mathbf{r}_s)]}{R^2} + \frac{1}{cR} \left[\frac{\partial [\mathbf{J}(\mathbf{r}_s)]}{\partial t} \right] \right) \times \hat{\mathbf{R}} \, dV_s \tag{1.134}$$

where $\hat{\mathbf{R}}$ is a unit vector in the direction from the source point at $\mathbf{r}_s$ to the field point at **r**, and $[\mathbf{J}(\mathbf{r}_s)]$ is the current density at $\mathbf{r}_s$ at the retarded time $(t - R/c)$. Equation (1.134) will be derived in Section 2.7 of Chapter 2 using the retarded potentials. We shall also derive equation (1.134) in Section 6.5

of Chapter 6 from the expression for the magnetic field due to a moving and accelerating classical point charge, where we shall show that equation (1.134) is valid in differential form. We shall apply equation (1.134) to the oscillating electric dipole in Section 2.7. It is of interest to note that the vacuum displacement current term $\varepsilon_0\dot{\mathbf{E}}$ does not appear as one of the sources of the magnetic field in equation (1.134).

Again by analogy with the derivation of the retarded potentials, we conclude that a solution of equation (1.125) giving the electric field $\mathbf{E}$ at a field point at $\mathbf{r}$ at the time of observation t is

$$\mathbf{E}(\mathbf{r},\, t) = -\frac{1}{4\pi\varepsilon_0}\int\left(\frac{[\boldsymbol{\nabla}_s\rho]}{R} + \frac{1}{Rc^2}\left[\frac{\partial\mathbf{J}}{\partial t}\right]\right)\mathrm{d}V_s \tag{1.135}$$

where the operation $\boldsymbol{\nabla}_s$ is given by equation (1.80) of Section 1.4.6. Jefimenko [16] has shown that equation (1.135) can be rewritten in the form

$$\mathbf{E}(\mathbf{r},\, t) = \frac{1}{4\pi\varepsilon_0}\int\left(\frac{[\rho]\hat{\mathbf{R}}}{R^2} + \frac{\hat{\mathbf{R}}}{Rc}\left[\frac{\partial\rho}{\partial t}\right] - \frac{1}{Rc^2}\left[\frac{\partial\mathbf{J}}{\partial t}\right]\right)\mathrm{d}V_s \tag{1.136}$$

where $[\rho]$ and $[\mathbf{J}]$ are the charge and current densities at the source point at $\mathbf{r}_s$ at the retarded time $(t - R/c)$. Equations (1.134) and (1.136) will be called Jefimenko's equations. Equation (1.136) will be derived using the retarded potentials in Section 2.7. An alternative expression for $\mathbf{E}$, which, unlike equation (1.136), is also valid in differential form, will be derived in Section 5.13 of Chapter 5 using the expression for the electric field due to a moving and accelerating classical point charge.

1.10. The Maxwell-Lorentz equations for the microscopic fields

Maxwell's equations were developed earlier in this chapter for continuous charge and current distributions in otherwise empty space. In practice, all macroscopic charge and current distributions are made up of large numbers of charged atomic particles such as electrons, protons and positive ions, whose charges are always on integral multiple of the electronic charge of

$$e = \pm 1.602 \times 10^{-19}\ \mathrm{C}.$$

It is now believed that atomic particles, such as protons, consist of tightly bound quarks, which have charges of $\pm e/3$ and $\pm 2e/3$. It will require enormous energies to produce free quarks, so that it is safe to assume that free quarks play no significant role in classical electromagnetism.

It will be assumed that individual atomic particles, such as protons and electrons, can be treated as classical point charges, that is as continuous charge distributions of finite but exceedingly small dimensions. As an example of a classical point charge we shall now assume that the charge distribution shown in Figure 1.8 is exceedingly small. Lorentz assumed that Maxwell equations (1.115), (1.116), (1.117) and (1.118) held for the microscopic electric field $\mathbf{e}$

and the microscopic magnetic field **b** at a field point, such as P in Figure 1.8 that is inside such a classical point charge, as well as at field points in the spaces between such classical point charges. Hence, according to Lorentz, at a field point inside a classical point charge

$$\nabla \cdot \mathbf{e} = \frac{\rho^{mic}}{\varepsilon_0} \tag{1.137}$$

$$\nabla \cdot \mathbf{b} = 0 \tag{1.138}$$

$$\nabla \times \mathbf{e} = -\frac{\partial \mathbf{b}}{\partial t} \tag{1.139}$$

$$\nabla \times \mathbf{b} = \mu_0 \left(\rho^{mic}\mathbf{u} + \varepsilon_0 \frac{\partial \mathbf{e}}{\partial t} \right) \tag{1.140}$$

where **e** and **b** are the resultant microscopic fields due to all the charges in the system and ρ^{mic} and $\mathbf{j}^{mic} = \rho^{mic}\,\mathbf{u}$ are the microscopic charge and current densities at a field point inside a classical point charge, that is moving with velocity **u**. Equations (1.137), (1.138), (1.139) and (1.140) will be called the **Maxwell-Lorentz equations.**

The Maxwell-Lorentz equations will be taken as axiomatic from now on. They are generally the starting point for the derivation of Maxwell's equations for the macroscopic fields **E** and **B** at field points inside stationary dielectrics and stationary magnetic materials. Until we reach Chapter 9, we shall only consider the electric and magnetic fields due to charge and current distributions in empty space for the special case when the relative permittivity ε_r and the relative permeability μ_r are both equal to unity everywhere. A brief outline of this special case will now be given.

1.11. Maxwell's equations for the macroscopic fields for the special case when $\epsilon_r = 1$ and $\mu_r = 1$ everywhere

When the scale of an electromagnetic phenomenon is very much greater than atomic dimensions, the enormous fluctuations in the microscopic fields **e** and **b** on the atomic scale average out and we often only need to know the values of the macroscopic variables, which are defined as the averages of the corresponding microscopic variables taken over a region of space that is much bigger than atomic dimensions, that is $\gg 10^{-10}$ m, but which is kept small on the laboratory scale, say less than one micron (10^{-6} m). For example, a sphere of a solid of diameter 10^{-6} m contains of the order of 10^{12} atoms.

Consider the field point P at a distance **r** from the origin O in Figure 1.12. We shall start with a simplified method of averaging a microscopic variable $f(x, y, z, t)$ to obtain the value of the corresponding macroscopic variable $F(x, y, z, t)$ at the field point P. Initially a microscopic variable will be averaged over the volume of a sphere of radius a having its centre at the field point P

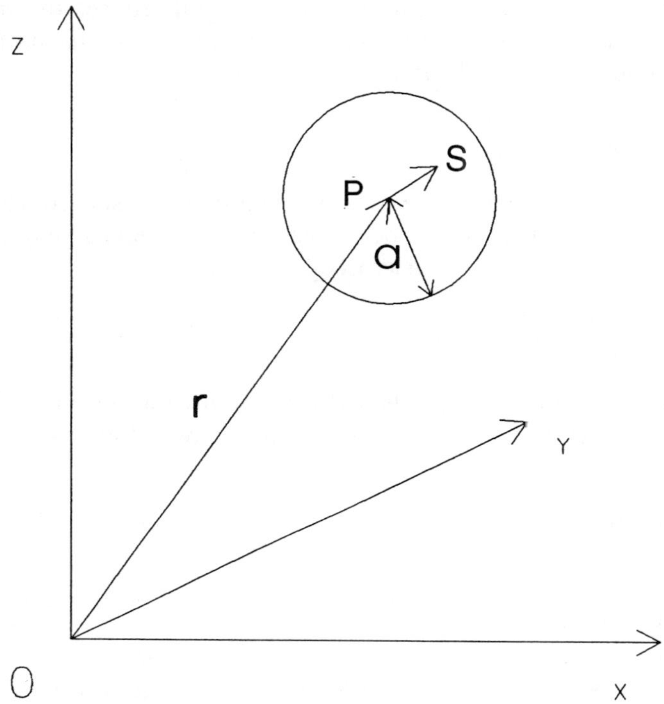

Figure 1.12. Determination of the macroscopic fields at the field point P by averaging the corresponding microscopic fields over a very small volume that surrounds P.

in Figure 1.12, where a is large on the atomic scale but small on the laboratory scale. Initially equal weights will be given to all the volume elements inside the sphere of radius a and volume $V_0 = 4\pi a^3/3$. Let $f(\mathbf{r} + \mathbf{s})$ be the value of the microscopic variable f at a point at a distance $\mathbf{s}$, having components s_x, s_y and s_z, from the field point at $\mathbf{r}$, at which the value of the macroscopic variable F is required. The value of the macroscopic variable F will be defined initially in terms of the microscopic variable f by the equation

$$F(\mathbf{r}, t) = \langle f \rangle = \frac{1}{V_0}\int f(x + s_x, y + s_y, z + s_z, t)\mathrm{d}^3\mathbf{s} \qquad (1.141)$$

where $\mathrm{d}^3\mathbf{s} = \mathrm{d}s_x\, \mathrm{d}s_y\, \mathrm{d}s_z$ and the integration is over the sphere of radius a, whose centre is at the field point P in Figure 1.12.

In the case of the macroscopic charge density ρ, equation (1.41) gives

$$\rho = \langle \rho^{\mathrm{mic}} \rangle = \frac{\int \rho^{\mathrm{mic}}\, \mathrm{d}^3\mathbf{s}}{V_0} \qquad (1.142)$$

where ρ^{mic} is the value of the microscopic charge density at the point $(\mathbf{r} + \mathbf{s})$.

For a system of atomic point charges the integral reduces to Σq_i, which is the sum of the charges of all the classical point charges inside the volume V_0, so that equation (1.142) becomes

$$\rho = \langle \rho^{\text{mic}} \rangle = \frac{\Sigma q_i}{V_0} . \tag{1.143}$$

Sometimes the macroscopic charge density ρ is defined using equation (1.143), as the average resultant charge per cubic metre. If the charge q_i inside V_0 is moving with velocity $\mathbf{u}_i$, the macroscopic current density $\mathbf{J}$ is given by

$$\mathbf{J} = \frac{\int \rho^{\text{mic}} \mathbf{u} \, \mathrm{d}^3\mathbf{s}}{V_0} = \frac{\Sigma q_i \mathbf{u}_i}{V_0} . \tag{1.144}$$

The macroscopic electric field $\mathbf{E}$ and the macroscopic magnetic field $\mathbf{B}$ are defined in terms of the corresponding microscopic fields $\mathbf{e}$ and $\mathbf{b}$ by the equations

$$\mathbf{E} = \langle \mathbf{e} \rangle = \frac{\int \mathbf{e} \, \mathrm{d}^3\mathbf{s}}{V_0} , \tag{1.145}$$

$$\mathbf{B} = \langle \mathbf{b} \rangle = \frac{\int \mathbf{b} \, \mathrm{d}^3\mathbf{s}}{V_0} . \tag{1.146}$$

The disadvantage of equation (1.141) when we come to dielectrics is that the surface of the sphere of radius a may cut through molecules. Russakoff [17] suggested using a spherically symmetric weighting function $w(\mathbf{s})$ that was constant out to the surface of the sphere of radius a in Figure 1.12 but which, instead of going to zero at $s = a$, decreases smoothly to zero over a distance that is small on the laboratory scale, but which is large on the atomic scale. The value of the macroscopic variable $F(\mathbf{r}, t)$ is then defined by the equation

$$F(\mathbf{r}, t) = \langle f \rangle = \int_0^\infty w(\mathbf{s}) \, f(\mathbf{r} + \mathbf{s}, t) \mathrm{d}^3\mathbf{s} \tag{1.147}$$

where $f(\mathbf{r} + \mathbf{s}, t)$ is the value of the microscopic variable f at a distance $\mathbf{s}$ from the field point P in Figure 1.12 at the time t. The weighting function $w(\mathbf{s})$ is normalised such that

$$\int_0^\infty w(\mathbf{s}) \, \mathrm{d}^3\mathbf{s} = 1. \tag{1.148}$$

Comparing equations (1.141) and (1.147), we see that the weighting function used previously, in equation (1.141) was

$$w(\mathbf{s}) = \frac{1}{V_0} \qquad \text{for } s \le a,$$

$$w(\mathbf{s}) = 0 \qquad \text{for } s > a.$$

According to equation (A2.4) of Appendix A2 for any microscopic variable f we have

$$\left\langle \frac{\partial f}{\partial x} \right\rangle = \frac{\partial \langle f \rangle}{\partial x} = \frac{\partial F}{\partial x} \tag{1.149}$$

where $F = \langle f \rangle$ is the macroscopic variable determined from the corresponding microscopic variable f using equation (1.147). Equation (1.149) can be summarised by saying that the operations of first averaging the microscopic variable f to determine the macroscopic variable F and then differentiating F partially with respect to x gives the same result as first differentiating the microscopic variable f partially with respect to x to determine $\partial f/\partial x$ and then averaging the differential coefficient $\partial f/\partial x$ to determine $\langle \partial f/\partial x \rangle$. For the x component of the electric field, equation (1.149) gives

$$\left\langle \frac{\partial e_x}{\partial x} \right\rangle = \frac{\partial E_x}{\partial x} . \tag{1.150}$$

Similar results hold for $\partial E_x/\partial y$, $\partial E_x/\partial z$, $\partial B_x/\partial x$ etc., and for $\partial E_x/\partial t$ etc.

According to the Maxwell-Lorentz equation (1.137)

$$\frac{\partial e_x}{\partial x} + \frac{\partial e_y}{\partial y} + \frac{\partial e_z}{\partial z} = \frac{\rho^{\text{mic}}}{\varepsilon_0} . \tag{1.151}$$

Averaging both sides of equation (1.151) we have

$$\left\langle \frac{\partial e_x}{\partial x} \right\rangle + \left\langle \frac{\partial e_y}{\partial y} \right\rangle + \left\langle \frac{\partial e_z}{\partial z} \right\rangle = \frac{\langle \rho^{\text{mic}} \rangle}{\varepsilon_0} .$$

Applying the general result given by equation (1.149) we have

$$\frac{\partial E_x}{\partial x} + \frac{\partial E_y}{\partial y} + \frac{\partial E_z}{\partial z} = \frac{\rho}{\varepsilon_0} .$$

Hence,

$$\nabla \cdot \mathbf{E} = \frac{\rho}{\varepsilon_0} . \tag{1.152}$$

where $\mathbf{E}$ is the macroscopic electric field and ρ is the macroscopic charge density, determined using equation (1.147). Proceeding in a similar way, the reader can show that the Maxwell-Lorentz equations (1.138), (1.139) and (1.140) lead to

$$\nabla \cdot \mathbf{B} = 0 \tag{1.153}$$

$$\nabla \times \mathbf{E} = -\dot{\mathbf{B}} \tag{1.154}$$

$$\nabla \times \mathbf{B} = \mu_0(\mathbf{J} + \varepsilon_0 \dot{\mathbf{E}}) \tag{1.155}$$

where $\mathbf{B}$ is the macroscopic magnetic field and $\mathbf{J}$ is the macroscopic current density, determined using equation (1.147). This analysis shows that the

Maxwell-Lorentz equations give Maxwell's equations as the appropriate relations between the macroscopic variables. **E**, **B**, **J** and ρ. A full discussion of macroscopic electromagnetism is given by Robinson [18].

References

1. Whittaker, E., *A History of the Theories of Aether and Electricity*, Vol. 1 (Thomas Nelson, Edinburgh, 1951).
2. Tricker, R. A. R., *Early Electrodynamics. The First Law of Circulation* (Pergamon Press, Oxford, 1965) and also *The Contributions of Faraday and Maxwell to Electrical Science* (Pergamon Press, Oxford, 1966).
3. Schaffner, K. W., *Nineteenth-Century Aether Theories* (Pergamon Press, Oxford, 1972).
4. King, J. G., *Phys. Rev. Letters*, Vol. 5, p. 562 (1960).
5. Jeans, J. J., *Electricity and Magnetism*, 5th Edition (Cambridge University Press, Cambridge, 1925).
6. Smythe, W. R., *Static and Dynamic Electricity* (McGraw-Hill, New York, 1939), Chapters 4 and 5.
7. Jackson, J. D., *Classical Electrodynamics*, 2nd Edition (John Wiley, New York, 1975), Chapters 2 and 3.
8. Weisskopf, V. F., *Amer. Journ. Phys.* Vol. 11, p. 1 (1943).
9 Rosser, W. G. V., *Introduction to the Theory of Relativity* (Butterworths, London, 1964), p. 337. Hint: Take the low velocity limits of equations (8.113) and (8.114) on p. 337.
10. Whittaker, E., see ref. 1, pp. 83–88.
11. McKinnon, W. R., McAlister, S. P. and Hurd, C. M., *Amer. Journ. Phys.*, Vol. 49, p. 493 (1981).
12. Scott, W. T., *The Physics of Electricity and Magnetism*, 2nd Edition (John Wiley, New York 1966), p. 286.
13. Rosser, W. G. V., *Contempory Physics*, Vol. 1, p. 134 (1959).
14. Maxwell, J. C., *The Scientific Papers of James Clerk Maxwell*, Vol. 1 (Dover, New York, 1965), pp. 451–525.
15. Tricker, R. A. R., *The Contributions of Faraday and Maxwell to Electrical Science* (Pergamon Press, Oxford, 1966).
16. Jefimenko, O. D., *Electricity and Magnetism*, 2nd Edition (Electric Scientific, Star City, W.V., 1989), p. 516 or see the 1st Edition (Appleton Century Crofts, 1966), p. 516.
17. Russakoff, G., *Amer. Journ. Phys.* Vol. 38, p. 1188 (1970).
18. Robinson, F. N. H., *Macroscopic Electromagnetism* (Pergamon Press, Oxford, 1973).

CHAPTER 2

The scalar potential ϕ and the vector potential **A**

2.1. Introduction

Though it is possible to determine the electric and magnetic fields **E** and **B** due to varying charge and current distributions by solving the differential equations (1.125) and (1.122) for **E** and **B**, for example using the Jefimenko solutions given by equations (1.136) and (1.134), it is sometimes more convenient to solve problems and to interpret electromagnetism using the scalar potential ϕ and the vector potential **A**. Our starting point in this chapter will be Maxwell's equations for continuous charge and current distributions in otherwise empty space. For these conditions, Maxwell's equations at a field point inside a charge and current distribution are

$$\nabla \cdot \mathbf{E} = \frac{\rho}{\varepsilon_0} \tag{2.1}$$

$$\nabla \cdot \mathbf{B} = 0 \tag{2.2}$$

$$\nabla \times \mathbf{E} = -\dot{\mathbf{B}} \tag{2.3}$$

$$\nabla \times \mathbf{B} = \mu_0(\mathbf{J} + \varepsilon_0\dot{\mathbf{E}}) \tag{2.4}$$

where ρ is the charge density and **J** is the current density at the field point. One dot above a variable denotes partial differentiation once with respect to time, two dots above a variable denote partial differentiation twice with respect to time etc. It will be assumed throughout this chapter that there are no dielectrics or magnetic materials so that the relative permittivity ε_r and the relative permeability μ_r are both equal to unity everywhere.

2.2. The differential equations for ϕ and A

The vector potential **A** was introduced in Section 1.4.6 of Chapter 1, where we showed, using the Biot-Savart law, that the magnetic field due to a steady current distribution (magnetostatics) could be expressed in terms of a vector potential **A** by the equation

$$\mathbf{B} = \boldsymbol{\nabla} \times \mathbf{A} \tag{2.5}$$

where **A** was given by equation (1.74). We shall now go on to consider varying charge and current distributions using Maxwell's equations as our starting point. It is consistent with equation (2.2) to assume that **B** can also be related to a vector potential **A** by equation (2.5) in the general case of a varying current distribution. To check this, take the divergence of both sides of equation (2.5) and then use the result that, according to equation (A1.25) of Appendix A1.6, the divergence of the curl of any vector is zero, to show that equation (2.5) leads to equation (2.2). The divergence of **A** has yet to be specified.

The scalar potential ϕ was first introduced in electrostatics in Section 1.2.9 of Chapter 1, where, in the context of electrostatics, ϕ was related to the electrostatic field **E** by equation (1.23), according to which

$$\mathbf{E} = -\boldsymbol{\nabla}\phi. \tag{1.23}$$

According to equation (A1.26) of Appendix A1.6 the curl of the gradient of any scalar function of position is zero. Hence it follows by taking the curl of both sides of equation (1.23) that equation (1.23) can only be applied in conditions where $\boldsymbol{\nabla} \times \mathbf{E}$ is zero, that is in electrostatics. According to equation (2.3), which expresses Faraday's law of electromagnetic induction,

$$\boldsymbol{\nabla} \times \mathbf{E} = -\dot{\mathbf{B}}. \tag{2.3}$$

Substituting for **B** using equation (2.5), we have

$$\boldsymbol{\nabla} \times \mathbf{E} = -\frac{\partial}{\partial t}(\boldsymbol{\nabla} \times \mathbf{A}) = -\boldsymbol{\nabla} \times \frac{\partial \mathbf{A}}{\partial t}.$$

Rearranging,

$$\boldsymbol{\nabla} \times (\mathbf{E} + \dot{\mathbf{A}}) = 0. \tag{2.6}$$

Since, according to equation (A1.26) of Appendix A1.6, the curl of the gradient of any scalar function of position is zero it is consistent with equation (2.6) to try putting $(\mathbf{E} + \dot{\mathbf{A}})$ equal to $-\boldsymbol{\nabla}\phi$ in the general case, when the charge and current distributions are varying, giving

$$\mathbf{E} = -\boldsymbol{\nabla}\phi - \frac{\partial \mathbf{A}}{\partial t}. \tag{2.7}$$

As a check, integrate equation (2.7) around any closed loop. Since, according to equation (A1.11) of Appendix A1.2, $\oint \boldsymbol{\nabla}\phi \cdot \mathbf{dl}$ is always zero, we have

$$\oint \mathbf{E} \cdot \mathbf{dl} = -\oint \boldsymbol{\nabla}\phi \cdot \mathbf{dl} - \oint \frac{\partial \mathbf{A}}{\partial t} \cdot \mathbf{dl} = -\frac{\partial}{\partial t}\oint \mathbf{A} \cdot \mathbf{dl}.$$

Applying Stokes' theorem, equation (A1.34) of Appendix A1.8, to $\oint \mathbf{A} \cdot \mathbf{dl}$ and putting $\boldsymbol{\nabla} \times \mathbf{A}$ equal to **B** we obtain

$$\oint \mathbf{E} \cdot d\mathbf{l} = -\frac{\partial}{\partial t}\int \nabla \times \mathbf{A} \cdot d\mathbf{S} = -\frac{\partial}{\partial t}\int \mathbf{B} \cdot d\mathbf{S} \tag{2.8}$$

which is the same as equation (1.105), which is Faraday's law of electromagnetic induction. It can be seen that the contribution of the $-\dot{\mathbf{A}}$ term in equation (2.7) to the total field $\mathbf{E}$ represents the contribution of electromagnetic induction to the total electric field in the case of a varying current distribution. For static conditions, $\dot{\mathbf{A}}$ is zero and equations (2.7) and (2.8) reduce to equations (1.23) and (1.28) of electrostatics respectively.

So far, we have only used the Maxwell equations (2.2) and (2.3) to develop equations (2.5) and (2.7), which relate the fields $\mathbf{B}$ and $\mathbf{E}$ to the potentials ϕ and $\mathbf{A}$. The other two Maxwell equations, namely equations (2.1) and (2.4), will now be used to develop the differential equations which relate ϕ and $\mathbf{A}$ to the charge and current distributions. According to equation (2.4),

$$\nabla \times \mathbf{B} = \mu_0(\mathbf{J} + \varepsilon_0\dot{\mathbf{E}}) \tag{2.4}$$

Substituting for $\mathbf{E}$ and $\mathbf{B}$ using equations (2.7) and (2.5) respectively, we have

$$\begin{aligned}\nabla \times (\nabla \times \mathbf{A}) &= \mu_0\mathbf{J} + \mu_0\varepsilon_0\frac{\partial}{\partial t}(-\nabla\phi - \dot{\mathbf{A}}) \\ &= \mu_0\mathbf{J} - \mu_0\varepsilon_0\nabla\left(\frac{\partial\phi}{\partial t}\right) - \mu_0\varepsilon_0\frac{\partial^2\mathbf{A}}{\partial t^2}.\end{aligned} \tag{2.9}$$

Notice that the $\mu_0\varepsilon_0\ddot{\mathbf{A}}$ term in equation (2.9) comes directly from the vacuum displacement current term $\varepsilon_0\dot{\mathbf{E}}$ in equation (2.4). From equation (A1.27) of Appendix A1.6

$$\nabla \times (\nabla \times \mathbf{A}) = \nabla(\nabla \cdot \mathbf{A}) - \nabla^2\mathbf{A}.$$

Substituting in equation (2.9), using $\mu_0\varepsilon_0 = 1/c^2$ and rearranging, we obtain

$$\nabla^2\mathbf{A} - \frac{1}{c^2}\frac{\partial^2\mathbf{A}}{\partial t^2} - \nabla\left(\nabla \cdot \mathbf{A} + \frac{1}{c^2}\frac{\partial\phi}{\partial t}\right) = -\mu_0\mathbf{J}. \tag{2.10}$$

The divergence of $\mathbf{A}$ has yet to be specified. It will be shown in Section 2.8 that there is flexibility in the choice of $\nabla \cdot \mathbf{A}$. In this section we shall specify $\nabla \cdot \mathbf{A}$ using the equation

$$\nabla \cdot \mathbf{A} + \frac{1}{c^2}\frac{\partial\phi}{\partial t} = 0. \tag{2.11}$$

Equation (2.11) is known as the **Lorentz condition**. This choice is sometimes called the Lorentz gauge or the covariant gauge. Using the Lorentz condition, equation (2.10) reduces to

$$\nabla^2\mathbf{A} - \frac{1}{c^2}\frac{\partial^2\mathbf{A}}{\partial t^2} = -\mu_0\mathbf{J}. \tag{2.12}$$

The reader can check that, if the vacuum displacement current term $\varepsilon_0\dot{\mathbf{E}}$

were absent from equation (2.4), then equations (2.12) and (2.11) would reduce to

$$\nabla^2\mathbf{A} = -\mu_0\mathbf{J}; \qquad \nabla \cdot \mathbf{A} = 0$$

which are equations (1.92) and (1.83) of magnetostatics. This shows that, just as was the case with the differential equation (1.122) for **B**, it is the presence of the vacuum displacement current term $\varepsilon_0\dot{\mathbf{E}}$ in equation (2.4) which converts the Laplacian of magnetostatics into the D'Alembertian in equation (2.12). Taking the curl of both sides of equation (2.12) gives equation (1.122), which is the partial differential equation relating **B** to the current distributions.

Substituting for **E** from equation (2.7) into equation (2.1), we obtain

$$\nabla \cdot (\nabla\phi + \dot{\mathbf{A}}) = -\frac{\rho}{\varepsilon_0}. \tag{2.13}$$

Now

$$\nabla \cdot \dot{\mathbf{A}} = \frac{\partial}{\partial t}(\nabla \cdot \mathbf{A}).$$

Using the Lorentz condition to substitute for $\nabla \cdot \mathbf{A}$, we have

$$\nabla \cdot \dot{\mathbf{A}} = \frac{\partial}{\partial t}\left(-\frac{1}{c^2}\frac{\partial\phi}{\partial t}\right) = -\frac{1}{c^2}\frac{\partial^2\phi}{\partial t^2}.$$

Substituting in equation (2.13) for $\nabla \cdot \mathbf{A}$ and putting $\nabla \cdot (\nabla\phi)$ equal to $\nabla^2\phi$, we find that equation (2.13) reduces to

$$\nabla^2\phi - \frac{1}{c^2}\frac{\partial^2\phi}{\partial t^2} = -\frac{\rho}{\varepsilon_0}. \tag{2.14}$$

Collecting the other equations for ϕ and **A**, in the Lorentz gauge we have

$$\nabla^2\mathbf{A} - \frac{1}{c^2}\frac{\partial^2\mathbf{A}}{\partial t^2} = -\mu_0\mathbf{J}. \tag{2.15}$$

$$\nabla \cdot \mathbf{A} + \frac{1}{c^2}\frac{\partial\phi}{\partial t} = 0. \qquad \text{(The Lorentz condition)} \tag{2.16}$$

The electric field **E** and the magnetic field **B** are given in terms of ϕ and **A** by the equations

$$\mathbf{E} = -\nabla\phi - \frac{\partial\mathbf{A}}{\partial t}. \tag{2.17}$$

$$\mathbf{B} = \nabla \times \mathbf{A}. \tag{2.18}$$

The choice of the Lorentz gauge decouples the differential equations for ϕ and **A** allowing us to solve the differential equations for ϕ and **A** separately. Notice that in equation (2.14), the scalar potential ϕ depends only on the charge density ρ and that in equation (2.12) the vector potential **A** depends only on the current density $\mathbf{J}$. The choice of the Lorentz gauge also gives equations that are relativistically invariant under a Lorentz transformation.

2.3. The retarded potentials

According to equation (2.14), in the Lorentz gauge the scalar potential ϕ due to a varying, continuous charge distribution is related to the charge density ρ by the partial differential equation

$$\nabla^2\phi - \frac{1}{c^2}\frac{\partial^2\phi}{\partial t^2} = -\frac{\rho}{\varepsilon_0}. \tag{2.19}$$

Equation (2.19) is an example of D'Alembert's equation. In this section, a simplified derivation of the solution of equation (2.19) will be given, which illustrates how the solution of equation (2.19) can be applied in practice. Readers interested in more rigorous solutions are referred to Panofsky and Phillips [1], Ferraro [2], or Hauser [3].

Consider the finite continuous charge distribution shown in Figure 2.1(a). The charge distribution is moving and varying in an arbitrary way such that the charge density $\rho(\mathbf{r}_s)$ and the current density $\mathbf{J}(\mathbf{r}_s)$ at the fixed source point at $\mathbf{r}_s$ are functions of time. Divide the charge distribution shown in Figure 2.1(a) into a large number of infinitesimal volume elements. Consider the variations in the charge, that is inside the volume element dV_s at the fixed position $\mathbf{r}_s$ in Figure 2.1(a), in isolation from the rest of the charge distribution, as shown in Figure 2.1(b). Choose a new coordinate system with a new origin O' at the position of the volume element dV_s as shown in Figure 2.1(b). Let the distance from the volume element dV_s at O' to the field point P in Figure 2.1(b) be denoted by $\mathbf{R}$. The total charge $dQ = \rho\, dV_s$ inside the volume element dV_s will be treated as a point charge. For a system consisting of only the varying charge inside the fixed volume element dV_s equation (2.19) becomes

$$\nabla^2\phi - \frac{1}{c^2}\frac{\partial^2\phi}{\partial t^2} = -\frac{\rho}{\varepsilon_0}\,\delta(R) \tag{2.20}$$

where the Dirac delta function $\delta(R)$ is zero unless $R = 0$ in which case $\delta(R) = 1$. Outside the volume element dV_s in Figure 2.1(b), equation (2.20) reduces to

$$\nabla^2\phi - \frac{1}{c^2}\frac{\partial^2\phi}{\partial t^2} = 0. \tag{2.21}$$

In the case of a point charge of magnitude $dQ = \rho dV_s$, the scalar potential ϕ should be spherically symmetric, so that using equation (A1.42) of Appendix A1.10, equation (2.21) becomes

$$\frac{1}{R^2}\frac{\partial}{\partial R}\left(R^2\frac{\partial\phi}{\partial R}\right) - \frac{1}{c^2}\frac{\partial^2\phi}{\partial t^2} = 0. \tag{2.22}$$

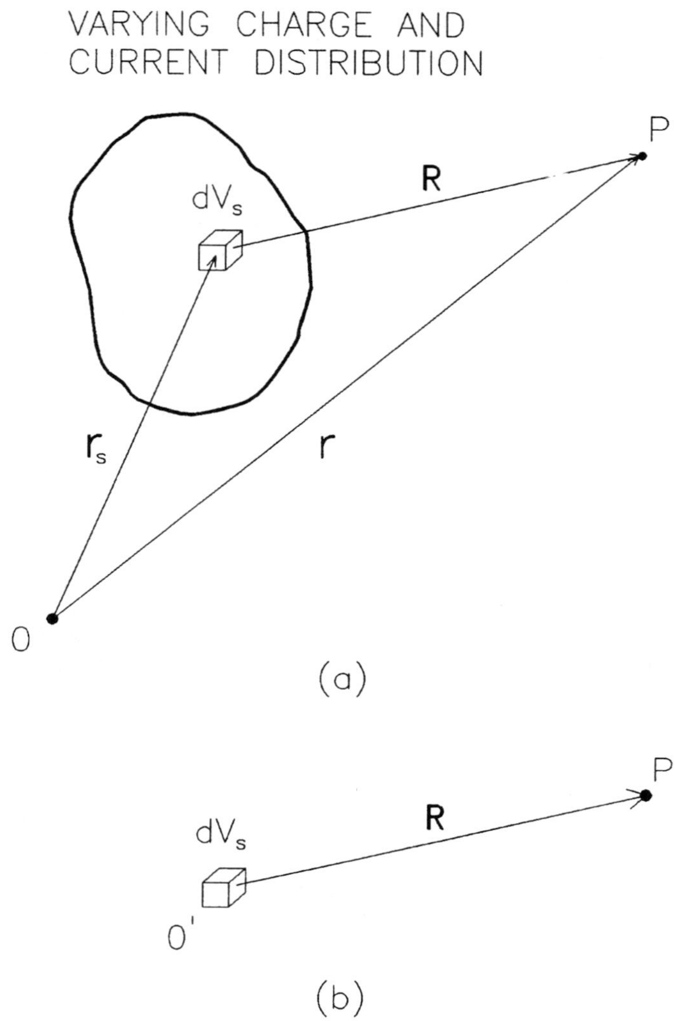

Figure 2.1. (a) Determination of the retarded potentials at the field point P due to a varying charge and current distribution (b) Derviation of the contribution of the charge and current inside the volume element dV_s to the retarded potentials at the field point P.

The solution of equation (2.22), that is valid at the field point P in Figure 2.1(b) at the time of observation t, is

$$\phi = \frac{f_1(t - R/c)}{R} + \frac{f_2(t + R/c)}{R} \tag{2.23}$$

where f_1 and f_2 are, so far, unspecified functions of $(t - R/c)$ and $(t + R/c)$ respectively. The reader can check this solution of equation (2.22) by substituting for ϕ from equation (2.23) into equation (2.22). The solution $(1/R)f_2(t + R/c)$ is normally rejected, since it corresponds to the advanced potentials. Hence outside the element of charge inside dV_s in Figure 2.1(b), the solution of equation (2.21) should be of the form

$$\phi = \frac{f_1(t - R/c)}{R} . \tag{2.24}$$

To specify the function $f_1(t - R/c)$ consider what happens very close to the origin O' in Figure 2.1(b). The solution given by equation (2.24) must satisfy equation (2.20) inside the volume element dV_s. Since ϕ is proportional to $1/R$, ϕ tends to infinity as R tends to zero in Figure 2.1(b). Hence near the origin O' in Figure 2.1(b), ϕ varies much more rapidly with distance at a fixed time than ϕ varies with time at a fixed value of R, so that near $R = 0$ the $\ddot{\phi}/c^2$ term in equation (2.20) can be neglected in comparison to the $\nabla^2\phi$ term. Hence near $R = 0$, equation (2.19) can be approximated by

$$\nabla^2\phi = -\frac{\rho}{\varepsilon_0} . \tag{2.25}$$

This is Poisson's equation. The solution of equation (2.25) in the case of a point charge is Coulomb's law. Hence when R tends to zero

$$\phi = \frac{f_1(t - R/c)}{R} \rightarrow \frac{dQ}{4\pi\varepsilon_0 R} = \frac{\rho dV_s}{4\pi\varepsilon_0 R} . \tag{2.26}$$

The general solution, given by equation (2.24), must go over to this form when R is very small. This suggests that in the general case

$$\phi = \frac{\rho(t - R/c)dV_s}{4\pi\varepsilon_0 R} \tag{2.27}$$

where $\rho(t - R/c)$ is a function of $(t - R/c)$. According to equation (2.27), the value of the scalar potential ϕ at the time of observation t at the field point P, at a distance $\mathbf{R}$ from the origin O' in Figure 2.1(b), depends on the magnitude of the charge inside the volume element dV_s at the time $(t - R/c)$, that is at a time R/c before the value of ϕ is required at the field point P. The time $(t - R/c)$ will be called the retarded time and denoted by t^*. Quantities measured at the retarded time t^* will be placed inside square brackets. For example the $\rho(t - R/c)$ term in equation (2.27) will be denoted by $[\rho]$, so that equation (2.27) can be rewritten in the form

$$\phi = \frac{[\rho]dV_s}{4\pi\varepsilon_0 R} . \tag{2.28}$$

Consider now the finite varying charge distribution shown in Figure 2.1(a). According to equation (2.28) the contribution, now denoted by $d\phi$, of the element of charge $[\rho]dV_s$ at $\mathbf{r}_s$ in Figure 2.1(a) to the total scalar potential ϕ at the field point P at $\mathbf{r}$ at the time of observation t is

$$d\phi = \frac{[\rho]dV_s}{4\pi\varepsilon_0 R}$$

where $\mathbf{R} = (\mathbf{r} - \mathbf{r}_s)$. Summing over all the volume elements making up the charge distribution shown in Figure 2.1(a), we find that the total scalar potential ϕ due to the continuous, varying charge distribution is given by

$$\phi(\mathbf{r}, t) = \frac{1}{4\pi\varepsilon_0} \int \frac{[\rho(\mathbf{r}_s)]\mathrm{d}V_s}{|\mathbf{r} - \mathbf{r}_s|} . \tag{2.29}$$

Since the various volume elements in the integral in equation (2.29) are at different distances $\mathbf{R} = |\mathbf{r} - \mathbf{r}_s|$ from the field point P in Figure 2.1(a), the appropriate retarded times are different for the various volume elements.

The solution of the equation

$$\nabla^2\mathbf{A} - \frac{1}{c^2}\frac{\partial^2\mathbf{A}}{\partial t^2} = -\mu_0\mathbf{J}. \tag{2.12}$$

can be obtained by solving equation (2.12) for the cartesian components A_x, A_y and A_z separately and then combining them to give

$$\mathbf{A}(\mathbf{r}, t) = \left(\frac{\mu_0}{4\pi}\right) \int \frac{[\mathbf{J}(\mathbf{r}_s)]\mathrm{d}V_s}{|\mathbf{r} - \mathbf{r}_s|} . \tag{2.30}$$

where $[\mathbf{J}(\mathbf{r}_s)]$ is the value of the current density at $\mathbf{r}_s$ at the retarded time $t^* = (t - R/c)$. Equations (2.29) and (2.30) are known as the **retarded potentials**. Notice that in the Lorentz gauge the vector potential **A** depends only on the current density **J**, showing that in the Lorentz gauge the vacuum displacement current density $\varepsilon_0\dot{\mathbf{E}}$ does not appear as one of the sources of the magnetic field in equation (2.30).

It is sometimes useful, when applying the retarded potentials, to introduce, for purposes of exposition only, an imaginary **information collecting sphere** whose centre is at the field point P in Figure 2.1(a) and which collapses with a velocity c in empty space, such that the information collecting sphere arrives at the field point P at the time of observation t, when the potentials ϕ and **A** are required at the field point P. The information collecting sphere passes the various volume elements $\mathrm{d}V_s$ in equations (2.29) and (2.30) at the appropriate retarded times. It is useful to imagine that this information collecting sphere collects information about charge density, current density and position at the appropriate retarded times. The data collected in this way can then be used to calculate ϕ and **A** using equations (2.29) and (2.30). The application of the retarded potentials will be illustrated in Section 2.4 by solving the example of the oscillating electric dipole.

So far in this section, we have only considered idealized continuous charge and current distributions in otherwise empty space. In practice all charge and current distributions are made up of moving atomic particles, such as electrons, protons and ions. In Section 1.11 of Chapter 1, the macroscopic variables **E**, **B**, ρ and **J** were defined in terms of the corresponding microscopic variables using equation (1.147). Maxwell's equations for the macroscopic fields **E** and **B** were derived from the Maxwell-Lorentz equations for the microscopic fields, namely from equations (1.137), (1.138), (1.139) and (1.140). This leads us to equations (1.152), (1.153), (1.154) and (1.155) of Chapter 1. These equations for the macroscopic fields have the same mathematical form as Maxwell's equations for an idealized continuous charge and current distribution, namely equations (2.1), (2.2), (2.3) and (2.4). It follows by the same mathematical steps as were used earlier in Sections 2.2 and 2.3 that the macroscopic fields **E** and **B** can be related to a macroscopic scalar potential ϕ and a macroscopic vector potential **A** by the equations

$$\mathbf{E} = -\nabla\phi - \dot{\mathbf{A}}; \qquad \mathbf{B} = \nabla \times \mathbf{A}$$

where ϕ and **A** are again given by the retarded potentials, namely equations (2.29) and (2.30) respectively, provided that ρ is now the macroscopic charge density and **J** is the macroscopic current density calculated using equation (1.147).

2.4. The oscillating electric dipole

2.4.1. *Introduction*

To illustrate the application and interpretation of the retarded potentials, we shall now give an account of the calculation of the electric and magnetic fields due to a stationary, oscillating, electric dipole. We shall assume that the electric dipole consists of two varying point charge distributions of magnitudes $+Q$ and $-Q$ respectively at a fixed infinitesimal distance dl apart, as shown in Figure 2.2. The instantaneous value of the electric dipole moment **p** is

$$\mathbf{p} = Q\mathrm{d}\mathbf{l}. \tag{2.31}$$

The mid-point of the electric dipole is fixed at the origin of a co-ordinate system, with **p** and **dl** pointing in the $+z$ direction from the negative to the positive charge as shown in Figure 2.2. The charges are joined by a straight conducting wire of infinitesimal length dl. If a charge $+\mathrm{d}Q$ flows along the wire in a time dt from the negative to the positive charge the charges are changed to $+(Q + \mathrm{d}Q)$ and $-(Q + \mathrm{d}Q)$ respectively, and the current flowing in the connecting wire is $I = \mathrm{d}Q/\mathrm{d}t$. Differentiating equation (2.31) with respect to time, for fixed dl we have

$$\dot{p} = \left(\frac{\mathrm{d}Q}{\mathrm{d}t}\right) \mathrm{d}l = I\, \mathrm{d}l. \tag{2.32}$$

It will be assumed that the electric dipole moment p varies sinusoidally with time, that is

$$p = p_0 \sin \omega t \tag{2.33}$$

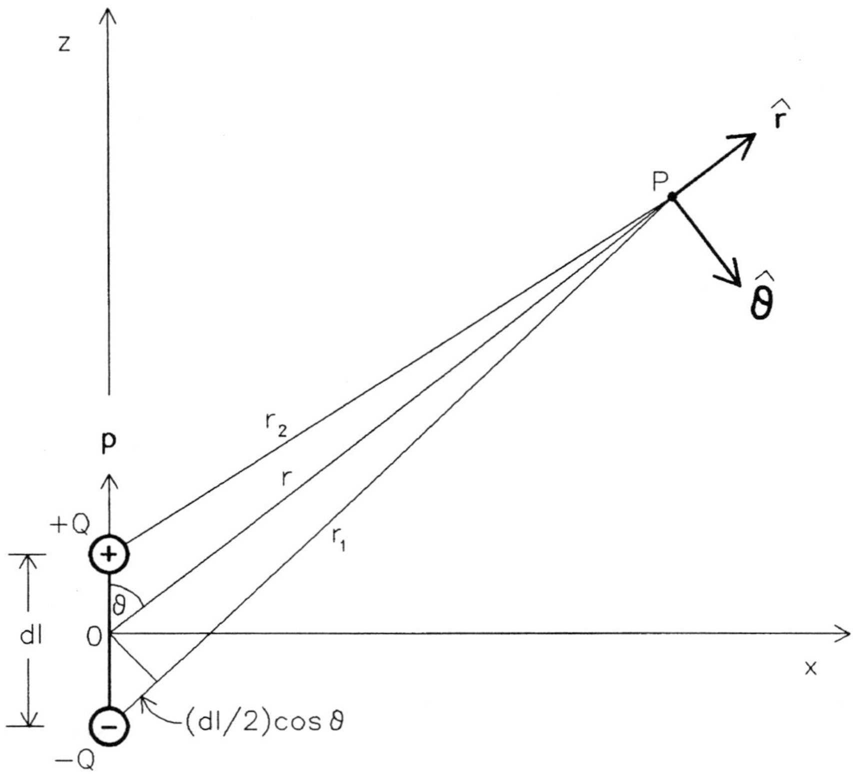

Figure 2.2. Determination of the electric and magnetic fields at the field point P which is at a distance $\mathbf{r}$ from the oscillating electric dipole at the origin. The dipole moment points along the z axis. The orientation of the x axis is chosen such that the field point P is in the xz plane. The polar angle θ of the spherical polar coordinate system is measured from the z axis. The directions of the unit vectors $\hat{\mathbf{r}}$ and $\hat{\boldsymbol{\theta}}$ at the field point P are shown. The unit vector $\hat{\boldsymbol{\phi}}$ is in the direction of $\hat{\mathbf{r}} \times \hat{\boldsymbol{\theta}}$, which is downwards into the paper.

where ω is the angular frequency. Using equation (2.32) we have

$$I = \frac{\dot{p}}{\mathrm{d}l} = \left(\frac{\omega p_0}{\mathrm{d}l}\right) \cos \omega t = I_0 \cos \omega t \tag{2.34}$$

where $I_0 = (\omega p_0/\mathrm{d}l)$ is the maximum electric current that flows in the wire of length $\mathrm{d}l$.

Consider a field point P at a distance $\mathbf{r}$ from the electric dipole that is at the origin in Figure 2.2. If the time of observation of the fields $\mathbf{E}$ and $\mathbf{B}$ at the field point P in Figure 2.2 is t, then the corresponding retarded time at the oscillating dipole is $t^* = (t - r/c)$. The value $[p]$ of the electric dipole moment at the retarded time t^* is

$$[p] = p_0 \sin \omega t^* = p_0 \sin \omega \left(t - \frac{r}{c}\right). \tag{2.35}$$

We shall now derive some of the formulae for the partial differential coefficients of the value of $[p]$ given by equation (2.35). Differentiating equation (2.35) partially with respect to time, we have

$$[\dot{p}] = \omega p_0 \cos \omega \left(t - \frac{r}{c}\right). \tag{2.36}$$

Using equations (2.32) and (2.34), we have

$$[\dot{p}] = [I]\mathrm{d}l = I_0 \,\mathrm{d}l \cos \omega \left(t - \frac{r}{c}\right). \tag{2.37}$$

Differentiating equation (2.36) and (2.37) partially with respect to time we obtain

$$[\ddot{p}] = -\omega^2 p_0 \sin \omega \left(t - \frac{r}{c}\right), \tag{2.38}$$

$$[\ddot{p}] = [\dot{I}]\,\mathrm{d}l = -\omega I_0 \,\mathrm{d}l \sin \omega \left(t - \frac{r}{c}\right). \tag{2.39}$$

Differentiating equation (2.35) partially with respect to r we have

$$\begin{aligned} \frac{\partial [p]}{\partial r} &= -\frac{\omega}{c} p_0 \cos \omega \left(t - \frac{r}{c}\right), \\ &= -\frac{[\dot{p}]}{c} = -\frac{[I]\,\mathrm{d}l}{c}. \end{aligned} \tag{2.40}$$

Differentiating equation (2.36) partially with respect to r and using equation (2.38) we have

$$\frac{\partial [\dot{p}]}{\partial r} = -\frac{[\ddot{p}]}{c} = -\frac{[\dot{I}]\,\mathrm{d}l}{c}. \tag{2.41}$$

2.4.2. *Determination of the magnetic field*

Consider again the field point P, that is at a distance $r \gg \mathrm{d}l$ from the electric dipole at the origin in Figure 2.2. According to the expression for the retarded vector potential, equation (2.30), the vector potential $\mathbf{A}$ at the field point P at the time of observation t is in the $+z$ direction in Figure 2.2 and, since in this example $[\mathbf{J}]\,\mathrm{d}V_s = [I]\,\mathbf{dl}$ and $R = r$, the magnitude of A_z is

$$A_z = \frac{\mu_0 [I]\,\mathrm{d}l}{4\pi r} = \frac{\mu_0 [\dot{p}]}{4\pi r} \tag{2.42}$$

where $[I] = I_0 \cos \omega(t - r/c)$ is the value of the current in the oscillating electric dipole at the retarded time $t^* = (t - r/c)$. Notice that the vacuum displacement current density $\varepsilon_0 \dot{\mathbf{E}}$ at various points in space should not be included as one of the sources of the vector potential in the Lorentz gauge.

Introduce spherical polar coordinates in Figure 2.2, measuring r from the origin, the polar angle θ from the z axis and the azimuthal angle ϕ from the x axis towards the y axis in the xy plane. Let $\hat{\mathbf{r}}$, $\hat{\boldsymbol{\theta}}$ and $\hat{\boldsymbol{\phi}}$ be unit vectors in the directions of increasing r, θ and ϕ respectively. The use of the same symbol ϕ for both the scalar potential and the azimuthal angle in spherical polar coordinates should not lead to any confusion since it should be clear from the text which quantity the symbol ϕ stands for.

Resolving the value of A_z given by equation (2.42) into components A_r, A_θ and A_ϕ in the spherical polar coordinate system we obtain

$$A_r = \frac{\mu_0[I]\,dl\cos\theta}{4\pi r} = \frac{\mu_0[\dot{p}]\cos\theta}{4\pi r} \tag{2.43}$$

$$A_\theta = -\frac{\mu_0[I]\,dl\sin\theta}{4\pi r} = -\frac{\mu_0[\dot{p}]\sin\theta}{4\pi r} \tag{2.44}$$

$$A_\phi = 0. \tag{2.45}$$

The magnetic field $\mathbf{B}$ at the field point P in Figure 2.2 at the time of observation t is given by $\nabla\times\mathbf{A}$, where in spherical polar coordinates $\nabla\times\mathbf{A}$ is given by equation (A1.41) of Appendix A1.10. It is straight forward for the reader to show that, since $A_\phi = 0$ and A_r and A_θ are independent of the azimuthal angle ϕ, then

$$B_r = 0$$

$$B_\theta = 0$$

$$B_\phi = \frac{1}{r}\frac{\partial}{\partial r}(rA_\theta) - \frac{1}{r}\frac{\partial A_r}{\partial\theta}$$

$$= \frac{\mu_0}{4\pi}\left\{\frac{1}{r}\frac{\partial}{\partial r}(-[\dot{p}]\sin\theta) - \frac{1}{r}\frac{\partial}{\partial\theta}\left(\frac{[\dot{p}]\cos\theta}{r}\right)\right\}.$$

From equation (2.41), $\partial[\dot{p}]/\partial r = -[\ddot{p}]/c$. Since $[p]$ is independent of θ, then $\partial[\dot{p}]/\partial\theta = 0$. Hence

$$\mathbf{B} = \frac{\mu_0\sin\theta}{4\pi}\left(\frac{[\dot{p}]}{r^2} + \frac{[\ddot{p}]}{rc}\right)\hat{\boldsymbol{\phi}}. \tag{2.46}$$

Using equation (2.36), (2.37), (2.38) and (2.39), the expression for $\mathbf{B}$ can be expressed in the alternative form

$$\mathbf{B} = \frac{\mu_0[I]\,dl\sin\theta}{4\pi r^2}\hat{\boldsymbol{\phi}} + \frac{\mu_0[\dot{I}]\,dl\sin\theta}{4\pi rc}\hat{\boldsymbol{\phi}}, \tag{2.47}$$

$$= \frac{\mu_0 I_0\cos\omega(t - r/c)\,dl\sin\theta}{4\pi r^2}\hat{\boldsymbol{\phi}} - \frac{\mu_0\omega I_0\sin\omega(t - r/c)\,dl\sin\theta}{4\pi rc}\hat{\boldsymbol{\phi}}. \tag{2.48}$$

The first term on the right hand side of equation (2.48) is proportional to $1/r^2$. Comparing equation (2.48) with equation (1.64), we see that the first term on the right hand side of equation (2.48) is equal to the magnetic field that would be predicted by the differential form of the Biot-Savart law, if the value $[I] = I_0 \cos \omega(t - r/c)$ of the current in the electric dipole at the retarded time $(t - r/c)$ is used in the differential form of the Biot-Savart law. The second term on the right hand side of equation (2.48) is proportional to $1/r$ and is called the radiation term. It predominates at large distances from the oscillating electric dipole in Figure 2.2. The maximum amplitudes of the two harmonically varying terms on the right hand side of equation (2.48) are numerically equal when $1/r = \omega/c$, that is when the distance from the oscillating electric dipole to the field point is

$$r = \frac{c}{\omega} = \frac{\lambda}{2\pi} = \frac{1}{k} \tag{2.49}$$

where λ is the wavelength of the electromagnetic variations and $k = 2\pi/\lambda$ is the wave number. The region, where $r < \lambda/2\pi$ and the first term on the right hand side of equation (2.48) predominates, is called the near zone. The region where $r > \lambda/2\pi$ and the second term on the right hand side of equation (2.48) predominates is called the far zone. When r is very much greater than $\lambda/2\pi$, the second term on the right hand side of equation (2.48), namely the radiation term, is very much bigger than the first term. The region where $r \gg \lambda/2\pi$ is called the radiation (or wave) zone.

Notice that, since according to equation (2.48) the only component of the magnetic field due to the oscillating electric dipole is in the direction of $\hat{\boldsymbol{\phi}}$, the magnetic field lines are closed circles having constant values of r and θ.

2.4.3. *Determination of the electric field*

Now that we have derived the expressions for the magnetic field **B** due to the oscillating electric dipole in Figure 2.2 we could determine the electric field **E** using the Maxwell equation

$$\nabla \times \mathbf{B} = \frac{1}{c^2}\frac{\partial \mathbf{E}}{\partial t}$$

in the way described later in Section 2.6.5. However, it will be useful in some of our discussions in later chapters, if we derive the expression for the electric field directly from the potentials ϕ and **A** using the equation

$$\mathbf{E} = -\nabla\phi - \frac{\partial \mathbf{A}}{\partial t}\,. \tag{2.17}$$

In addition to the expression for the vector potential **A** given by equations (2.43), (2.44) and (2.45), to determine **E** we must also derive the expression for the scalar potential ϕ from the charge distributions using equation (2.29). To determine ϕ, we shall assume that the negative and positive charges are at distances $r_1 = (r + \frac{1}{2}\mathrm{d}l \cos \theta)$ and $r_2 = (r - \frac{1}{2}\mathrm{d}l \cos \theta)$ from the field point P in Figure 2.2. We shall assume that, when the information collecting sphere passes the negative charge at the retarded time $(t - r_1/c)$, the magnitude of the negative charge is $[-Q]$. We shall also assume that the magnitude of the positive charge is varying at the rate $[\dot{Q}]$ when the information collecting sphere is passing the oscillating electric dipole in Figure 2.2. Since the positive charge is at a distance $(\mathrm{d}l \cos\theta)$ closer to the field point P than the negative charge, it takes the information collecting sphere a time $(\mathrm{d}l \cos\theta)/c$ to cross the oscillating dipole so that the magnitude of the positive charge recorded by the information collecting sphere is $[Q + \dot{Q}(\mathrm{d}l \cos\theta)/c]$. Substituting in the expression for the retarded scalar potential, given by equation (2.29), we have

$$\phi = \frac{[Q + \dot{Q}(\mathrm{d}l \cos\theta)/c]}{4\pi\varepsilon_0 r[1 - (\mathrm{d}l \cos\theta)/2r]} - \frac{[Q]}{4\pi\varepsilon_0 r[1 + (\mathrm{d}l \cos\theta)/2r]}.$$

Expanding $[1 \pm (\mathrm{d}l \cos\theta)2r)]^{-1}$ using the binomial theorem, then multiplying out and ignoring terms of order $(\mathrm{d}l)^2$ we finally obtain

$$\phi = \frac{[Q]\,\mathrm{d}l \cos\theta}{4\pi\varepsilon_0 r^2} + \frac{[\dot{Q}]\,\mathrm{d}l \cos\theta}{4\pi\varepsilon_0 rc}. \tag{2.50}$$

From equations (2.31) and (2.32), $[Q]\,\mathrm{d}l = [p]$ and $[\dot{Q}]\,\mathrm{d}l = [I]\,\mathrm{d}l = [\dot{p}]$. Substituting in equation (2.50), we find that the scalar potential ϕ is given by

$$\phi = \frac{\cos\theta}{4\pi\varepsilon_0}\left(\frac{[p]}{r^2} + \frac{[\dot{p}]}{rc}\right). \tag{2.51}$$

Alternatively, the expression for ϕ can be determined from the Lorentz condition, equation (2.11), in the way described later in Section 2.5. It is left as an exercise for the reader to show, using the expression for $\nabla\phi$ given by equation (A1.39) of Appendix 1.10 and using equations (2.40) and (2.41), that

$$-\nabla\phi = \frac{\cos\theta}{4\pi\varepsilon_0}\left(\frac{2[p]}{r^3} + \frac{2[\dot{p}]}{r^2 c} + \frac{[\ddot{p}]}{rc^2}\right)\hat{\mathbf{r}} + \frac{\sin\theta}{4\pi\varepsilon_0}\left(\frac{[p]}{r^3} + \frac{[\dot{p}]}{r^2 c}\right)\hat{\boldsymbol{\theta}}. \tag{2.52}$$

Differentiating equations (2.43) and (2.44) partially with respect to time and using the relation $\mu_0 = 1/\varepsilon_0 c^2$ we find that

$$-\frac{\partial \mathbf{A}}{\partial t} = -\frac{\cos\theta[\ddot{p}]}{4\pi\varepsilon_0 rc^2}\hat{\mathbf{r}} + \frac{\sin\theta[\ddot{p}]}{4\pi\varepsilon_0 rc^2}\hat{\boldsymbol{\theta}}. \tag{2.53}$$

Adding equations (2.52) and (2.53) we finally obtain

$$\mathbf{E} = \frac{\cos\theta}{4\pi\varepsilon_0}\left(\frac{2[p]}{r^3} + \frac{2[\dot{p}]}{r^2 c}\right)\hat{\mathbf{r}} + \frac{\sin\theta}{4\pi\varepsilon_0}\left(\frac{[p]}{r^3} + \frac{[\dot{p}]}{r^2 c} + \frac{[\ddot{p}]}{rc^2}\right)\hat{\boldsymbol{\theta}}. \tag{2.54}$$

Notice that the term proportional to $[\ddot{p}]/r$ in the $\hat{\mathbf{r}}$ direction in equation (2.52) for $-\nabla\phi$ cancels the term in the $\hat{\mathbf{r}}$ direction proportional to $-[\ddot{p}]/r$ in equation

(2.53) for $-\dot{\mathbf{A}}$ leaving only the component proportional to $[\ddot{p}]/r$ that is in the direction of $\hat{\boldsymbol{\theta}}$, which is in a direction perpendicular the vector **r** from the electric dipole to the field point. It is useful to rewrite equation (2.54) in the form

$$\mathbf{E} = \mathbf{E}_1 + \mathbf{E}_2 + \mathbf{E}_3 \tag{2.55}$$

where

$$\mathbf{E}_1 = \frac{2[p]\cos\theta}{4\pi\varepsilon_0 r^3}\hat{\mathbf{r}} + \frac{[p]\sin\theta}{4\pi\varepsilon_0 r^3}\hat{\boldsymbol{\theta}} \tag{2.56}$$

$$\mathbf{E}_2 = \frac{2[\dot{p}]\cos\theta}{4\pi\varepsilon_0 r^2 c}\hat{\mathbf{r}} + \frac{[\dot{p}]\sin\theta}{4\pi\varepsilon_0 r^2 c}\hat{\boldsymbol{\theta}} \tag{2.57}$$

$$\mathbf{E}_3 = \frac{[\ddot{p}]\sin\theta}{4\pi\varepsilon_0 r c^2}\hat{\boldsymbol{\theta}}. \tag{2.58}$$

Notice that E_1, E_2 and E_3 are proportional to $1/r^3$, $1/r^2c$ and $1/rc^2$ respectively.

Examples of the electric field due to an oscillating electric dipole at successive instants of time are shown in Figures 2.3. The $\mathbf{E}_1$ term, given by equation (2.56), is similar to the expression for the electrostatic field due to an electric dipole of dipole moment $[p] = p_0 \sin \omega(t - r/c)$. The $\mathbf{E}_1$ term is proportional to $1/r^3$ and predominates close to the oscillating electric dipole, where the electric field resembles the electric field due to an electrostatic dipole, as shown in the examples in Figure 2.3. The direction of the electric field reverses every half period. The $\mathbf{E}_2$ term, given by equation (2.57), is proportional to $1/r^2$. It depends on $[\dot{p}] = [I]\,dl$. It is the $\mathbf{E}_2$ contribution to **E** that gives rise to the induction electric field, that gives the main contribution to the induced emf in a stationary coil in the near zone ($r < \lambda/2\pi$). The $\mathbf{E}_3$ term, given by equation (2.58) is proportional to $1/r$ and predominates at very large distances from the oscillating electric dipole. It is the $\mathbf{E}_3$ term that gives the radiation electric field.

It follows from equations (2.54), (2.38) and (2.46) that the expressions for **E** and **B** in the radiation zone where $r \gg \lambda/2\pi$ are

$$\mathbf{E} = E_0 \sin\omega\left(t - \frac{r}{c}\right)\hat{\boldsymbol{\theta}}. \tag{2.59}$$

$$\mathbf{B} = B_0 \sin\omega\left(t - \frac{r}{c}\right)\hat{\boldsymbol{\phi}}. \tag{2.60}$$

where

$$E_0 = -\frac{\omega^2 p_0 \sin\theta}{4\pi\varepsilon_0 r c^2}; \qquad B_0 = -\frac{\omega^2 p_0 \sin\theta}{4\pi\varepsilon_0 r c^3} = \frac{E_0}{c}. \tag{2.61}$$

Notice that in the radiation zone **E** and **B** are in phase and that **E** and **B** are perpendicular to each other. Equations (2.59) and (2.60) represent electromagnetic waves travelling outwards from the oscillating electric dipole with the

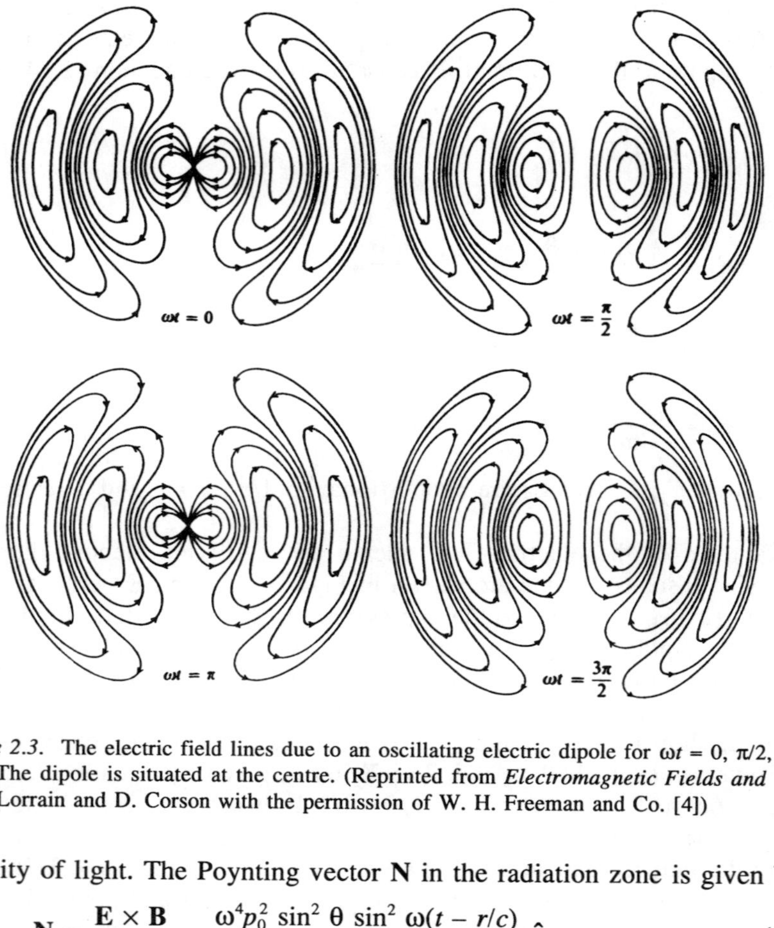

Figure 2.3. The electric field lines due to an oscillating electric dipole for $\omega t = 0$, $\pi/2$, π and $3\pi/2$. The dipole is situated at the centre. (Reprinted from *Electromagnetic Fields and Waves* by P. Lorrain and D. Corson with the permission of W. H. Freeman and Co. [4])

velocity of light. The Poynting vector **N** in the radiation zone is given by

$$\mathbf{N} = \frac{\mathbf{E} \times \mathbf{B}}{\mu_0} = \frac{\omega^4 p_0^2 \sin^2 \theta \sin^2 \omega(t - r/c)}{16\pi^2 \varepsilon_0 r^2 c^3} \hat{\mathbf{r}}. \tag{2.62}$$

The direction of the Poynting vector **N** in the radiation zone is radially outwards from the oscillating electric dipole.

2.4.4. *Relation of the fields* E *and* B *to experimental measurements*

The fields **E** and **B** at any field point in empty space, due to the oscillating electric dipole shown in Figure 2.2, could, in principle, be related to experiments by using stationary and moving test charges. If the values of **E** and **B** were known then the forces on the test charges could be calculated using the Lorentz force law, which according to equation (1.1) is

$$\mathbf{F} = \frac{\mathrm{d}}{\mathrm{d}t}\left(\frac{m_0\mathbf{u}}{(1 - u^2/c^2)^{1/2}}\right) = q\mathbf{E} + q\mathbf{u} \times \mathbf{B} \tag{2.63}$$

where m_0 is the rest mass and **u** is the velocity of a test charge of magnitude

q. Alternatively, if the values of **E** and **B** were unknown, their values could be determined from the forces on a stationary and on moving test charges, in the way described in Section 1.4.2 of Chapter 1, using equations (1.56) and (1.58).

In practice, in the case of the oscillating electric dipole shown in Figure 2.2, it is generally easier experimentally to get observable effects by placing a stationary antenna at the field point in empty space, for example of the type shown in Figure 2.4(a), which is a simple dipole. Typically such a dipole antenna consists of a metal rod, split in the middle, and connected to a high resistance R as shown in Figure 2.4(a). If at a particular instant the electric field **E** due to the oscillating electric dipole is in the direction shown in Figure 2.4(a), the electric field **E** gives an electric force on each of the conduction electrons in the receiving antenna leading to a current flow in the antenna, which in turn gives rise to a potential difference V across the resistor R. This potential difference can be measured using electronic methods. It can be shown that, if the length l of the receiving dipole is much less than

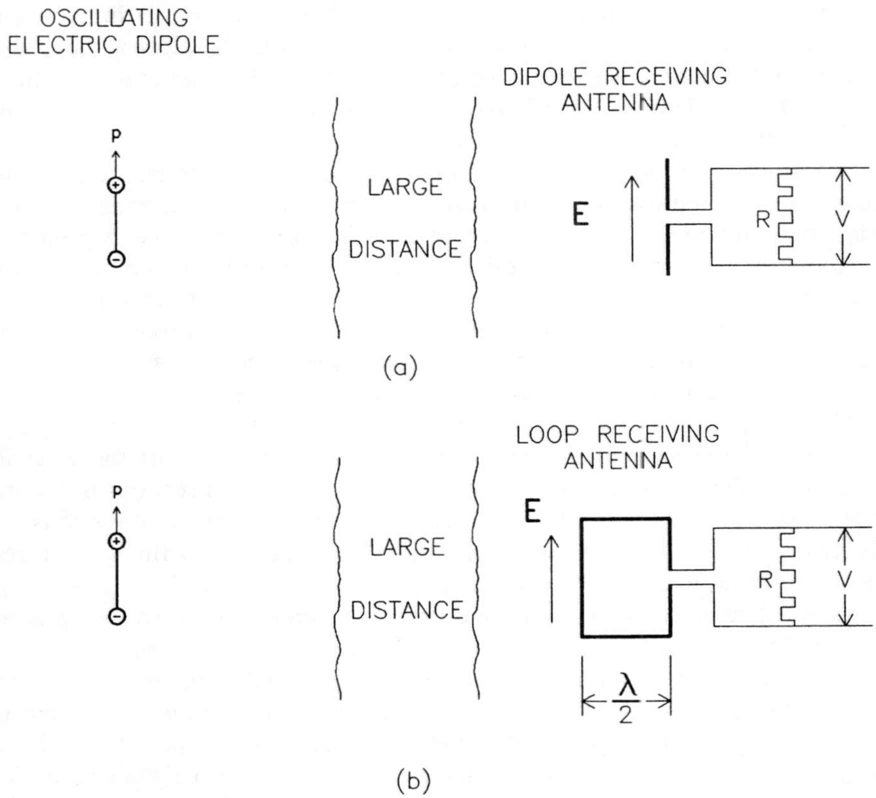

Figure 2.4. Measurement of the electric field due to an oscillating electric dipole (a) using a dipole receiving antenna and (b) using a loop receiving antenna.

the wavelength λ, the potential difference across the resistor R is approximately $E_p l$, where E_p is the component of $\mathbf{E}$ parallel to the dipole antenna. In practice, in order to increase the signal strength, the total length of the receiving dipole is generally made equal to $\lambda/2$, so that each half section is of length $\lambda/4$. The full theory of a dipole antenna of finite length is rather complicated. The interested reader is referred to a text book such as Lorrain and Corson [4].

In the example shown in Figure 2.4(a), the oscillating electric dipole is equivalent to a radio transmitter and the receiving dipole antenna and associated electronic circuits corresponds to a radio receiver. If the distance from the transmitter to the receiver is $\gg \lambda/2\pi$, the receiver is in the radiation (or wave) zone, and responds mainly to the radiation electric field $\mathbf{E}_3$ given by equation (2.58).

It is assumed in the idealized case shown in Figure 2.4(a) that there is nothing in the space between the transmitter and the receiver. If there were isolated metallic conductors present, the varying electric field due to the oscillating electric dipole would give varying conduction current flows in the conductors, which in turn would give rise to electric and magnetic fields which would be superimposed on the fields due to the oscillating electric dipole. Any dielectrics present would be polarized in the electric field and any magnetic materials present would be magnetized in the magnetic field due to the oscillating electric dipole and would also give contributions to the total electric and magnetic fields.

Advantage can be taken of the effects of induced electric currents in conductors to improve the designs of receiving antennae. For example, in a Yagi type antenna a rod is placed at an appropriate distance behind the receiving dipole, but it is not connected electrically to the receiving dipole. This extra rod acts as a reflector. A series of rods, called directors, are sometimes placed at appropriate distances in front of the receiving dipole to increase the signal strength and to improve the directional properties of the antenna.

Another type of receiving antenna is the loop antenna of the type shown in Figure 2.4(b). In this case the magnitude of the electrical signal in the receiving antenna circuit depends on the spatial variations of the electric field due to the transmitter. For example, if the two vertical sections in Figure 2.4(b) were $\lambda/2$ apart, the electric field would be in opposite directions on the two vertical sides of the receiving antenna, leading to a finite value for $\oint \mathbf{E} \cdot d\mathbf{l}$, which gives rise to a voltage V across the resistor R. According to Faraday's law of electromagnetic induction, this emf is numerically equal to the rate of change of the magnetic flux through the loop antenna.

The example of the oscillating electric dipole shown in Figures 2.4(a) and 2.4(b) can be used to illustrate the retarded potentials. Assume that there is an oscillating electric dipole transmitter on a spaceship that is in the vicinity of the planet Jupiter. Experiments, for example using radio signals from the Pioneer 10 spaceship, have confirmed that it takes about 30 minutes for radio signals to reach the Earth from such a spaceship. This shows that the value of the electric field of a radio signal reaching the Earth from a spaceship near Jupiter depends on the value of the electric current in the transmitting antenna on the spaceship at the retarded time, which is approximately 30 minutes before the time the radio signal reaches the Earth. If there were a series of radio transmitters in space, we could imagine the radio signals coming in with the information collecting sphere to reach the radio receiver on the Earth at the time of observation.

2.5. Use of the Lorentz condition to determine the scalar potential ϕ from the vector potential A

In practice we do not always need to know all four of the variables ϕ, A_x, A_y and A_z to determine the fields **E** and **B**. In the Lorentz gauge which we have been using, ϕ and **A** are related by the Lorentz condition, equation (2.11). In some problems it is possible to derive the scalar potential ϕ from the vector potential **A** using the Lorentz condition. Integrating the Lorentz condition, equation (2.11), with respect to time, we have

$$\phi = -c^2 \int \nabla \cdot \mathbf{A}\, dt + \phi_0(x, y, z) \tag{2.64}$$

where $\phi_0(x, y, z)$ is a scalar function of x, y, z that is independent of time. When the electric field is determined using equation (2.7), the $-\nabla\phi_0$ contribution gives rise to a time independent, that is an electrostatic contribution to the total electric field. By substituting the values of A_r, A_θ and A_ϕ given by equations (2.43), (2.44) and (2.45) respectively into the expression for the divergence of **A** given by equation (A1.40) of Appendix A1.10, the reader can show that, in the case of the oscillating electric dipole shown in Figure 2.2, $\nabla \cdot \mathbf{A}$ is given by

$$\nabla \cdot \mathbf{A} = -\frac{\mu_0 \cos\theta}{4\pi}\left(\frac{[\dot{p}]}{r^2} + \frac{[\ddot{p}]}{rc}\right). \tag{2.65}$$

Substituting for $\nabla \cdot \mathbf{A}$ in equation (2.64), then integrating with respect to time and using $\mu_0 = 1/\varepsilon_0 c^2$ we find that

$$\phi = \frac{\cos\theta}{4\pi\varepsilon_0}\left(\frac{[p]}{r^2} + \frac{[\dot{p}]}{rc}\right) + \phi_0(x, y, z) \tag{2.66}$$

where ϕ_0 is the time independent constant of integration. Since there is no electrostatic field in the case of the oscillating electric dipole shown in Figure 2.2, the ϕ_0 term is zero in equation (2.66), which then reduces to equation (2.51). This shows that in the case of an oscillating electric dipole that has no resultant total charge, the six components of **E** and **B** can all be determined from the three components of the vector potential **A**. If there were an additional electrostatic charge Q_0 at the origin in Figure 2.2 there would be an additional electrostatic contribution $\phi_0 = Q_0/4\pi\varepsilon_0 r$ to ϕ, and the scalar potential ϕ could not be determined completely from the vector potential **A**.

2.6. Application of Maxwell's equations to the electric and magnetic fields due to an oscillating electric dipole

2.6.1. *Introduction*

Since the four variables A_x, A_y, A_z and ϕ, given as functions of the variables x, y, z and t, are sufficient to specify the varying electromagnetic field at a field point in empty space, it is not necessary to use as many as the six variables E_x, E_y, E_z, B_x, B_y and B_z to specify the electromagnetic field due to the oscillating electric dipole shown in Figure 2.2. Hence it is reasonable to find that there are relations between the six components that specify the fields **E** and **B** given by equations (2.54) and (2.46).

Problem. A plane transverse wave, that is propagating in the $+x$ direction and is linearly polarized in the y direction is described by the equation

$$\mathbf{Y} = \hat{\mathbf{j}}Y_0 \cos \omega \left(t - \frac{x}{c} \right)$$

where c is the velocity of the wave. A typical example would be an elastic S wave of seismology, in which case **Y** could stand for the displacement of a point from its equilibrium position. Now, for the fun of it, introduce new variables defined by

$$\mathbf{F} = -\dot{\mathbf{Y}}; \qquad \mathbf{G} = \nabla \times \mathbf{Y}.$$

Show that it follows from the definitions of **F** and **G** that

$$\nabla \times \mathbf{F} = -\dot{\mathbf{G}}.$$

Also show that

$$\mathbf{F} = \hat{\mathbf{j}}\omega Y_0 \sin \omega \left(t - \frac{x}{c} \right); \qquad \mathbf{G} = \hat{\mathbf{k}} \left(\frac{\omega}{c} \right) Y_0 \sin \omega \left(t - \frac{x}{c} \right)$$

and that the new variables **F** and **G** are related by the equation

$$\nabla \times \mathbf{G} = \frac{1}{c^2} \frac{\partial \mathbf{F}}{\partial t} = \hat{\mathbf{j}} \left(\frac{\omega^2 Y_0}{c^2} \right) \cos \omega \left(t - \frac{x}{c} \right).$$

This is a relation between the new variables **F** and **G**. It illustrates how, if we use more variables than is necessary, we can end up with relations between the variables.

One of the main aims of this book is to develop the interpretation of the roles of the various terms in Maxwell's equations, which we shall do in Chapter 4, where we shall start from the expressions for the electric and magnetic fields due to an accelerating classical point charge. In this section we shall give a brief introduction to the ideas, that we shall develop more fully in Chapter 4, by seeing how Maxwell's equations apply to the fields **E** and **B** due to the oscillating electric dipole in Figure 2.2.

2.6.2. *The equation* $\nabla \cdot \mathbf{E} = 0$

According to equation (A1.40) of Appendix A1.10, in the spherical polar co-ordinate system (r, θ, ϕ) shown in Figure 2.2, the divergence of **E** is given by

$$\nabla \cdot \mathbf{E} = \frac{1}{r^2}\frac{\partial}{\partial r}(r^2 E_r) + \frac{1}{r \sin\theta}\frac{\partial}{\partial \theta}(E_\theta \sin\theta) + \frac{1}{r\sin\theta}\left(\frac{\partial E_\phi}{\partial \phi}\right). \tag{2.67}$$

According to equation (2.54), for the oscillating electric dipole shown in Figure 2.2

$$E_r = \frac{1}{4\pi\varepsilon_0}\left[\frac{2p}{r^3} + \frac{2\dot{p}}{r^2 c}\right]\cos\theta \tag{2.68}$$

$$E_\theta = \frac{1}{4\pi\varepsilon_0}\left[\frac{p}{r^3} + \frac{\dot{p}}{r^2 c} + \frac{\ddot{p}}{rc^2}\right]\sin\theta \tag{2.69}$$

$$E_\phi = 0. \tag{2.70}$$

Throughout this section, all the quantities inside square brackets are measured at the retarded time $t^* = (t - r/c)$. The reader can show, using equations (2.68), (2.69) and (2.70), that

$$\begin{aligned}\frac{1}{r^2}\frac{\partial}{\partial r}(r^2 E_r) &= \frac{1}{4\pi\varepsilon_0 r^2}\frac{\partial}{\partial r}\left[\frac{2p}{r} + \frac{2\dot{p}}{c}\right]\cos\theta \\ &= -\frac{2\cos\theta}{4\pi\varepsilon_0}\left[\frac{p}{r^4} + \frac{\dot{p}}{r^3 c} + \frac{\ddot{p}}{r^2 c^2}\right]\end{aligned} \tag{2.71}$$

$$\begin{aligned}\frac{1}{r\sin\theta}\frac{\partial}{\partial\theta}(E_\theta \sin\theta) &= \left(\frac{1}{4\pi\varepsilon_0}\right)\frac{1}{r\sin\theta}\left[\frac{p}{r^3} + \frac{\dot{p}}{r^2 c} + \frac{\ddot{p}}{rc^2}\right]\frac{\partial}{\partial\theta}\sin^2\theta \\ &= \frac{2\cos\theta}{4\pi\varepsilon_0}\left[\frac{p}{r^4} + \frac{\dot{p}}{r^3 c} + \frac{\ddot{p}}{r^2 c^2}\right]\end{aligned} \tag{2.72}$$

$$\frac{1}{r\sin\theta}\frac{\partial E_\phi}{\partial\phi} = 0. \tag{2.73}$$

Adding equations (2.71), (2.72) and (2.73), we find that

$$\nabla \cdot \mathbf{E} = 0.$$

In cartesian coordinates we have

$$\nabla \cdot \mathbf{E} = \frac{\partial E_x}{\partial x} + \frac{\partial E_y}{\partial y} + \frac{\partial E_z}{\partial z} = 0. \tag{2.74}$$

Equation (2.74) is a relation between the components E_x, E_y and E_z. According to equation (2.74), the divergence of the electric field due to an oscillating electric dipole is zero at every field point in empty space. Integrating equation (2.74) over any arbitrary finite volume that does not enclose any part of the oscillating electric dipole, and applying Gauss' theorem of vector analysis, which is equation (A1.30) of Appendix A1.7, we find that at any fixed time

$$\int \nabla \cdot \mathbf{E}\, \mathrm{d}V = \int \mathbf{E} \cdot \mathrm{d}\mathbf{S} = 0. \tag{2.75}$$

To illustrate equation (2.75), consider a Gaussian surface in any of the examples in Figure 2.3 that does not enclose any part of the oscillating electric dipole. At a fixed instant of time, as many lines of **E** enter such a Gaussian surface as leave it. The values of **E** at different points on the Gaussian surface are at different distances from the oscillating electric dipole and have different retarded times at the oscillating electric dipole.

Well away from the oscillating electric dipole in Figure 2.2, that is in the radiation (wave) zone, the terms proportional to $1/r^3$ and $1/r^2$ are very much smaller than the term proportional to $1/r$, and to an excellent approximation equation (2.54) reduces to

$$\mathbf{E} = \mathbf{E}_{\text{rad}} = \frac{[\ddot{p}] \sin\theta}{4\pi\varepsilon_0 r c^2}\,\hat{\boldsymbol{\theta}}. \tag{2.76}$$

It is tempting to assume that equation (2.74) applies to $\mathbf{E}_{\text{rad}}$ in the radiation zone. However if we apply equation (2.67) to the radiation field only, we find that, since E_r and E_ϕ do not contribute to $\mathbf{E}_{\text{rad}}$ which only has a component in the $\hat{\boldsymbol{\theta}}$ direction, then

$$\nabla \cdot \mathbf{E}_{\text{rad}} = \frac{1}{r\sin\theta}\frac{\partial}{\partial\theta}(E_{\text{rad}} \sin\theta) = \frac{2[\ddot{p}]\cos\theta}{4\pi\varepsilon_0 r^2 c^2} \tag{2.77}$$

showing that the divergence of $\mathbf{E}_{\text{rad}}$ is finite. The reader can check back that the contribution to $\nabla \cdot \mathbf{E}$, given by equation (2.77), is cancelled in equation (2.74) by the third term inside the square brackets on the right hand side of equation (2.71), which arises from one of the contributions of the E_r term to $\nabla \cdot \mathbf{E}$, where E_r is given by equation (2.68). Notice E_r does not contribute to the radiation electric field. This result illustrates how, in the general case, Maxwell's equations are only valid when they are applied to the total electric and the total magnetic fields, and cannot always be applied to only the radiation fields even in the radiation zone. To illustrate this result consider one of the closed electric field lines, well away from the oscillating electric dipole in any one of the examples in Figure 2.3. According to equation (2.76), the radiation electric field has only a component in the direction of $\hat{\boldsymbol{\theta}}$. Hence if the radiation electric field given by equation (2.76) were the only contribution to the electric field, then the electric field lines would have to be circles of constant radii and, comparing equation (2.77) with equation (2.1) we see that there would have to be a volume charge distribution at field points in the radiation zone given by

$$\rho = \varepsilon_0 \nabla \cdot \mathbf{E}_{\text{rad}} = \frac{\varepsilon_0 \, 2\cos\theta \, [\ddot{p}]}{4\pi\varepsilon_0 r^2 c^2}. \tag{2.78}$$

There is no such charge distribution in empty space. It can be seen from Figure 2.3 that the change in the magnitude of E_θ with θ is not due to the termination of electric field lines on a charge distribution, but is due to the deviation of continuous electric field lines from circles due to the radial component E_r given by equation (2.68). It can be seen from the examples in Figure 2.3 that the electric field lines in the radiation zone are continuous so that $\nabla \cdot \mathbf{E}$, where $\mathbf{E}$ is the total electric field, is zero in this region. As r tends to infinity, the value of $\nabla \cdot \mathbf{E}_{\text{rad}}$, which according to equation (2.77) is proportional to $1/r^2$, tends to zero and, in the limit of the idealized case of a plane wave, the divergence of the radiation electric field is zero at field points in empty space.

2.6.3. *The equation* $\nabla \cdot \mathbf{B} = 0$

According to equation (2.46) the magnetic field $\mathbf{B}$, due to the oscillating electric dipole shown in Figure 2.2, has only a B_ϕ component which is independent of ϕ. Hence substituting in equation (A1.40) of Appendix A1.10 we find that

$$\nabla \cdot \mathbf{B} = \frac{1}{r^2}\frac{\partial}{\partial r}(r^2 B_r) + \frac{1}{r\sin\theta}\frac{\partial}{\partial\theta}(B_\theta \sin\theta) + \frac{1}{r\sin\theta}\frac{\partial B_\phi}{\partial\phi} = 0.$$

In cartesian coordinates we have

$$\frac{\partial B_x}{\partial x} + \frac{\partial B_y}{\partial y} + \frac{\partial B_z}{\partial z} = 0. \tag{2.79}$$

Equation (2.79) is a relation between the components B_x, B_y and B_z. According to equation (2.79), the divergence of the magnetic field $\mathbf{B}$ due to the oscillating electric dipole in Figure 2.2 is always zero. Integrating equation (2.79) at a fixed instant of time and applying Gauss' theorem of vector analysis, we have

$$\int \nabla \cdot \mathbf{B} \, \mathrm{d}V = \int \mathbf{B} \cdot \mathrm{d}\mathbf{S} = 0. \tag{2.80}$$

According to equation (2.80), as many lines of $\mathbf{B}$ should enter a closed surface as leave it. In the case of the oscillating electric dipole in Figure 2.2, the magnetic field lines are closed circles having constant values of r and θ, so that as many magnetic field lines enter any closed surface in Figure 2.2 as leave it.

2.6.4. *The equation* $\nabla \times \mathbf{E} = -\dot{\mathbf{B}}$ *at a field point in empty space*

According to equation (2.54), in the case of the oscillating electric dipole shown in Figure 2.2, E_ϕ is zero and E_r and E_θ are independent of ϕ. Hence the expression for $\nabla \times \mathbf{E}$, given by equation (A1.41) of Appendix A1.10, reduces to

$$\nabla \times \mathbf{E} = \left(\frac{1}{r}\frac{\partial}{\partial r}(rE_\theta) - \frac{1}{r}\frac{\partial E_r}{\partial \theta}\right)\hat{\boldsymbol{\phi}}. \tag{2.81}$$

Using equations (2.68) and (2.69) and remembering that, according to equation (2.40), $\partial[p]/\partial r = -[\dot{p}]/c$ etc., the reader can show that at any field point in empty space

$$\begin{aligned}\frac{1}{r}\frac{\partial}{\partial r}(rE_\theta) &= \frac{\sin\theta}{4\pi\varepsilon_0 r}\frac{\partial}{\partial r}\left[\frac{p}{r^2} + \frac{\dot{p}}{rc} + \frac{\ddot{p}}{c^2}\right]\\ &= -\frac{\sin\theta}{4\pi\varepsilon_0}\left[\frac{2p}{r^4} + \frac{2\dot{p}}{r^3c} + \frac{\ddot{p}}{r^2c^2} + \frac{\dddot{p}}{rc^3}\right]\end{aligned} \tag{2.82}$$

$$\begin{aligned}-\frac{1}{r}\frac{\partial E_r}{\partial\theta} &= \left(-\frac{1}{r}\right)\left(\frac{1}{4\pi\varepsilon_0}\right)\left[\frac{2p}{r^3} + \frac{2\dot{p}}{r^2c}\right]\frac{\partial}{\partial\theta}(\cos\theta)\\ &= \frac{\sin\theta}{4\pi\varepsilon_0}\left[\frac{2p}{r^4} + \frac{2\dot{p}}{r^3c}\right].\end{aligned} \tag{2.83}$$

Adding equations (2.82) and (2.83), we find that

$$\nabla \times \mathbf{E} = \left(-\frac{\sin\theta}{4\pi\varepsilon_0}\right)\left[\frac{\ddot{p}}{r^2c^2} + \frac{\dddot{p}}{rc^3}\right]\hat{\boldsymbol{\phi}}. \tag{2.84}$$

According to equation (2.46), for an oscillating electric dipole

$$\mathbf{B} = \frac{\sin\theta}{4\pi\varepsilon_0 c^2}\left[\frac{\dot{p}}{r^2} + \frac{\ddot{p}}{rc}\right]\hat{\boldsymbol{\phi}}.$$

Hence at any fixed field point

$$\frac{\partial\mathbf{B}}{\partial t} = \frac{\sin\theta}{4\pi\varepsilon_0}\left[\frac{\ddot{p}}{r^2c^2} + \frac{\dddot{p}}{rc^3}\right]\hat{\boldsymbol{\phi}}. \tag{2.85}$$

Comparing equations (2.84) and (2.85) we see that at any field point

$$\nabla \times \mathbf{E} = -\frac{\partial\mathbf{B}}{\partial t}. \tag{2.86}$$

Equation (2.86) is a relation we have derived which relates the spatial variations of **E** evaluated at a fixed time to the time variations of **B** evaluated at a fixed field point. For example, the x component of equation (2.86) is

$$\frac{\partial E_z}{\partial y} - \frac{\partial E_y}{\partial z} = -\frac{\partial B_x}{\partial t}$$

which is a relation between the components E_y, E_z and B_x.

Integrating equation (2.86) over a finite area at a fixed time t and applying Stokes' theorem of vector analysis, we have

$$\int \nabla \times \mathbf{E} \cdot d\mathbf{S} = \oint \mathbf{E} \cdot d\mathbf{l} = \int \left(-\frac{\partial \mathbf{B}}{\partial t} \right) \cdot d\mathbf{S}$$
$$= -\frac{\partial}{\partial t} \int \mathbf{B} \cdot d\mathbf{S} = -\frac{\partial \Phi}{\partial t} \tag{2.87}$$

where Φ is the magnetic flux through the area. Equation (2.87) is valid at any instant of time and the values of **E** and **B** are their values at that instant of time.

Consider one of the closed electric field lines, well away from the oscillating electric dipole in any one of the examples in Figure 2.3. It can be seen that $\oint \mathbf{E} \cdot d\mathbf{l}$ evaluated around the field line, which implies integrating at a fixed time, is finite, showing that $\nabla \times \mathbf{E}$ is finite. The magnetic field **B**, which is perpendicular to the paper in Figure 2.3, varies harmonically with time leading to finite values for $\dot{\mathbf{B}}$ and $\dot{\Phi}$. According to equation (2.87) $\oint \mathbf{E} \cdot d\mathbf{l}$ evaluated around a closed electric field line in any one of the examples in Figure 2.3 is equal to minus the rate of change of the magnetic flux $\Phi = \int \mathbf{B} \cdot d\mathbf{S}$ that goes through the closed electric field line. Different points on the closed electric field line are at different distances from the oscillating electric dipole in Figure 2.3 and correspond to different retarded times at the oscillating electric dipole. The line integral $\oint \mathbf{E} \cdot d\mathbf{l}$ is evaluated at a fixed time of observation.

Since equation (2.3) is a relation between the fields **E** and **B**, equation (2.3) can be used to determine **B** if **E** is known. Integrating equation (2.3) with respect to time we have

$$\mathbf{B} = -\int \nabla \times \mathbf{E}\, dt + \mathbf{B}_0(x, y, z) \tag{2.88}$$

where $\mathbf{B}_0(x, y, z)$ is a time independent, that is a magnetostatic contribution to the magnetic field, which is zero in the case of the oscillating electric dipole shown in Figure 2.2. To apply equation (2.88) to the oscillating electric dipole, take the curl of equation (2.54) to give equation (2.84). Then substitute in equation (2.88) and integrate with respect to time to obtain equation (2.46).

2.6.5. *The equation $\nabla \times \mathbf{B} = \dot{\mathbf{E}}/c^2$ at a field point in empty space*

According to equation (2.46), for the oscillating electric dipole in Figure 2.2

$$\mathbf{B} = \frac{\sin\theta}{4\pi\varepsilon_0 c^2} \left[\frac{\dot{p}}{r^2} + \frac{\ddot{p}}{rc} \right] \hat{\boldsymbol{\phi}}. \tag{2.89}$$

Since **B** has only the B_ϕ component, the expression for $\nabla \times \mathbf{B}$ given by equation (A1.41) of Appendix A1.10, reduces to

$$\nabla \times \mathbf{B} = \frac{1}{r \sin\theta} \frac{\partial}{\partial \theta} (B_\phi \sin\theta)\hat{\mathbf{r}} - \frac{1}{r}\frac{\partial}{\partial r}(rB_\phi)\hat{\boldsymbol{\theta}}. \tag{2.90}$$

It is left as an exercise for the reader to show that

$$\frac{1}{r \sin\theta}\frac{\partial}{\partial\theta}(B_\phi \sin\theta) = \frac{2\cos\theta}{4\pi\varepsilon_0 c^2}\left[\frac{\dot{p}}{r^3} + \frac{\ddot{p}}{r^2 c}\right] \tag{2.91}$$

$$-\frac{1}{r}\frac{\partial}{\partial r}(rB_\phi) = \frac{\sin\theta}{4\pi\varepsilon_0 c^2}\left[\frac{\dot{p}}{r^3} + \frac{\ddot{p}}{r^2 c} + \frac{\dddot{p}}{rc^2}\right]. \tag{2.92}$$

Substituting in equation (2.90) using equations (2.91) and (2.92) we have

$$\nabla \times \mathbf{B} = \left(\frac{2\cos\theta}{4\pi\varepsilon_0 c^2}\right)\left[\frac{\dot{p}}{r^3} + \frac{\ddot{p}}{r^2 c}\right]\hat{\mathbf{r}} + \left(\frac{\sin\theta}{4\pi\varepsilon_0 c^2}\right)\left[\frac{\dot{p}}{r^3} + \frac{\ddot{p}}{r^2 c} + \frac{\dddot{p}}{rc^2}\right]\hat{\boldsymbol{\theta}}. \tag{2.93}$$

Since according to equation (2.54)

$$\mathbf{E} = \left(\frac{2\cos\theta}{4\pi\varepsilon_0}\right)\left[\frac{p}{r^3} + \frac{\dot{p}}{r^2 c}\right]\hat{\mathbf{r}} + \left(\frac{\sin\theta}{4\pi\varepsilon_0}\right)\left[\frac{p}{r^3} + \frac{\dot{p}}{r^2 c} + \frac{\ddot{p}}{rc^2}\right]\hat{\boldsymbol{\theta}}$$

it follows that

$$\frac{1}{c^2}\frac{\partial \mathbf{E}}{\partial t} = \left(\frac{2\cos\theta}{4\pi\varepsilon_0 c^2}\right)\left[\frac{\dot{p}}{r^3} + \frac{\ddot{p}}{r^2 c}\right]\hat{\mathbf{r}} + \left(\frac{\sin\theta}{4\pi\varepsilon_0 c^2}\right)\left[\frac{\dot{p}}{r^3} + \frac{\ddot{p}}{r^2 c} + \frac{\dddot{p}}{rc^2}\right]\hat{\boldsymbol{\theta}}. \tag{2.94}$$

Comparing equation (2.93) and (2.94), we see that

$$\nabla \times \mathbf{B} = \frac{1}{c^2}\frac{\partial \mathbf{E}}{\partial t} = \mu_0\left(\varepsilon_0 \frac{\partial \mathbf{E}}{\partial t}\right). \tag{2.95}$$

Equation (2.95) is a relation we have derived between the electric and magnetic fields due to the oscillating electric dipole in Figure 2.2, which is valid at any field point in empty space. We shall go on in Section 4.8 of Chapter 4 to show that in the general case, when there is a current distribution at the field point, we must add the $\mu_0\mathbf{J}$ term to the right hand side of equation (2.95).

Integrating equation (2.95) over any surface in Figure 2.3 at a fixed time of observation t, and applying Stokes' theorem of vector analysis we obtain

$$\int \nabla \times \mathbf{B} \cdot d\mathbf{S} = \oint \mathbf{B} \cdot d\mathbf{l} = \frac{1}{c^2}\int \frac{\partial \mathbf{E}}{\partial t} \cdot d\mathbf{S} = \frac{1}{c^2}\frac{\partial}{\partial t}\int \mathbf{E} \cdot d\mathbf{S}$$
$$= \frac{1}{c^2}\left(\frac{\partial \Psi}{\partial t}\right) \tag{2.96}$$

where $\Psi = \int \mathbf{E} \cdot d\mathbf{S}$ is the electric flux through the surface. The magnetic field lines due to the oscillating electric dipole are closed circles having constant values of r and θ, so that $\oint \mathbf{B} \cdot d\mathbf{l}$ evaluated around one of these magnetic field lines is finite, showing that $\nabla \times \mathbf{B}$ is finite. Since the electric field $\mathbf{E}$ at all points on any surface bounded by the chosen magnetic field line varies harmonically with time, $\dot{\mathbf{E}}$ and hence $\dot{\Psi}$ are finite. According to equation (2.95) at any field point in empty space $\nabla \times \mathbf{B}$ is equal to $\dot{\mathbf{E}}/c^2$, and, according to equation (2.96), $\oint \mathbf{B} \cdot d\mathbf{l}$ evaluated around any closed loop is equal to $(1/c^2)$ times the rate of change of the electric flux Ψ through the loop. These are relations between the field vectors **E** and **B**.

At large distances from the oscillating electric dipole in Figure 2.2, the radiation fields predominate and at these large distances the other contributions to **E** and **B** are often neglected. According to equations (2.54) and (2.46), the radiation fields due to the oscillating dipole are given by

$$\mathbf{E}_{\text{rad}} = \frac{[\ddot{p}]\sin\theta}{4\pi\varepsilon_0 rc^2}\,\hat{\boldsymbol{\theta}}; \qquad \mathbf{B}_{\text{rad}} = \frac{[\ddot{p}]\sin\theta}{4\pi\varepsilon_0 rc^3}\,\hat{\boldsymbol{\phi}}.$$

It is straightforward for the reader to show, remembering that $\partial[\ddot{p}]/\partial r$ is equal to $-[\dddot{p}]/c$, that

$$\nabla \times \mathbf{B}_{\text{rad}} = \frac{2[\ddot{p}]\cos\theta}{4\pi\varepsilon_0 r^2 c^3}\,\hat{\mathbf{r}} + \frac{[\dddot{p}]\sin\theta}{4\pi\varepsilon_0 rc^4}\,\hat{\boldsymbol{\theta}} \tag{2.97}$$

$$\frac{1}{c^2}\frac{\partial \mathbf{E}_{\text{rad}}}{\partial t} = \frac{[\dddot{p}]\sin\theta}{4\pi\varepsilon_0 rc^4}\,\hat{\boldsymbol{\theta}} \tag{2.98}$$

Comparing equations (2.97) and (2.98), we see that $\nabla \times \mathbf{B}_{\text{rad}}$ is not equal to $\dot{\mathbf{E}}_{\text{rad}}/c^2$, showing that equation (2.95) is a relation between the total fields **B** and **E** due to the oscillating electric dipole and cannot be applied to the radiation fields on their own. In the limit when r tends to infinity the component of $\nabla \times \mathbf{B}_{\text{rad}}$ in the direction of $\hat{\mathbf{r}}$, which is proportional to $1/r^2$, becomes very much smaller than the other terms in equations (2.97) and (2.98) so that in the limit of a plane wave equation (2.95) can be applied to the radiation fields.

Integrating equation (2.95) with respect to time, we have

$$\mathbf{E} = c^2\int \nabla \times \mathbf{B}\,\mathrm{d}t + \mathbf{E}_0(x, y, z) \tag{2.99}$$

where $\mathbf{E}_0(x, y, z)$ is a time independent, that is an electrostatic contribution to the electric field. The $\mathbf{E}_0$ term in equation (2.99) is zero in the case of the oscillating electric dipole, shown in Figure 2.2, since there is no resultant total electric charge to give an electrostatic field. To determine **E** from **B** for the oscillating electric dipole, we start by taking the curl of the expression for **B**, given by equation (2.46), to give equation (2.93). Then substitute for $\nabla \times \mathbf{B}$ in equation (2.99) and integrate with respect to time to obtain equation (2.54). Using this method, there is no need, in the case of the oscillating electric dipole shown in Figure 2.2, to determine the scalar potential ϕ to determine **E**, since **B** can be determined from the vector potential **A** and **E** can then be determined using equation (2.99). If the oscillating electric dipole had a resultant electric charge we would have to include the $\mathbf{E}_0$ term in equation (2.99), which would arise from a $-\nabla\phi$ contribution to the total electric field **E**.

2.6.6. *Discussion of Maxwell's equations*

By deriving equations (2.74), (2.79), (2.86) and (2.95) from the expressions for the electric field **E** and the magnetic field **B** due to the oscillating electric dipole shown in Figure 2.2, both of which can be determine independently from the vector potential in the way described in Section 2.4, we have illustrated how, **at a field point in empty space**, Maxwell's equations are relations between the components E_x, E_y, E_z, B_x, B_y, and B_z of the field vectors **E** and **B**. We shall defer until Chapter 4, our discussion of how equations (2.74) and (2.95) must be extended when there is a charge and current distribution at the field point.

2.7. Derivation of the Jefimenko formulae for E and B from the retarded potentials

Consider again the charge and current distribution shown previously in Figure 2.1(a). Consider the field point P at position **r** having coordinates (x, y, z) and a source point at position $\mathbf{r}_s$ having coordinates (x_s, y_s, z_s). Let

$$\mathbf{R} = (\mathbf{r} - \mathbf{r}_s)$$

$$\hat{\mathbf{R}} = \mathbf{R}/R$$

$$R = \{(x - x_s)^2 + (y - y_s)^2 + (z - z_s)^2\}^{1/2}.$$

According to equation (2.17), the electric field at the field point P is given by

$$\mathbf{E} = -\nabla\phi - \dot{\mathbf{A}} \tag{2.17}$$

where

$$\nabla = \hat{\mathbf{i}}\frac{\partial}{\partial x} + \hat{\mathbf{j}}\frac{\partial}{\partial y} + \hat{\mathbf{k}}\frac{\partial}{\partial z}\,. \tag{2.100}$$

According to equation (2.29) the retarded scalar potential at the field point P is

$$\phi = \frac{1}{4\pi\varepsilon_0}\int\frac{[\rho]}{R}\,\mathrm{d}V_s$$

where $[\rho] = \rho(t - R/c)$ is the value of the charge density ρ at the retarded time $t^* = (t - R/c)$. Since differentiating partially with respect to x, y and z and integrating with respect to x_s, y_s and z_s are independent linear operations

$$\nabla\phi = \frac{1}{4\pi\varepsilon_0}\nabla\int\frac{[\rho]}{R}\,\mathrm{d}V_s = \frac{1}{4\pi\varepsilon_0}\int\nabla\left(\frac{[\rho]}{R}\right)\mathrm{d}V_s.$$

Using equation (A1.17) of Appendix A1.6, we have

$$\nabla\phi = \frac{1}{4\pi\varepsilon_0}\int [\rho]\nabla\left(\frac{1}{R}\right) dV_s + \frac{1}{4\pi\varepsilon_0}\int \frac{\nabla[\rho]}{R} dV_s. \tag{2.101}$$

According to equation (1.20) of Chapter 1

$$\nabla\left(\frac{1}{R}\right) = -\frac{\mathbf{R}}{R^3} = -\frac{\hat{\mathbf{R}}}{R^2}. \tag{2.102}$$

The x component of $\nabla[\rho]$ is

$$\frac{\partial\rho(t - r/c)}{\partial x} = \frac{\partial[\rho]}{\partial(t - R/c)}\frac{\partial(t - R/c)}{\partial x} = -\left[\frac{\partial\rho}{\partial t}\right]\frac{(x - x_s)}{Rc}.$$

Hence

$$\nabla[\rho] = -\left[\frac{\partial\rho}{\partial t}\right]\frac{\mathbf{R}}{Rc} = -\left[\frac{\partial\rho}{\partial t}\right]\frac{\hat{\mathbf{R}}}{c}. \tag{2.103}$$

Substituting from equations (2.102) and (2.103) into equation (2.101) we have

$$\nabla\phi = -\frac{1}{4\pi\varepsilon_0}\int\left(\frac{[\rho]\hat{\mathbf{R}}}{R^2} + \left[\frac{\partial\rho}{\partial t}\right]\frac{\hat{\mathbf{R}}}{Rc}\right) dV_s. \tag{2.104}$$

Differentiating equation (2.30) partially with respect to time, for fixed R we have

$$\frac{\partial\mathbf{A}}{\partial t} = \frac{1}{4\pi\varepsilon_0 c^2}\frac{\partial}{\partial t}\int\frac{[\mathbf{J}]}{R} dV_s = \frac{1}{4\pi\varepsilon_0 c^2}\int\left[\frac{\partial\mathbf{J}}{\partial t}\right]\frac{dV_s}{R}. \tag{2.105}$$

Substituting from equations (2.104) and (2.105) into equation (2.17) we finally obtain

$$\mathbf{E} = \frac{1}{4\pi\varepsilon_0}\int\left(\frac{[\rho]\hat{\mathbf{R}}}{R^2} + \left[\frac{\partial\rho}{\partial t}\right]\frac{\hat{\mathbf{R}}}{Rc} - \left[\frac{\partial\mathbf{J}}{\partial t}\right]\frac{1}{Rc^2}\right) dV_s. \tag{2.106}$$

This is the same as equation (1.136) of Chapter 1.

Using equations (2.18) and (2.30) we have

$$\mathbf{B} = \nabla\times\mathbf{A} = \frac{1}{4\pi\varepsilon_0 c^2}\nabla\times\int\frac{[\mathbf{J}]}{R} dV_s = \frac{1}{4\pi\varepsilon_0 c^2}\int\nabla\times\left(\frac{[\mathbf{J}]}{R}\right) dV_s.$$

Using equation (A1.23) of Appendix A1.6, we find that

$$\mathbf{B} = \frac{1}{4\pi\varepsilon_0 c^2}\int\nabla\left(\frac{1}{R}\right)\times[\mathbf{J}]\, dV_s + \frac{1}{4\pi\varepsilon_0 c^2}\int\frac{\nabla\times[\mathbf{J}]}{R} dV_s. \tag{2.107}$$

The x component of $\nabla\times[\mathbf{J}]$ is

$$(\nabla\times[\mathbf{J}])_x = \frac{\partial[J_z]}{\partial y} - \frac{\partial[J_y]}{\partial z}$$

where $[J_z]$ amd $[J_y]$ are function of $(t - R/c)$. Hence

$$(\nabla \times [\mathbf{J}])_x = \left[\frac{\partial J_z}{\partial t^*}\right]\left(\frac{\partial t^*}{\partial y}\right) - \left[\frac{\partial J_y}{\partial t^*}\right]\left(\frac{\partial t^*}{\partial z}\right)$$

$$= -\frac{(y - y_s)}{Rc}\left[\frac{\partial J_z}{\partial t}\right] + \frac{(z - z_s)}{Rc}\left[\frac{\partial J_y}{\partial t}\right]$$

$$= \left(\left[\frac{\partial \mathbf{J}}{\partial t}\right] \times \frac{\mathbf{R}}{Rc}\right)_x = \left(\left[\frac{\partial \mathbf{J}}{\partial t}\right] \times \frac{\hat{\mathbf{R}}}{c}\right)_x. \qquad (2.108)$$

with similar expressions for the other components of $(\nabla \times [\mathbf{J}])$. Using equations (2.102) and (2.108) in equation (2.107) we finally obtain

$$\mathbf{B} = \frac{1}{4\pi\varepsilon_0 c^2}\int\left(\left[\frac{[\mathbf{J}]}{R^2}\right] + \frac{1}{Rc}\left[\frac{\partial \mathbf{J}}{\partial t}\right]\right) \times \hat{\mathbf{R}}\, \mathrm{d}V_s. \qquad (2.109)$$

This is the same as equation (1.134) of Chapter 1.

For the example of the oscillating electric dipole shown in Figure 2.2, **[J]** is zero except at the dipole, where using equations (2.37) and (2.39) we have

$$[\mathbf{J}]\, \mathrm{d}V_s = [I]\, \mathrm{d}\mathbf{l} = [\dot{\mathbf{p}}]$$

$$\left[\frac{\partial \mathbf{J}}{\partial t}\right] \mathrm{d}V_s = [\dot{I}]\mathrm{d}\mathbf{l} = [\ddot{\mathbf{p}}].$$

Substituting in equation (2.109) we find that, for the oscillating electric dipole shown in Figure 2.2,

$$\mathbf{B} = \frac{1}{4\pi\varepsilon_0 c^2}\left(\frac{[\dot{\mathbf{p}}]}{R^2} + \frac{[\ddot{\mathbf{p}}]}{Rc}\right) \times \hat{\mathbf{R}}. \qquad (2.110)$$

This is in agreement with equation (2.46). Having derived equation (2.109), this is probably the quickest way of determining the expression for the magnetic field due to the oscillating electric dipole. Once we know **B**, we can determine **E** using the Maxwell equation (2.95) in the integral form given by equation (2.99).

2.8. Gauge transformations and the Coulomb gauge

So far in this chapter, we have only used the Lorentz gauge, in which the divergence of **A** satisfies the Lorentz condition

$$\nabla \cdot \mathbf{A} + \frac{1}{c^2}\frac{\partial \phi}{\partial t} = 0. \qquad (2.16)$$

Among the advantages of the Lorentz gauge are that its choice decouples the partial differential equations (2.14) and (2.15) for ϕ and **A** and leads to the retarded potentials, which are easy to interpret and apply. Furthermore, the equations for ϕ and **A** in the Lorentz gauge are Lorentz covariant, that is they have the same mathematical form in all intertial reference frames, when the coordinates and time are transformed using the Lorentz transformations.

In classical electromagnetism, it is the field vectors **E** and **B** that appear in the expression ($q\mathbf{E} + q\mathbf{u} \times \mathbf{B}$) for the Lorentz force acting on a test charge of magnitude q moving with velocity **u**. The potentials ϕ and **A** can be modified provided the values of **E** and **B** are unchanged. For example, consider the following transformation

$$\mathbf{A}' = \mathbf{A} + \nabla\psi \tag{2.111}$$

$$\phi' = \phi - \frac{\partial\psi}{\partial t} \tag{2.112}$$

where ψ is a scalar function of position and of time. Since according to equation (A1.26) of Appendix A1.6, $\nabla \times \nabla\psi$ is zero, using equations (2.17) and (2.18), we have

$$\nabla \times \mathbf{A}' = \nabla \times \mathbf{A} + \nabla \times (\nabla\psi) = \nabla \times \mathbf{A} = \mathbf{B} \tag{2.111}$$

$$-\nabla\phi' - \frac{\partial \mathbf{A}'}{\partial t} = -\nabla\phi + \nabla\left(\frac{\partial\psi}{\partial t}\right) - \frac{\partial \mathbf{A}}{\partial t} - \frac{\partial}{\partial t}(\nabla\upsilon)$$

$$= -\nabla\phi - \frac{\partial \mathbf{A}}{\partial t} = \mathbf{E}.$$

Hence the transformation given by equations (2.111) and (2.112) leaves the values of **E** and **B** unchanged. Such a transformation is called a **gauge transformation**. Gauge transformations can be used to give flexibility in the choice of $\nabla \cdot \mathbf{A}$. One popular choice is the Coulomb gauge, in which the divergence of the vector potential is put equal to zero.

Let the scalar potential and the vector potential in the Coulomb gauge be denoted by ϕ^* and $\mathbf{A}^*$ respectively. The discussion will again be confined to the case when $\varepsilon_r = 1$ and $\mu_r = 1$ everywhere. It will be assumed that the charge and current distributions are given as functions of position and time. Substituting for **E** in the Maxwell equation $\nabla \cdot \mathbf{E} = \rho/\varepsilon_0$, using the relation

$$\mathbf{E} = -\nabla\phi^* - \dot{\mathbf{A}}^*$$

and using the result that $\nabla \cdot \nabla\phi^* = \nabla^2\phi^*$, we have

$$\nabla \cdot \mathbf{E} = -\nabla \cdot (\nabla\phi^*) - \nabla \cdot \frac{\partial \mathbf{A}^*}{\partial t} = -\nabla^2\phi^* - \frac{\partial}{\partial t}(\nabla \cdot \mathbf{A}^*) = \frac{\rho}{\varepsilon_0}. \tag{2.113}$$

It is assumed in the Coulomb gauge that

$$\nabla \cdot \mathbf{A}^* = 0. \tag{2.114}$$

Substituting in equation (2.113), we find that in the Coulomb gauge

$$\nabla^2\phi^* = -\frac{\rho}{\varepsilon_0}. \tag{2.115}$$

Comparing equation (2.115) with equation (2.14) in the Lorentz gauge, we see that the $-\ddot{\phi}/c^2$ term is absent from equation (2.115). Equation (2.115) is the same as Poisson's equation of electrostatics, which is equation (1.29) of Chapter 1. By analogy with equation (1.26), we conclude that the solution of equation (2.115) that gives the value of ϕ^* at a field point at position $\mathbf{r}$ at the time of observation t, is

$$\phi^*(\mathbf{r}, t) = \frac{1}{4\pi\varepsilon_0}\int \frac{\rho(\mathbf{r}_s, t)}{|\mathbf{r} - \mathbf{r}_s|}\,\mathrm{d}V_s \tag{2.116}$$

where $\rho(\mathbf{r}_s, t)$ is the charge density at the source point at $\mathbf{r}_s$ at the time of observation t. Equation (2.116) suggests that the scalar potential ϕ^* in the Coulomb gauge is propagated from the varying charge distributions at an infinite speed. Consequently the potential ϕ^* in the Coulomb gauge is often called the instantaneous scalar potential.

Substituting $\mathbf{B} = \nabla \times \mathbf{A}^*$ into the Maxwell equation

$$\nabla \times \mathbf{B} = \mu_0(\mathbf{J} + \varepsilon_0\dot{\mathbf{E}})$$

where $\mathbf{J}$ is the electric current density due to the motion of free charges, and $\varepsilon_0\dot{\mathbf{E}}$ is the vacuum displacement current density, we have

$$\nabla \times (\nabla \times \mathbf{A}^*) = \mu_0(\mathbf{J} + \varepsilon_0\dot{\mathbf{E}}) = \mu_0\mathbf{C} \tag{2.117}$$

where

$$\mathbf{C} = \mathbf{J} + \varepsilon_0\dot{\mathbf{E}} \tag{2.118}$$

is what Maxwell called the "true current on which the electromagnetic phenomena depend". Using equation (A1.27) of Appendix A1.6 and putting $\nabla \cdot \mathbf{A}^*$ equal to zero we obtain

$$\nabla^2\mathbf{A}^* = -\mu_0\mathbf{C} = -\mu_0(\mathbf{J} + \varepsilon_0\dot{\mathbf{E}}). \tag{2.119}$$

Maxwell [5] wrote the solution of equation (2.119), which is similar to Poisson's equation, in the form

$$\mathbf{A}^*(\mathbf{r}, t) = \frac{\mu_0}{4\pi}\int \frac{\{\mathbf{J}(\mathbf{r}_s, t) + \varepsilon_0\dot{\mathbf{E}}(\mathbf{r}_s, t)\}}{|\mathbf{r} - \mathbf{r}_s|}\,\mathrm{d}V_s. \tag{2.120}$$

The integration in equation (2.120) is carried out over the whole of space at the time of observation t. To evaluate the integral, we must know the values of both $\mathbf{J}(\mathbf{r}_s, t)$ and $\varepsilon_0\dot{\mathbf{E}}(\mathbf{r}_s, t)$ at all points of space at the time of observation t.

Following Maxwell, equation (2.120) was interpreted in the nineteenth century by saying that both the conduction current density $\mathbf{J}$ and the displacement current density $\varepsilon_0\dot{\mathbf{E}}$ contributed to the magnetic field, and that the electromagnetic interaction was propagated at infinite speed from the source to the field point. The vector potential $\mathbf{A}^*$, given by equation (2.120), is sometimes called the instantaneous vector potential. In the nineteenth century, ideas based on equations (2.116) and (2.120) fitted in with the then prevailing ideas of instantaneous action at a distance and Newtonian mechanics. The position at the end of the nineteenth century is illustrated by the following quotation from Poincaré [6].

> In calculating **A** Maxwell takes into account the currents of conduction and those of displacement; and he supposes that the attraction takes place according to Newton's law i.e. instantaneously. But in calculating [the retarded potential] on the contrary we take account only of conduction currents and we suppose that the attraction is propagated with the velocity of light . . . . It is a matter of indifference whether we make this hypothesis [of a propagation in time] and consider only the induction due to conduction currents, or whether like Maxwell, we retain the old law of [instantaneous] induction and consider both conduction and the displacement currents.

This quotation illustrates how the use of different gauges in the nineteenth century lead to very different interpretations of the field equations of classical electromagnetism.

Equation (2.120) is really a little bit of an illusion. In order to evaluate the integral in equation (2.120) to determined $\mathbf{A}^*$, we need to know both $\mathbf{J}$ and $\varepsilon_0\dot{\mathbf{E}}$ at all points of space at the time of observation t, when the vector potential $\mathbf{A}^*$ is determined at the field point. In cases of practical importance, the vacuum displacement current term $\varepsilon_0\dot{\mathbf{E}}$ is not given but has to be determined from the given charge and current distributions. In order to determine $\mathbf{E}$ using the equation

$$\mathbf{E} = -\nabla\phi^* - \dot{\mathbf{A}}^* \tag{2.121}$$

we need to know both the value of ϕ^*, obtained by solving equation (2.116), and the unknown vector potential $\mathbf{A}^*$. However, to determine $\mathbf{A}^*$ using equation (2.120) we would have to know $\dot{\mathbf{E}}$, but to determine $\mathbf{E}$ to determine $\dot{\mathbf{E}}$ we would have to know the value of $\mathbf{A}^*$ to use in equation (2.121), but $\mathbf{A}^*$ is what we are trying to determine using equation (2.120). This circular argument shows clearly that we cannot use equation (2.120) to determine $\mathbf{A}^*$ from the charge and current distributions. Equation (2.120) can only be used if we have already solved the problem to determine $\mathbf{E}$ and hence $\varepsilon_0\dot{\mathbf{E}}$ from the charge and current distributions. When the Coulomb gauge is used nowadays, the vacuum displacement current term is eliminated from equation (2.120) as follows. Differentiating equation (2.121) partially with respect to time, we have

$$\dot{\mathbf{E}} = \frac{\partial}{\partial t}\left(-\nabla\phi^* - \frac{\partial \mathbf{A}^*}{\partial t}\right) = -\nabla\left(\frac{\partial\phi^*}{\partial t}\right) - \frac{\partial^2\mathbf{A}^*}{\partial t^2}\,.$$

Eliminating $\varepsilon_0\dot{\mathbf{E}}$ from equation (2.119), we obtain

$$\nabla^2\mathbf{A}^* - \frac{1}{c^2}\frac{\partial^2\mathbf{A}^*}{\partial t^2} = -\mu_0\mathbf{J} + \mu_0\varepsilon_0\nabla\left(\frac{\partial\phi^*}{\partial t}\right)\,. \tag{2.122}$$

Equation (2.122) can be solved after we have determined ϕ^* using equation

(2.116). Notice the presence of the $-\ddot{\mathbf{A}}^*/c^2$ term in equation (2.122). This shows that the vector potential $\mathbf{A}^*$ in the Coulomb gauge depends on the conditions at the source point at the appropriate retarded time $(t - |\mathbf{r} - \mathbf{r}_s|/c)$. A reader interested in a discussion of the solution of equation (2.122) is referred to Jackson [7]. A full analysis of the general case was given by Brill and Goodman [8], who showed that, when the contribution of the $-\nabla\phi^*$ term to the total electric field $\mathbf{E}$ is combined with the contribution of the $-\dot{\mathbf{A}}^*$ term, where ϕ^* and $\mathbf{A}^*$ are the solutions of equations (2.115) and (2.122) respectively, the resultant electric field $\mathbf{E}$ depends on the values of the charge and current densities at the appropriate retarded times and the value of $\mathbf{E}$ is exactly the same as the value of $\mathbf{E}$ determined using the retarded potentials in the Lorentz gauge. The effect of a change of gauge is to change the magnitudes of the contributions of the $-\nabla\phi$ and $-\dot{\mathbf{A}}$ terms in equation (2.121), but to leave the sum of their contributions to the electric field $\mathbf{E}$ unchanged. Generally, the retarded potentials in the Lorentz gauge, are simpler to use in classical electromagnetism than the equations in the Coulomb gauge, but the Coulomb gauge has some advantages in the quantum theory of radiation. Reference: Heitler [9].

Instead of using the potentials ϕ and $\mathbf{A}$ we can introduce alternative mathematical functions, such as the Hertz vectors. References: Stratton [10] and Heading [11]. If we try to interpret what may happen in the space between the source and field point using different gauges for ϕ and $\mathbf{A}$, or using different mathematical functions such as the Hertz vectors, we can end up with apparently very different models of how the various contributions to the resultant values of $\mathbf{E}$ and $\mathbf{B}$ arise. The important thing to realize is that, in the context of classical electromagnetism all the methods give the same values for the total electric field $\mathbf{E}$ and the total magnetic field $\mathbf{B}$ and hence for the observable force on a moving test charge. The attitude we have tried to cultivate is that, in the context of classical electromagnetism, there is no need in any of the methods to say anything about what may or may not happen in the empty space between the charge and current distributions and the field point. The choice of which method to use to solve a particular problem is a matter of mathematical convenience.

References

1. Panofsky, W. K. H. and Phillips, M., *Classical Electricity and Magnetism*, 2nd Edition (Addison-Wesley, Reading, Mass., 1962), p. 242.
2. Ferraro, V. C. A., *Electromagnetic Theory* (The Athlone Press, London, 1956), p. 524.
3. Hauser, W., *Introduction to the Principles of Electromagnetism* (Addison-Wesley, Reading, Mass., 1971), p. 524.
4. Lorrain, P. and Corson, D., *Electromagnetic Fields and Waves*, 2nd Edition (W. H. Freeman, San Francisco, 1970), p. 461.
5. Maxwell, J. C., *A Treatise on Electricity and Magnetism*, Vol. 2 (Dover, New York, 1954), p. 255.

6. O'Rahilly, A., *Electromagnetics* (Longmans Green, London, 1938), p. 176.
7. Jackson, J. D., *Classical Electrodynamics*, 2nd Edition (John Wiley, New York, 1975), p. 221.
8. Brill, O. W. and Goodman, B., *Amer. Journ. Phys.* Vol 35, p. 832 (1967).
9. Heitler, W., *Quantum Theory of Radiation*, 3rd Edition (Oxford University Press, London, 1954).
10. Stratton, J. A., *Electromagnetic Theory* (McGraw-Hill, New York, 1941), p. 28.
11. Heading, J., *Electromagnetic Theory and Special Relativity* (University Tutorial Press, London, 1964), p. 103.

CHAPTER 3

The electric and magnetic fields due to an accelerating classical point charge

3.1. The Liénard-Wiechert potentials

The retarded potentials will now be used to derive the potentials ϕ and $\mathbf{A}$ due to an accelerating classical point charge of magnitude q. We shall again assume that the classical point charge is a continuous charge of finite, but exceedingly small dimensions. Due to the finite dimensions of the classical point charge, when we apply the retarded potentials we must allow for the motion of its finite charge distribution, while the information collecting sphere is crossing the classical point charge. To simplify the discussion, we shall assume initially that the accelerating classical point charge is moving directly towards the field point P in Figure 3.1 with a velocity $[\mathbf{u}]$ at its retarded position. The information collecting sphere, that reaches the field point P in

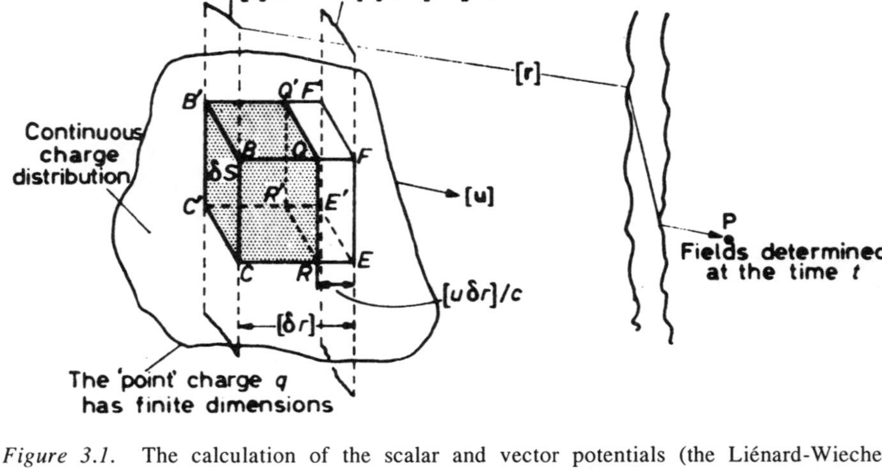

Figure 3.1. The calculation of the scalar and vector potentials (the Liénard-Wiechert potentials) at the field point P due to an accelerating classical point charge q moving with non-uniform velocity. In Figure 3.1, $[\mathbf{u}]$ is parallel to $[\mathbf{r}]$. It is assumed that the classical point charge q has finite dimensions and is made up from a continuous charge distribution.

Figure 3.1 at the time of observation t, is at a distance $[r]$ from the field point at the retarded time $t^* = (t - r/c)$. Consider the element of area δS of the information collecting sphere, that is crossing the continuous charge distribution shown in Figure 3.1. Let the surface area δS be at the position $CBB'C'$ at the time $(t - r/c)$ as shown in Figure 3.1. Let δS be at the position $EFF'E'$ at a time $\delta t = \delta r/c$ later, where δr is equal to the distance from C to E in Figure 3.1. The volume element δV_s swept out by the area δS in the time interval δt is given by

$$\delta V_s = (\delta S)(\delta r).$$

The information collecting sphere will record a charge density $[\rho]$ in this volume element δV_s where $[\rho]$ is the charge per unit volume measured at the fixed time $(t - r/c)$. Since the charge distribution is moving directly towards the field point P in Figure 3.1, the amount of charge actually passed by the element of area δS of the information collecting sphere in a time δt is less than $[\rho]\ \delta V_s$. In the time the information collecting sphere takes to move from $CBB'C'$ to $EFF'E'$. in Figure 3.1, charge on the surface $RQQ'R'$ moves to $EFF'E'$. Hence the total quantity of charge, denoted δq, actually passed by the element of area δS of the information collecting sphere inside δV_s is equal to the charge, that at the time $(t - r/c)$ was inside the volume between $CBB'C'$ and $RQQ'R'$ in Figure 3.1. Since the distance from R to E is equal to $[u]\delta t = [u]\delta r/c$, the distance from C to R is equal to $\delta r - [u]\delta r/c = \delta r[1 - u/c]$. Hence the volume between $CBB'C'$ and $RQQ'R'$ is equal to $(\delta S)(\delta r)[1 - u/c] = \delta V_s[1 - u/c]$. Since the charge density at the time $(t - r/c)$ is $[\rho]$, the total quantity of charge δq passed by the area δS of the information collecting sphere inside δV_s is

$$\delta q = [\rho]\left[1 - \frac{u}{c}\right]\delta V_s. \tag{3.1}$$

In general, the direction of the velocity $[\mathbf{u}]$ of the classical point charge at its retarded position is not directly towards the field point P in Figure 3.1. In the general case, $[u]$ must be replaced in equation (3.1) by the component of $[\mathbf{u}]$ in the direction of the field point, which can be expressed in the form $[\mathbf{u}\cdot\mathbf{r}/r]$, where $\mathbf{r}$ is a vector from the position of the volume element δV_s to the field point. Hence in the general case, equation (3.1) becomes

$$\delta q = [\rho]\delta V_s\left[1 - \frac{\mathbf{u}\cdot\mathbf{r}}{rc}\right].$$

Rearranging, we have

$$[\rho]\delta V_s = \frac{\delta q}{[1 - \mathbf{u}\cdot\mathbf{r}/rc]}. \tag{3.2}$$

Substituting in equation (2.29), we obtain

$$\phi = \frac{1}{4\pi\varepsilon_0}\int\frac{[\rho]}{r}\,\mathrm{d}V_s = \frac{1}{4\pi\varepsilon_0}\int\frac{\mathrm{d}q}{[r - \mathbf{r}\cdot\mathbf{u}/c]}. \tag{3.3}$$

It will now be assumed that the dimensions of the charge distribution are exceedingly small, so that it corresponds to our model of a classical point charge. The variation of r in equation (3.3) over the dimensions of the classical point charge can then be neglected, so that equation (3.3) can be rewritten in the form

$$\phi = \frac{1}{4\pi\varepsilon_0[r - \mathbf{r}\cdot\mathbf{u}/c]}\int \mathrm{d}q$$

where [**r**] is a vector from the retarded position of the classical point charge to the field point. The integral $\int \mathrm{d}q$ is the total quantity of charge passed by the information collecting sphere, which is equal to the total charge q of the classical point charge. Hence

$$\phi = \frac{q}{4\pi\varepsilon_0[r - \mathbf{u}\cdot\mathbf{r}/c]} = \frac{q}{4\pi\varepsilon_0 s} \tag{3.4}$$

where

$$s = \left[r - \frac{\mathbf{u}\cdot\mathbf{r}}{c}\right]. \tag{3.5}$$

For the continuous charge distribution shown in Figure 3.1, the current density [**J**] at the retarded time $(t - r/c)$ is equal to [ρ**u**]. Using equation (3.2), we have

$$[\mathbf{J}]\delta V_s = [\rho\mathbf{u}]\delta V_s = \left[\frac{\mathbf{u}\delta q}{1 - \mathbf{u}\cdot\mathbf{r}/rc}\right].$$

Substituting in equation (2.30) and proceeding as for the determination of the scalar potential ϕ we find that the vector potential **A** due to the accelerating classical point charge is

$$\mathbf{A} = \frac{\mu_0}{4\pi}\left[\frac{q\mathbf{u}}{r - \mathbf{u}\cdot\mathbf{r}/c}\right] = \frac{\mu_0 q[\mathbf{u}]}{4\pi s}. \tag{3.6}$$

Equations (3.4) and (3.6) are known as the **Liénard-Wiechert potentials.** They are valid in the Lorentz gauge. All the quantities inside the square brackets are measured at the retarded position of the charge at the appropriate retarded time. Alternative mathematical derivations of equations (3.4) and (3.6) are given by Jackson [1] and Hauser [2].

3.2. The formulae for the electric and magnetic fields due to an accelerating classical point charge

The expressions for the electric and magnetic fields due to an accelerating classical point charge will now be derived from the Liénard-Wiechert potentials. Consider the accelerating classical point charge shown in Figure 3.2. The expressions for the fields **E** and **B** will be determined at the field point P at

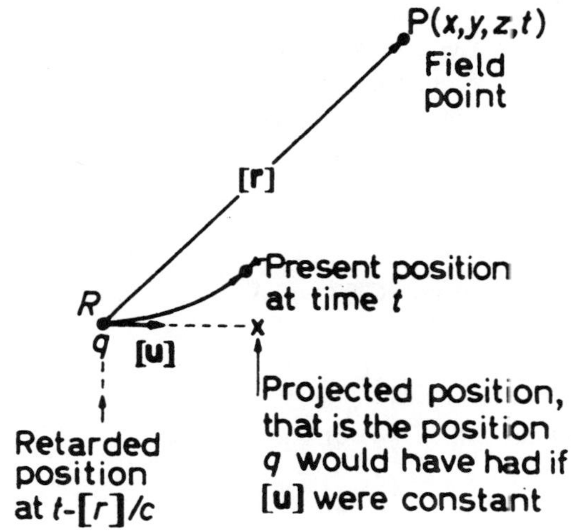

Figure 3.2. The retarded, present and projected positions of an accelerating classical point charge.

the position (x, y, z) a the time of observation t. The accelerating charge q is at the retarded position R at (x^*, y^*, z^*) in Figure 3.2 at the retarded time $t^* = (t - [r]/c)$. As in Section 3.1 we are again assuming that $[\mathbf{r}]$ is a vector from the retarded position R of the charge to the field point P in Figure 3.2. We shall assume that the velocity $[\mathbf{u}]$ and the acceleration $[\mathbf{a}]$ of the charge at its retarded position R at the retarded time t^* are known. The Liénard-Wiechert potentials are given by equations (3.4) and (3.6) and are

$$\phi = \frac{q}{4\pi\varepsilon_0 s}; \qquad \mathbf{A} = \frac{\mu_0 q[\mathbf{u}]}{4\pi s} = \frac{q[\mathbf{u}]}{4\pi\varepsilon_0 c^2 s}$$

where according to equation (3.5)

$$s = \left[r - \frac{\mathbf{r}\cdot\mathbf{u}}{c}\right]. \tag{3.5}$$

The expressions for $\mathbf{E}$ and $\mathbf{B}$ can be determined using the relations

$$\mathbf{E} = -\nabla\phi - \frac{\partial \mathbf{A}}{\partial t} \tag{3.7}$$

$$\mathbf{B} = \nabla \times \mathbf{A} \tag{3.8}$$

where in cartesian coordinates

$$\nabla = \hat{\mathbf{i}}\frac{\partial}{\partial x} + \hat{\mathbf{j}}\frac{\partial}{\partial y} + \hat{\mathbf{k}}\frac{\partial}{\partial z} \tag{3.9}$$

and where $\hat{\mathbf{i}}$, $\hat{\mathbf{j}}$ and $\hat{\mathbf{k}}$ are unit vectors in the directions of increasing x, y and z respectively. The partial differential coefficients in equations (3.7) and (3.8) are with respect to changes in the coordinates (x, y, z) of the field point P at the fixed time of observation t. However, the velocity $[\mathbf{u}]$ in the Liénard-Wiechert potentials is given in terms of the coordinates (x^*, y^*, z^*) of the retarded position of the charge at the retarded time $t^* = (t - [r]/c)$. This makes the calculation of $\mathbf{E}$ and $\mathbf{B}$ in terms of $[\mathbf{r}]$, $[\mathbf{u}]$ and $[\mathbf{a}]$ both long and tedious. The full calculation is given in Appendix C, where it is shown that the electric field $\mathbf{E}$ can be expressed in the form:

$$\mathbf{E} = \mathbf{E}_V + \mathbf{E}_A \tag{3.10}$$

where according to equations (C.33) of Appendix C

$$\mathbf{E}_V = \frac{q}{4\pi\varepsilon_0 s^3}\left[\mathbf{r} - \frac{r\mathbf{u}}{c}\right]\left[1 - \frac{u^2}{c^2}\right] \tag{3.11}$$

and according to equation (C.34) of Appendix C

$$\mathbf{E}_A = \frac{q}{4\pi\varepsilon_0 c^2 s^3}[\mathbf{r}] \times \left\{\left[\mathbf{r} - \frac{r\mathbf{u}}{c}\right] \times [\mathbf{a}]\right\} \tag{3.12}$$

where according to equation (3.5)

$$s = \left[r - \frac{\mathbf{r}\cdot\mathbf{u}}{c}\right]. \tag{3.5}$$

The magnetic field $\mathbf{B}$ can be expressed in the form:

$$\mathbf{B} = \mathbf{B}_V + \mathbf{B}_A \tag{3.13}$$

where according to equations (C.41), (C.42) and (C.43)

$$\mathbf{B}_V = \frac{q[\mathbf{u}] \times [\mathbf{r}]}{4\pi\varepsilon_0 c^2 s^3}\left[1 - \frac{u^2}{c^2}\right] \tag{3.14}$$

$$\mathbf{B}_A = \frac{q}{4\pi\varepsilon_0 c^3 s^3}[\mathbf{r}] \times \left[-\mathbf{u}\left(\frac{\mathbf{r}\cdot\mathbf{a}}{c}\right) - s\mathbf{a}\right] \tag{3.15}$$

$$= \frac{q}{4\pi\varepsilon_0 c^3 s^3}\left[\frac{\mathbf{r}}{r}\right] \times \left\{[\mathbf{r}] \times \left(\left[\mathbf{r} - \frac{r\mathbf{u}}{c}\right] \times [\mathbf{a}]\right)\right\}. \tag{3.16}$$

The quantities q, $[\mathbf{u}]$ and $[\mathbf{a}]$ are the values of the charge, velocity and acceleration of the classical point charge at its retarded position R in Figure 3.2, $[\mathbf{r}]$ is a vector from the retarded position R of the charge to the field point P, and s is given by equation (3.5).

It can be seen from equations (3.11), (3.12), (3.14) and (3.16) that

$$\mathbf{B} = \frac{[\mathbf{r}] \times \mathbf{E}}{[rc]}. \tag{3.17}$$

According to equation (3.17), the resultant magnetic field $\mathbf{B}$ is perpendicular to both the resultant electric field $\mathbf{E}$ and the vector $[\mathbf{r}]$ from the retarded position R of the charge to the field point P in Figure 3.2.

The Liénard-Wiechert potentials and equations (3.10) and (3.13), which give the electric and magnetic fields due to an accelerating classical point charge, are the summit of our development of classical electromagnetism at the atomistic level. These results will be used in later chapters to interpret macroscopic classical electromagnetism in terms of the electric and magnetic fields due to moving and accelerating classical point charges.

3.3. The electric and magnetic fields due to a classical point charge moving with uniform velocity

Before going on to discuss the general case of an accelerating classical point charge, we shall consider first the special case of a classical point charge of magnitude q, that is moving with uniform velocity $\mathbf{u}$ as shown in Figure 3.3. We shall also assume that the charge q has always been moving with the same uniform velocity $\mathbf{u}$ throughout its past history. Since the acceleration $[\mathbf{a}]$ of the charge at its retarded position R in Figure 3.3 in zero, the expression for the total electric field $\mathbf{E}$, given by equation (3.10), reduces to

$$\mathbf{E} = \mathbf{E}_V = \frac{q}{4\pi\varepsilon_0 s^3}\left[\mathbf{r} - \frac{r\mathbf{u}}{c}\right][1 - \beta^2] \tag{3.18}$$

where $\beta = u/c$. In the time interval $[r]/c$ it takes the information collecting sphere to go from R, the retarded position of the charge in Figure 3.3, to the field point P, the charge q moves a distance $[\mathbf{u}][r/c] = [\mathbf{u}r/c]$ at uniform velocity $\mathbf{u}$ to reach O, the position of the charge q at the time of observation t, when

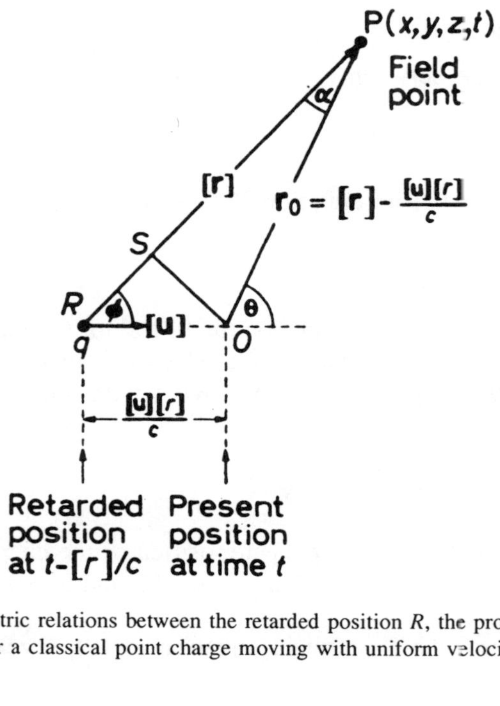

Figure 3.3. Geometric relations between the retarded position R, the projected position O and the field point P for a classical point charge moving with uniform velocity.

the fields are determined at the field point P. The position of the charge q at the time of observation t will be called the **present position** of the charge. Let $\mathbf{r}_0$ be a vector from O, the present position of the charge, to the field point P in Figure 3.3. In the triangle RPO in Figure 3.3, the lengths of the sides RP, OP and RO are $[\mathbf{r}]$, $\mathbf{r}_0$ and $[\mathbf{u}r/c]$ respectively. It follows from the law of vector addition that

$$\mathbf{r}_0 = [\mathbf{r}] - \left[\frac{\mathbf{u}r}{c}\right] = \left[\mathbf{r} - \frac{r\mathbf{u}}{c}\right]. \tag{3.19}$$

Draw a perpendicular OS to the line RP in Figure 3.3. Let ϕ be the angle between the lines RP and RO in Figure 3.3. The distance from R to S in Figure 3.3. is equal to $[ur/c]\cos\phi$, which can be written in the form $[\mathbf{u}\cdot\mathbf{r}/c]$. Hence the distance from S to P in Figure 3.3 is given by

$$[r] - \left[\frac{\mathbf{u}\cdot\mathbf{r}}{c}\right] = \left[r - \frac{\mathbf{u}\cdot\mathbf{r}}{c}\right] = s. \tag{3.20}$$

where s is defined by equation (3.5). In the triangle OSP in Figure 3.3, since s is equal to the distance from S to P and α is the angle between the lines RP and OP, we have

$$s = r_0 \cos\alpha = r_0(1 - \sin^2\alpha)^{1/2}. \tag{3.21}$$

Applying the standard trigonmetrical result that $A/\sin\ a = B/\sin b$ to the triangle ROP in Figure 3.3, we have

$$\frac{r}{\sin(\pi - \theta)} = \frac{[\beta]\mathrm{r}}{\sin\alpha} \tag{3.22}$$

where θ is the angle between $\mathbf{r}_0$ and $\mathbf{u}$, as shown in Figure 3.3. Equation (3.22) gives

$$\sin\alpha = [\beta]\sin\theta. \tag{3.23}$$

Substituting for $\sin\alpha$ from equation (3.23) into equation (3.21), we find that

$$s = r_0(1 - \beta^2\sin^2\theta)^{1/2}. \tag{3.24}$$

Substituting for $[\mathbf{r} - r\mathbf{u}/c]$ using equation (3.19) and for s using equation (3.24) into equation (3.18), we find that the total electric field $\mathbf{E}$, due to the charge q in Figure 3.3, that has always been moving with uniform velocity $\mathbf{u}$, is given by

$$\mathbf{E} = \frac{q\mathbf{r}_0(1 - \beta^2)}{4\pi\varepsilon_0 r_0^3(1 - \beta^2\sin^2\theta)^{3/2}} \tag{3.25}$$

where $\mathbf{r}_0$ is a vector from O, the present position of the charge q at the time of observation t, to the field point P in Figure 3.3 and θ is the angle between $\mathbf{u}$ and $\mathbf{r}_0$.

When $[\mathbf{a}]$ is zero, the $\mathbf{B}_A$ term in equation (3.13) is zero and equation

(3.13) for the total magnetic field due to a classical point charge moving with uniform velocity $\mathbf{u}$ reduces to

$$\mathbf{B} = \mathbf{B}_V = \frac{q[\mathbf{u}] \times [\mathbf{r}](1 - \beta^2)}{4\pi\varepsilon_0 c^2 s^3} . \tag{3.26}$$

Since $[\mathbf{u}] \times [\mathbf{u}]$ is zero the $[\mathbf{u} \times \mathbf{r}]$ term can be rewritten in form

$$[\mathbf{u}] \times [\mathbf{r}] = [\mathbf{u}] \times \left[\mathbf{r} - \frac{r\mathbf{u}}{c}\right] = \mathbf{u} \times \mathbf{r}_0. \tag{3.27}$$

where $\mathbf{r}_0$ is given by equation (3.19). Substituting in equation (3.26) and using equation (3.24) to substitute for s, we finally obtain

$$\mathbf{B} = \frac{q\mathbf{u} \times \mathbf{r}_0(1 - \beta^2)}{4\pi\varepsilon_0 c^2 r_0^3(1 - \beta^2 \sin^2 \theta)^{3/2}} \tag{3.28}$$

where $\mathbf{u}$ is the uniform velocity of the classical point charge. Notice that in the special case of a classical point charge moving with uniform velocity $\mathbf{u}$

$$\mathbf{B} = \frac{\mathbf{u} \times \mathbf{E}}{c^2} . \tag{3.29}$$

When $\mathbf{u}$, the uniform velocity of the classical point charge in Figure 3.3, is zero, that is when $\beta = 0$, equation (3.25) reduces to

$$E = \frac{q}{4\pi\varepsilon_0 r_0^2} \tag{3.30}$$

in agreement with Coulomb's law. When $\mathbf{u} = 0$, the electric field $\mathbf{E}$ is the same in all directions, as illustrated in Figure 3.4(a). The number of lines of $\mathbf{E}$ is limited in both Figures 3.4(a) and 3.4(b), such that the number of lines of $\mathbf{E}$ per square metre perpendicular to $\mathbf{E}$ is equal to (or proportional to) the magnitude of $\mathbf{E}$. This gives a visual picture of both the strength and direction of $\mathbf{E}$. The lines of $\mathbf{E}$ are closest together where the magnitude of the electric field is greatest.

According to equation (3.25), provided the velocity $\mathbf{u}$ of a positive charge has been constant in the past, the electric field lines diverge radially from the present position of the charge, that is from the position of the positive charge at the time of observation, when $\mathbf{E}$ is determined at the field point. This is illustrated in Figure 3.4(b). According to equation (3.25), the magnitude of the electric field $\mathbf{E}$ is still proportional to $1/r_0^2$, but unlike the electrostatic case given by equation (3.30) the magnitude of the electric field $\mathbf{E}$ is not the same in all directions when the charge is moving with uniform velocity, though $\mathbf{E}$ is still symmetric about $\theta = \pi/2$, where θ is the angle between $\mathbf{u}$ and $\mathbf{r}_0$. When $\theta = 0$ or $\theta = \pi$, equation (3.25) reduces to

$$E = \frac{q(1 - \beta^2)}{4\pi\varepsilon_0 r_0^2} . \tag{3.31}$$

Thus the electric field $\mathbf{E}$ is reduced in the direction of $\mathbf{u}$, the direction of motion

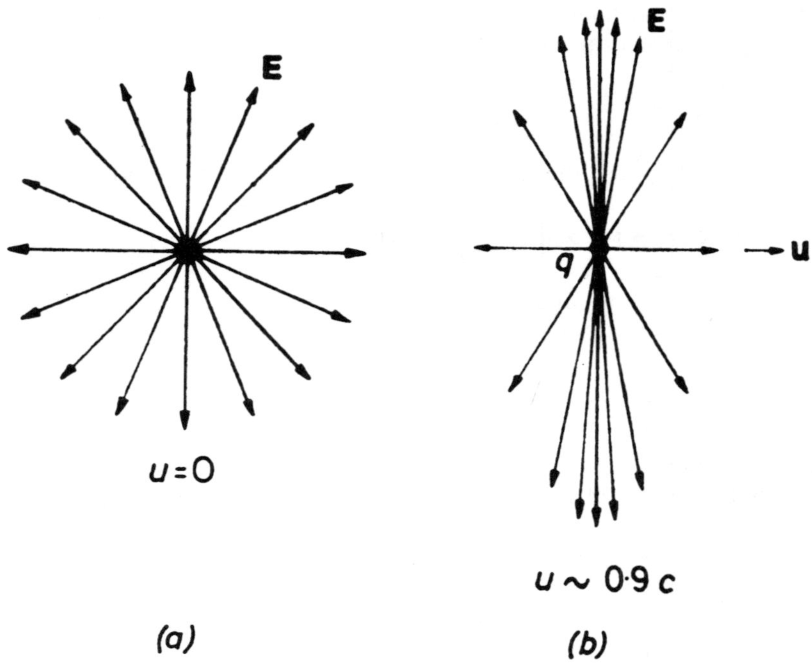

Figure 3.4. (a) The electric field of a stationary positive charge is spherically symmetric. (b) If the charge is moving with uniform velocity, the electric field diverges radially from the present position of the charge. The electric field is increased in the direction perpendicular to **u**, but decreased in the directions parallel to and antiparallel to **u**.

of the charge, and in the direction opposite to **u** to $(1 - \beta^2)$ times the electrostatic value given by equation (3.30). When $\theta = \pi/2$, equation (3.25) reduces to

$$E = \frac{q}{4\pi\varepsilon_0 r_0^2(1 - \beta^2)^{1/2}} \,. \tag{3.32}$$

According to equation (3.32), the electric field **E** is increased in the direction perpendicular to **u** to $1/(1 - \beta^2)^{1/2}$ times the electrostatic value given by equation (3.30). It will be confirmed in Section 4.2 of Chapter 4, that the total flux of the electric field **E** from a classical point charge of magnitude q that is moving with uniform velocity **u**, is always equal to q/ε_0, so that the total number of electric field lines is the same as in the electrostatic case.

As a typical example, the electric field due to a positive point charge, moving with uniform velocity $u = 0.9c$, is sketched in Figure 3.4(b). The diagram illustrates how, though the total number of lines of **E** is the same as in the electrostatic case, the lines of **E** are bunched towards the direction perpendicular to **u**, the uniform velocity of the charge. For $\beta = 0.9$, the electric field for $\theta = 0$ is 0.19 times the electrostatic value and for $\theta = \pi/2$ it is 2.3 times the electrostatic value. For $\beta = 0.99$, the corresponding ratios are 0.02 and 7.1 respectively. In the extreme relativistic case, when u tends to c, nearly all the lines of **E** are almost perpendicular to **u**.

According to equation (3.29), for a classical point charge q moving with uniform velocity **u**, the magnetic field lines are perpendicular to both **u** and the lines of **E**, which diverge radially from the present position of the charge. Thus the lines of **B** are closed circles in the plane perpendicular to the direction of motion of the charge. These circles are concentric with the direction of **u**. The lines of **B** in two planes are sketched in Figure 3.5. The magnitude of **B** decreases as $1/r_0^2$. For a given value of r_0, the magnitude of **B** is the same for $\theta = \alpha$ and $\theta = (\pi - \alpha)$ and, according to equation (3.29), the sense of rotation of the lines of **B** around the direction of **u** is the same in both cases. For a positive charge, the direction of **B** is the direction a right-handed corkscrew would have to be rotated, if the corkscrew is to advance in the direction of **u**, the velocity of the charge.

When $u \ll c$, $\beta \ll 1$, equations (3.25) and (3.28) reduce to

$$\mathbf{E} = \frac{q\mathbf{r}_0}{4\pi\varepsilon_0 r_0^3} \tag{3.33}$$

$$\mathbf{B} = \frac{q\mathbf{u} \times \mathbf{r}_0}{4\pi\varepsilon_0 c^2 r_0^3} \, . \tag{3.34}$$

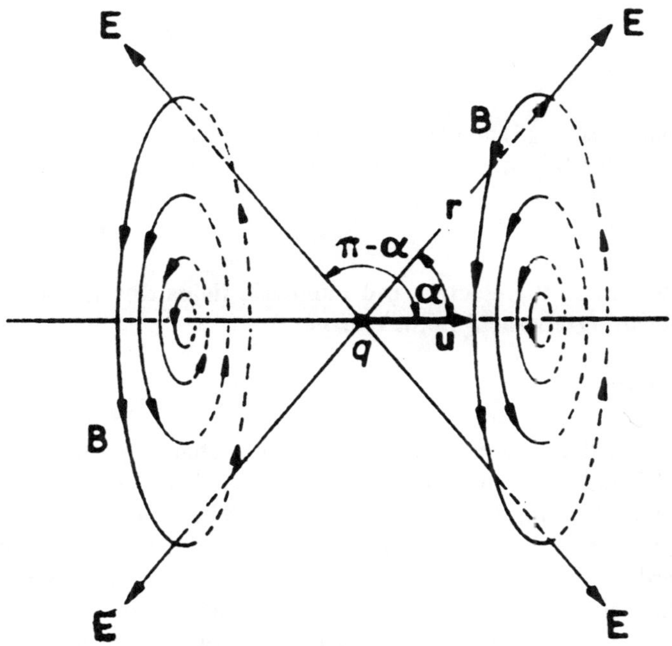

Figure 3.5. The magnetic field lines of a charge moving with uniform velocity are circles concentric with the direction of **u**. The direction of **B** is given by the right-handed corkscrew rule. The field is in the same direction for values of $\theta = \alpha$ and $\theta = \pi - \alpha$, where θ is the angle between $\mathbf{r}_0$ and **u**.

Equation (3.33) is similar to Coulomb's law, but in the case of a charge moving with a uniform velocity $u \ll c$, $\mathbf{r}_0$ must be measured from the present position of the moving charge, which is the position of the charge at the time of observation. This position changes with time due to the motion of the charge. Because of its similarity to equation (1.64) of Chapter 1, the expression given by equation (3.34) is sometimes called the Biot-Savart approximation for the magnetic field due to a classical point charge that is moving with a uniform velocity $u \ll c$.

Substituting for s from equation (3.24) into the Liénard-Wiechert potentials, namely equations (3.4) and (3.6), we find that the potentials ϕ and $\mathbf{A}$ due to a classical point charge q that is and always has been moving with uniform velocity $\mathbf{u}$ are

$$\phi = \frac{q}{4\pi\varepsilon_0 r_0(1 - \beta^2 \sin^2 \theta)^{1/2}} \tag{3.35}$$

$$\mathbf{A} = \frac{\mu_0 q\mathbf{u}}{4\pi r_0(1 - \beta^2 \sin^2 \theta)^{1/2}} \,. \tag{3.36}$$

where $\mathbf{r}_0$ is the distance from the present position of the charge to the field point at the time of observation and θ is the angle between $\mathbf{u}$ and $\mathbf{r}_0$. It is left as an exercise for the reader to show that application of the equations

$$\mathbf{E} = -\nabla\phi - \frac{\partial \mathbf{A}}{\partial t}\,; \qquad \mathbf{B} = \nabla \times \mathbf{A}$$

leads to equations (3.25) and (3.28). (Hint: Use the equation (4.24) of Chapter 4 to replace $\partial/\partial t$ by $-\mathbf{u} \cdot \nabla$ in the case of a charge moving with uniform velocity).

3.4. Discussion of the electric and magnetic fields due to an accelerating classical point charge

3.4.1. *Introduction*

Consider again the accelerating classical point charge q, shown previously in Figure 3.2 and again in Figure 3.6. The fields $\mathbf{E}$ and $\mathbf{B}$ are determined at the field point P at the time of observation t. The accelerating charge is at the appropriate retarded position R, at a distance $[r]$ from the field point P at the retarded time $t^* = (t - [r]/c)$. Draw a sphere of radius $[r]$, with its centre at the retarded position R, as shown in Figure 3.6. The point R is the appropriate retarded position for all field points on the surface of the sphere of radius $[r]$ at the time of observation t.

According to equation (3.10)

$$\mathbf{E} = \mathbf{E}_V + \mathbf{E}_A$$

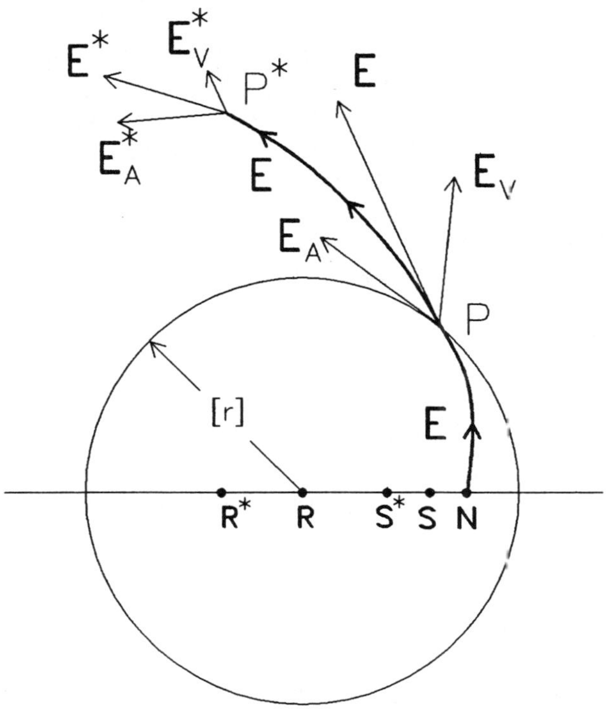

Figure 3.6. A sketch of an electric field line of a classical point charge, that is accelerating in its direction of motion. The electric field line starts from *N*, the position of the charge, and curves to pass through *P* and *P**.

where according to equation (3.11)

$$\mathbf{E}_V = \frac{q}{4\pi\varepsilon_0 s^3}\left[\mathbf{r} - \frac{r\mathbf{u}}{c}\right]\left[1 - \frac{u^2}{c^2}\right]. \tag{3.37}$$

It was shown in Section 3.3 that equation (3.37) leads to equation (3.25) for the electric field due to a charge moving with uniform velocity [**u**]. If the accelerating charge q were moving with uniform velocity [**u**] it would travel a distance [$\mathbf{u}r/c$] to reach the point S in the time $[r]/c$ it would take light to go from the retarded position R to the field point P in Figure 3.6. The point S will be called the **projected position** of the charge, which is the position the accelerating charge would have reached, if it had carried on with uniform velocity [**u**]. Since [u] is always less than c, the projected position S is always inside the sphere of radius [r] in Figure 3.6. Since equation (3.37) leads to equation (3.25), we conclude that the value of $\mathbf{E}_V$ at all points on the surface of the sphere of radius [r] in Figure 3.6 at the time of observation t is given by the equation

$$\mathbf{E}_V = \frac{q\mathbf{r}_0(1 - \beta^2)}{4\pi\varepsilon_0 r_0^3(1 - \beta^2 \sin^2\theta)^{3/2}} \tag{3.38}$$

where $\mathbf{r}_0$ is measured from the projected position S of the charge q and θ is the angle between $[\mathbf{u}]$ and $\mathbf{r}_0$ in Figure 3.6. The contribution of $\mathbf{E}_V$ to the total electric field at every point on the sphere of radius $[r]$ in Figure 3.6, at the time of observation t, is in the direction radially outwards from S, the projected position of the charge. The magnitude of $\mathbf{E}_V$ is proportional to $1/r_0^2$, where r_0 is the distance from the projected position S to the field point P. The magnitude of $\mathbf{E}_V$ also depends on θ, the angle between $[\mathbf{u}]$ and $\mathbf{r}_0$.

According to equation (3.12), the acceleration dependent contribution $\mathbf{E}_A$ to the total electric field $\mathbf{E}$ at the field point P in Figure 3.6 at the time of observation t is given by

$$\mathbf{E}_A = \frac{q}{4\pi\varepsilon_0 c^2 s^3}\,[\mathbf{r}] \times \left[\left(\mathbf{r} - \frac{r\mathbf{u}}{c}\right) \times \mathbf{a}\right]. \tag{3.39}$$

Since the direction of the vector product of two vectors is perpendicular to the plane containing the two vectors, it follows from equation (3.39) that $\mathbf{E}_A$ is always perpendicular to the vector $[\mathbf{r}]$ from R, the retarded position of the charge, to the field point P which is on the surface of the sphere of radius $[r]$ in Figure 3.6. The direction of $\mathbf{E}_A$, at the time of observation t, is therefore trangential to the sphere of radius $[r]$ in Figure 3.6, but in the general case $\mathbf{E}_A$ is not necessarily in the plane of the paper. To simply the discussion, we shall consider the special case when the acceleration $[\mathbf{a}]$ is **uniform** and when $[\mathbf{u}]$ and $[\mathbf{a}]$ are always in the same direction, so that the charge q moves in a straight line in Figure 3.6. In this special case $[\mathbf{u} \times \mathbf{a}]$ is zero, and equation (3.39) reduces to

$$\mathbf{E}_A = \frac{q}{4\pi\varepsilon_0 c^2 s^3}\,[\mathbf{r}] \times [\mathbf{r} \times \mathbf{a}]. \tag{3.40}$$

Consider the field point P in Figure 3.6. Since the direction of $[\mathbf{r} \times \mathbf{a}]$ is downwards into the paper in Figure 3.6, the direction of the vector $[\mathbf{r}] \times [\mathbf{r} \times \mathbf{a}]$ is in the plane of the paper in Figure 3.6 in a direction perpendicular to the line jointing R and P, which is in the direction of the tangent to the circle of radius $[r]$ at the field point P. The resultant electric field $\mathbf{E}$ is the vector sum of $\mathbf{E}_V$ and $\mathbf{E}_A$, as shown in Figure 3.6. We are assuming in Figure 3.6 that $\mathbf{E}_V$ and $\mathbf{E}_A$ have same magnitudes at the field point P. The vector $\mathbf{E}$ gives the direction of the resultant electric field line passing through the field point P in Figure 3.6.

We shall now discuss the shape of the electric field line that goes through the field point P in Figure 3.6. If the accelerating charge q is at the position N at the time of observation t, then the electric field line must start from N. Near the accelerating charge the $\mathbf{E}_V$ term predominates, and so the electric field line starts radially outwards from N. It then curves to the left to pass through the field point P in Figure 3.6 in the direction of the electric field at P. Now consider another field point P^*, which is further from the accelerating charge than P. The retarded position of the charge corresponding to the measurement of the total electric field at P^* at the same time of observation t, is R^*, as shown in Figure 3.6. When the acceleration [**a**] is positive, the velocity of the charge is less at the retarded position R^* than at the retarded position R so that S^*, the projected position of the charge corresponding to the retarded position R^* is to the left of S, as shown in Figure 3.6. The direction of the contribution of the $\mathbf{E}_V$ term at the new field point P^* is radially outwards from the appropriate projected position S^*, as shown in Figure 3.6. The direction of the contribution of the $\mathbf{E}_A$ term at the field point P^* is perpendicular to the line jointing R^* to P^*, as shown in Figure 3.6. Since the directions of both the $\mathbf{E}_V$ and $\mathbf{E}_A$ contributions to the electric field are different at P and P^*, the changes in both $\mathbf{E}_V$ and $\mathbf{E}_A$ contribute to the curvature of the electric field line when the charge is accelerating. The magnitudes of both $\mathbf{E}_V$ and $\mathbf{E}_A$ are less at P^* than at P, but the acceleration dependent term E_A decreases less rapidly with increasing distance from the accelerating charge than the E_V term and at large distances from the charge the $\mathbf{E}_A$ term predominates, in which case the electric field line is almost perpendicular to the vector from the appropriate retarded position to the field point. We shall consider a typical example in detail in Section 3.4.2. If the charge q were decelerating the electric field line starting from N would be curved in the opposite direction.

According to equation (3.29), the magnetic field **B** due to the accelerating charge is given, at the field point P in Figure 3.6 at the time of observation t, by

$$\mathbf{B} = \frac{[\mathbf{r}] \times \mathbf{E}}{[rc]} . \tag{3.41}$$

The total magnetic field **B** at the field point P in Figure 3.5 is perpendicular to [**r**] and is therefore tangential to the surface of the sphere of radius $[r]$ in a direction perpendicular to the total electric field **E**. In the special case when [**a**] and [**u**] are in the same direction, the electric field **E** is in the plane of the paper in Figure 3.6. Hence in this special case **B**, which is in the direction of $[\mathbf{r}] \times \mathbf{E}$, is vertically upwards from the paper in Figure 3.6. When [**a**] and [**u**] have always been in the same direction, there is rotational symmetry around the direction of [**u**] and in this special case the lines of **B**, at the time of observation t, are closed circles with centres on the line of motion of the charge.

In the case of a negative charge, the directions of both **E** and **B** are reversed.

3.4.2. *Example of the electric field due to an accelerating charge*

An example of a computer computation of the electric field due to an accelerating classical point charge is given in Figure 3.7. Reference: Hamilton and Schwartz [3]. The electric field lines in Figure 3.7 represent the magnitude and direction of the electric field at one instant of time. Different field points correspond to different retarded positions of the charge. In the example shown in Figure 3.7, a positive charge that has been moving from left to right with a uniform velocity of $0.9c$ in Figure 3.7 is brought to rest with uniform deceleration and then remains at rest. In the vicinity of the stationary

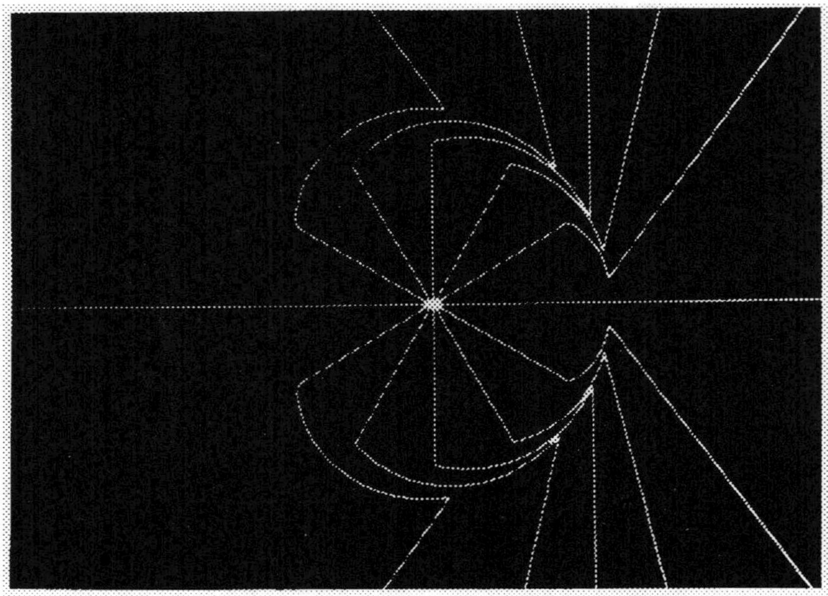

Figure 3.7. A classical point charge, which is moving with a velocity of 0.9*c*, is brought to rest with uniform deceleration. (References: Hamilton and Schwartz [3]. Reproduced by the permission of the *American Journal of Physics*.)

charge in Figure 3.7, the electric field lines diverge isotropically from the charge and the electric field in this region is given by Coulomb's law. Moving outwards along an electric field line, we reach a kink in the electric field line corresponding to the time when the charge was finally brought to rest and the deceleration stopped. Next, moving outwards along the electric field line we come to a curved section of field line, which has a large non-radial component, and which is associated with the period when the charge was decelerating at its retarded positions and when both $\mathbf{E}_V$ and $\mathbf{E}_A$ contributed to the electric field. Continuing to move outwards along a field line we reach a second kink corresponding to the time when the charge started decelerating at its retarded position. Beyond this kink the charge was moving with a uniform velocity of 0.9c at its retarded positions. In this region the $\mathbf{E}_A$ term is zero and the electric field $\mathbf{E}$ is given by the $\mathbf{E}_V$ term. Since in this region all the projected positions coincide with the position that the charge would have reached if it had carried on with uniform velocity 0.9*c*, the electric field lines in this region diverge radially from the position the charge would have reached, if it had carried on with uniform velocity 0.9*c*, as shown in Figure 3.7. The electric field lines in this region are bunched towards a direction perpendicular to the direction of the initial velocity of the charge. In this region, the electric field lines are the same as the appropriate section of the electric field lines due to a charge that is and always has been moving with uniform velocity as shown previously in Figure 3.4(b).

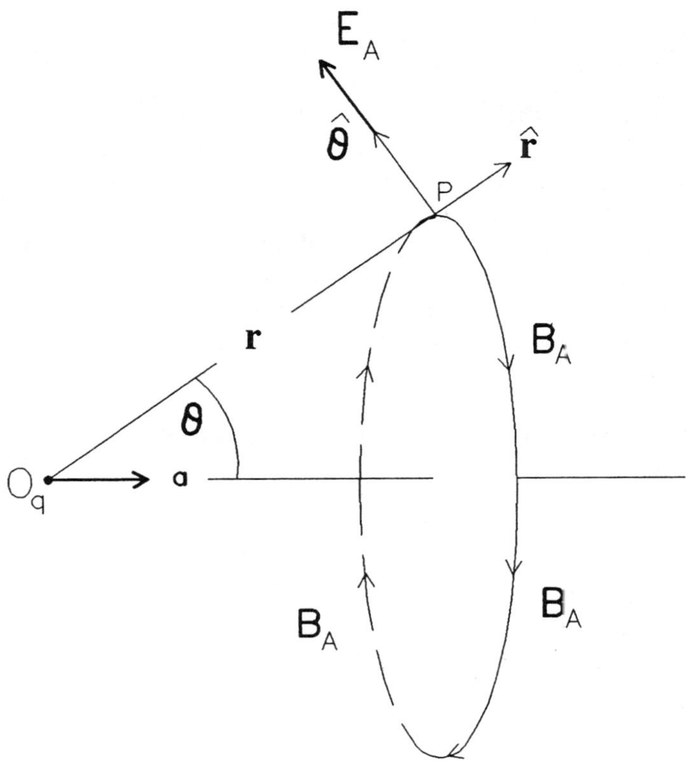

Figure 3.8. The directions of the acceleration dependent contributions $\mathbf{E}_A$ and $\mathbf{B}_A$ to the total electric and magnetic fields at the field point P due to the charge q which is moving with a velocity $\ll c$ and which is accelerating in its direction of motion.

3.4.3. *Discussion of the radiation fields due to an accelerating charge*

Assume that the accelerating charge is at the origin of the spherical polar coordinate system shown in Figure 3.8. The polar angle θ is measured from the direction of the acceleration of the charge. Let $\hat{\mathbf{r}}$, $\hat{\boldsymbol{\theta}}$ and $\hat{\boldsymbol{\phi}}$ be unit vectors in the directions of increasing r, θ and ϕ at the field point P that is at the position $\mathbf{r}$. Since $\hat{\boldsymbol{\phi}} = \hat{\mathbf{r}} \times \hat{\boldsymbol{\theta}}$, it is probably easiest for the reader to determine the direction of the unit vector $\hat{\boldsymbol{\phi}}$ at the field point P from the direction $\hat{\mathbf{r}} \times \hat{\boldsymbol{\theta}}$. In the zero-velocity limit the expressions for $\mathbf{E}_A$ and $\mathbf{B}_A$, given by equations (3.12) and (3.15) respectively, reduce to

$$\mathbf{E}_A = E_\theta \hat{\boldsymbol{\theta}} = \frac{qa \sin\theta}{4\pi\varepsilon_0 c^2 r}\,\hat{\boldsymbol{\theta}}, \tag{3.42}$$

$$\mathbf{B}_A = B_\phi \hat{\boldsymbol{\phi}} = \frac{qa \sin\theta}{4\pi\varepsilon_0 c^3 r}\,\hat{\boldsymbol{\phi}}. \tag{3.43}$$

Strictly r is the distance from the retarded position of the charge to the field point, but in the zero-velocity limit the change in the position of the charge in the time $[r]/c$ it takes the radiation field to reach the field point P is very small and, to a good approximation, we can assume that r is the distance from the charge to the field point at the time of observation. The angle θ in equations (3.42) and (3.43) is the angle between **r** and the acceleration **a** of the charge. According to equations (3.42) and (3.43), $E_A = cB_A$ and E_A and B_A are proportional to $1/r$, whereas E_V and B_V are both proportional to $1/r^2$. Hence the radiation fields $\mathbf{E}_A$ and $\mathbf{B}_A$ predominate at large distances from the charge. According to classical electromagnetism, the rate at which electromagnetic energy is crossing an area of 1 m^2 perpendicular to the direction of energy flow at the field point P in Figure 3.8 is given by the Poynting vector $\mathbf{N} = \mathbf{E} \times \mathbf{B}/\mu_0$. Using equations (3.42) and (3.43) we find that at very large distances from the accelerating charge, where E_V and B_V can be neglected, in the limit when $\beta \ll 1$ we have

$$\mathbf{N} = \frac{\mathbf{E}_A \times \mathbf{B}_A}{\mu_0} = \frac{q^2 a^2 \sin^2 \theta}{16\pi^2 \varepsilon_0 c^3 r^2}\,\hat{\mathbf{r}}. \tag{3.44}$$

The Poynting vector is proportional to $1/r^2$. Equation (3.44) gives the angular distribution of the radiation emitted, which in the zero velocity limit is symmetrical about $\theta = \pi/2$, which is the direction perpendicular to **a**. The total rate of emission of electromagnetic radiation by an accelerating charge in the zero-velocity limit can be obtained by integrating equation (3.44) over the surface of the sphere of radius r, to give

$$-\frac{\mathrm{d}W}{\mathrm{d}t} = \frac{q^2 a^2}{6\pi\varepsilon_0 c^3}\,. \tag{3.45}$$

Equation (3.44) and (3.45) must be modified when the charge is moving at a high velocity [**u**]. We shall consider again the special case when [**a**] is parallel to [**u**]. According to equations (3.40) and (3.41),

$$\mathbf{E}_A = \frac{q[r^2]\,[a]\sin\theta}{4\pi\varepsilon_0 c^2 s^3}\,\hat{\boldsymbol{\theta}} \tag{3.46}$$

$$\mathbf{B}_A = \frac{q[r^2]\,[a]\sin\theta}{4\pi\varepsilon_0 c^3 s^3}\,\hat{\boldsymbol{\phi}} \tag{3.47}$$

where [**r**] is now a vector from the retarded position of the charge to the field point and θ is the angle between [**r**] and both [**u**] and [**a**]. The difference between equations (3.46) and (3.47) on the one hand and equations (3.42) and (3.43) on the other is the change of $1/r$ into r^2/s^3 in equations (3.46) and (3.47).

According to equation (3.5),

$$s = \left[r - \frac{\mathbf{r}\cdot\mathbf{u}}{c}\right] = [r][1 - \beta\cos\theta] \qquad [3.48]$$

where $[\beta] = [u]/c$ and θ is the angle between $[\mathbf{u}]$ and $[\mathbf{r}]$. The effect of replacing $1/r$ by r^2/s^3 is to increase the value of $\mathbf{E}_A$ in the forward direction, when $\theta < \pi/2$ and to decrease $\mathbf{E}_A$ in the backward direction, when $\theta > \pi/2$. The Poynting vector $\mathbf{N}$ at large distances from the accelerating charge is

$$\mathbf{N} = \frac{\mathbf{E} \times \mathbf{B}}{\mu_0} = \frac{q^2[r^4][a^2] \sin^2 \theta}{16\pi^2\varepsilon_0 c^3 s^6} [\hat{\mathbf{r}}]. \qquad (3.49)$$

The effect of the $s^6 = [r(1 - \beta \cos \theta)]^6$ term in the denominator of equation (3.49) is to increase the amount of radiation emitted in the forward direction where $\theta < \pi/2$ and to decrease the amount of radiation emitted in the backward direction where $\theta > \pi/2$ compared with the zero velocity case given by equation (3.44). In the zero-velocity limit s tends to r and equation (3.49) reduces to equation (3.44).

The case when the acceleration $[\mathbf{a}]$ of a charge at its retarded position is perpendicular to its velocity $[\mathbf{u}]$ at its retarded position is another very important example, as it gives the rate of emission of synchrotron radiation by a charge moving in a magnetic field. A reader interested in a full discussion of the classical theory of the emission of radiation by an accelerating charge, is referred to text books on classical electromagnetism such as Panofsky and Phillips [4], Jackson [1] and Laud [5].

References

1. Jackson, J. D., *Principles of Electrodynamics*, 2nd Edition (John Wiley, New York, 1975), Chapter 14.
2. Hauser, W., *Introduction to the Principles of Electromagnetism* (Addison-Wesley, Reading, Mass., 1971), p. 412.
3. Hamilton, J. C. and Schwartz, J. L., *Amer. Journ. Phys.*, Vol. 32, p. 1541 (1971).
4. Panofsky, W. K. H. and Phillips, M., *Classical Electricity and Magnetism*, 2nd Edition (Addison-Wesley, Reading, Mass., 1962), Chapter 20.
5. Laud, B. B., *Electromagnetics* (Wiley Eastern, New Delhi, 1983), Chapter 9.

CHAPTER 4

Development of Maxwell's equations from the expressions for the electric and magnetic fields due to a moving classical point charge

4.1. Introduction

Before going on in later chapters to illustrate how individual macroscopic electromagnetic phenomena can be interpreted directly in terms of the electric and magnetic fields due to moving and accelerating classical point charges, we shall in this chapter develop Maxwell's equations from the expressions for the electric and magnetic field due to an accelerating classical point charge. Since we used Maxwell's equations to develop the differential equations for ϕ and **A** which were then used in Chapter 2 to derive the Liénard-Wiechert potentials and which were then used in turn in Chapter 3 to derive the fields **E** and **B** due to an accelerating charge, given by equations (3.10) and (3.13) respectively, it is only to be expected that these fields **E** and **B** due to an accelerating charge obey Maxwell's equations. However, by going in the reverse direction and starting with the fields **E** and **B** due to an accelerating charge and deriving Maxwell's equations as relations between these fields, we shall be able to interpret the origins and roles of the various terms in each of Maxwell's equations. We shall confine our discussions in this chapter to the special case where $\varepsilon_r = 1$ and $\mu_r = 1$ everywhere. We shall go on to discuss the general case when $\varepsilon_r > 1$ and $\mu_r > 1$ in Chapter 9.

To simplify the mathematics we shall confine most of our discussions to the simpler case of a classical point charge that is and always has been moving with uniform velocity, in which case the acceleration dependent contributions $\mathbf{E}_A$ and $\mathbf{B}_A$ to the electric and magnetic fields are zero and the total fields **E** and **B** are given by equations (3.25) and (3.28) of Chapter 3 respectively.

Consider an isolated classical point charge of magnitude q, that is and always has been moving with uniform velocity **u** along the x axis in Figure 4.1. The charge q is at the origin O at the time of observation t, when the fields are determined at the field point P. We shall use the spherical polar coordinate system, shown in Figure 4.1. The radial coordinate r is the distance from the origin O to the field point P. In the coordinate system used in Figure 4.1 the polar θ is the angle between the x axis (which is the direction of **u**) and

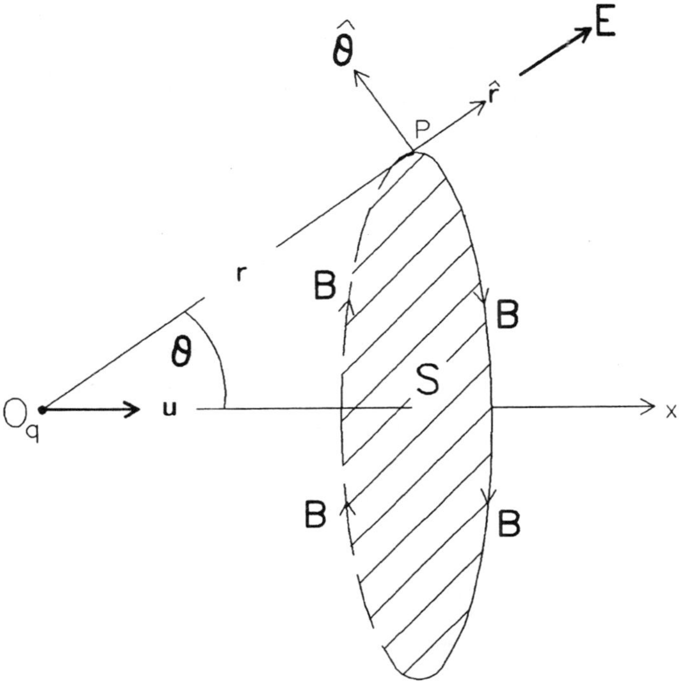

Figure 4.1. The charge q, which is moving with uniform velocity $\mathbf{u}$ along the x axis, is at the origin at the time $t = 0$. The stationary circular disk shaped surface S is perpendicular to the x axis, with its centre on the x axis. In the spherical polar coordinate system shown, the polar angle θ is the angle between $\mathbf{u}$ and the position vector $\mathbf{r}$ of the field point P and is measured from the x axis. The unit vectors $\hat{\mathbf{r}}$ and $\hat{\boldsymbol{\theta}}$ at the field point P are shown. The unit vector $\hat{\boldsymbol{\phi}}$ is in the direction of $\hat{\mathbf{r}} \times \hat{\boldsymbol{\theta}}$, which, with the choice of polar angle is in the directions upwards from the paper.

the vector $\mathbf{r}$ from the origin O to the field point P. The azimuthal angle ϕ is measured from the y axis in the yz plane in the direction towards the z axis. The unit vector $\hat{\mathbf{r}}$, $\hat{\boldsymbol{\theta}}$ and $\hat{\boldsymbol{\phi}}$ at the field point P at (r, θ and ϕ) are in the directions of increasing r, θ and ϕ in Figure 4.1. The determination of the directions of $\hat{\mathbf{r}}$, and $\hat{\boldsymbol{\theta}}$ is straightforward. Since $\hat{\boldsymbol{\phi}} = \hat{\mathbf{r}} \times \boldsymbol{\theta}$ it is probably easiest for the reader to determine the direction of the unit vector $\hat{\boldsymbol{\phi}}$ at the field point P from the direction of $\hat{\mathbf{r}} \times \hat{\boldsymbol{\theta}}$.

Putting $\mathbf{r}_0 = \mathbf{r}$ in equation (3.25) and with $\beta = u/c$, we find that when the charge q is at the origin O in Figure 4.1, the electric field $\mathbf{E}$ at the field point P at (r, θ, ϕ) is given by

$$\mathbf{E} = E_r\hat{\mathbf{r}} + E_\theta\hat{\boldsymbol{\theta}} + E_\phi\hat{\boldsymbol{\phi}} \tag{4.1}$$

where the components E_r, E_θ and E_ϕ are given by

$$E_r = \frac{q(1 - \beta^2)}{4\pi\varepsilon_0 r^2(1 - \beta^2 \sin^2 \theta)^{3/2}}; \qquad E_\theta = 0; \qquad E_\phi = 0. \tag{4.2}$$

The electric field lines emanate radially from the present position of a charge moving with uniform velocity, that is from the position of the charge at the time of observation t when the fields $\mathbf{E}$ and $\mathbf{B}$ are determined at the field point P, as shown in Figure 3.4(b) of Chapter 3.

According to equation (3.28) the magnetic field $\mathbf{B}$ at the field point P at (r, θ, ϕ) in Figure 4.1, at the time of observation t, is given by

$$\mathbf{B} = B_r\hat{\mathbf{r}} + B_\theta\hat{\boldsymbol{\theta}} + B_\phi\hat{\boldsymbol{\phi}} \tag{4.3}$$

where the components B_r, B_θ and B_ϕ are given by

$$B_r = 0; \quad B_\theta = 0; \quad B_\phi = \frac{qu(1-\beta^2)\sin\theta}{4\pi\varepsilon_0 c^2 r^2(1-\beta^2\sin^2\theta)^{3/2}}\,. \tag{4.4}$$

The magnetic field lines, due to a classical point charge moving with uniform velocity, are closed circles concentric with the direction of motion of the charge, as shown in Figure 3.5 of Chapter 3.

4.2. The equation $\nabla\cdot\mathbf{E} = \rho/\epsilon_0$

Consider the isolated, classical point charge of magnitude q, that is and has always been moving with uniform velocity $\mathbf{u}$ along the x axis in Figure 4.1. The charge is at the origin O at the time of observation t. According to equation (A1.40) of Appendix A1.10 the divergence of $\mathbf{E}$ is given in spherical polar coordinates by

$$\nabla\cdot\mathbf{E} = \frac{1}{r^2}\frac{\partial}{\partial r}(r^2E_r) + \frac{1}{r\sin\theta}\frac{\partial}{\partial\theta}(E_\theta\sin\theta) + \frac{1}{r\sin\theta}\frac{\partial E_\phi}{\partial\phi}\,. \tag{4.5}$$

According to equation (4.2), when the charge q is at the origin in Figure 4.1, $E_\theta = 0$, $E_\phi = 0$ and E_r is proportional to $1/r^2$. Substituting in equation (4.5), we find that

$$\nabla\cdot\mathbf{E} = 0. \tag{4.6}$$

Equation (4.6) shows that the divergence of the electric field $\mathbf{E}$ due to a charge moving with uniform velocity is zero at any field point in empty space. Integrating equation (4.6) over a finite volume, that does not enclose the moving charge and then applying Gauss' theorem of vector analysis, which is equation (A1.30) of Appendix A1.7, we have

$$\int\nabla\cdot\mathbf{E}\,\mathrm{d}V = \int\mathbf{E}\cdot\mathrm{d}\mathbf{S} = 0. \tag{4.7}$$

We shall now assume that the moving charge q, which is moving with uniform velocity $\mathbf{u}$, is inside the Gaussian surface shown in Figure 4.2. We shall choose O, the position of the moving charge at the time of observation t, as the origin of the same spherical polar coordinate system as shown in Figure 4.1. Consider an element of area $\mathrm{d}\mathbf{S}$ at a field point P at (r, θ, ϕ) on the arbitrary Gaussian surface, as shown in Figure 4.2. Since, according to equations (4.2),

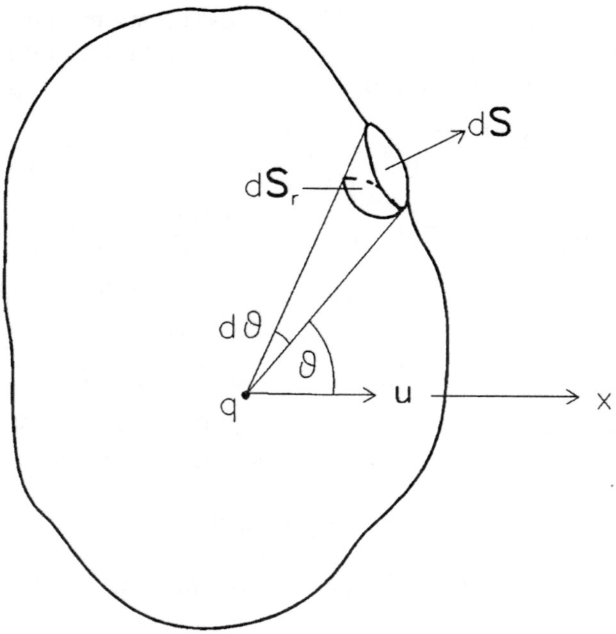

Figure 4.2. Calculation of the total electric flux from a classical point charge moving with uniform velocity **u**.

$E_\theta = 0$ and $E_\phi = 0$, the electric flux $d\Psi$ through $d\mathbf{S}$ at the time of observation t is

$$d\Psi = \mathbf{E} \cdot d\mathbf{S} = E_r \, dS_r \tag{4.8}$$

where dS_r is the component of the element of area $d\mathbf{S}$ in the direction of increasing r. The magnitude of dS_r is equal to the area of the projection of the element of surface area $d\mathbf{S}$ on to a sphere of radius r, with centre at O. Consider an element of area of part of the sphere of radius r which has dimensions $r \, d\theta$ by $r \sin\theta \, d\phi$ and is of area $r^2 \sin\theta \, d\theta \, d\phi$. Substituting for E_r from equations (4.2) into equation (4.8), we find that

$$d\Psi = E_r \, dS_r = \frac{q(1-\beta^2)}{4\pi\varepsilon_0 r^2 (1-\beta^2 \sin^2\theta)^{3/2}} \, r^2 \sin\theta \, d\theta \, d\phi. \tag{4.9}$$

After cancelling r^2, we find that the expression for $d\Psi$ is independent of r. Similar expressions for $d\Psi$, which are all independent of r, are obtained for all the elements of area of the arbitrary Gaussian surface in Figure 4.2. To integrate $d\Psi$ over the arbitrary Gaussian surface in Figure 4.2, we can integrate equation (4.9) first with respect to ϕ from $\phi = 0$ to $\phi = 2\pi$, and then integrate with respect to θ from $\theta = 0$ to $\theta = \pi$. Since $\int d\phi = 2\pi$ integrating first over ϕ and then over θ we obtain

$$\Psi = \frac{q(1-\beta^2)}{2\varepsilon_0} \int_0^\pi \frac{\sin\theta \, d\theta}{(1-\beta^2 \sin^2\theta)^{3/2}} \, . \tag{4.10}$$

Rewriting $(1 - \beta^2 \sin^2\theta)$ as $[(1 - \beta^2) + \beta^2 \cos^2\theta]$ and making substitutions $w = \beta \cos\theta$, $dw = -\beta \sin\theta\, d\theta$ and $(1 - \beta^2) = a^2$, the integral I in equation (4.10) can be rewritten in the form

$$\mathrm{I} = \int \frac{\sin\theta\, d\theta}{(a^2 + \beta^2 \cos^2\theta)^{3/2}} = -\int \frac{dw}{\beta(a^2 + w^2)^{3/2}}$$

$$= -\frac{w}{\beta a^2(a^2 + w^2)^{1/2}} = -\frac{\beta \cos\theta}{\beta(1 - \beta^2)(1 - \beta^2 \sin^2\theta)^{1/2}} .$$

Substituting for the integral in equation (4.10), we finally find that

$$\Psi = \int \mathbf{E} \cdot d\mathbf{S} = -\frac{q}{2\varepsilon_0}\left[\frac{\cos\theta}{(1 - \beta^2 \sin^2\theta)^{1/2}}\right]_{\theta=0}^{\theta=\pi} = \frac{q}{\varepsilon_0} . \tag{4.11}$$

According to equation (4.11), the total electric flux $\Psi = \int \mathbf{E} \cdot d\mathbf{S}$ from a classical point charge moving with uniform velocity is the same as the total electric flux from a stationary charge of the same magnitude, even though in the case of the moving charge the magnitude of the electric field **E** varies with direction. Equations (4.11) and (4.7) are also valid for the electric fields of accelerating charges, illustrating how Gauss' flux law is valid for moving and accelerating classical point charges.

It was shown in Section 1.2.5 of Chapter 1 that, if the number of electric field lines per square metre is equal to the magnitude E of the electric field, then $\mathbf{E} \cdot d\mathbf{S}$ is equal to the number of electric field lines crossing the element of area $d\mathbf{S}$ of the Gaussian surface shown in Figure 4.2. It follows from equation (4.11), that in the case of the charge moving with uniform velocity inside the Gaussian surface in Figure 4.2, the total number of electric field lines crossing the Gaussian surface in an outward direction is equal to q/ε_0. Since this is true whatever the shape and dimensions of the Gaussian surface, it follows that there are q/ε_0 electric field lines from a classical point charge moving with uniform velocity and and these electric field lines carry on radially outwards from the position of the charge to infinity as shown in Figure 4.3(a). If the moving charge q is outside the Gaussian surface as shown in Figure 4.3(b), then as many electric field lines enter the Gaussian surface as leave it, in agreement with equation (4.7). The same results, leading up to equations (4.11) and (4.7) are true for the electric field lines due to an accelerating classical point charge, though in the case of an accelerating classical point charge the electric field lines are curved corresponding to periods when the charge was accelerating at its retarded position.

We shall now consider a field point inside a moving, continuous, macroscopic charge distribution of the type shown previously in Figure 1.8 of Chapter 1. Equations (4.11) and (4.7) have the same mathematical forms as equations (1.13) and (1.14). Hence we can use the same mathematical steps as we used in Section 1.2.7 of Chapter 1. Consider a Gaussian surface of volume V_0 inside the moving charge distribution. If we divide the moving, continuous charge distribution into infinitesimal volume elements we can treat the charge inside each volume element as the equivalent of a moving

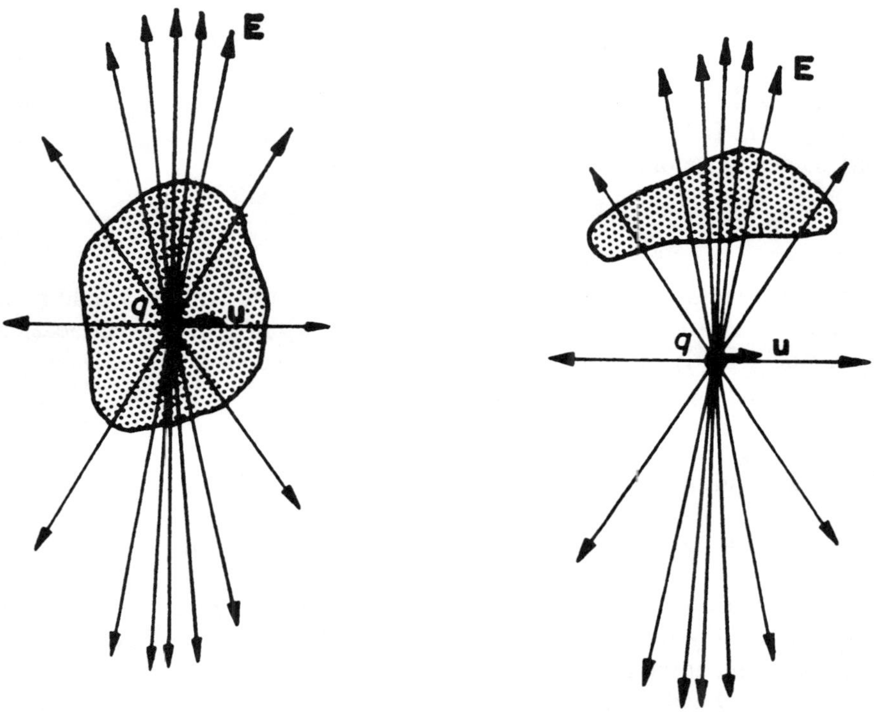

Figure 4.3. (a) The flux of **E** through a Gaussian surface, due to a charge moving with uniform velocity, is equal to q/ε_0 if the charge is inside the surface, and (b) equal to zero if the charge is outside the Gaussian surface. The lines of **E** for an isolated classical point charge, which is and always has been, moving with uniform velocity, continue radially outwards to infinity.

classical point charge. Using equations (4.11) and (4.7) as appropriate and following the method we used in Section 1.2.7 of Chapter 1, we find that after applying Gauss' theorem of vector analysis we have for the volume V_0 inside the charge distribution

$$\int \mathbf{E} \cdot \mathrm{d}\mathbf{S} = \int \nabla \cdot \mathbf{E}\, \mathrm{d}V = \int \frac{\rho \mathrm{d}V}{\varepsilon_0}$$

where **E** is the resultant electric field due to the complete charge distribution. If we now make the volume V_0 small enough for the variations of $\nabla \cdot \mathbf{E}$ and ρ inside V_0 to be negligible then, after cancelling V_0, we have

$$\nabla \cdot \mathbf{E} = \frac{\rho}{\varepsilon_0} \tag{4.12}$$

where **E** is the total electric field at a field point inside the moving continuous charge distribution where the charge density is ρ.

We shall now assume that the dimensions of the moving continuous charge distribution are exceedingly small so that it corresponds to our model of a classical point charge. At a field point inside such a charge distribution we can rewrite equation (4.12) in the form.

$$\nabla \cdot \mathbf{e} = \frac{\rho^{\text{mic}}}{\varepsilon_0} \tag{4.13}$$

where **e** is the microscopic electric field and ρ^{mic} is the microscopic charge density at the field point. After averaging, in the way described in Section 1.11 of Chapter 1, we obtain

$$\nabla \cdot \mathbf{E} = \frac{\rho}{\varepsilon_0} \tag{4.14}$$

where in equation (4.14) **E** is the macroscopic electric field at a field point inside a macroscopic charge distribution made up of moving and accelerating atomic particles and ρ is the macroscopic charge density at the field point, defined using equation (1.147) of Chapter 1.

4.3. The equation $\nabla \cdot \mathbf{B} = 0$

4.3.1. *A classical point charge moving with uniform velocity*

Consider the isolated classical point charge, of magnitude q, that is and always has been moving with uniform velocity **u** along the x axis in Figure 4.1 and is at the origin at the time of observation. Since, according to equations (4.4) $B_r = 0$, $B_\theta = 0$ and B_ϕ is independent of ϕ, if follows from equation (A1.40) of Appendix A1.10 that, at any field point in empty space,

$$\nabla \cdot \mathbf{B} = 0. \tag{4.15}$$

For the special case of a classical point charge moving with uniform velocity, the magnetic field lines are closed circles concentric with the direction of motion of the charge, as shown in Figure 4.1. Since the magnetic field lines are closed, as many lines of **B** enter any closed surface as leave it. This is true whether the moving charge is inside or outside the closed surface. Hence, in general, for a classical point charge moving with uniform velocity

$$\int \mathbf{B} \cdot \mathrm{d}\mathbf{S} = 0. \tag{4.16}$$

if the surface integral is evaluated, at a fixed time, over any closed surface, whether the moving charge is inside or outside the closed surface. Equations (4.15) and (4.16) are also valid for the magnetic field due to an accelerating classical point charge.

4.3.2. *A field point inside a moving continuous charge distribution*

It is straight forward for the reader to show, using equation (4.15) and the methods used in the case of the electric field in Section 4.2, that $\nabla \cdot \mathbf{B}$ is zero at a field point inside a moving, macroscopic, continuous charge distribution.

The corresponding Maxwell-Lorentz equation at a field point inside a moving charged atomic particle is

$$\nabla \cdot \mathbf{b} = 0 \tag{4.17}$$

where **b** is the microscopic magnetic field. Averaging equation (4.17) at a field point inside a distribution of moving atomic charged particles in the way described in Section 1.11 of Chapter 1, we obtain

$$\nabla \cdot \mathbf{B} = 0 \tag{4.18}$$

where **B** is now the macroscopic magnetic field, defined using equation (1.147).

4.3.3. *Use of magnetic field lines*

Some authors conclude erroneously on the basis of equation (4.16), according to which $\int \mathbf{B} \cdot d\mathbf{S}$ evaluated over any closed surface is zero, that all magnetic field lines are closed. This is true in the special case of an isolated classical point charge moving with uniform velocity, where the magnetic field lines are closed circles concentric with the direction of the velocity of the charge. The magnetic field lines are also closed in some of the idealized symmetrical examples treated in elementary text books. For example, the magnetic field lines representing the magnetic field due to the steady current in an extremely long, straight, thin wire are closed circles. If, however, there is also a magnetic field parallel to the long straight wire produced for example by the current in a circular coil with its centre on the wire, then the magnetic field lines representing the resultant magnetic field are helices which return to the other end of the very long straight wire at large distances from the wire. It is extremely unlikely that a particular magnetic field line will join up with itself, particularly if the small coil is not circular or if there are kinks in the wire connecting the two ends of the very long straight wire. This illustrates how in practical cases the magnetic field lines are not necessarily closed. In such cases, when the magnetic field lines are not closed, there are difficulties when using the density of magnetic field lines to represent the strength of the magnetic field quantitatively. A reader interested in a full discussion of this point is referred to Slepian [1] McDonald [2] and Iona [3].

4.4. Relation between the spatial and time derivatives of the fields of a classical point charge moving with uniform velocity

Consider the electric field **E** due to the isolated classical point charge, of magnitude q, that is and always has been moving with uniform velocity **u** along x axis in Figure 4.4(a). The charge q is at the position shown in Figure 4.4(a) at the time of observation t of the electric field **E** at the field point P, which has coordinates (x, y, z). The electric field lines diverge radially from the position of the charge q at the time of observation t, as shown in Figure 4.4(a). Let an experimenter at rest at the field point P in Figure 4.4(a) measure

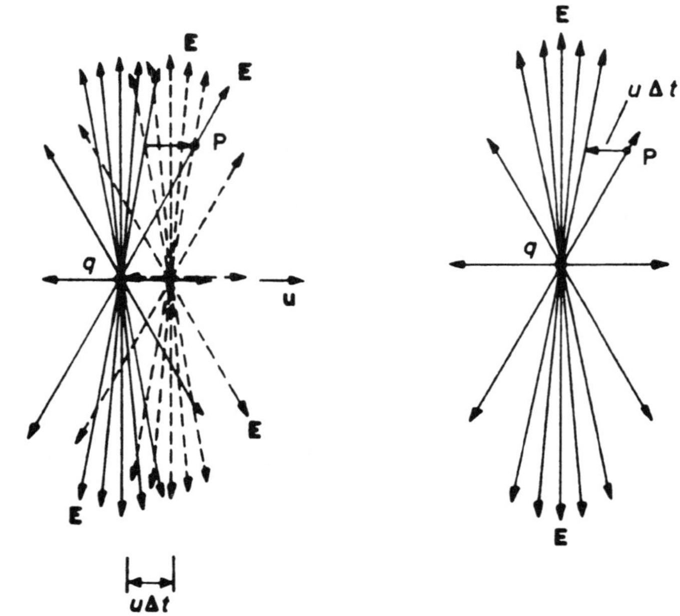

Figure 4.4. (a) The field point P is fixed and the change in the electric field $\mathbf{E}$ in a time Δt due to the motion of the charge is measured. (b) The field point P is moved a distance $u\Delta t$ to the left at a fixed time. The same change in $\mathbf{E}$ is measured in both (a) and (b).

an increase $\Delta\mathbf{E}$ in the electric field at the field point P in a short time interval Δt. Since x, y and z are constant

$$\Delta\mathbf{E} = \left(\frac{\partial \mathbf{E}}{\partial t}\right)_{x,\, y,\, z} \Delta t. \tag{4.19}$$

If at a fixed time t, that is considering the charge q as fixed in space, the experimenter at P moved a distance $\Delta x = u\,\Delta t$ to the left parallel to the x axis, as shown in Figure 4.4(b), the experimenter would measure the same increase $\Delta\mathbf{E}$ in $\mathbf{E}$. For the conditions shown in Figure 4.4(b) y, z and t are constant, so that

$$\Delta\mathbf{E} = \left(\frac{\partial \mathbf{E}}{\partial x}\right)_{y,\, z,\, t} \Delta x. \tag{4.20}$$

Equating the right hand sides of equations (4.19) and (4.20) and putting $\Delta x = -u\Delta t$, we find that for a charge moving with uniform velocity along the x axis

$$\left(\frac{\partial \mathbf{E}}{\partial t}\right)_{x,\, y,\, z} = -u\left(\frac{\partial \mathbf{E}}{\partial x}\right)_{y,\, z,\, t}. \tag{4.21}$$

Similarly

$$\left(\frac{\partial \mathbf{B}}{\partial t}\right)_{x,\,y,\,z} = -u\left(\frac{\partial \mathbf{B}}{\partial x}\right)_{y,\,z,\,t}. \tag{4.22}$$

In general, for the fields **E** and **B** due to a classical point charge moving with uniform velocity **u** along the x axis

$$\frac{\partial}{\partial t} = -u\frac{\partial}{\partial x}. \tag{4.23}$$

If the uniform velocity **u** is not parallel to the x axis, then

$$\frac{\partial}{\partial t} = -\mathbf{u} \cdot \boldsymbol{\nabla}. \tag{4.24}$$

4.5. The equation $\boldsymbol{\nabla} \times \mathbf{E} = -\dot{\mathbf{B}}$

4.5.1. *A classical point charge moving with uniform velocity*

Consider again the isolated classical point charge of magnitude q, that is and always has been moving with uniform velocity **u** along the x axis in Figure 4.1, and is at the origin O at the time of observation t, when the fields **E** and **B** are determined at the field point P, which has coordinates (x, y, z) and is at a distance **r** from the origin O. Choose the spherical polar coordinate system, shown in Figure 4.1, measuring the polar angle θ from the x axis, which is the direction of motion of the charge. Let $\hat{\mathbf{r}}$, $\hat{\boldsymbol{\theta}}$ and $\hat{\boldsymbol{\phi}}$ be unit vectors in the directions of increasing r, θ and ϕ respectively, as shown in Figure 4.1.

According to equations (4.2), the electric field **E** at the field point P in Figure 4.1 has the components

$$E_r = \frac{q(1-\beta^2)}{4\pi\varepsilon_0 r^2(1-\beta^2\sin^2\theta)^{3/2}}\,;\; E_\theta = 0; \qquad E_\phi = 0. \tag{4.25}$$

Since $E_\theta = 0$, $E_\phi = 0$ and $\partial E_r/\partial\phi = 0$, the expression for $\boldsymbol{\nabla} \times \mathbf{E}$ in spherical polar coordinates, which is given by equation (A1.41) of Appendix A1.10, reduces to

$$\boldsymbol{\nabla} \times \mathbf{E} = \left(-\frac{1}{r}\frac{\partial E_r}{\partial\theta}\right)\hat{\boldsymbol{\phi}}. \tag{4.26}$$

It can be seen from Figure 3.4(b) of Chapter 3 and from equations (4.25) that, for fixed values of r and ϕ, when θ is less than 90°, E_r increases with increasing θ so that $\partial E_r/\partial\theta$ is positive. Hence, according to equation (4.26), for the position of the field point P in Figure 4.1, for which $\theta < 90°$, $\boldsymbol{\nabla} \times \mathbf{E}$ has a component in the direction of $-\hat{\boldsymbol{\phi}}$ Substituting for E_r in equation (4.26) and carrying out the partial differentiation, we find that

$$\nabla \times \mathbf{E} = -\frac{3q(1-\beta^2)\beta^2 \sin\theta \cos\theta}{4\pi\varepsilon_0 r^3(1-\beta^2 \sin^2\theta)^{5/2}}\hat{\boldsymbol{\phi}}. \tag{4.27}$$

The magnetic field at the field point P in Figure 4.1 is upwards out of the paper, that is in the direction of the unit vector $\hat{\boldsymbol{\phi}}$. Since the magnitude of **B** at the fixed field point P increases with time as the charge gets closer to the field point, $\dot{\mathbf{B}}$ is positive and is in the direction of $\hat{\boldsymbol{\phi}}$. The value of $\dot{\mathbf{B}}$ will now be determined. The magnetic field **B** at the field point P in Figure 4.1 at the time of observation t has the components:

$$B_r = 0; \quad B_\theta = 0; \quad B_\phi = \frac{qu(1-\beta^2)\sin\theta}{4\pi\varepsilon_0 c^2 r^2(1-\beta^2\sin^2\theta)^{3/2}}. \tag{4.28}$$

Since the field point P has coordinates (x, y, z) and the distance form O to P in Figure 4.1 is r we have

$$r^2 = x^2 + y^2 + z^2. \tag{4.29}$$

Since θ is the angle between the x axis and the line joining O and P

$$\cos\theta = \frac{x}{r}; \quad \sin\theta = (1-\cos^2\theta)^{1/2} = \frac{(y^2+z^2)^{1/2}}{r}. \tag{4.30}$$

Hence the expression for B_ϕ can be expressed in cartesian coordinates as follows

$$B_\phi = \frac{qu(1-\beta^2)(y^2+z^2)^{1/2}}{4\pi\varepsilon_0 c^2[x^2+(1-\beta^2)(y^2+z^2)]^{3/2}}. \tag{4.31}$$

According to equation (4.23), for the classical point charge that is moving along the x axis with uniform velocity **u** in Figure 4.1

$$\frac{\partial \mathbf{B}}{\partial t} = -u\frac{\partial \mathbf{B}}{\partial x} = -u\frac{\partial}{\partial x}(B_\phi\hat{\boldsymbol{\phi}}). \tag{4.32}$$

The direction of the unit vector $\hat{\boldsymbol{\phi}}$ does not change when the x coordinate of the field point P in Figure 4.1 is changed keeping y, z and t constant. Hence equation (4.32) becomes

$$\frac{\partial \mathbf{B}}{\partial t} = -u\left(\frac{\partial B_\phi}{\partial x}\right)\hat{\boldsymbol{\phi}}. \tag{4.33}$$

Substituting for B_ϕ from equation (4.31) and carrying out the partial differentiation, we find that

$$-\frac{\partial \mathbf{B}}{\partial t} = -\frac{3qu^2(1-\beta^2)(y^2+z^2)^{1/2}\,x}{4\pi\varepsilon_0 c^2[x^2+(1-\beta^2)(y^2+z^2)]^{5/2}}\hat{\boldsymbol{\phi}}. \tag{4.34}$$

Using equation (4.29) and (4.30), equation (4.34) can be rewritten in the form

$$-\frac{\partial \mathbf{B}}{\partial t} = -\frac{3q\beta^2(1-\beta^2)\sin\theta\cos\theta}{4\pi\varepsilon_0 r^3[1-\beta^2\sin^2\theta]^{5/2}}\hat{\boldsymbol{\phi}}. \tag{4.35}$$

Comparing equations (4.27) and (4.35), we can see that the values of **E** and **B** at the field point P in Figure 4.1, which are given by equations (4.2) and (4.4) respectively, are related by the equation

$$\nabla \times \mathbf{E} = -\frac{\partial \mathbf{B}}{\partial t}. \tag{4.36}$$

Equation (4.36) is one of Maxwell's equations. It was derived in this section from the values for the electric and magnetic fields due to a classical point charge which is moving with uniform velocity, which we derived in Section 3.3 of Chapter 3 using the Liénard-Wiechert potentials. Equation (4.36) is a relation between the field vectors **E** and **B**, that is valid at any field point. If we know how **B** is varying with time at a fixed field point, such as P in Figure 4.1, then the value of $\nabla \times \mathbf{E}$ at that field point can be calculated using equation (4.36). Conversely if the value of $\nabla \times \mathbf{E}$ at any field point is known, then equation (4.36) gives the value of $\dot{\mathbf{B}}$ at that field point. There is no need to enquire what the velocity and position of the moving charge are when we apply equation (4.36).

Consider any closed curve in Figure 4.1. Integrating equation (4.36) over a surface bounded by the closed curve at a fixed time t, we have

$$\int \nabla \times \mathbf{E} \cdot \mathrm{d}\mathbf{S} = -\int \left(\frac{\partial \mathbf{B}}{\partial t}\right) \cdot \mathrm{d}\mathbf{S} = -\frac{\partial}{\partial t}\int \mathbf{B} \cdot \mathrm{d}\mathbf{S}.$$

Applying Stokes' theorem, which is equation (A1.34) of Appendix A1.8, we have

$$\oint \mathbf{E} \cdot \mathrm{d}\mathbf{l} = -\frac{\partial}{\partial t}\int \mathbf{B} \cdot \mathrm{d}\mathbf{S} = -\frac{\partial \Phi}{\partial t} \tag{4.37}$$

where

$$\Phi = \int \mathbf{B} \cdot \mathrm{d}\mathbf{S} \tag{4.38}$$

is the magnetic flux through the surface bounded by the closed curve. Equation (4.37) is an example of Faraday's law of electromagnetic induction. Equations (4.36) and (4.37) will now be illustrated by showing how a classical point charge that is moving with uniform velocity can give rise to an induced emf in a stationary circuit.

4.5.2. *Induced emf due to a classical point charge moving with uniform velocity*

Consider a positive classical point charge, of magnitude $+q$, that is moving with uniform velocity **u**, as shown in Figure 4.5. The line integral $\oint \mathbf{E} \cdot \mathrm{d}\mathbf{l}$ will be evaluated around the closed curve $ABCD$ in Figure 4.5 at the instant the charge q is at the point O. The electric field lines diverge radially from O, which is the present position of the moving charge at the time of observation t. The lines of **E** bunch towards a direction perpendicular to the velocity

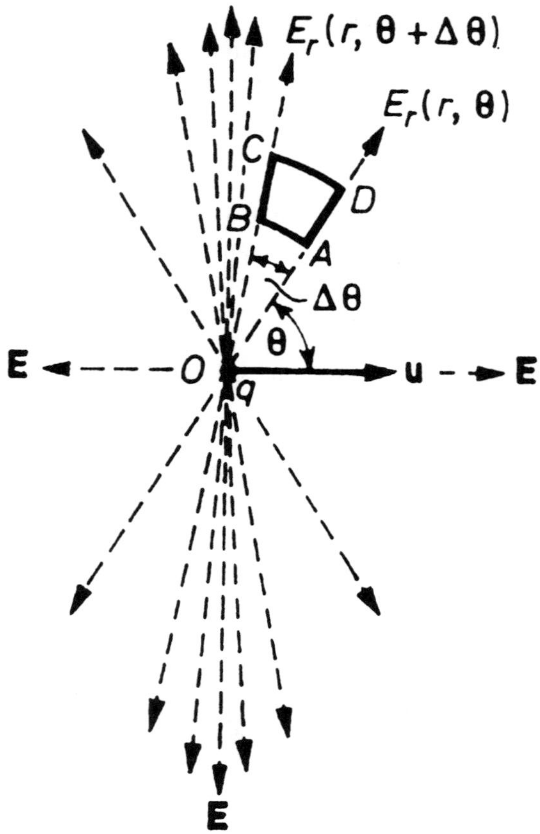

Figure 4.5. The calculation of the curl of the electric field **E** due to a charge q moving with uniform velocity. Since the electric field is greater along BC than AD, $\oint \mathbf{E} \cdot d\mathbf{l}$ evaluated around $ABCD$ is finite, so that $\nabla \times \mathbf{E}$ is finite. It is shown that $\nabla \times \mathbf{E} = -(\partial \mathbf{B}/\partial t)$. Both **E** and **B** arise from the moving charge.

u of the charge, as shown in Figure 4.5. The arc AB is at a radial distance r from O and the arc CD is at a radial distance $(r + \Delta r)$ from O, where $\Delta r \ll r$. Since the electric field **E** is radial, the electric field has no component along either of the sections AB or CD, so that $\int \mathbf{E} \cdot d\mathbf{l}$ is zero along both of the sections $A\,B$ and $C\,D$ in Figure 4.5. For the conditions shown in Figure 4.5, the value of E_r, given by equation (4.2), at any point along the section BC is greater than the value of E_r at a point along the section AD at the same radial distance from O, so that $\int \mathbf{E} \cdot d\mathbf{l}$ evaluated from B to C is greater than evaluated from A to D by an amount $\{(\partial E_r/\partial\theta)\Delta\theta\}\Delta r$. Using equation (4.2) for E_r and differentiating, the reader can show that $\int \mathbf{E} \cdot d\mathbf{l}$ evaluated from A to D to C to B to A in Figure 4.5 is given by

$$\oint \mathbf{E} \cdot d\mathbf{l} = -\left(\frac{\partial E_r}{\partial \theta}\right)\Delta\theta\Delta r = -\frac{q(1-\beta^2)}{4\pi\varepsilon_0 r^2}\left[\frac{3\beta^2 \sin\theta\cos\theta}{(1-\beta^2\sin^2\theta)^{5/2}}\right]\Delta\theta\Delta r. \tag{4.39}$$

We shall now assume that a stationary conducting coil of wire coincides with the closed curve *ABCD* in Figure 4.5, as shown in Figure 4.6. As the charge q approaches the coil *ABCD* from the left, as illustrated in Figure 4.6, the magnetic flux through the coil is in the direction upwards towards the reader and is increasing in magnitude. At the same time, when the charge q is approaching the coil *ABCD* in Figure 4.6, the electric field is greater along *BC* then along *AD* giving a finite value for the emf $\oint \mathbf{E} \cdot \mathbf{dl}$, which is given by equation (4.39). It follows from the constitutive equation $\mathbf{J} = \sigma\mathbf{E}$, that, for the experimental conditions shown in Figure 4.6, a conduction current will flow around the stationary coil *ABCD* from A to B to C to D to A. This conduction current flow in the stationary coil is due to the electric field due to the moving charge. According to the right-handed corkscrew rule, the conduction current flow from A to B to C to D to A gives rise to an extra contribution to the magnetic field in a direction that is downwards into the paper inside the coil *ABCD* in Figure 4.6. This is in such a direction as to "tend to oppose" the increase in the magnetic flux through the stationary coil *ABCD* as the classical point charge approaches the coil *ABCD* from the left in Figure 4.6. This illustrates Lenz's law.

The reader can show, using equations (4.22) and (4.35), that the rate of change of the magnetic flux Φ through the coil *ABCD*, which has area $\Delta S = r\,\Delta\theta\,\Delta r$ due to the motion of the charge q in Figure 4.6 is given by

$$
\begin{aligned}
-\frac{\partial \Phi}{\partial t} &= -\left(\frac{\partial \mathbf{B}}{\partial t}\right) \cdot \Delta \mathbf{S} = -\left(\frac{\partial B_\phi}{\partial t}\right) \Delta S \\
&= -\left(\frac{\partial B_\phi}{\partial t}\right) r\Delta\theta\Delta r = u\left(\frac{\partial B_\phi}{\partial x}\right) r\Delta\theta\Delta r \\
&= -\frac{q(1-\beta^2)}{4\pi\varepsilon_0 r^2}\left[\frac{3\beta^2 \sin\theta\cos\theta}{(1-\beta^2\sin^2\theta)^{5/2}}\right]\Delta\theta\Delta r.
\end{aligned}
\tag{4.40}
$$

The right hand sides of equations (4.39) and (4.40) are the same in agreement with Faraday's law of electromagnetic induction, according to which $\oint \mathbf{E} \cdot \mathbf{dl}$ is equal to $-\dot{\Phi}$. After the charge q has moved to the right of the stationary coil *ABCD* in Figure 4.6, the magnetic flux Φ through the coil due to the moving charge q is still upwards towards the reader in Figure 4.6, but the magnetic flux Φ through the coil is then decreasing in magnitude as q moves away from the coil *ABCD*. In this case the electric field $\mathbf{E}$ due to the moving charge q is less on the left hand side of the coil *ABCD* in Figure 4.6 than on the right hand side of the coil, so that in this case the conduction current flows from A to D to C to B to A and gives a contribution to the magnetic flux through the coil *ABCD* in a direction upwards towards the reader. This extra magnetic flux in the upward direction "tends to oppose" the reduction in the magnetic flux in the upward direction through the coil *ABCD* in Figure 4.6, as the charge q moves away from the coil *ABCD*. This is again in agreement with Lenz's law.

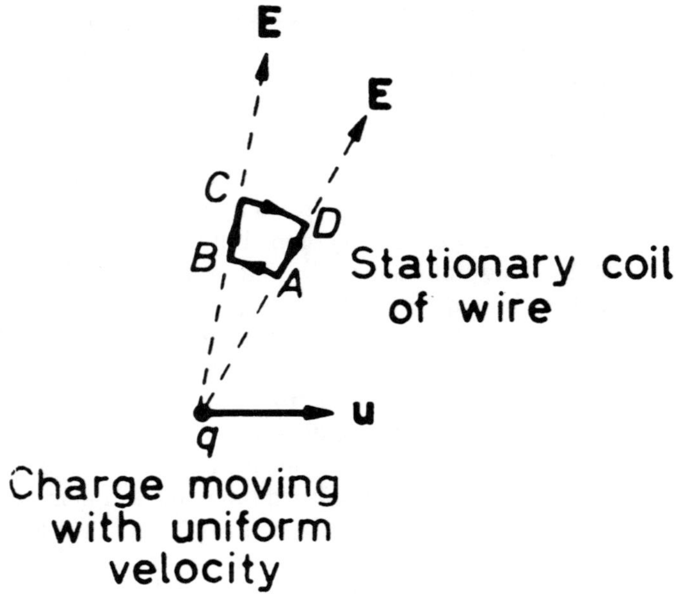

Figure 4.6. The electric field due to the charge q, which is moving with uniform velocity, is stronger along BC than along AD, which drives a conduction current around the coil from B to C to D to A to B. This direction of current flow is consistent with Lenz's law.

4.5.3. *An accelerating classical point charge*

In the case of an accelerating classical point charge, the expressions for **E** and **B**, determined from the Liénard-Wiechert potentials, are given in terms of the velocity and acceleration of the charge at its appropriate retarded position by equations (3.10) and (3.13) of Section 3.2 of Chapter 3. According to Maxwell's equations, at any field point the curl of the electric field **E** due to an isolated accelerating classical point charge, is related to the rate of change of its magnetic field **B** by the relation

$$\nabla \times \mathbf{E} = -\frac{\partial \mathbf{B}}{\partial t}. \tag{4.41}$$

An example of the electric field due to an accelerating classical point charge was shown in Figure 3.7 of Chapter 3. It can be seen from Figure 3.7 that the electric field lines are curved and have kinks in places, associated with periods when the charge was accelerating at its retarded position. The magnetic field in Figure 3.7 is in a direction perpendicular to the paper. The reader can see from the shape and number of electric field lines per square metre in Figure 3.7 that the value of the integral $\int \mathbf{E} \cdot d\mathbf{l}$ evaluated around any closed curve in Figure 3.7 at a fixed time t is finite. The rate of change of the magnetic flux $\Phi = \int \mathbf{B} \cdot d\mathbf{S}$ through that closed curve is also finite due to the motion of the accelerating classical point charge relative to the closed curve, and according to Maxwell's equations is related to the line integral of **E** around the curve by the equation

$$\oint \mathbf{E} \cdot d\mathbf{l} = -\frac{\partial \Phi}{\partial t}. \tag{4.42}$$

In the case of an isolated accelerating classical point charge, equation (4.41) is again a relation between the field vectors **E** and **B**, which in the case of an accelerating classical point charge are given by equations (3.10) and (3.13) respectively.

4.5.4. *A system of moving and accelerating classical point charges*

Consider a system of N moving and accelerating classical point charges, which build up a macroscopic charge and current distribution. Consider a field point in empty space. According to equation (4.36) the electric field $\mathbf{E}_i$ and the magnetic field $\mathbf{B}_i$ at any field point in empty space due to the ith moving charge q_i, are related by the equation

$$\nabla \times \mathbf{E}_i = -\frac{\partial \mathbf{B}_i}{\partial t}. \tag{4.43}$$

An equation, such as equation (4.43) is valid at the field point in empty space for the fields due to every one of the classical point charges in the system at the time of observation, so that we can add up the contribution of all the N charges in the macroscopic charge and current distribution, to give at a field point outside the charge distribution

$$\nabla \times \mathbf{E}_1 + \nabla \times \mathbf{E}_2 + \ldots + \nabla \times \mathbf{E}_N = -\frac{\partial \mathbf{B}_1}{\partial t} - \frac{\partial \mathbf{B}_2}{\partial t} - \ldots - \frac{\partial \mathbf{B}_N}{\partial t}.$$

Since for any vectors **C**, **D**, . . .

$$\nabla \times \mathbf{C} + \nabla \times \mathbf{D} + \ldots = \nabla \times (\mathbf{C} + \mathbf{D} + \ldots)$$

it follows that

$$\nabla \times \mathbf{E}_1 + \nabla \times \mathbf{E}_2 + \ldots + \nabla \times \mathbf{E}_N = \nabla \times (\mathbf{E}_1 + \mathbf{E}_2 + \ldots + \mathbf{E}_N).$$

Also

$$\frac{\partial \mathbf{B}_1}{\partial t} + \frac{\partial \mathbf{B}_2}{\partial t} + \ldots + \frac{\partial \mathbf{B}_N}{\partial t} = \frac{\partial}{\partial t}(\mathbf{B}_1 + \mathbf{B}_2 + \ldots + \mathbf{B}_N).$$

Hence

$$\nabla \times \mathbf{E} = -\frac{\partial \mathbf{B}}{\partial t} \tag{4.44}$$

where $\mathbf{E} = (\mathbf{E}_1 + \mathbf{E}_2 + \ldots + \mathbf{E}_N)$ is the resultant electric field and $\mathbf{B} = (\mathbf{B}_1 + \mathbf{B}_2 + \ldots + \mathbf{B}_N)$ is the resultant magnetic field at the external field point due to all the N classical point charges in the charge and current distribution.

As a typical example of a macroscopic current distribution, consider a varying conduction current in a stationary arbitrarily shaped coil and consider an external field point in empty space. The stationary positive ions in the coil give rise to electrostatic fields which contribute to the resultant electric field **E** at an external field point, but the stationary positive ions give no contribution to the resultant magnetic field **B**. Since, according to equation (1.27), the curl of any electrostatic field is zero, the stationary positive ions make no contribution to either $\nabla \times \mathbf{E}$ or $\dot{\mathbf{B}}$. The resultant values of $\nabla \times \mathbf{E}$ and $\dot{\mathbf{B}}$ at the external field point are due to the moving and accelerating conduction electrons. Equation (4.43) is a relation between the values of the resultant total fields **E** and **B** at the external field point. If the accelerations of the conduction electrons in the coil were negligible, it could be assumed, as a first approximation, that the conduction electrons were moving with uniform velocities, in which case, each individual moving conduction electron in the coil would give, in this approximation, an electric field distribution of the type shown in Figure 4.3(b), except that for electrons the charge q would be negative and the electric field lines would converge on the individual conduction electrons. Each moving conduction electron would give a contribution to $\nabla \times \mathbf{E}$ at the external field point due to the bunching of its electric field lines towards a direction perpendicular to its velocity. Each of the conduction electrons would also give a contribution to $\dot{\mathbf{B}}$ due to its motion relative to the fixed field point. In practice, the conduction electrons are continually accelerated and decelerated. They also have centripetal accelerations in regions where the coil is curved. When these accelerations are important, the electric field due to the individual accelerating conduction electrons are more complicated than in Figure 3.4(b) of Chapter 3 and the electric field lines would be curved and have kinks in them, as shown for example in Figure 3.7 of Chapter 3. It is the sums of the electric and magnetic fields of this type due to individual moving and accelerating conduction electrons, that give the resultant values of $\nabla \times \mathbf{E}$ and $\dot{\mathbf{B}}$ at the external field point. We shall return in Chapter 5 to interpret the origin of the induced emf due to a varying current in an electrical circuit using equation (3.10), which gives the electric field due to an accelerating classical point charge.

4.5.5. *Critique of an obsolete interpretation of Faraday's law of electromagnetic induction*

The reader may find in many text books that the equation

$$\nabla \times \mathbf{E} = -\dot{\mathbf{B}}. \tag{4.45}$$

is interpreted by saying that a varying magnetic field produces an electric field. The following quotation, taken from Grant and Phillips [4] is typical:

> A changing magnetic field produces an electric field according to Faraday's law.

The integral form of equation (4.45), namely

$$\oint \mathbf{E} \cdot d\mathbf{l} = -\frac{\partial \Phi}{\partial t} \tag{4.46}$$

is often interpreted by saying that a varying magnetic flux through a stationary coil produces an induced emf in the coil. Such interpretations are relics of the nineteenth century aether theories. The values of **E** and **B** in the general case can be derived independently from the values of ϕ and **A** which can be determined using the retarded potentials. We have shown that equation (4.45) is just a relation between these values of **E** and **B**. That equation (4.45) is a relation between **E** and **B** can be illustrated by introducing the potentials ϕ and **A**. According to equation (2.17) of Chapter 2

$$\mathbf{E} = -\nabla\phi - \dot{\mathbf{A}}. \tag{2.17}$$

By taking the curl of both sides of equation (2.17) and using the result that the curl of the gradient of any scalar function of position is zero, we find that

$$\nabla \times \mathbf{E} = \nabla \times \left(-\nabla\phi - \frac{\partial \mathbf{A}}{\partial t}\right) = -\nabla \times \frac{\partial \mathbf{A}}{\partial t}. \tag{4.47}$$

According to equation (2.18)

$$\mathbf{B} = \nabla \times \mathbf{A}. \tag{2.18}$$

Differentiating equation (2.18) partially with respect to time, we have

$$-\frac{\partial \mathbf{B}}{\partial t} = -\frac{\partial}{\partial t}(\nabla \times \mathbf{A}). \tag{4.48}$$

Notice immediately that both $\nabla \times \mathbf{E}$ and $\dot{\mathbf{B}}$ can be determined from the vector potential **A**. Using equations (4.47) and (4.48) we can rewrite equation (4.45) in the form

$$-\nabla \times \left(\frac{\partial \mathbf{A}}{\partial t}\right) = -\frac{\partial}{\partial t}(\nabla \times \mathbf{A}). \tag{4.49}$$

Since the operations of taking the curl of a vector and differentiating partially with respect to time are independent linear operations they commute, so that equation (4.49) is always valid for any vector. There is no need to interpret equation (4.49) by saying that the right hand side produces the left hand side or vice versa. One side of equation (4.49) is just a rearrangement of the other. Hence it is best to interpret equations (4.45) and (4.49) as relations between the components of **E** and the components of **B** and not as cause-effect relations, in which $\dot{\mathbf{B}}$ is interpreted as the cause of **E**.

When teaching electromagnetism, we can interpret equation (4.45) by saying that, if we set up the experimental conditions such that we get a varying magnetic field, then the same charge and current distributions that give rise to the varying magnetic field also give rise to an electric field whose curl is equal to $-\dot{\mathbf{B}}$. When equation (4.46) is applied to a stationary electrical circuit, if we set up the experimental conditions to get a varying magnetic flux through the stationary circuit, then the varying charge and current distributions that give rise to the varying magnetic flux also give rise to an induction electric field, which gives rise to an induced emf $\oint \mathbf{E} \cdot \mathbf{dl}$ in the stationary circuit, which is numerically equal to minus the rate of change of the magnetic flux through the circuit. We shall show in Chapter 7 that we generally only need to know the value of the emf $\oint \mathbf{E} \cdot \mathbf{dl}$ in AC circuit theory and not the value of $\mathbf{E}$ at every point in the circuit.

4.6. The equation $\nabla \times \mathbf{B} = \dot{\mathbf{E}}/c^2$ at a field point in empty space

4.6.1. *A classical point charge moving with uniform velocity*

Consider again the isolated classical point charge of magnitude q, that is and always has been moving with uniform velocity $\mathbf{u}$ along the x axis of Figure 4.1. Let $\beta = u/c$. The charge q is at the origin O in Figure 4.1 at the time of observation t, when the fields $\mathbf{E}$ and $\mathbf{B}$ are determined at the field point P, which is at a distance $\mathbf{r}$ from the origin. Choose again the spherical polar coordinate system (r, θ, ϕ) shown in Figure 4.1. The polar angle θ is the angle between $\mathbf{u}$ and $\mathbf{r}$, measured at the time of observation t. According to equations (4.4),

$$\mathbf{B} = B_\phi \hat{\boldsymbol{\phi}} = \frac{qu(1 - \beta^2) \sin \theta}{4\pi\varepsilon_0 c^2 r^2 (1 - \beta^2 \sin^2 \theta)^{3/2}} \hat{\boldsymbol{\phi}}. \tag{4.50}$$

It can be seen from equation (4.50) and Figure 3.5 of Chapter 3 that, in the special case of a classical point charge moving with uniform velocity in Figure 4.1, the magnetic field lines are closed circles with centres on the x axis, which is the direction of the uniform velocity $\mathbf{u}$ of the charge. Hence $\oint \mathbf{B} \cdot \mathbf{dl}$ evaluated around a magnetic field line is finite, showing that $\nabla \times \mathbf{B}$ is finite on a surface bounded by a magnetic field line. Since, according to equation (4.50) both B_r and B_θ are zero, the expression for $\nabla \times \mathbf{B}$ in spherical polar coordinates, given by equation (A1.41) of Appendix A1.10, reduces to

$$\nabla \times \mathbf{B} = \frac{1}{r \sin \theta} \left(\frac{\partial}{\partial \theta} (B_\phi \sin \theta) \right) \hat{\mathbf{r}} - \frac{1}{r} \frac{\partial}{\partial r} (r B_\phi) \hat{\boldsymbol{\theta}}. \tag{4.51}$$

According to equation (4.50), B_ϕ varies with θ at fixed values of r and ϕ giving a component of $\nabla \times \mathbf{B}$ in the direction of $\hat{\mathbf{r}}$, and B_ϕ varies with r at fixed values of θ and ϕ giving a component of $\nabla \times \mathbf{B}$ in the direction of $\hat{\boldsymbol{\theta}}$. Substituting for B_ϕ from equation (4.50) into equation (4.51) and carrying out the partial differentiations, the reader can show that

$$\nabla \times \mathbf{B} = \frac{qu(1-\beta^2)}{4\pi\varepsilon_0 c^2}\left[\frac{2\cos\theta(1+{}^1\!/_2\beta^2\sin^2\theta)}{r^3(1-\beta^2\sin^2\theta)^{5/2}}\,\hat{\mathbf{r}} + \frac{\sin\theta}{r^3(1-\beta^2\sin^2\theta)^{3/2}}\,\hat{\boldsymbol{\theta}}\right]. \tag{4.52}$$

According to equation (4.2)

$$\mathbf{E} = \frac{q(1-\beta^2)}{4\pi\varepsilon_0 r^2(1-\beta^2\sin^2\theta)^{3/2}}\,\hat{\mathbf{r}}. \tag{4.53}$$

The electric field **E** at the fixed field point P in Figure 4.1 varies in both magnitude and direction with time due to the motion of the charge q relative to the fixed field point. According to equation (4.24), for a classical point charge moving with uniform velocity **u**

$$\frac{\partial \mathbf{E}}{\partial t} = -(\mathbf{u}\cdot\nabla)\mathbf{E}.$$

Expressing the velocity **u** of the charge q in Figure 4.1 in terms of the unit vectors $\hat{\mathbf{r}}$ and $\hat{\boldsymbol{\theta}}$ at the field point P we have

$$\mathbf{u} = (u\cos\theta)\hat{\mathbf{r}} - (u\sin\theta)\hat{\boldsymbol{\theta}}.$$

According to equation (A1.39) of Appendix A1.10

$$\nabla = \hat{\mathbf{r}}\frac{\partial}{\partial r} + \hat{\boldsymbol{\theta}}\frac{1}{r}\frac{\partial}{\partial\theta} + \hat{\boldsymbol{\phi}}\frac{1}{r\sin\theta}\frac{\partial}{\partial\phi}.$$

Hence, since $\mathbf{E} = E_r\hat{\mathbf{r}}$, after using equation (A1.1) of Appendix A1.1 to expand $\mathbf{u}\cdot\nabla$, we find that

$$\frac{1}{c^2}\frac{\partial \mathbf{E}}{\partial t} = -\frac{1}{c^2}(\mathbf{u}\cdot\nabla)\mathbf{E} = -\frac{u\cos\theta}{c^2}\frac{\partial}{\partial r}(E_r\hat{\mathbf{r}}) + \frac{u\sin\theta}{c^2 r}\frac{\partial}{\partial\theta}(E_r\hat{\mathbf{r}}).$$

Since

$$\frac{\partial\hat{\mathbf{r}}}{\partial r} = 0; \qquad \frac{\partial\hat{\mathbf{r}}}{\partial\theta} = \hat{\boldsymbol{\theta}}$$

we have

$$\frac{1}{c^2}\frac{\partial \mathbf{E}}{\partial t} = \left(-\frac{u\cos\theta}{c^2}\frac{\partial E_r}{\partial r} + \frac{u\sin\theta}{c^2 r}\frac{\partial E_r}{\partial\theta}\right)\hat{\mathbf{r}} + \frac{u\sin\theta}{c^2 r}E_r\hat{\boldsymbol{\theta}}.$$

By substituting for E_r from equation (4.53) and carrying out the partial differentiatons, the reader can show that the above expression for $\dot{\mathbf{E}}/c^2$ reduces to the right hand side of equation (4.52), showing that at a field point in empty space

$$\nabla \times \mathbf{B} = \frac{1}{c^2}\frac{\partial \mathbf{E}}{\partial t} = \mu_0\varepsilon_0\frac{\partial \mathbf{E}}{\partial t} \tag{4.54}$$

where we have used $c^2 = 1/\mu_0\varepsilon_0$. Equation (4.54) is a relation we have derived from the expressions for the electric field **E** and the magnetic field **B** due to a classical point charge moving with uniform velocity, which we had previously derived in Chapter 3 using the Liénard-Wiechert potentials.

In the case of an isolated accelerating classical point charge, the electric and magnetic fields are given by equations (3.10) and (3.13) of Chapter 3. These electric and magnetic fields are more complicated than is the case when the charge is moving with uniform velocity. For example the electric field lines, due to an accelerating classical point charge, can be curved and have kinks in them associated with periods when the charge was accelerating at its retarded position. However equation (4.54) is still valid at any field point in empty space. It is again a relation between the field vectors **E** and **B**.

4.6.2. *The low velocity approximation*

It is of interest to note that, whereas there is a factor $(1/c^2)$ on the right hand side of equation (4.54) relating $\nabla \times \mathbf{B}$ and $\dot{\mathbf{E}}$, there are no factors involving c in equation (4.36) which relates $\nabla \times \mathbf{E}$ and $\dot{\mathbf{B}}$. Consider again the example of the charge q moving with uniform velocity $\mathbf{u}$ shown in Figure 4.1. Expanding the denominators in equations (3.25) and (3.28) using the binomial theorem, we find that at the field point P at the time of observation we have

$$\mathbf{E} = \frac{q\mathbf{r}}{4\pi\varepsilon_0 r^3}\left\{1 + \beta^2\left(\frac{3}{2}\sin^2\theta - 1\right) + \ldots\right\}, \tag{4.55}$$

$$\mathbf{B} = \frac{q\mathbf{u}\times\mathbf{r}}{4\pi\varepsilon_0 r^3}\left(\frac{1}{c^2}\right)\left\{1 + \beta^2\left(\frac{3}{2}\sin^2\theta - 1\right) + \ldots\right\}. \tag{4.56}$$

According to equation (3.29)

$$\mathbf{B} = \frac{\mathbf{u}\times\mathbf{E}}{c^2}. \tag{4.57}$$

In the low velocity limit when $\beta \ll 1$, the zero order terms are

$$\mathbf{E}_0 = \frac{q\mathbf{r}}{4\pi\varepsilon_0 r^3}; \qquad \mathbf{B}_0 = \frac{q\mathbf{u}\times\mathbf{r}}{4\pi\varepsilon_0 r^3}\left(\frac{1}{c^2}\right). \tag{4.58}$$

It is straightforward for the reader to show using the methods used in Section 4.6.1 that when $\beta \ll 1$

$$\nabla \times \mathbf{B}_0 = \frac{q}{4\pi\varepsilon_0}\left(\frac{1}{c^2}\right)\left(\frac{2u\cos\theta}{r^3}\hat{\mathbf{r}} + \frac{u\sin\theta}{r^3}\hat{\boldsymbol{\theta}}\right) \tag{4.59}$$

$$\frac{\partial\mathbf{E}_0}{\partial t} = \frac{q}{4\pi\varepsilon_0}\left(\frac{2u\cos\theta}{r^3}\hat{\mathbf{r}} + \frac{u\sin\theta}{r^3}\hat{\boldsymbol{\theta}}\right). \tag{4.60}$$

Comparing equations (4.59) and (4.60) we see that

$$\nabla \times \mathbf{B}_0 = \frac{1}{c^2}\frac{\partial\mathbf{E}_0}{\partial t} \tag{4.61}$$

showing that equation (4.54) is valid for the zero order terms. The zero order contribution $\mathbf{B}_0$ to the magnetic field gives a finite curl since the magnetic field lines are closed circles in the special case of the charge moving with uniform velocity shown in Figure 4.1. The zero order contribution $\mathbf{E}_0$ to the electric field gives a finite value for $\dot{\mathbf{E}}_0$, due to the motion of the charge relative to the fixed field point. Since according to equations (4.58), B_0 is of order uE_0/c^2 it is reasonable to find that we must multiply $\dot{\mathbf{E}}_0$ by $(1/c^2)$ to be equal to the value of $\nabla \times \mathbf{B}_0$ given by equation (4.59).

When we come to calculate the value of $\nabla \times \mathbf{E}$ at the field point P in Figure 4.1 at the fixed time t, $\nabla \times \mathbf{E}_0$ is zero since, at a fixed instant of time the expression for $\mathbf{E}_0$ is the same as Coulomb's law. Hence the first term that contributes to $\nabla \times \mathbf{E}$ in the low velocity limit is $\nabla \times \mathbf{E}_2$, where

$$\mathbf{E}_2 = \frac{q\mathbf{r}}{4\pi\varepsilon_0 r^3}\beta^2\left(\frac{3}{2}\sin^2\theta - 1\right).$$

Evaluating $\nabla \times \mathbf{E}_2$ using the methods used in Section 4.5.1, the reader can show that

$$\nabla \times \mathbf{E}_2 = -\beta^2\left(\frac{q}{4\pi\varepsilon_0 r^3}\right) 3\cos\theta\sin\theta\,\hat{\boldsymbol{\phi}}. \tag{4.62}$$

The zero order contribution $\mathbf{B}_0$ to the magnetic field gives a finite contribution to $\dot{\mathbf{B}}$ due to the motion of the charge q relative to the field point in Figure 4.1. Using the method used in Section 4.5.1, the reader can show that

$$\frac{\partial \mathbf{B}_0}{\partial t} = -u\frac{\partial \mathbf{B}_0}{\partial x} = \beta^2\left(\frac{q}{4\pi\varepsilon_0 r^3}\right) 3\cos\theta\sin\theta\,\hat{\boldsymbol{\phi}}. \tag{4.63}$$

Comparing equations (4.62) and (4.63) we see that, in the low velocity limit

$$\nabla \times \mathbf{E}_2 = -\frac{\partial \mathbf{B}_0}{\partial t}. \tag{4.64}$$

Since $\nabla \times \mathbf{E}$ depends on $\mathbf{E}_2$ not $\mathbf{E}_0$ in the low velocity limit and E_2 is proportional to E_0/c^2, it is reasonable to find that we do not need any factors of c in equation (4.64) since, according to equation (4.58) B_0 is also proportional to E_0/c^2 so that the factors of β^2 in equations (4.62) and (4.63) cancel out in equation (4.64).

4.6.3. *A system of moving and accelerating classical point charges*

Consider the system of N moving and accelerating classical point charges that build up the macroscopic charge and current distribution, which we discussed in Section 4.5.4. Corresponding to equation (4.43) we now have

$$\nabla \times \mathbf{B}_i = \frac{1}{c^2}\frac{\partial \mathbf{E}_i}{\partial t}. \tag{4.65}$$

Adding up the contributions of all the N charges we find that at a field point in empty space

$$\nabla \times \mathbf{B} = \frac{1}{c^2}\frac{\partial \mathbf{E}}{\partial t} \tag{4.66}$$

where $\mathbf{B} = (\mathbf{B}_1 + \mathbf{B}_2 + \ldots + \mathbf{B}_N)$ is resultant magnetic field and $\mathbf{E} = (\mathbf{E}_1 + \mathbf{E}_2 + \ldots + \mathbf{E}_N)$ is the resultant electric field at the external field point in empty space. Equation (4.66) is a relation between the resultant electric and magnetic fields at a field point in empty space. We shall now go on to show that equation (4.66) must be extended by adding the $\mu_0\mathbf{J}$ term to the right hand side when there is a current distribution of current density $\mathbf{J}$ at the field point.

4.7. Application of the equation $\oint \mathbf{B} \cdot d\mathbf{l} = \dot{\Psi}/c^2$ to the fields of a point charge of zero dimensions moving with uniform velocity

In this section we shall assume that the moving charge, of magnitude q, in Figure 4.1 is an idealized point charge of zero dimensions, which is and always has been moving with uniform velocity $\mathbf{u}$ along the x axis of Figure 4.1. The charge q is at the origin of the spherical polar coordinate system at the time of observation t. Let $\hat{\mathbf{r}}$, $\hat{\boldsymbol{\theta}}$ and $\hat{\boldsymbol{\phi}}$ be unit vectors in the directions of increasing r, θ and ϕ respectively. To simplify the discussions we shall assume that $u \ll c$ so that $\beta = u/c \ll 1$. According to equation (3.34) of Chapter 3, when $\beta \ll 1$, the magnetic field at the field point P in Figure 4.1, at a distance $\mathbf{r}$ from the position of the charge at the time of observation t, is given to a good approximation by the Biot-Savart approximation, which is

$$\mathbf{B} = \frac{q\mathbf{u} \times \mathbf{r}}{4\pi\varepsilon_0 c^2 r^3} = \frac{qu \sin\theta}{4\pi\varepsilon_0 c^2 r^2}\hat{\boldsymbol{\phi}}. \tag{4.67}$$

The angle θ in equation (4.67) is the angle between $\mathbf{u}$ and $\mathbf{r}$.

According to equation (3.33) of Chapter 3, when $\beta \ll 1$ the electric field $\mathbf{E}$ at the field point P in Figure 4.1 at the time of observation t is given to a good approximation by

$$\mathbf{E} = \frac{q\mathbf{r}}{4\pi\varepsilon_0 r^3} = \frac{q}{4\pi\varepsilon_0 r^2}\hat{\mathbf{r}} \tag{4.68}$$

where $\mathbf{r}$ is again measured from the position of the point charge at the time of observation t.

It is important for the reader to realize that there is an important difference between the electric field $\mathbf{E}$ and the magnetic field $\mathbf{B}$ due to a point charge moving with uniform velocity $\mathbf{u}$. The electric field lines diverge radially from the position of the charge at the time of observation t as shown in Figure 3.4(b). The electric field lines on the x axis in Figure 4.1, which is the line of motion of the charge, are in opposite directions on either side of the moving charge. However, the magnetic field lines just off the x axis are in the same direction, given by the right handed corkscrew rule, on both sides of the moving charge, as shown in Figure 3.5.

Consider a circular disk shaped surface S of radius a, with its centre on the x axis at a distance x from the origin and with the field point P on its circumference as shown in Figure 4.1. We shall start by evaluating the line integral $\oint \mathbf{B} \cdot d\mathbf{l}$ around the circumference of the surface S in the direction of the magnetic field due to the moving charge in Figure 4.1 at the instant the charge is at the origin. The positive direction for an element of area $d\mathbf{S}$ of the surface S is then the direction in which a right-handed corkscrew would advance if it were rotated in the direction of $d\mathbf{l}$. This is in the $+x$ direction from the right hand side of the surface S in Figure 4.1. Using equation (4.67) we have

$$\oint \mathbf{B} \cdot d\mathbf{l} = 2\pi a B_\phi = \frac{qua \sin\theta}{2\varepsilon_0 c^2 r^2} = \frac{qua^2}{2\varepsilon_0 c^2 (a^2 + x^2)^{3/2}} . \tag{4.69}$$

The electric flux $\Psi = \int \mathbf{E} \cdot d\mathbf{S}$ through the surface S, when the charge q is at the origin in Figure 4.1, can be obtained by evaluating the integral in equation (4.10) from $\theta = 0$ to $\theta = \theta$. Neglecting terms in β^2 compared with unity in equation (4.10), we find that

$$\Psi = \frac{q}{2\varepsilon_0} \int_0^\theta \sin\theta \, d\theta = \frac{q}{2\varepsilon_0} (1 - \cos\theta) \tag{4.70}$$

$$= \frac{q}{2\varepsilon_0} \left(1 - \frac{x}{(a^2 + x^2)^{1/2}} \right) . \tag{4.71}$$

Applying equation (4.23) to equation (4.71) we find that

$$\frac{\partial \Psi}{\partial t} = - u \frac{\partial \Psi}{\partial x} = \frac{qua^2}{2\varepsilon_0 (a^2 + x^2)^{3/2}} . \tag{4.72}$$

Comparing equations (4.69) and (4.72), we see that

$$\oint \mathbf{B} \cdot d\mathbf{l} = \frac{1}{c^2} \frac{\partial \Psi}{\partial t} . \tag{4.73}$$

Equation (4.73) was derived from the values of **E** and **B** given by equations (4.68) and (4.67) respectively. It is left as an exercise for the reader to show that equation (4.73) can also be applied after the charge of zero dimensions has passed through the surface S in Figure 4.1.

Problem. Use equation (4.11) to show that, if u tends to c, the electric flux through the surface S in Figure 4.1 is

$$\Psi = \frac{q}{2\varepsilon_0} \left[1 - \frac{\cos\theta}{(1 - \beta^2 \sin^2\theta)^{1/2}} \right] . \tag{4.74}$$

Express $\cos\theta$ and $\sin\theta$ in cartesian coordinates and apply equation (4.23) to determine $\dot{\Psi}$. Use equation (4.4) to evaluate $\oint \mathbf{B} \cdot d\mathbf{l}$ to show that

$$\oint \mathbf{B} \cdot d\mathbf{l} = \frac{1}{c^2} \frac{\partial \Psi}{\partial t} = \frac{qu(1 - \beta^2) \sin^2\theta}{2\varepsilon_0 r c^2 (1 - \beta^2 \sin^2\theta)^{3/2}} \tag{4.75}$$

where r is the distance from the charge to a point on the circumference of the surface S. This is in agreement with equation (4.73).

We shall now show that equation (4.73) is not valid and must be extended when the idealized point charge is actually crossing the surface S in Figure 4.1. According to equation (4.71), when the charge q of zero dimensions is approaching the surface S in Figure 4.1 from the left, the electric flux Ψ through the surface S increases continuously until, just before the charge reaches the surface S, the electric flux reaches the value $+q/2\varepsilon_0$, as illustrated in Figure 4.7(a). After the point charge of zero dimensions has passed completely through the surface S, the electric field lines go through the surface S in the opposite direction to previously, as shown in Figure 4.7(b). According to equation (4.71), just after the charge has passed through the surface S, $\Psi = -q/2\varepsilon_0$. The variation of $\Psi = \int \mathbf{E} \cdot d\mathbf{S}$ through the stationary surface S is shown for various values of x_0 in Figure 4.8, where x_0 is the distance from the centre of the circular disk shaped surface S to the position of the moving charge in Figure 4.1. The distance x_0 is negative for the position of the moving charge in Figure 4.1. It can be seen from Figure 4.8 that there is a discontinuity of $-q/\varepsilon_0$ in the electric flux Ψ through the surface S in Figure 4.1, when the idealized point charge of zero dimensions and of magnitude q is crossing the surface S at $x_0 = 0$, so that $\dot{\Psi}$ is equal to $-\infty$ at that instant. However,

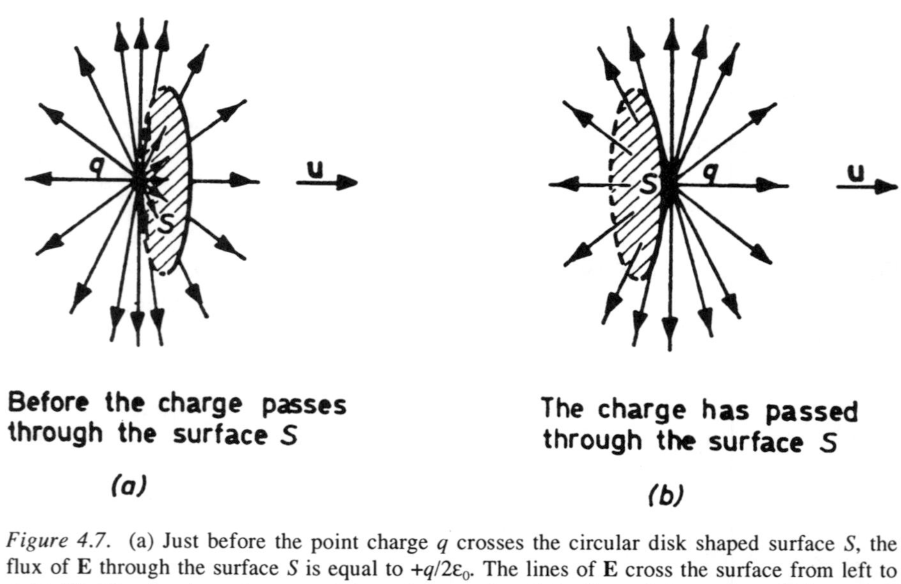

Figure 4.7. (a) Just before the point charge q crosses the circular disk shaped surface S, the flux of $\mathbf{E}$ through the surface S is equal to $+q/2\varepsilon_0$. The lines of $\mathbf{E}$ cross the surface from left to right. (b) After the charge has passed through the surface S, the lines of $\mathbf{E}$ pass through the surface from right to left. There is a discontinuity in the flux of $\mathbf{E}$, when the point charge passes through the surface S, which is equal to $-q/\varepsilon_0$, whatever the speed of the charge. There is however no discontinuity in the magnitudes and directions of $\mathbf{B}$ and $\nabla \times \mathbf{B}$ at a field point on the circumference of the surface S when the charge q crosses the surface S.

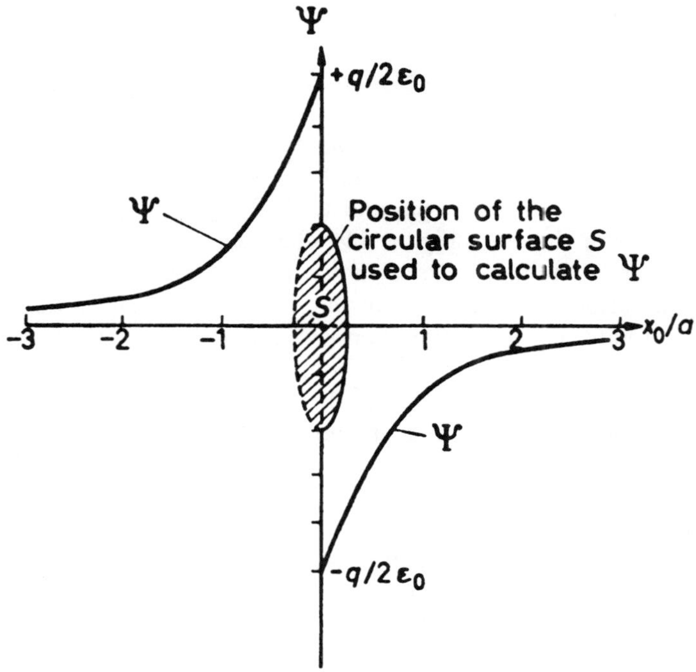

Figure 4.8. The variation of Ψ the flux of **E** through the circular disk surface S in Figure 4.1 with the position of the moving point charge, x_0 is the distance of the charge from the surface S. There is a discontinuity of $-q/\varepsilon_0$ in Ψ when the point charge crosses the surface. (For a charge of finite size the variation of Ψ with x_0 is similar to that in Figure 4.10.)

according to equation (4.69) $\oint \mathbf{B} \cdot d\mathbf{l}$ is still finite and equal to $qu/2\varepsilon_0 c^2 a$. Since the total flux of **E** from a moving point charge of magnitude q is equal to q/ε_0, whatever the speed and acceleration of the charge, the discontinuity in Ψ is $-q/\varepsilon_0$ wherever the idealized point charge of zero dimensions crosses the surface S in Figure 4.1. If the charge q passes outside the disk-shaped surface S in Figure 4.1, the electric flux through the surface S is zero when the charge q is in the plane of the surface S, so that there is no discontinuity in the electric flux Ψ through the surface S, and equation (4.73) is valid at all times in this case.

Following Lorentz, we have been assuming in the text that a classical point charge is a continuous charge distribution of finite but exceedingly small dimensions. To illustrate how equation (4.73) must be extended when a charge distribution is crossing the surface S in Figure 4.1, we shall now extend the analysis to a charge distribution of finite dimensions, by considering the example of a line of continuous charge, that is moving with a uniform velocity $u \ll c$ and which crosses the surface S, as shown in Figure 4.9. We can then proceed to the limit of a moving line of charge of finite but exceedingly small length. This will be a simplified version of our model of a moving classical point charge.

4.8. A moving line of charge

4.8.1. *A moving line of charge of finite length*

Consider a thin line of charge that is of finite length $2L$ and has a charge of λ coulombs per metre length. Let the line of charge move with a uniform velocity $u \ll c$ in a direction parallel to its length along the x axis of the coordinate system shown in Figure 4.9, The field point P is on the y axis at a distance a from the origin O, as shown in Figure 4.9. Consider a circular disk-shaped surface S of radius a in the x = 0 plane, with its centre at the origin O and with the field point P on its circumference, as shown in Figure 4.9. Let the mid-point of the line of charge have coordinates (x_0, 0, 0). Note that in this section the origin O is at the centre of the circular disk-shaped surface S in Figure 4.9 and not at the position of the charge.

The magnetic field $\mathbf{B}$ at the field point P in Figure 4.9 can be derived using the Biot-Savart law, and is given by the standard expression derived in elementary text books for the magnetic field due to the current flowing in a thin straight wire of finite length. For the experimental conditions shown in Figure 4.9, with $I = \lambda u$ we have from the Biot-Savart law

$$\mathbf{B} = \frac{\lambda u}{4\pi\varepsilon_0 c^2 a}(\cos\theta_1 + \cos\theta_2). \tag{4.76}$$

The magnetic field lines due to the moving line of charge are closed circles with centres on the x axis and direction given by the right-handed corkscrew rule, as shown in Figure 4.9. Integrating around the circumference of the

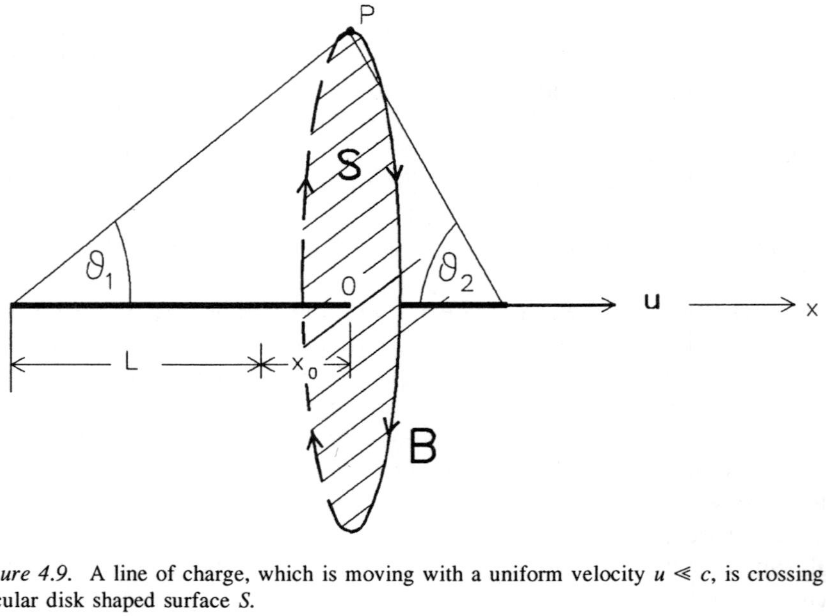

Figure 4.9. A line of charge, which is moving with a uniform velocity $u \ll c$, is crossing the circular disk shaped surface S.

circular disk-shaped surface S in Figure 4.9 in the direction of **B**, we have

$$\oint \mathbf{B} \cdot d\mathbf{l} = 2\pi a B = \frac{\lambda u}{2\varepsilon_0 c^2}(\cos\theta_1 + \cos\theta_2). \tag{4.77}$$

When the line integral is evaluated in the direction of the magnetic field, the positive direction for an element of area d**S** of the surface S is in the $+x$ direction from the surface S in Figure 4.9. It follows from the right-handed corkscrew rule that the direction of the magnetic field lines are as shown in Figure 4.9, whatever the position of the moving line of charge. Equations (4.76) and (4.77) can be applied for all positions of the moving line of charge, provided θ_1 and θ_2 are chosen appropriately.

The magnitude of the electric flux Ψ through the surface S in Figure 4.9 is sketched for various positions of the moving line of charge in Figure 4.10. Before the line of charge reaches the surface S from the left, the electric field lines go through the surface S in Figure 4.9 in the $+x$ direction so that $\Psi = \int \mathbf{E} \cdot d\mathbf{S}$ is positive and is increasing in magnitude as the line of charge approaches the surface S. The variation of Ψ with x_0, the x coordinate of the mid-point of the line of charge in the region $x_0 < -L$ is given by the section RV of the curve in Figure 4.10. In this region, Ψ is positive. By applying

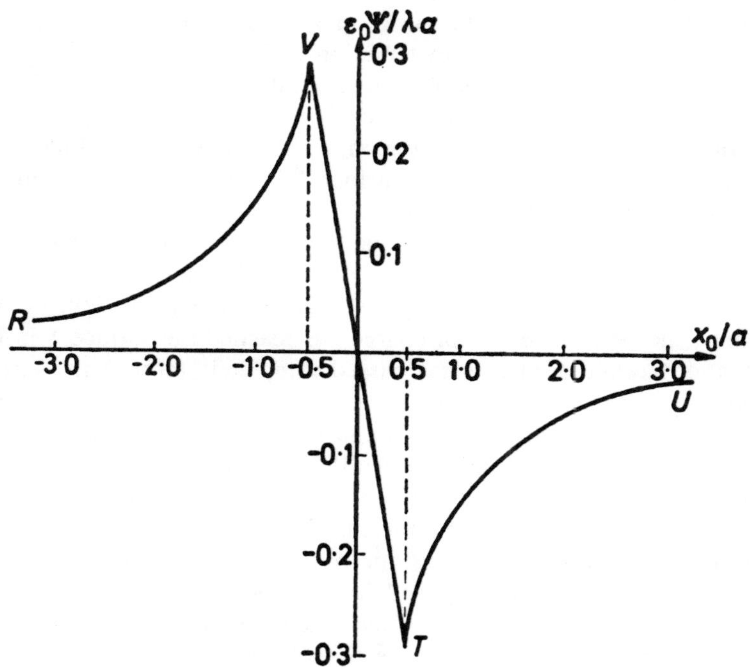

Figure 4.10. The variation of Ψ, the flux of **E** through the circular disk shaped surface S in Figure 4.9 for various positions of the moving line of charge. (Actually $\varepsilon_0\Psi/\lambda a$ is plotted against x_0/a, where λ = charge/unit length, and a, the distance of the field point from the x axis, is equal to the length of the charge.)

equation (4.54) to each element of the line of charge, then summing and following the method of Section 4.6.3, the reader can show that before the line of charge reaches the surface S in Figure 4.9 and after the line of charge has passed completely through the surface S, the equation

$$\nabla \times \mathbf{B} = \frac{1}{c^2}\frac{\partial \mathbf{E}}{\partial t} \tag{4.78}$$

is valid at the field point P, so that

$$\oint \mathbf{B} \cdot \mathbf{dl} = \frac{1}{c^2}\frac{\partial \Psi}{\partial t} . \tag{4.79}$$

When the line of charge is crossing the surface S, as shown in Figure 4.9, the charge to the left of the surface S still gives an electric field through the surface S in the $+x$ direction and hence gives a positive contribution of the electric flux Ψ through the surface S. However, the electric field due to the charge to the right of the surface S in Figure 4.9 is in the $-x$ direction and gives a negative contribution to Ψ. Hence Ψ starts decreasing, when the line of charge starts crossing the surface S, reaching zero for the case when $x_0 = 0$. The variation of Ψ with x_0, the position of the mid-point of the moving line of charge, when the line of charge is crossing the surface S in Figure 4.9 is given by the section VT of the curve in Figure 4.10. In this region $\dot{\Psi}$ is negative. However, according to equation (4.77) $\oint \mathbf{B} \cdot \mathbf{dl}$ is still positive, showing that equation (4.79) is no longer valid and must be extended when the line of charge is crossing the surface S.

The variation of Ψ with x_0 in the region $x_0 > L$, that is after the line of charge has passed completely through the surface S in Figure 4.9, is given by the section TU of the curve in Figure 4.10. It can be seen that in this region Ψ is negative, but Ψ is getting less negative with increasing x_0 so that $\dot{\Psi}$ is positive and in this region equation (4.79) is valid again.

Since we have shown earlier using equation (4.54) that equations (4.78) and (4.79) are valid before the line of charge has reached the surface S in Figure 4.9 and after the line of charge has passed completely through the surface S, we need only consider in detail the case shown in Figure 4.9 when the line of charge is actually passing through the surface S. Let the line of charge move an infinitesimal distance $\mathrm{d}x = u\,\mathrm{d}t$ in an infinitesimal time $\mathrm{d}t$. The change in the electric flux through the surface S in Figure 4.9 in the time $\mathrm{d}t$ is the same as removing, at a fixed time, an infinitesimal section of length $\mathrm{d}x$ and of charge $\lambda\,\mathrm{d}x$ from the left hand end of the line of charge and adding it to the right hand end. It follows from equation (4.70) that the change $\mathrm{d}\Psi_1$ in the electric flux through the surface S in Figure 4.9, when the element of charge $\lambda\,\mathrm{d}x$ is removed from the left hand end of the line of charge is

$$\mathrm{d}\Psi_1 = -\frac{\lambda \mathrm{d}x}{2\varepsilon_0}(1 - \cos\theta_1) \tag{4.80}$$

where θ_1 is shown in Figure 4.9. When the element of charge $\lambda\,\mathrm{d}x$ is added to the right hand end of the line of charge in Figure 4.9, it gives an electric field in the $-x$ direction and its contribution to the electric flux through the surface S is negative. According to equation (4.70) this change $d\Psi_2$ in the electric flux through the surface S is

$$d\Psi_2 = -\frac{\lambda dx}{2\varepsilon_0}(1 - \cos\theta_2). \tag{4.81}$$

Adding equations (4.80) and (4.81), putting $dx = u\,dt$ and then dividing by c^2dt we obtain

$$\begin{aligned}\frac{1}{c^2}\frac{\partial\Psi}{\partial t} &= -\frac{\lambda u}{2\varepsilon_0 c^2}(1 - \cos\theta_1 + 1 - \cos\theta_2)\\ &= \frac{\lambda u}{2\varepsilon_0 c^2}(\cos\theta_1 + \cos\theta_2 - 2).\end{aligned} \tag{4.82}$$

According to equation (4.77), for the conditions shown in Figure 4.9

$$\oint \mathbf{B}\cdot d\mathbf{l} = \frac{\lambda u}{2\varepsilon_0 c^2}(\cos\theta_1 + \cos\theta_2). \tag{4.83}$$

Comparing equations (4.82) and (4.83), we see that, for the case shown in Figure 4.9,

$$\oint \mathbf{B}\cdot d\mathbf{l} = \frac{1}{c^2}\frac{\partial\Psi}{\partial t} + \frac{\lambda u}{\varepsilon_0 c^2} = \mu_0\left\{\varepsilon_0\frac{\partial\Psi}{\partial t} + I\right\} \tag{4.84}$$

where $I = \lambda u$ is the electric current crossing the surface S and $1/c^2 = \mu_0\varepsilon_0$. We see that if we start with equations (4.78) and (4.79) as relations between the field vectors **E** and **B**, then the need for consistency when there is a moving charge distribution crossing the surface S leads us to equation (4.84) which is the integral form of equation (1.118) of Chapter 1.

The negative value of $\dot{\Psi}/c^2$ in the region $-L < x_0 < +L$ is due to the fact that, when an infinitesimal element of length dx and of charge $dq = \lambda\, dx$ crosses the surface S in Figure 4.9 in an infinitesimal time dt, its contribution to the total electric field changes from the $+x$ to the $-x$ direction and its contribution to the electric flux through the surface S changes from $+dq/2\varepsilon_0$ to $-dq/2\varepsilon_0$. It was shown in Section 4.7 that when a point charge of magnitude q and zero dimensions crosses the surface S in Figure 4.1 there is no discontinuity in $\oint \mathbf{B}\cdot d\mathbf{l}$, even though there is a discontinuity of $-q/\varepsilon_0$ in the flux of **E** through the surface. Hence, when the infinitesimal elements of charge $dq = \lambda\, dx$ cross the surface S in Figure 4.9, though the changes of $-dq/\varepsilon_0$ in Ψ contribute to the value of $\dot{\Psi}/c^2$ given by equation (4.82), the changes of $-dq/\varepsilon_0$ in Ψ are not associated with any corresponding discontinuities in $\oint \mathbf{B}\cdot d\mathbf{l}$. Hence it is reasonable to find that, if the contributions of the successive changes of $-dq/\varepsilon_0$ in the electric flux Ψ, when successive elements of charge dq cross the surface S in Figure 4.9, are included in the values of $\dot{\Psi}/c^2$, to which they make a contribution of $-dq/\varepsilon_0 \div c^2dt = -I/\varepsilon_0 c^2 = -\lambda u/\varepsilon_0 c^2$ then, since the elements of charge dq crossing the surface S do not give corresponding changes in $\oint \mathbf{B} \cdot d\mathbf{l}$, to get consistency we must compensate for the inclusion of their contributions to $\dot{\Psi}/c^2$ by adding $+\lambda u/\varepsilon_0 c^2$ to $\dot{\Psi}/c^2$ to give equation (4.84). The terms in equation (4.82) involving $\cos\theta_1$ and $\cos\theta_2$ arise from changes in the positions of the ends of the ends of the moving line of charge in Figure 4.9, when the line of charge is crossing the surface S.

As an exercise the reader can apply the method we used to determine the value of $\dot{\Psi}/c^2$ given by equation (4.82) to determine $\dot{\Psi}/c^2$ for the cases before the line of charge has reached the surface S in Figure 4.9 and after the line of charge has passed completely through the surface S. Then, by comparing the values obtained with the values of $\oint \mathbf{B} \cdot d\mathbf{l}$ the reader can confirm that equation (4.79) is valid in these cases.

It will now be assumed that the moving line of charge in Figure 4.9 is accelerating and moving in an arbitrary direction when it is crossing the surface S. It was pointed out in Section 4.2 that the total flux of $\mathbf{E}$ from a classical point charge of magnitude q is always equal to q/ε_0, whatever the velocity and acceleration of the charge. Hence, it does not matter where the line of charge crosses the surface S in Figure 4.9 and what its velocity and acceleration are at that instant, when the element of charge of magnitude dq crosses the surface S its contribution to Ψ, the electric flux through the surface S, changes by $-dq/\varepsilon_0$. This gives a contribution to $\dot{\Psi}$ equal to $-dq/\varepsilon_0 \div dt = I/\varepsilon_0$ where $I = dq/dt$ is the current crossing the surface S. Hence the compensation term $I/\varepsilon_0 c^2$ in equation (4.84) is the same wherever the line of charge is crossing the surface S and whatever the velocity and acceleration of the line of charge are at that instant. Hence in the general case, when the line of charge is crossing the surface S in Figure 4.9

$$\oint \mathbf{B} \cdot d\mathbf{l} = \frac{1}{c^2}\frac{\partial \Psi}{\partial t} + \frac{I}{\varepsilon_0 c^2} = \mu_0 \left\{ \varepsilon_0 \frac{\partial \Psi}{\partial t} + I \right\} \tag{4.85}$$

where $\Psi = \int \mathbf{E} \cdot d\mathbf{S}$ is evaluated over this surface S and $I = \int \mathbf{J} \cdot d\mathbf{S}$ is the total current crossing the surface S. If the line of charge does not cross the surface S

$$\oint \mathbf{B} \cdot d\mathbf{l} = \frac{1}{c^2}\frac{\partial \Psi}{\partial t} . \tag{4.86}$$

If the line of charge in Figure 4.9 were of infinite length, $\dot{\Psi}$ would be zero and equation (4.85) would reduce to

$$\int \mathbf{B} \cdot d\mathbf{l} = \mu_0 I$$

which is an example of Ampère's circuital theorem.

Problem. Divide the moving line of charge in Figure 4.9 into infinitesimal elements of length. The electric flux due to each element of charge is given by equation (4.71). Integrate to show that for $x_0 < -L$

$$\Psi = \frac{\lambda}{2\varepsilon_0}\left[2L - \{(x_0 - L)^2 + a^2\}^{1/2} + \{(x_0 + L)^2 + a^2\}^{1/2}\right] \tag{4.87}$$

and for $x_0 > L$

$$\Psi = -\frac{\lambda}{2\varepsilon_0}\left[2L - \{(x_0 + L)^2 + a^2\}^{1/2} + \{(x_0 - L)^2 + a^2\}^{1/2}\right]. \tag{4.88}$$

When $-L < x_0 < L$, the contributions of the charge on opposite sides of the surface S to Ψ, the electric flux through S, have opposite signs. By adapting the expressions for Ψ for $x_0 < -L$ and $x_0 > L$ for the sections to the left and to the right of S and then combining the results, show that for $-L < x_0 < L$

$$\Psi = -\frac{\lambda}{2\varepsilon_0}\left[+2x_0 + \{(x_0 - L)^2 + a^2\}^{1/2} - \{(x_0 + L)^2 + a^2\}^{1/2}\right]. \tag{4.89}$$

Plot variation of Ψ against x_0 for the case when $a = 2L$ to obtain Figure 4.10. Adapt equation (4.23) to determine the values of $\dot{\Psi}/c^2$. Then by comparing your values of $\dot{\Psi}/c^2$ with the values of $\oint \mathbf{B} \cdot d\mathbf{l}$ given by equation (4.77), show that equation (4.79) is valid for $x_0 < L$ and $x_0 > L$ and show that equation (4.84) is valid for $-L < x_0 < L$.

4.8.2. *A moving classical point charge*

If the length of the moving line of charge in Figure 4.9 is kept finite but made exceedingly small, it corresponds to a special case of our model of a classical point charge. In the limit, when the length $2L$ of the moving line of charge tends to zero, the variation of Ψ with x_0 in Figure 4.10 tends to the variation of Ψ with x_0 for a point charge of zero dimensions shown in Figure 4.8. However, since, on our model of a classical point charge, $2L$ remains finite, the decrease of Ψ between the points V and T in Figure 4.10 does not quite become discontinuous, whereas when an idealised point charge of zero dimensions is crossing the surface S in Figure 4.1, $\dot{\Psi}$ is equal to $-\infty$ at the instant the charge is crossing the surface. It can be seen that we have removed this infinity in $\dot{\Psi}$ by treating a classical point charge as the limiting case of a line of charge of length $2L$, when $2L$ is made exceedingly small but is kept finite to fit in with our model of a classical point charge. It is equation (4.85) that is valid when the classical point charge is crossing the surface S in Figure 4.9.

It is of interest to see why, in the absence of magnetic monopoles, it is not necessary to add any extra terms to the equations

$$\nabla \times \mathbf{E} = -\dot{\mathbf{B}} \tag{4.90}$$

$$\oint \mathbf{E} \cdot d\mathbf{l} = -\dot{\mathbf{\Phi}} \tag{4.91}$$

when there is a charge and current distribution at the field point. Consider the idealized point charge of zero dimensions, that is approaching the surface

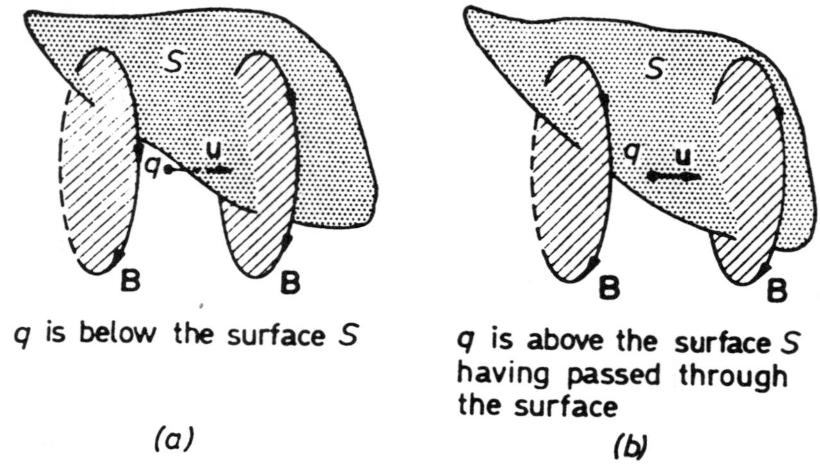

Figure 4.11. The charge q passes through the surface S. The magnetic field lines are upwards out of the surface both before and after the charge crosses the surface.

S in Figure 4.11(a) with uniform velocity. The charge has passed completely through the surface S in Figure 4.11(b). Typical magnetic field lines before and after the charge has passed through the surface S are shown in Figures 4.11(a) and 4.11(b) respectively. It can be seen that the lines of **B** are closed circles which, according to the right-handed corkscrew rule, are in the same direction for field points in front of and behind the moving charge in Figures 4.11(a) and 4.11(b). Hence there is no change in the direction of the flux of **B** through the surface S, when the point charge of zero dimensions crosses the surface S and there is no discontinuity in the flux of **B** to compensate for when evaluating $\dot{\Phi}$, where $\Phi = \int \mathbf{B} \cdot d\mathbf{S}$ is the flux of **B** through the surface S. If magnetic monopoles played a significant role in classical electromagnetism, equations (4.90) and (4.91) would have to be extended. The interested reader is referred to Rosser [5].

4.8.3. *A field point inside a macroscopic charge distribution*

Consider the continuous macroscopic charge distribution that is moving with uniform velocity **u** in Figure 4.12. Divide the current distribution into a number of lines of charge parallel to **u**. Consider an arbitrary surface S inside the moving charge distribution in Figure 4.12. By applying to the surface S, equation (4.85) for the lines of charge crossing S and equation (4.86) for the lines of charge not crossing S and then adding, we have

$$\Sigma \oint_S \mathbf{B}_i \cdot d\mathbf{l} = \mu_0 \Sigma' I_i + \mu_0 \varepsilon_0 \Sigma \int_S \dot{\mathbf{E}}_i \cdot d\mathbf{S} \tag{4.92}$$

where Σ' means summing only over the lines of charge actually crossing the surface S. Equation (4.92) can be expressed in the form

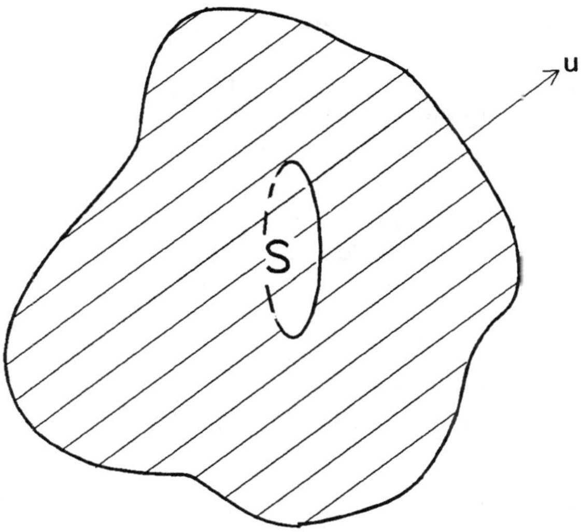

Figure 4.12. The continuous charge distribution, which is moving with uniform velocity **u**, is divided into a series of convection line currents parallel to **u** and used to develop the equation $\nabla \times \mathbf{B} = \mu_0(\mathbf{J} + \varepsilon_0\dot{\mathbf{E}})$ at an internal field point.

$$\oint \mathbf{B} \cdot d\mathbf{l} = \mu_0 \int \mathbf{J} \cdot d\mathbf{S} + \mu_0\varepsilon_0 \int \dot{\mathbf{E}} \cdot d\mathbf{S}$$

where $\mathbf{B} = \Sigma\mathbf{B}_i$ is the resultant magnetic field $\mathbf{E} = \Sigma\mathbf{E}_i$ is the resultant electric field and **J** is the current density. Applying Stokes' theorem of vector analysis to $\oint \mathbf{B} \cdot d\mathbf{l}$ and rearranging we have

$$\int (\nabla \times \mathbf{B} - \mu_0\mathbf{J} - \mu_0\varepsilon_0\dot{\mathbf{E}}) \cdot d\mathbf{S} = 0.$$

If the surface S in Figure 4.12 is made small enough for the variations of **B**, **J** and **E** over the surface S to be negligible, then

$$(\nabla \times \mathbf{B} - \mu_0\mathbf{J} - \mu_0\varepsilon_0\dot{\mathbf{E}}) \cdot \mathbf{S} = 0. \tag{4.93}$$

Since equation (4.93) must be true for all orientations of the surface **S** we conclude that at a field point inside a continuous charge and current distribution

$$\nabla \times \mathbf{B} = \mu_0(\mathbf{J} + \varepsilon_0\dot{\mathbf{E}}). \tag{4.94}$$

Thus if we start with equation (4.54) as a relation between the field vectors **E** and **B** at a field point in empty space, then the need for consistency, when there is a current distribution at the field point, leads us to equation (4.94) which is the same as equation (1.118). An alternative development of equation (4.94) from equation (4.54) is given in Appendix D.

4.8.4. *The Maxwell-Lorentz equations*

We shall now assume that the moving, continuous charge distribution, shown in Figure 4.12 is exceedingly small so that it corresponds to our model of an atomic charged particle. If $\mathbf{e}$ is the microscopic electric field, $\mathbf{b}$ is the microscopic magnetic field and $\mathbf{j}^{mic} = \rho^{mic}\mathbf{u}$ is the microscopic current density at a field point inside the moving charged atomic particle then, if we assume that equation (4.94) is valid inside the moving charged atomic particle, we have

$$\nabla \times \mathbf{b} = \mu_0(\mathbf{j}^{mic} + \varepsilon_0\dot{\mathbf{e}}). \tag{4.95}$$

This is the same as equation (1.140). We can now apply the method of averaging the microscopic fields given in Section 1.11 of Chapter 1 to derive equation (1.155) which is

$$\nabla \times \mathbf{B} = \mu_0(\mathbf{J} + \varepsilon_0\dot{\mathbf{E}}). \tag{4.96}$$

where $\mathbf{E}$ is now the macroscopic electric field, $\mathbf{B}$ is the macroscopic magnetic field and $\mathbf{J}$ is the macroscopic current density at a field point inside a macroscopic current distribution made up of moving and accelerating charged atomic particles. The macroscopic vectors $\mathbf{E}$, $\mathbf{B}$ and $\mathbf{J}$ are defined using equation (1.147) of Chapter 1.

It follows from equation (4.90) that at a field point inside a moving charged atomic particle we have

$$\nabla \times \mathbf{e} = -\dot{\mathbf{b}}. \tag{4.97}$$

Following the method leading from equation (4.95) to equation (4.96), we now have

$$\nabla \times \mathbf{E} = -\dot{\mathbf{B}} \tag{4.98}$$

where $\mathbf{E}$ and $\mathbf{B}$ are the macroscopic fields.

Summarizing, equations (4.13), (4.17), (4.97) and (4.95) are the Maxwell-Lorentz equations inside a moving and accelerating classical charged atomic particle and equations (4.14), (4.18), (4.98) and (4.96) are the corresponding Maxwell's equations relating the macroscopic fields at a field point inside a charge and current distribution made up of moving and accelerating classical charged atomic particles.

4.9. A charging capacitor in the quasi-stationary approximation

In order to illustrate the application of the Maxwell equation (4.96), we shall consider the example of the parallel plate capacitor shown in Figure 4.13. We shall assume that the capacitor plates are in a vacuum and that the plates are small enough for the fringing electric fields to be very significant, as shown in Figure 4.13. We shall assume that the frequency of the alternating current

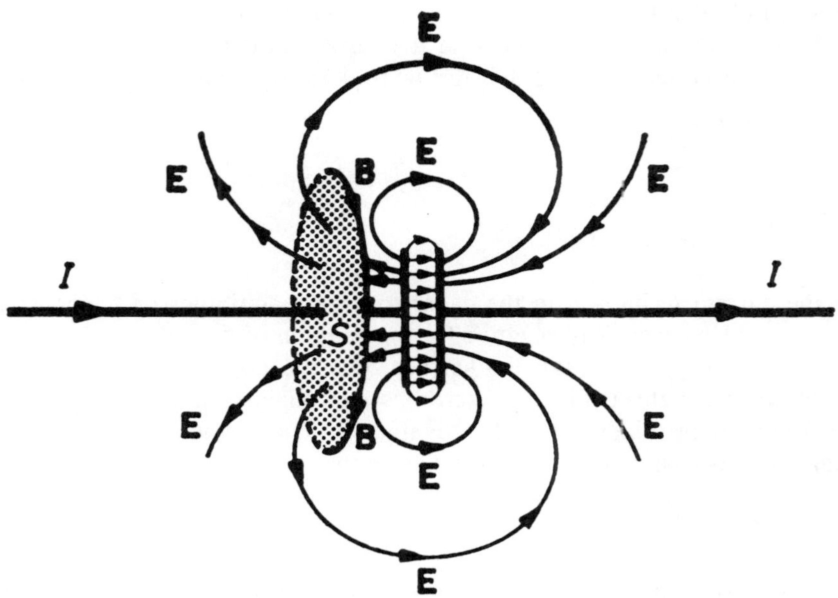

Figure 4.13. The charging of a capacitor. In this case the fringing electric field is important.

I is low enough for the quasi-stationary approximations to be valid, so that the current I at a given instant can be assumed to have the same value in all parts of the connecting leads. At large distances from the capacitor, the effects due to the presence of the capacitor are negligible and the magnetic field **B** at a perpendicular distance r from the long straight wire, calculated using the Biot-Savart law, is given, to a very good approximation, by

$$B = \frac{\mu_0 I}{2\pi r}. \tag{4.99}$$

As we approach the capacitor from the left in Figure 4.13, the effect of the presence of the parallel plate capacitor becomes more and more important, when we calculate the magnetic field outside the capacitor by applying the Biot-Savart law to all the currents in the system, including the currents in the capacitor plates that give rise to the changes in the charge densities on the distant parts of the capacitor plates, and the value of **B** deviates more and more from that given by equation (4.99). We shall now show how these results can be interpreted using equation (4.96). Applying equation (4.96) to the circular disk shaped surface S of radius r and then applying Stokes' theorem of vector analysis at a fixed time we have

$$\int \nabla \times \mathbf{B} \cdot d\mathbf{S} = \oint \mathbf{B} \cdot d\mathbf{l} = \mu_0 \int \mathbf{J} \cdot d\mathbf{S} + \mu_0 \varepsilon_0 \dot{\Psi} \tag{4.100}$$

where $\Psi = \int \mathbf{E} \cdot d\mathbf{S}$ is the flux of **E** through the surface S due to the fringing electric field due to the capacitor. Since by symmetry B has the same value at all points on the circumference of the surface S, evaluating $\oint \mathbf{B} \cdot \mathrm{d}\mathbf{l}$ in the direction of the magnetic field $\mathbf{B}$ in Figure 4.13 we have

$$2\pi rB = \mu_0 I + \mu_0 \varepsilon_0 \dot{\Psi}$$

so that

$$B = \frac{\mu_0}{2\pi r}[I + \varepsilon_0 \dot{\Psi}]. \tag{4.101}$$

As the current I charges up the capacitor, the magnitudes of $\mathbf{E}$ and Ψ are increasing. However Ψ is negative for the position of the surface S in Figure 4.13, since $\mathbf{E}$ goes through the surface S in the direction opposite to the direction of a right-handed corkscrew would advance if it were rotated in the direction in which $\oint \mathbf{B} \cdot \mathrm{d}\mathbf{l}$ is evaluated. Thus, when the capacitor in Figure 4.13 is charging up, $\dot{\Psi}$ is negative so that according to equation (4.101)

$$B < \frac{\mu_0 I}{2\pi r}. \tag{4.102}$$

Well away from the capacitor, $\dot{\Psi}$ tends to zero and equation (4.99) is a very good approximation. The nearer the surface S is to the capacitor the bigger the numerical values of $|\Psi|$ and $|\dot{\Psi}|$ and according to equation (4.101) the smaller the magnetic field B becomes.

If the surface S in Figure 4.13 were between the plates of the capacitor the current density $\mathbf{J}$ would be zero everywhere on the surface S and equation (4.101) would reduce to

$$B < \frac{\mu_0 \varepsilon_0 \dot{\Psi}}{2\pi r}. \tag{4.103}$$

However in this case Ψ and $\dot{\Psi}$ would both be positive, when the current I in the connecting leads was increasing, and the direction of I and hence of $\mathbf{B}$ were as shown previously in Figure 4.13. As an example, we shall assume that the plates of the capacitor in Figure 4.13 are so big that the fringing electric field can be neglected. It is shown in text books on electromagnetism that, if Q is the total charge on the positively charged circular capacitor plate, which is of radius a and area A, then the electric field E between the capacitor plates is equal to $Q/\varepsilon_0 A$. If the circular disk shaped surface S of radius r is between the plates and $r > a$, then the value of the electric flux Ψ crossing the surface S is $(Q/\varepsilon_0 A) \times A = Q/\varepsilon_0$, so that differentiating with respect to time we have

$$\varepsilon_0 \dot{\Psi} = \frac{\mathrm{d}Q}{\mathrm{d}t} = I. \tag{4.104}$$

Substituting in equation (4.103) we find that in this special case

$$B = \frac{\mu_0 I}{2\pi r}. \tag{4.105}$$

The effect of a finite fringing electric field would be to reduce the value of the electric flux Ψ crossing the surface S which, according to equation(4.103) would give a lower value for B than given by equation (4.105).

When the surface S in Figure 4.13 is on the right-hand side of the capacitor the current I crossing the surface S is finite in equation (4.101), but now Ψ and $\dot{\Psi}$ are negative again, when I is in the direction shown in Figure 4.13 and I is increasing in magnitude, showing that equation (4.102) is again satisfied.

So far we have assumed in this section that the frequency of the varying current I in Figure 4.13 is low enough for the quasi-stationary approximations to be valid. Equation (4.100) is still valid in the general case at a fixed time, when the current to and from the capacitor is varying at very high frequencies and whatever the shapes of the capacitor plates and the surface S. In these conditions we would have to use the retarded potentials or equation (1.134) of Chapter 1 to determine the magnetic field at a field point from the current distributions in the leads and the capacitor plates.

4.10. The displacement current and the continuity equation

Many text books still imply that the vacuum displacement current produces a magnetic field. The following quotation taken from Grant and Phillips [4] is typical:

> The inclusion of displacement current in the electromagnetic field equations restores a degree of symmetry to electricity and magnetism. A changing magnetic field produces an electric field according to Faraday's law. Now we see that a changing electric field produces a magnetic field.

We showed in Section 1.9.3 of Chapter 1 that, despite its name, the vacuum displacement current density $\varepsilon_0\dot{\mathbf{E}}$ should not be included as one of the sources of the magnetic field when the Jefimenko equation (1.134), which is the solution of the differential equation (1.122) for $\mathbf{B}$, is used to determine the magnetic field due to a varying current distribution. Neither should the vacuum displacement current be included as one of the sources of the magnetic field, when calculating the vector potential $\mathbf{A}$ in the Lorentz gauge using the retarded vector potential, namely equation (2.30). The values of the current density $\mathbf{J}$ at the appropriate retarded times are sufficient to determine $\mathbf{A}$ and hence $\mathbf{B}$ in the general case. Furthermore, if there is a current distribution, of current density $\mathbf{J}$, in an external magnetic field $\mathbf{B}$, there is a magnetic force $(\mathbf{J} \times \mathbf{B})$ N m^{-3} on the current distribution, so that, if the vacuum displacement current did behave like a conduction current, then by analogy we would expect that there would be a force equal to $(\varepsilon_0\dot{\mathbf{E}} \times \mathbf{B})$ N m^{-3} on empty space. However, it is assumed nowadays that there is no such force on empty space. Hence, unlike a conduction current, neither does the vacuum displacement current $\varepsilon_0\dot{\mathbf{E}}$ produce a magnetic field nor is it acted upon by a magnetic field, showing that the vacuum displacement current has none of the properties of an electric current. This shows that the roles of the **J** and $\varepsilon_0\dot{\mathbf{E}}$ terms are completely different in classical electromagnetism. We showed in Section 2.6.5 of Chapter 2 and in Section 4.6.1 that the equation

$$\nabla \times \mathbf{B} = \mu_0\varepsilon_0\dot{\mathbf{E}} \tag{4.106}$$

at a field point in empty space can be interpreted as a relation between the field vectors **E** and **B**. We showed in Section 4.8.1 that the need for consistency when interpreting equation (4.106) as a relation between the field vectors **E** and **B** requires the addition of the $\mu_0\mathbf{J}$ term to the right hand side of equation (4.106), giving

$$\nabla \times \mathbf{B} = \mu_0 (\mathbf{J} + \varepsilon_0\dot{\mathbf{E}}) \tag{4.107}$$

It was shown in Section 1.9.1 of Chapter 1 that it was the addition of the vacuum displacement current term $\varepsilon_0\dot{\mathbf{E}}$ to Maxwell's equations that converted the Laplacians of electrostatics and magnetostatics into D'Alembertians. For example, according to equation (1.122)

$$\nabla^2\mathbf{B} - \mu_0\varepsilon_0\ddot{\mathbf{B}} = -\mu_0\nabla \times \mathbf{J}. \tag{4.108}$$

It can be seen from the Jefimenko equation (1.134) that the **J** term in equation (4.108) is the source of the magnetic field, whereas it follows from the derivation of equation (4.108) that the vacuum displacement current term is associated with the propagation of the electromagnetic interaction at the speed of light in empty space.

In view of all the above evidence it is a little surprising to find that the vacuum displacement current term $\varepsilon_0\dot{\mathbf{E}}$ is still treated in some text books as an electric current which produces a magnetic field. One reason is the historical one. When Maxwell introduced the vacuum displacement current term the aether theories were generally accepted and Maxwell used a precise model of the aether. Maxwell also used the Coulomb gauge leading up to our equation (2.120) of Section 2.8. After the idea that the $\varepsilon_0\dot{\mathbf{E}}$ term in Maxwell's equations behaved like a current was introduced into the teaching of electromagnetism it has remained as such in the minds of many people and has not yet been thoroughly expurgated from some text books. Another reason why the $\varepsilon_0\dot{\mathbf{E}}$ term is still treated as a current is the way the $\varepsilon_0\dot{\mathbf{E}}$ term is introduced. In our typical approach to electromagnetism in Chapter 1, the vacuum displacement current term was introduced in Section 1.7 by seeing how the equation

$$\nabla \times \mathbf{B} = \mu_0\mathbf{J}$$

of magnetostatics had to be modified if it was to become consistent with equation (1.49), the continuity equation of charge and current densities. This led to equation (1.113). In many introductory courses the vacuum displacement current is introduced following a discussion of the example shown in Figure 4.13 for the special case when there is no fringing electric field.

Since according to equation (4.104)

$$I = \frac{dQ}{dt} = \int (\varepsilon_0 \dot{\mathbf{E}}) \cdot d\mathbf{S} = \varepsilon_0 \dot{\Psi} \tag{4.109}$$

many authors conclude that the conduction current I in the input lead in Figure 4.13 is carried on between the plates of the capacitor by a displacement current $\varepsilon_0\dot{\Psi}$ between the plates so that in their view the current is continuous. There is no need to assume that the displacement current $\varepsilon_0\dot{\Psi}$ is an electric current to interpret equation (4.109).

Consider the distribution of moving classical point charges shown in Figure 4.14. Each charge has magnitude q. Consider the volume V_0 shown in Figure 4.14. According to the equation of continuity, equation (1.49)

$$\nabla \cdot \mathbf{J} + \frac{\partial \rho}{\partial t} = 0. \tag{4.110}$$

Integrating equation (4.110) over the volume V_0, and applying Gauss' theorem of vector analysis, which is equation (A1.30) of Appendix A1.7, we have

$$\int \mathbf{J} \cdot d\mathbf{S} = -\frac{\partial}{\partial t} \int \rho dV. \tag{4.111}$$

Since the divergence of the curl of any vector is zero, it follows by taking the divergence of equation (4.107) that

$$\nabla \cdot (\mathbf{J} + \varepsilon_0 \dot{\mathbf{E}}) = \nabla \cdot \left(\nabla \times \frac{\mathbf{B}}{\mu_0}\right) = 0. \tag{4.112}$$

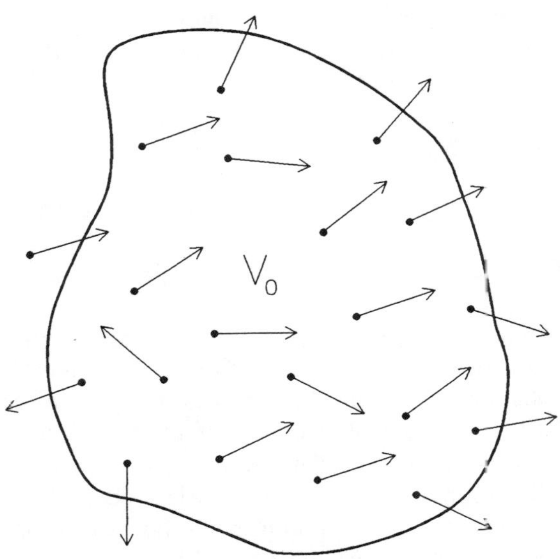

Figure 4.14. The system of moving classical point charges is used to illustrate the close relation between the displacement current and the continuity equation for charge and current densities.

Integrating equation (4.112) over the volume V_0 in Figure 4.14 and applying Gauss' theorem of vector analysis, we obtain

$$\int \mathbf{J} \cdot d\mathbf{S} = -\varepsilon_0 \frac{\partial}{\partial t} \int \mathbf{E} \cdot d\mathbf{S}. \tag{4.113}$$

Comparing equations (4.111) and (4.113) we see that

$$\int \mathbf{J} \cdot d\mathbf{S} = -\frac{\partial}{\partial t} \int \rho dV = -\varepsilon_0 \frac{\partial}{\partial t} \int \mathbf{E} \cdot d\mathbf{S}. \tag{4.114}$$

We can go directly from equation (4.111) to equation (4.113) by putting $\rho = \varepsilon_0 \nabla \cdot \mathbf{E}$ in equation (4.111) and applying Gauss' theorem of vector analysis.

To illustrate equation (4.114) we shall assume that N classical point charges, each of magnitude q, leave the surface of the volume V_0 in Figure 4.14 per second. The total charge crossing the surface of the volume V_0 per second is

$$\int \mathbf{J} \cdot d\mathbf{S} = Nq. \tag{4.115}$$

Assuming that the total electric charge in the system is conserved, it follows that the rate of increase of the total charge inside the volume V_0 in Figure 4.14 is equal to $-Nq$. Hence

$$\frac{\partial}{\partial t} \int \rho dV = -Nq. \tag{4.116}$$

Every one of the N charges, each of magnitude q, leaving the surface of the volume V_0 in Figure 4.14 per second takes an electric flux of q/ε_0 with it, so that the rate of increase of the total electric flux coming from the volume V_0 in Figure 4.14 is equal to $-N(q/\varepsilon_0)$, so that

$$\varepsilon_0 \frac{\partial}{\partial t} \int \mathbf{E} \cdot d\mathbf{S} = -Nq. \tag{4.117}$$

Comparing equations (4.115), (4.116) and (4.117) we see that

$$\int \mathbf{J} \cdot d\mathbf{S} = Nq = -\frac{\partial}{\partial t} \int \rho dV = -\varepsilon_0 \int \dot{\mathbf{E}} \cdot d\mathbf{S} = -\varepsilon_0 \dot{\Psi}. \tag{4.118}$$

There is no need to assume that the vacuum displacement current is an electric current to interpret equation (4.109). Equations (4.118) merely show that we can equate the total electric current $\int \mathbf{J} \cdot d\mathbf{S}$ crossing the surface of the volume V_0 in Figure 4.14 to the rate of decrease of the total charge inside V_0 or, since each classical point charge of magnitude q takes a flux of q/ε_0 with it, when it leaves the volume V_0, we can equate $\int \mathbf{J} \cdot d\mathbf{S}$ to ε_0 times the rate of decrease of the total electric flux coming from the surface of the volume V_0, due to the decrease in the number of classical point charges left inside V_0.

In the example of a capacitor in an AC circuit in Figure 4.13 we can interpret equation (4.109) as follows. We can either equate the conduction current I in the connecting wire leading to the positive place of the capacitor in Figure 4.13 to $\mathrm{d}Q/\mathrm{d}t$ the rate of increase of positive charge on the positive place, or to ε_0 times the consequential rate of increase of the total electric flux between the plates of the capacitor, which arises from the changes in the total charge on the capacitor plates due to the conduction current in the connecting lead.

The erroneous idea that the vacuum displacement current term always behaves like an electric current is reinforced in the minds of many students by the continued use of the phrase vacuum displacement current to label the $\varepsilon_0\dot{\mathbf{E}}$ term in Maxwell's equations. The author has tried for many years to have it called something else, such as the Maxwell term, but old habits die hard and so to avoid confusion we have followed standard practice in the text and used the phrase vacuum displacement current.

Notice that we have always included the word vacuum in the phrase vacuum displacement current to stress that we have only been considering moving charge distributions in otherwise empty space. We shall show in Chapter 9 that, in the presence of dielectrics we must add a $\dot{\mathbf{P}}$ term to the current density $\mathbf{J}$ in equation (4.107) where $\mathbf{P}$ is the polarization vector. The varying polarization vector $\dot{\mathbf{P}}$ arises from changes in the positions of atomic charges inside molecules and consequently behaves like an electric current and gives a contribution to the magnetic field.

4.11. Discussion of Maxwell's equations

We have now completed our discussions of Maxwell's equations for the case of charge and current distributions in empty space, for which $\varepsilon_r = 1$ and $\mu_r = 1$ everywhere, though we shall return in Chapter 9 to discuss the form Maxwell's equations take at field points inside dielectrics and magnetic materials. In Chapter 1 we gave an account of how, in introductory courses, Maxwell's equations are developed on the basis of very limited experimental evidence. In Section 2.6 of Chapter 2, we showed how Maxwell's equations should be interpreted by applying them to the fields of an oscillating electric dipole. That interpretation was consolidated in this chapter by deriving Maxwell's equations from the expressions for the fields $\mathbf{E}$ and $\mathbf{B}$ due to a moving classical point charge. The equation

$$\nabla \cdot \mathbf{E} = \frac{\rho}{\varepsilon_0} \tag{4.14}$$

was interpreted in terms of the result that the total flux of $\mathbf{E}$ from a moving classical point charge of magnitude q is always equal to q/ε_0. At field points in empty space, equation (4.14) means that there is no net flux of $\mathbf{E}$ from any surface that does not enclose an electric charge. The equation

$$\nabla \cdot \mathbf{B} = 0 \tag{4.18}$$

can be interpreted by saying that there are no sources of **B**, such as magnetic monopoles, so that there is no net flux of **B** from any closed surface.

We interpreted the equations

$$\nabla \times \mathbf{E} = -\frac{\partial \mathbf{B}}{\partial t} \tag{4.44}$$

$$\nabla \times \mathbf{B} = \frac{1}{c^2}\frac{\partial \mathbf{E}}{\partial t} \tag{4.66}$$

at a field point in empty space as relations between the vectors **E** and **B** and not as cause–effect relations. We showed in Section 4.8.3 that, for consistency, we must add the $\mu_0\mathbf{J}$ term to the right hand side of equation (4.66) to give

$$\nabla \times \mathbf{B} = \frac{1}{c^2}\frac{\partial \mathbf{E}}{\partial t} + \mu_0\mathbf{J} = \mu_0(\mathbf{J} + \varepsilon_0\dot{\mathbf{E}}) \tag{4.96}$$

when there is a current distribution at the field point. The reader can check back to show that it is the $\mu_0\mathbf{J}$ term in equation (4.96) that ends up as the source of the magnetic field **B** in equation (1.134) of Chapter 1 and as the source of the retarded vector potential **A** in equation (2.30) of Chapter 2.

In elementary classical electromagnetism, Maxwell's equations are generally applied individually, but in more advanced work Maxwell's equations are often used collectively, for example to derive equations (1.125) and (1.122) of Chapter 1 which are

$$\nabla^2\mathbf{E} - \mu_0\varepsilon_0\frac{\partial^2\mathbf{E}}{\partial t^2} = \nabla\left(\frac{\rho}{\varepsilon_0}\right) + \mu_0\frac{\partial \mathbf{J}}{\partial t} \tag{1.125}$$

$$\nabla^2\mathbf{B} - \mu_0\varepsilon_0\frac{\partial^2\mathbf{B}}{\partial t^2} = -\mu_0\nabla \times \mathbf{J} \tag{1.122}$$

and are the partial differential equations for **E** and **B** respectively. It is these equations that are used to interpret electromagnetic waves and which lead to Jefimenko's equations (1.136) and (1.134) which relate the fields **E** and **B** to the charge and current distributions.

It was pointed out in Section 1.9.1 that historically it was the addition of the displacement current term $\varepsilon_0\dot{\mathbf{E}}$ to Maxwell's equations that converted the Laplacians of electrostatics and magnetostatics into the D'Alembertians in equations (1.125) and (1.122). It was the differential equations (1.125) and (1.122) which we used in Section 1.9.2 to predict the existence of electromagnetic waves in empty space. It is also the differential equations (1.125) and (1.122) which can be used to determine the fields **E** and **B** from the charge and current distributions, which are the sources of the electromagnetic field. Maxwell's equations were used in Chapter 2 to derive the differential equations for the potentials ϕ and **A**, the solutions of which are given in the Lorentz gauge by the retarded potentials.

When we have stressed that, in the context of classical electromagnetism, there is no need to say anything about what happens in the empty space between moving charges, we were emphasising the fact that there is no need to introduce any mechanical aether models. We were trying to break the chain in which students are still taught these days by somebody who was taught by somebody . . . who was taught by somebody who was taught by somebody when the mechanical aether theories were in fashion. In emphasising that we need not say anything about what may or may not happen in the space between moving charges, we have probably gone too far, as we must remember that wherever there is a moving test charge in an electromagnetic field, it experiences a force given by the Lorentz force law, equation (1.1). This shows that there is in the electromagnetic field a capacity to impart energy and momentum to test charges wherever they are. We shall therefore find it convenient in Chapter 8 to attribute energy and in some cases momentum to the electromagnetic field. For example, it is assumed in classical electromagnetism that the radio waves emitted by a transmitting antenna travel outwards in all directions and are present in empty space, whether there is a test charge there or not, and even if the transmitting antenna has been subsequently destroyed. We must also remember that there is a more comprehensive theory than classical electromagnetism, namely quantum electrodynamics in which the electromagnetic interaction is interpreted in terms of the exchange of real and virtual photons.

Some readers may like to adopt a similar attitude to classical electromagnetism, to that which most people have towards quantum mechanics, where the wave function is treated as a quantity that can be determined by solving Schrödinger's equation subject to the appropriate boundary and continuity conditions. The wave function can then be used to make predictions about observable quantities such as the energy and momentum of a particle. By analogy with quantum mechanics, we could, in the context of classical electromagnetism, treat the fields **E** and **B** as theoretical quantities that could be determined by solving equations (1.125) and (1.122) respectively. The values of **E** and **B** could then be related to observable quantities, for example using the Lorentz force law. However, unlike the wave function of quantum mechanics, we can give operational definitions for the field vectors **E** and **B** using the Lorentz force law, equation (1.1), which was used in Section 1.2.4 of Chapter 1 to give an operational definition of **E** and in Section 1.4.2 of Chapter 1 to give an operational definition of **B**. Consequently many people prefer to interpret Maxwell's equations as relations between operationally defined quantities. However in this interpretation also, the operationally defined vectors **E** and **B** must ultimately be related back to experiments using the Lorentz force law or relations derived from it. At this stage it is worth reminding ourselves of the role of force laws in classical physics by considering a simple example based on Newton's law of universal gravitation.

We shall only consider the low velocity limit, so that Newton's laws of motion can be applied. At first sight, Newton's second law in the form $\mathbf{F} = m\mathbf{a}$ might appear to be no more than a definition of force. In practice, when $\mathbf{F} = m\mathbf{a}$ is applied, it is assumed that the force $\mathbf{F}$ is known from a force law, such as Newton's law of universal gravitation. As an example, consider two isolated particles 1 and 2 of known masses m_1 and m_2 respectively, situated a distance r apart. The gravitational force of attraction between the particles accelerates the particles. These accelerations can be determined from the changes in the motions of the particles. If the masses of the particles are known, the force acting on each particle at the separation r can be calculated using $\mathbf{F} = m\mathbf{a}$. The experiment can be repeated for different values of r, the separation of the particles. A general pattern will emerge, which can be summarized in Newton's law of universal gravitation, according to which the gravitational force of attraction between the two particles is given by

$$F = \frac{Gm_1m_2}{r^2} \tag{4.119}$$

where the gravitational constant G has the experimental value of 6.67×10^{-11} N m^2 kg^{-2}. According to equation (4.119) every particle in the Universe attracts every other particle with a force given by this equation. If we are later presented with a gravitational problem, such as the motion of a particle in the Earth's gravitational field, then, on the basis of our previous experimental investigations, summarized in equation (4.119), we can use that equation to determine the total gravitational force $\mathbf{F}$ acting on a particle of mass m from the positions of the other particles. According to Newton's second law, the acceleration of the particle is equal to $\mathbf{F}/m$, and the subsequent motion of the particle can be calculated. This illustrates how, when Newton's second law of motion is generally applied in practice, it is assumed that the force acting on the particle is known from a force law, such as Newton's law of universal gravitation. In the case of electrostatics, on the basis of previous experiments we can, if the charge distributions are known, use Coulomb's law, or equations derived from it, to make quantitative predictions of what the electric force on a test charge will be, which can then be used to predict the subsequent motion of the charge. Similarly, on the basis of previous experiments, we can, if the current distributions are given, determine $\mathbf{B}$ quantitatively, for example using equation (1.134) of Chapter 1, or using the retarded vector potential given by equation (2.30) of Chapter 2. This value of $\mathbf{B}$ can then be used in the Lorentz force law to predict the magnetic force on a moving charge, which can then be used to predict the subsequent motion of the charge.

Up to radio frequencies, we can generally treat problems in electromagnetism using only classical electromagnetism, and the theory can be related to experiments using the Lorentz force law, or relations derived from it. However, when we reach optical frequencies it is best to interpret the interaction of electromagnetic waves with matter in terms of the absorption of individual photons from the electromagnetic wave, so that quantum theory must be applied to the interactions. For example, it is the absorption of individual photons that affects the silver bromide grains in a photographic plate such that the grains turn to silver when the plate is developed. However, even at optical frequencies, Maxwell's equations or the equations for the potentials

ϕ and $\mathbf{A}$ can still be used to estimate the probabilities that photons will interact at various positions on a screen, for example in a diffraction experiment. We shall return to discuss this probabilistic interpretation in Section 4.13.5.

4.12. Comparison of the use of the potentials ϕ and A with the use of the fields E and B

It is often more convenient to use the potentials ϕ and $\mathbf{A}$ rather than $\mathbf{E}$ and $\mathbf{B}$. In Chapter 2, we used Maxwell's equations to derive the equations:

$$\nabla^2\phi - \frac{1}{c^2}\frac{\partial^2\phi}{\partial t^2} = -\frac{\rho}{\varepsilon_0} \tag{4.120}$$

$$\nabla^2\mathbf{A} - \frac{1}{c^2}\frac{\partial^2\mathbf{A}}{\partial t^2} = -\mu_0\mathbf{J} = -\frac{\mathbf{J}}{\varepsilon_0 c^2} \tag{4.121}$$

for the scalar potential ϕ and the vector potential $\mathbf{A}$ in the Lorentz gauge, in which the Lorentz condition

$$\nabla \cdot \mathbf{A} + \frac{1}{c^2}\frac{\partial\phi}{\partial t} = 0$$

is used to specify $\nabla \cdot \mathbf{A}$. It is more economic in the number of variables if we use the potentials ϕ and $\mathbf{A}$ rather than $\mathbf{E}$ and $\mathbf{B}$, since in the case of the potentials the four variables ϕ, A_x, A_y and A_z suffice whereas we need six components when we use $\mathbf{E}$ and $\mathbf{B}$. Hence a strong case can be made for taking the equations for the potentials rather than Maxwell's equations as our axioms for the theory of classical electromagnetism. For example we could assume that ϕ and $\mathbf{A}$ are variables that are the solutions of the partial differential equations (4.120) and (4.121) respectively, the solutions of which are given by the retarded potentials. The potentials ϕ and $\mathbf{A}$ can then be related to experiments using the Lorentz force equation (1.1) acting on a moving test charge q after defining $\mathbf{E}$ and $\mathbf{B}$ using the equations

$$\mathbf{E} = -\nabla\phi - \frac{\partial\mathbf{A}}{\partial t} \tag{4.122}$$

$$\mathbf{B} = \nabla \times \mathbf{A}. \tag{4.123}$$

If we did not want to mention the fields $\mathbf{E}$ and $\mathbf{B}$ we could rewrite the Lorentz force law in the form

$$\mathbf{F} = q(-\nabla\phi - \dot{\mathbf{A}}) + q\mathbf{u} \times (\nabla \times \mathbf{A}). \tag{4.124}$$

Alternatively we could avoid the use of the field vectors $\mathbf{E}$ and $\mathbf{B}$ by using either Lagrange's equations or Hamilton's equations. It can be shown, see for example Rosser [6], that if the **Lagrangian** L of a classical point charge of magnitude q and (rest) mass m_0, which is moving with a relativistic velocity **u** in an electromagnetic field described by potentials ϕ and **A** is defined to be

$$L = m_0c^2\left\{1 - \left(1 - \frac{u^2}{c^2}\right)^{1/2}\right\} - q\phi + q(\mathbf{u} \cdot \mathbf{A}) \tag{4.125}$$

then the application of Lagrange's equations gives the Lorentz force law. In the low velocity limit when $u \ll c$, we can use the Lagrangian

$$L = \frac{1}{2}m_0u^2 - q\phi + q(\mathbf{u} \cdot \mathbf{A}). \tag{4.126}$$

Similarly if the **Hamiltonian** H is defined to be

$$H = q\phi + c\{m_0^2c^2 + (\mathbf{P} - q\mathbf{A})^2\}^{1/2} \tag{4.127}$$

where **P** is the generalized momentum, then application of Hamilton's equations also leads to the Lorentz force law. The use of equations (4.125) and (4.127) can simplify the solution of many problems. The expression for the Hamiltonian given by equation (4.127) is often used to set up the appropriate differential equation for the wave function in quantum mechanics.

Summarizing, instead of taking Maxwell's equations as axiomatic, we could just as well choose the equations for the potentials ϕ and **A** as our axioms which would be checked a *posteriori* by comparing the predictions of the theory based on the potentials with the experimental results. In such an approach we could treat the potentials ϕ and **A** as theoretical quantities which are the solutions of equations (4.119) and (4.120) subject to the appropriate boundary and continuity conditions. These values of ϕ and **A** can be related to experiments using, for example, the Lorentz force law in the form given by equation (4.124) or using the Lagrangian given by equation (4.125) and Lagrange's equations or using the Hamiltonian given by equation (4.127) and Hamilton's equations. On the other hand some readers may prefer to think of ϕ and **A** as quantities that can be defined operationally. In Section 1.2.10 of Chapter 1, we defined ϕ operationally in terms of the potential energy of a test charge using equation (1.35). In Section 8.8.3 of Chapter 8 we shall show that when a charge q is at rest at a field point where the vector potential is **A**, there is a contribution of $q\mathbf{A}$ to the total "electromagnetic momentum" of the field. This "potential momentum" can appear as an increase in the momenta of the charges in the system, if the experimental conditions are changed. The vector potential will be defined operationally in Section 8.8.3 as the ratio of this "potential momentum" to the magnitude of the test charge q in the limit as q tends to zero. Reference: Konopinski [7]. The differential equations (4.120) and (4.121) can therefore be interpreted as relations between operationally defined quantities ϕ and **A** if the reader so wishes, but again the operationally defined quantities ϕ and **A** must ultimately be related back to experiments using equations (4.124), (4.125) or (4.127).

If we did take the equations for the potentials ϕ and **A** as our axioms, we could then reverse the arguments of Section 2.2 to derive Maxwell's equations as follows. If we defined **E** and **B** in terms of ϕ and **A** using equations (4.122) and (4.123) then, since the divergence of the curl of any vector is zero by taking the divergence of both sides of equation (4.123) we would have

$$\nabla \cdot \mathbf{B} = \nabla \cdot (\nabla \times \mathbf{A}) = 0. \tag{4.128}$$

Taking the curl of both sides of equation (4.122) and using the fact that $\nabla \times (\nabla\phi)$ is always zero, we would have

$$\nabla \times \mathbf{E} = -\nabla \times (\nabla\phi) - \nabla \times \dot{\mathbf{A}} = 0 - \frac{\partial}{\partial t}(\nabla \times \mathbf{A}) = -\frac{\partial \mathbf{B}}{\partial t}. \tag{4.129}$$

Thus two of Maxwell's equations, namely equations (4.128) and (4.129) would follow directly from the definitions of **E** and **B** in terms of ϕ and **A**. Using equation (A1.27) of Appendix A1.6 to substitute for $\nabla^2\mathbf{A}$ in equation (4.120) we would obtain

$$-\nabla \times (\nabla \times \mathbf{A}) + \nabla(\nabla \cdot \mathbf{A}) - \frac{1}{c^2}\frac{\partial^2 \mathbf{A}}{\partial t^2} = -\mu_0 \mathbf{J}. \tag{4.130}$$

From the Lorentz condition $\nabla \cdot \mathbf{A} = -\dot{\phi}/c^2$ and from equation (4.123) $\nabla \times \mathbf{A} = \mathbf{B}$. Hence equation (4.130) would become

$$\begin{aligned}\nabla \times \mathbf{B} &= \nabla\left(-\frac{1}{c^2}\frac{\partial \phi}{\partial t}\right) - \frac{1}{c^2}\frac{\partial}{\partial t}\left(\frac{\partial \mathbf{A}}{\partial t}\right) + \mu_0 \mathbf{J} \\ &= \frac{1}{c^2}\frac{\partial}{\partial t}\left(-\nabla\phi - \frac{\partial \mathbf{A}}{\partial t}\right) + \mu_0 \mathbf{J} = \mu_0\left(\varepsilon_0 \frac{\partial \mathbf{E}}{\partial t} + \mathbf{J}\right).\end{aligned} \tag{4.131}$$

Notice that the vacuum displacement current term would come out naturally from the $\ddot{\mathbf{A}}/c^2$ term in equation (4.121) plus the Lorentz condition. Since $\nabla^2\phi = \nabla \cdot (\nabla\phi)$ and since by differentiating the Lorentz condition we would have

$$\frac{\ddot{\phi}}{c^2} = -\frac{\partial}{\partial t}(\nabla \cdot \mathbf{A}) = -\nabla \cdot \dot{\mathbf{A}}$$

then equation (4.120) could be rewritten in the form

$$\nabla \cdot (-\nabla\phi - \dot{\mathbf{A}}) = \nabla \cdot \mathbf{E} = \frac{\rho}{\varepsilon_0}. \tag{4.132}$$

This would complete the derivation of Maxwell's equations, if the equations for the potentials ϕ and **A** were taken as axiomatic.

It does not matter, in the strict context of classical electromagnetism, whether we use the fields **E** and **B** and take Maxwell's equations as axiomatic or whether we use the potentials ϕ and **A** and take the equations (4.120) and (4.121) for the potentials as axiomatic. The two approaches are just alternative ways of expressing the laws of classical electromagnetism. Whichever set of axioms that is adopted can be used to derive the other set of axioms. However it has been argued on the basis of experiments on the Bohm-Ahranov effect that, in the wider context of quantum mechanics, the potentials ϕ and **A** are more useful than the field vectors **E** and **B**. A reader interested in an introductory account of the Bohm-Ahranov effect is referred to Feynman, Leighton and Sands [8], who consider a two slit interference experiment using electrons.

4.13. Historical note on the development of classical electromagnetism and the nineteenth century aether theories

4.13.1. *The luminiferous aether*

After the initial success of Newtonian mechanics, it seemed plausible to try to explain all natural phenomena in terms of Newtonian mechanics, particularly as the concepts of mechanics were familiar to people in their daily lives. This approach proved very successful in the case of sound. Sound will not travel through a vacuum and must have a material medium to transmit it. Sound is now interpreted as an elastic wave propagating through a material medium which can be a solid, liquid or a gas.

In the same way, attempts were made to interpret light in terms of mechanical models, and two theories arose. One was the corpuscular theory, in which light was pictured as a stream of little corpuscles. It was assumed that these small corpuscles obeyed the laws of mechanics, and produced the sensation of light when they struck the eye. The other theory was the wave theory. The corpuscular theory remained pre-eminent until the beginning of the nineteenth century, when Young investigated the interference of two beams of light. Young and Fresnel were able to account for the newly observed phenomena of interference and diffraction on the wave theory, and from that time onwards the wave theory came to be accepted. Since sound will not travel through a vacuum and must have a material medium to transmit it, it seemed plausible in the early nineteenth century to assume that, if light was a form of wave motion, then there should be a light transmitting medium present in a vacuum, as well as inside a material medium, that was able to transmit the vibrations constituting light. This hypothetical light transmitting medium was called the aether. The idea of the aether arose originally in Greek science where it was introduced as an element in addition to the four elements of fire, earth, water and air. It was assumed in ancient cosmology that the aether filled the celestial regions.

In 1828, Poisson showed that both longitudinal and transverse elastic waves can be propagated in a solid. The velocities of longitudinal and transverse elastic waves in a solid are $[(K + 4G/3)/\rho_m]^{1/2}$ and $[G/\rho_m]^{1/2}$ respectively, where K is the bulk modulus, G is the rigidity modulus and ρ_m is the mass density of the solid. In order to account for the phenomenon of the polarization of light on the wave theory, light was pictured as a transverse wave motion. No longitudinal light waves have ever been observed. For transverse waves to exist in a solid, the rigidity modulus G must be finite. Since solids contract under pressure, the bulk modulus K is positive, so that if G is finite $(K + 4G/3)$ must be finite for a solid, so that longitudinal as well as transverse waves should propagate in an elastic solid. Since no longitudinal light waves have ever been observed, it was concluded that the hypothetical aether could not be a normal elastic solid. Neither could the hypothetical aether be a perfect fluid, since the rigidity modulus G of a gas or a liquid is zero, in which case there would be no transverse waves in the hypothetical aether. It was concluded that the hypothetical aether had to be a new type of elastic medium. A large number of extremely complicated mathematical and mechanical models were suggested in the nineteenth century. For example, in 1889 Kelvin suggested a mechanical model for an element of the aether which consisted of rotating gyroscopes. This model was able to resist all rotatory disturbances, but was unable to resist translatory movements. An aether constructed of such elements would be able to transmit transverse but not longitudinal waves. Reference: Schaffner [9].

4.13.2. *The electromagnetic aether*

At the beginning of the nineteenth century, it was suggested that a type of aether transmitted the electric forces between electric charges and the magnetic forces between permanent magnets, though people were not sure whether or not this aether was the same as the light transmitting aether. For example, Young wrote: "Whether the electric aether is to be considered the same with the luminous aether, if such a fluid exists, may perhaps at some future time be discovered."

Faraday pictured electric and magnetic forces in terms of lines of force. This picture was developed mathematically by Maxwell who used the concepts of electric and magnetic lines of force. A line of force was what we have called an electric or a magnetic field line. It was assumed that the electric lines of force were in a state of tension. On this model, it was assumed, for example, that the lines of force which started on a positive electric charge and ended on a negative charge behaved like stretched rubber bands pulling the charges towards each other. In order to satisfy the condition that the aether had to be in equilibrium under the influence of electrostatic forces, it was assumed that the electric field lines repelled each other in the transverse direction giving rise to a pressure at right angles to the lines of force. It was assumed that this pressure was transmitted by an aether. A similar interpretation was developed for the transmission of magnetic forces in terms of magnetic lines of force. This latter model is still often used in plasma physics where it is often assumed that the magnetic field lines are under a tension of (B^2/μ_0) newtons per square metre and that there is a magnetic pressure of $(B^2/2\mu_0)$ pascals.

Using a complicated model of the aether, in which he assumed that the aether consisted of vortices, idler wheels etc, and using mechanical ideas, Maxwell concluded in 1862 that the vacuum displacement current term $\varepsilon_0\dot{\mathbf{E}}$ should be added to equation (1.89). References: Maxwell [10], Tricker [11] and Rosenfeld [12]. This in turn led Maxwell to the theory of electromagnetic waves in empty space. After the identification of electromagnetic waves with light waves, it was generally assumed that the same aether transmitted electric forces, magnetic forces and light. Interpretations of electromagnetism were developed, in which it was assumed that, according to Faraday's law of electromagnetic induction, varying magnetic fields in the aether generated electric fields in the aether and that varying electric fields, that is displacement currents in the aether, generated magnetic fields. It was assumed that in electromagnetic waves, varying magnetic fields generated varying electric fields which in turn generated varying magnetic fields which generated varying electric fields and so on leading to wave propagation in the aether. Many complicated models of the aether were developed during this period. The mechanical aether theories became more and more complicated as they tried to account for a wider and wider range of phenomena. For example, Kelvin went so far as to suggest that atoms might be vortices in the aether. The interested reader is referred to Whittaker [13] or Schaffner [9]. These complicated models of the aether gave rise to no observable effects other than the electromagnetic forces they were meant to interpret. To quote Born [14]

> If we were to accept them literally, the aether would be a monstrous mechanism of invisible toothed wheels, gyroscopes and gears intergripping in the most complicated fashion, and of all this confused mess nothing would be observable but a few relatively simple forces which would present themselves as an electromagnetic field.

Towards the end of the nineteenth century, the view was beginning to arise that one should merely accept that the laws of electromagnetism describe the electromagnetic forces between moving charges, and one should not try to interpret the electromagnetic forces themselves in terms of a mechanical aether, whose properties could not be measured. For example, in his book on electric waves, Hertz [15] wrote

> To the question 'what is Maxwell's theory?' I know of no shorter or more definite answer than the following. Maxwell's theory is Maxwell's system of equations.

To this concise statement, the author would just add the statement that the variables in Maxwell's equations can be related to observable effects using the Lorentz force law.

4.13.3. *The rise of the theory of special relativity*

By the year 1900, the mechanical theories of the aether had been largely abandoned, but one difficulty still remained. It was still generally believed that the Galilean transformations, which were based on Newtonian mechanics and the concept of absolute time, were correct. Maxwell's equations and the laws of electromagnetism do not satisfy the principle of relativity, if the coordinates and time are transformed from one inertial reference frame to one that is moving with uniform velocity relative to the first using the Galilean transformations. It was concluded initially that Maxwell's equations could only be applied in one absolute reference frame, which was generally identified with the reference frame in which the hypothetical aether was at rest. This can be illustrated by considering the speed of light in empty space. For purposes of discussion we shall assume, by analogy with sound, that the speed of light is the same and equal to c in all directions in the reference frame in which the aether is at rest. Let the Earth move with velocity v relative to the aether. If the Galilean velocity transformations were applied the speed of light in empty space should vary in the laboratory frame, in which the Earth is at rest, from $(c - v)$ in the direction in which the Earth is moving relative to the aether to $(c + v)$in the opposite direction. Hence, if the Galilean transformations could be applied to light waves then, by measuring the speed of light in different directions in the laboratory reference frame, it should have been possible to determine the velocity of the Earth relative to the aether. The most famous of the experiments was carried out by Michelson and Morley [16] in 1887. All such experiments failed to determine the speed of the Earth relative to any absolute reference frame.

The way Einstein overcame this dilemma in 1905 was to assume that the laws of electromagnetism, namely Maxwell's equations, held in all inertial reference frames, even though this meant abandoning the Galilean transformations in favour of the Lorentz transformations. This in turn lead to a revision of the Newtonian concepts of absolute space and absolute time. Thus, according to special relativity there is no absolute reference frame in which and only in which Maxwell's equations are valid. The predictions of the theory of special relativity have been confirmed by experiments. Reference: Rosser [17]. We shall return to discuss special relativity in more detail in Chapter 10.

The view that prevailed after the rise of the theory of special relativity can be summarized by the following quotation from a book written by Born [14] in 1924.

> Light or electromagnetic forces are never observable except in connection with bodies. Empty space free of all matter is no object of observation at all. All that we can ascertain is that an action starts out from one material body and arrives at another material body some time later. What occurs in the interval is purely hypothetical, or, more precisely expressed, arbitrary. This signifies that theorists may use their own judgement in equipping a vacuum with phase quantities (denoting state), fields, or similar things, with the one restriction that these quantities serve to bring changes observed with respect to material things into clear and concise relationship.
>
> This view is a new step in the direction of higher abstraction and in

> releasing us from common ideas that are apparently necessary components of our world of thought. At the same time, however, it is an approach to the ideal of allowing only that to be valid as constructive elements of the physical world which is directly given by experience, all superfluous pictures and analogies which originate from a state of more primitive and more unrefined experience being eliminated.
>
> From now onwards aether as a substance vanishes from theory. In its place we have the abstract 'electromagnetic field' as a mere mathematical device for conveniently describing processes in matter and their regular relationship.

This is the approach we have been trying to cultivate. The equations of electromagnetism give the electric and magnetic fields due to a system of moving charges. From these fields the force that would act on a moving test charge at any field point in empty space can be calculated using the Lorentz force law. In the context of classical electromagnetism, it is the effects of the forces on charges that are observed experimentally, for example the conduction current produced in a receiving antenna by a radio wave.

4.13.4. *Discussion of the mechanical models of the aether and their aftermath*

The mechanical models of the aether helped in the historical development of Maxwell's equations. For example, Maxwell was using a mechanical model of the aether when he developed the idea of the displacement current. These mechanical models have been compared to the scaffolding used when building a house. When the house is finished the scaffolding can be taken down and forgotten. It does not matter how Maxwell's equations were originally arrived at, the important thing is that they can be applied to make reliable predictions in the context of classical electromagnetism.

In our approach to classical electromagnetism, we have eliminated all hypotheses about what may or may not happen in the empty space between a charge distribution and a field point. If hypotheses, such as a mechanical aether theory, are introduced which lead to no extra experimental result other than the electromagnetic forces themselves, we cannot rule out these hypotheses on the basis of the experimental evidence. Our attitude is that such mechanical hypotheses are superfluous. To quote Schaffner [18]:

> the nineteenth-century aether has been relegated to that Ideal Realm populated by Caloric, Phlogiston, Epicycles, and other scientific concepts that have done their work so well as to have forced science beyond them.

In retrospect the mechanical models of the aether introduced in the second half of the nineteenth century should not be viewed as attempts to produce exact replicas of the aether, reducing everything to the laws of mechanics. The prevailing attitude even then was that they were trying to devise possible mechanical models (or analogues) that could be used to illustrate the laws of electromagnetism in terms of the laws of Newtonian mechanics, which had been so successful in interpreting so many of the phenomena they encountered in their daily lives. They felt that they had a feel for and understood the laws of mechanics. For example, Kelvin wrote: "I am never content until I have constructed a mechanical model of the object I am studying. If I succeed in making one, I understand; otherwise I do not'. After using a precise mechanical model of the aether in the paper in which he first introduced the vacuum displacement current, Maxwell went on to comment later in his Treatise:

> The attempt which I then [in 1862] made to imagine a working model of this mechanism must be taken to be no more than it really is, a demonstration that mechanism may be imagined capable of producing a connexion mechanically equivalent to the actual connexion of the parts of the electromagnetic field. The problem of determining the mechanism required to establish a given species of connexion between the motions of the parts of a system always admits of an infinite number of solutions.

In 1865 Maxwell wrote:

> I have on a former occasion attempted to describe a particular kind of motion and a particular kind of strain, so arranged as to account for the phenomena. In the present paper I avoid any hypothesis of this kind; and in using such words as electric momentum and electric elasticity in reference to the known phenomena of the induction of currents and the polarization of dielectrics, I wish merely to direct the mind of the reader to mechanical phenomena which will assist him in understanding the electrical ones. All such phrases in the present paper are to be considered as illustrative, not as explanatory.

It is of interest to note that in the twentieth century there has been a complete reversal of roles. Nowadays, instead of trying to interpret electromagnetic forces in terms of mechanical models, we now try to interpret the mechanical properties of solids and fluids in terms of atomic theory and quantum mechanics using electromagnetic forces.

Some ideas based on the nineteenth century mechanical models of the aether are still prevalent in the teaching of some branches of classical electromagnetism, such as plasma physics. These models should now be looked upon simply as mechanical analogues. As an illustrative example, assume that a mechanical analogue of an oscillating *LCR* circuit is an oscillating simple pendulum with a damping term proportional to the velocity of the pendulum. The physical principles are completely different in the two cases, but both systems are described by the same mathematical equation, namely the differential equation of damped harmonic motion. If the solutions of the mechanical case are known, or can be developed from mechanical insight or experience, then these solutions can often be adapted to the electrical case by interchanging the corresponding terms in the mathematical equations describing the systems. It will be shown later in Section 8.2 of Chapter 8 that, starting with the Lorentz force giving the electromagnetic force on a finite current distribution and using Maxwell's equations we can rewrite the total force on a current distribution in a mathematical form that is the same as if the magnetic field lines were in a state of tension equal to (B^2/μ_0) newtons per square metre and as if there were a magnetic pressure equal to $(B^2/2\mu_0)$ pascals. If we know the solutions of these mechanical analogues from the properties of rubber bands and fluids, these solutions can often be adapted to the electromagnetic case. The use of such analogues need not imply a reality to magnetic field lines, that behave like rubber bands.

In the text, electric and magnetic fields are sometimes represented graphically by imaginary field lines, drawn at a fixed time such that the direction of the tangent to the field line at a point is in the direction of the electric field **E** (or the magnetic field **B**) at that point at that instant. The number of field lines is generally limited such that the number of field lines per square metre crossing a surface at right angles to the field line is equal to, or if more convenient is proportional to, the value of the electric (or magnetic) field at that point. The field lines are closest together where the field strengths are greatest. In this way, all we have done is to give a graphical, pictorial representation of the magnitude and direction of the electric (or magnetic) field at every point of space at one given instant of time. Many people find this a more helpful way of thinking about electromagnetic problems than using tabulated values of the field vectors **E** and **B**.

As an analogy consider an ordinance survey map that shows the contour lines. One could just as well use a table of geographic coordinates and their altitudes above sea level. This is all the information a walker needs to plan a suitable walk, but many walkers would prefer to have the same data presented in the form of a contour map, as, with experience, many walkers would find it easier to visualize and plan a walk using the contour map. In a similar way, many scientists find it easier to think about classical electromagnetism using electric and magnetic field line diagrams, rather than using tabulated values of the fields. Field line diagrams are used in the text whenever they help to illustrate the interpretation of Maxwell's equations etc, but we shall not attribute any independent reality, such as mechanical properties, to these field lines. Field line diagrams are just a convenient pictorial way of presenting the data about the direction and magnitude of the electric (or magnetic) field at one instant of time.

4.13.5. *The advent of quantum theory*

In 1900, Planck introduced the quantum theory of radiation and in 1905 Einstein interpreted the photoelectric effect in terms of the interactions with individual electrons of individual photons of energy $h\nu$, where $h = 6.625 \times 10^{-34}$ J s is Planck's constant and ν is the frequency of the light. Some readers may already like to think of a light beam as consisting of a stream of an extremely large number of photons moving from the light source with the speed of light. This goes beyond classical electromagnetism, which must be extended to incorporate the effects of quantization. The starting point for the quantization of the electromagnetic field is generally the equations for the potentials ϕ and $\mathbf{A}$. Reference: Heitler [19].

Photons are the quanta (or carriers) of the electromagnetic interaction. Photons are emitted when electric charges are accelerated, and they give the main contribution to the electromagnetic interaction in the radiation zone. In the near zone, that is for radial distances $r < \lambda/2\pi$ in the case of the oscillating electric dipole, where the electric and magnetic field contributions given by equations (2.46), (2.56) and (2.57), predominate and the acceleration dependent fields are negligible in comparison, the electromagnetic interaction between moving charges has been interpreted in terms of the exchange of **virtual photons**, which are the carriers of the electromagnetic interaction. In the case of a virtual photon, one electric charge emits a photon which is absorbed by another electric charge within a very short time interval Δt. According to the uncertainty principle, the uncertainty ΔE in energy in the time interval Δt can be of the order of $h/(2\pi\ \Delta t)$. The theory of quantum electrodynamics is a more comprehensive theory than classical electromagnetism. Having pointed out the existence of quantum electrodynamics and made a few comments, we shall not consider the theory further and we shall ignore all effects arising from the finite value of Planck's constant. A reader interested in a fuller account of the roles of real and virtual photons is referred to Lawson [20]. A reader interested in quantum electrodynamics is referred to Mandl and Shaw [21].

Even though light beams are composed of photons, Maxwell's equations or the equations for the potentials ϕ and $\mathbf{A}$ can still be used to give a comprehensive interpretation of physical optics. Reference: Born and Wolf [22]. The reason for this success is that photons are non-interacting bosons and obey Bose-Einstein statistics. There is no restriction on the number of photons that can be in a single-particle (photon) quantum state. In most laboratory experiments in optics and radio there are generally an exceedingly large number of photons in the same state. As an example, consider an idealized light source which emits 10W of monochromatic light of wavelength 600 nm. The energy of each photon is $h\nu = hc/\lambda = 3.3 \times 10^{-19}$ J = 2.07 eV. At a distance of 1m from such a light source, there are 2.4×10^{18} photons crossing 1 m^2 per second, which is an exceedingly large number. If such a monochromatic light source is used to produce an interference pattern on a screen, due to the persistence of vision of the eye we cannot see the effects of individual photons' hitting the screen, but we see what seems to the eye to be a complete interference pattern. If we photograph the interference pattern using a very weak source and a very short exposure time, we will not get a photograph of a complete, continuous interference pattern on the film but we will get a series of dots where individual photons have struck the photographic plate. The wave theory can be used to predict the probability that a photon will hit a certain region of the screen. If the source is a radio transmitter emitting 10W of radio waves of frequency 3×10^8 Hz and wavelength 1m, then at a distance of 100 km from the source there are still 4×10^{14} photons crossing 1 m^2 per second. Thus in practice, radio detectors generally measure the superimposed contributions of a very large number of photons in the electromagnetic wave, which add up to give the radiation electric and magnetic fields in the electromagnetic wave. The response of the radio detector can be interpreted classically in terms of the radiation electric and magnetic fields of the electromagnetic wave.

The reason why the laws of classical electromagnetism can be used to interpret physical optics is summarized in the following quotation from Feynman, Leighton and Sands [23].

> When we have the wave function of a single photon, it is the amplitude to find a photon somewhere. Although we haven't ever written it down there is an equation for the photon wave function analogous to the Schrödinger equation for the electron. The photon equation is just the same as Maxwell's equations for the electromagnetic field, and the wave function is the same as the vector potential **A**. The wave function turns out to be just the vector potential. The quantum physics is the same thing as the classical physics because photons are non-interacting Bose particles and many of them can be in the same state – as you know they *like* to be in the same state. The moment you have billions in the same state (that is, in the same electromagnetic wave), you can measure the wave function, which is the vector potential directly. Of course, it worked historically the other way. The first observations were on situations with many photons in the same state, and so we were able to discover the correct equation for a single photon by observing directly with our hands on a macroscopic level the nature of wave function.

Thus Maxwell's equations or the equations for the potentials ϕ and **A** can be used to interpret physical optics. Once ϕ and **A** or **E** and **B** are calculated from the charge and current distributions, the energy density of photons in empty space is generally estimated using the expression $(\varepsilon_0 E^2/2 + B^2/2\mu_0)$. That electric and magnetic radiation fields can be associated with individual photons is illustrated by the photo-disintegration of a deuteron by a high energy γ-ray. At photon energies a few keV above threshold, the photo-disintegration of the deuteron is due mainly to electric dipole absorption. The proton arising from the disintegration is emitted preferentially in the direction of the electric vector of the incident photon. Reference: Wilkinson [24]. It is the contributions of the large number of photons in an electromagnetic wave which add up to give the radiation electric and magnetic fields of classical electromagnetism.

According to quantum mechanics, the wave properties of individual photons are interpreted in a similar statistical way to the wave properties associated with individual electrons. It is natural to accept that just as electrons can travel through a vacuum, individual photons, such as γ-rays can also travel through a vacuum without a light transmitting medium such as an aether. In the nineteenth century, it was believed that light was a continuous wave motion in a continuous medium. It was reasonable to postulate a mechanical aether in such circumstances. No mechanical aether is needed to let photons, such as γ-rays, go through a vacuum.

References

1. Slepian, J., *Amer. Journ. Phys.*, Vol. 19, p. 87 (1951).
2. McDonald, K. L., *Amer. Journ. Phys.*, Vol. 22, p. 586, (1954).
3. Iona, M., *Amer. Journ. Phys.*, Vol. 31, p. 398 (1963).
4. Grant, I. S. and Phillips, W. R., *Electromagnetism*, 2nd Edition (John Wiley, Chichester, 1990), p. 353.
5. Rosser, W. G. V., *Classical Electromagnetism Via Relativity* (Butterworths, London, 1968) pp 123-125
6. Rosser, W. G. V., *Introduction to the Theory of Relativity* (Butterworths, London, 1964), p. 365.
7. Konopinski, E. J., *Amer. Journ. Phys.*, Vol. 46, p. 499 (1978).
8. Feynman, R. P., Leighton, R. B and Sands, M., *The Feynman Lectures on Physics* (Addison-Wesley, Reading, Mass., 1964), Vol. 2, Section 15-5.
9. Schaffner, K. W., *Nineteenth-Century Aether Theories* (Pergamon Press, Oxford, 1972), p. 68.
10. Maxwell, J. C., *The Scientific Papers of James Clerk Maxwell* (Dover, New York, 1965), Vol. 1, pp. 451–525.
11. Tricker, R. A. R., *The Contributions of Faraday and Maxwell to Electrical Science* (Pergamon Press, Oxford, 1966).
12. Rosenfeld, L., *Supplemento Al, Vol. 4, Series 10, Del Nuovo Cimento*, p. 1630 (1957).
13. Whittaker, E., *A History of the Theories of Aether and Electricity*, Vol. 1 (Thomas Nelson, Edinburgh, 1951), p. 294.
14. Born, M., *Einstein's Theory of Relativity* (Methuen, London, 1924), Chapter 5.
15. Hertz, H., *Electric Waves* (Macmillan, London, 1893), p. 21.
16. Michelson, A. A and Morley, E. W., *Amer. Journ. Science*, Vol. 34, p. 333 (1887).
17. Rosser, W. G. V., *Introductory Special Relativity* (Taylor and Francis, London, 1991).
18. Schaffner, K. W., see ref. 9, p. 117.
19. Heitler, W., *Quantum Theory of Radiation*, 3rd Edition (Oxford University Press, London, 1954).
20. Lawson, J. D., *Contemporary Physics*, Vol. 11, p. 575 (1970).
21. Mandl, F and Shaw, G., *Quantum Field Theory* (John Wiley, Chichester, 1984).
22. Born, M, and Wolf, W., *Principles of Optics* (Pergamon Press, Oxford, 1959).
23. Feynman, R. P., Leighton, R. B. and Sands, M., *The Feynman Lectures on Physics*, Vol. 3 (Addison-Wesley, Reading, Mass., 1964), Section 21-4.
24. Wilkinson, D. H., *Phil. Mag., Series* 7, Vol. 43, p. 659 (1952).

CHAPTER 5

Electric fields due to electrical circuits

5.1. Introduction

It is important to distinguish two different contributions to the total electric field due to the current flowing in an electrical circuit. First there are the electric fields, denoted $\mathbf{E}_s$ due to the surface and boundary charge distributions that give the appropriate value for the resultant electric field in the various parts of the circuit so as to give the same value of current in all parts of a series circuits when the current is steady or is varying at mains frequency. The properties of these electric fields due to surface and boundary charge distributions are discussed in detail in Appendix B. Secondly there are the electric fields due to the moving and accelerating conduction electrons and stationary positive ions inside the conductors making up the circuit. The formulae for the electric fields due to the moving conduction electrons and stationary positive ions will be derived in this chapter using the expression for the electric field due to an accelerating classical point charge, which is given by equation (3.10).

It is important to consider three frequency ranges, namely

(a) DC or steady current flow.
(b) AC variations at mains frequency when the quasi-stationary approximations are valid.
(c) Frequencies much higher than mains frequency when the contributions of the radiation electric fields are important.

If the mains frequency is 50 Hz, the wavelength of the electromagnetic variations is 6000 km so that $\lambda/2\pi$ is equal to about 1000 km. Hence, for laboratory experiments carried out at mains frequency, we are always in the near zone and we can generally ignore the radiation electric fields that arise from the $\mathbf{E}_A$ term in equation (3.10), which gives the electric field due to an accelerating classical point charge. Hence for DC circuits and at mains frequency we generally only need to use the $\mathbf{E}_V$ contribution given by equation (3.11) to determine the electric field due to a current element. Before going on to do so, we shall derive the expression for the number of moving conduction electrons counted by the information collecting sphere inside a current element.

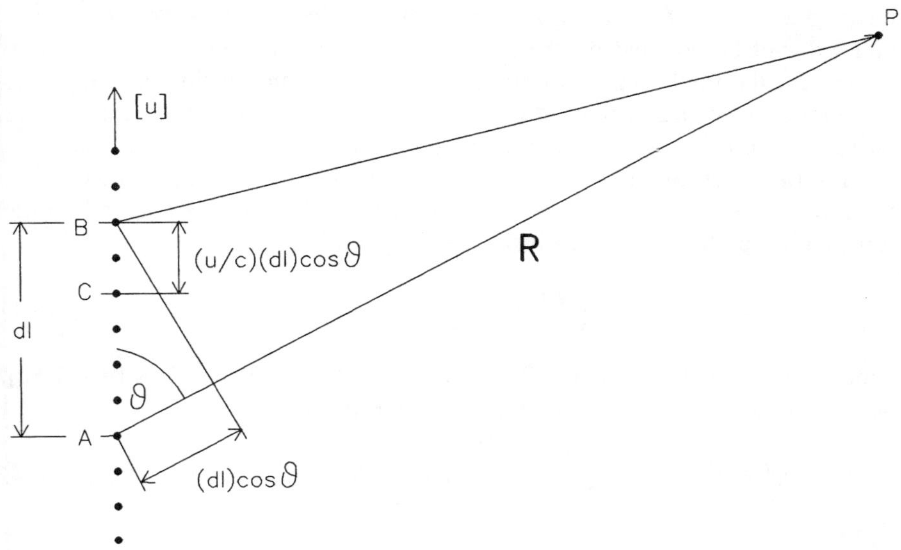

Figure 5.1. The counting of moving charges by the information collecting sphere that reaches the field point P at the time of observation t.

5.2. The counting of moving charges by the information collecting sphere

Consider the line of moving classical point charges shown in Figure 5.1. We shall determine the number of moving charges counted by the information collecting sphere that reaches the field point P at the time of observation t. Consider the section AB, which is of infinitesimal length $\mathrm{d}l$. The distance from A, which is at the bottom end of $\mathrm{d}l$, to the field point P is equal to $\mathbf{R}$. We shall assume that all the charges are moving upwards with the same velocity $[\mathbf{u}]$ at the retarded time $t^* = (t - R/c)$ when the information collecting sphere passes the point A in Figure 5.1. It can be seen from Figure 5.1 that the point B, which is at the top end of the section of length $\mathrm{d}l$ is at a distance $\mathrm{d}l \cos \theta$ closer to the field point P than the point A, where θ is as shown in Figure 5.1. In the case when the charges are moving upwards θ is equal to the angle between $[\mathbf{u}]$ and $\mathbf{R}$. The information collecting sphere passes B a time $(\mathrm{d}l \cos \theta)/c$ after it has passed A. In the time interval $(\mathrm{d}l \cos \theta)/c$, all the charges move upwards with a velocity $[\mathbf{u}]$ covering a distance of $[u/c]\, \mathrm{d}l \cos \theta$. Since $\mathrm{d}l$ is infinitesimal, we can neglect the effect of any accelerations of the charges on the total distances the charges travel in the infinitesimal time $(\mathrm{d}l \cos \theta)/c$. A charge that was at the position C in Figure 5.1 at a distance $[u/c](\mathrm{d}l)\cos \theta$ below B at the time $t^* = (t - R/c)$, when the information collecting sphere passes A, reaches B at the instant the information collecting sphere passes B. Charges which were at a distance less than $[u/c](\mathrm{d}l) \cos \theta$ below B when the information collecting sphere passes A will have gone beyond B by the time the information collecting sphere reaches B and will not be counted by the information collecting sphere in the section dl. Hence, the total number of charges actually counted inside dl is equal to the number of charges, that, at the time t^* the information collecting sphere passes A, were in the section AC, which is of length dl – $[u/c](\mathrm{d}l)\cos\theta$ = dl $[1-(u/c)\cos\theta]$. If the number of charges per metre length, counted at the fixed time t^*, is N_0, then δN, the total number of charges counted by the information collecting sphere inside dl, is given by

$$\delta N = N_0\,\mathrm{d}l\left[1-\left(\frac{u}{c}\right)\cos\theta\right]. \tag{5.1}$$

Since $[uR\cos\theta]$ is equal $[\mathbf{u}\cdot\mathbf{R}]$, we can rewrite $[(u/c)\cos\theta]$ as $[\mathbf{u}\cdot\mathbf{R}/Rc]$, so that equation (5.1) can be rewritten in the form

$$\delta N = N_0\left[1-\frac{\mathbf{u}\cdot\mathbf{R}}{Rc}\right]\mathrm{d}l = N_0\left(\frac{s}{R}\right)\mathrm{d}l \tag{5.2}$$

where

$$s = \left[R-\frac{\mathbf{u}\cdot\mathbf{R}}{c}\right] \tag{5.3}$$

is the same as the quantity s defined by equation (3.5) and which appears in the expressions for the Liénard-Wiechert potentials.

Assume now that the charges are moving downwards with velocity $[\mathbf{u}]$ in Figure 5.1, and that the angle θ is still as shown in Figure 5.1. In the time interval $(\mathrm{d}l\cos\theta)/c$ it takes the information collecting sphere to cross dl, a charge that is at a distance $[u/c]$ (dl) $\cos\theta$ above B at the time $t^* = (t - R/c)$, when the information collecting sphere passes A, will move downwards to reach B and be counted by the information collecting sphere when it passes B. In this case the number of charges counted by the information collecting sphere inside dl is equal to the number of charges in a length dl $[1+(u/c)\cos\theta]$ at the time t^*, so that

$$\delta N = N_0\left[1+\left(\frac{u}{c}\right)\cos\theta\right]\mathrm{d}l. \tag{5.4}$$

When the charges are moving downwards, the angle between $[\mathbf{u}]$ and $\mathbf{R}$ is equal to $(\pi-\theta)$ so that

$$[\mathbf{u}]\cdot\mathbf{R} = [u]R\cos(\pi-\theta) = -[uR\cos\theta].$$

Hence equation (5.4) can be rewritten in the form

$$\delta N = N_0\left[1-\frac{\mathbf{u}\cdot\mathbf{R}}{Rc}\right]\mathrm{d}l = N_0\left(\frac{s}{R}\right)\mathrm{d}l \tag{5.5}$$

which is the same as equation (5.2).

Assume now that the moving charges in Figure 5.1 are distributed over an infinitesimal area of cross section dS. If n is the number of moving charges per cubic metre, counted at the retarded time t^*, then the number of charges per metre length at the retarded time t^* is

$$N_0 = n \, \mathrm{d}S.$$

Substituting in equation (5.2), we obtain

$$\delta N = n \left(\frac{s}{R} \right) \mathrm{d}V_s \tag{5.6}$$

where the volume element $\mathrm{d}V_s$ is equal to $(\mathrm{d}S)(\mathrm{d}l)$. Equation (5.6) will be used in Section 6.7 of Chapter 6 to derive the retarded potentials from the Liénard-Wiechert potentials.

It will be assumed, that when equations (5.2), (5.5) and (5.6) are applied, the number of conduction electrons is so large that fluctuations in N_0 and n can be neglected. Since n is of the order of 8.3×10^{28} per cubic metre for a copper conductor, this assumption is reasonable for conduction current flow in a metallic conductor.

5.3. Induction electric field due to a current element that forms part of a complete circuit

5.3.1. *A list of basic assumptions*

To illustrate the method we shall use in Chapters 5 and 6, the reader should start at the field point P in Figure 5.1 at the time of observation t and then work out the appropriate retarded positions of each of the moving classical point charges in Figure 5.1. The reader should find that these retarded positions are precisely the positions of the charges, when they were passed by the imaginary information collecting sphere that reaches the field point P in Figure 5.1 at the time of observation t. For example, the charge whose retarded position is at the point B in Figure 5.1 would have been at the position C at the earlier time $t^* = t - R/c$ when the information collection sphere passed A. Hence the number δN of charges, whose retarded positions were between A and B in Figure 5.1, is equal to the number of charges that were between A and C, that is in a length $\mathrm{d}l(1 - (u/c)\cos\theta)$, at the time $t^* = t - R/c$. If there are N_0 charges per unit length at the time t^*, it follows that

$$\delta N = N_0 \, \mathrm{d}l \left[1 - \left(\frac{u}{c} \right) \cos\theta \right]$$

which is the same as equation (5.1) and leads on to equation (5.6), which now gives the number of retarded positions in a volume element $\mathrm{d}V_s$ at a distance R from the field point. We shall assume that the electric and magnetic fields due to each of the moving charges is given by equations (3.10) and (3.13) of Chapter 3 respectively. Writing out our assumptions explicitily, we shall assume that:

1. The number of moving and accelerating classical point charges, whose retarded positions are inside a volume element dV_s, which is the same as the number of charges passed by the information collection sphere inside dV_s, is given by equation (5.6).
2. The electric and magnetic fields due to each of these accelerating charges are given by equations (3.10) and (3.13) respectively.
3. The resultant force **F** on a test charge of magnitude q that is moving with velocity **u** at a field point in empty space is given by the vector sum of the Lorentz forces $\mathbf{F}_i$ on the test charge due to all the individual moving classical point charges making up the charge and current distributions that is

$$\begin{aligned}\mathbf{F} &= \Sigma\mathbf{F}_i = \Sigma(q\mathbf{E}_i + q\mathbf{u} \times \mathbf{B}_i) \\ &= q\Sigma\mathbf{E}_i + q\mathbf{u} \times (\Sigma\mathbf{B}_i) = q\mathbf{E} + q\mathbf{u} \times \mathbf{B}\end{aligned}$$

   where $\mathbf{E} = \Sigma\mathbf{E}_i$ and $\mathbf{B} = \Sigma\mathbf{B}_i$ are the resultant electric and magnetic fields at the field point, which can be determined by adding the fields due to individual classical point charges vectorially.
4. In practice there are interactions between the various charge and current distributions, which affect the values of the charge and current densities. We shall assume, for purposes of interpretation, that the charge and current densities at the appropriate retarded times are all known.

5.3.2. *The induction electric field due to the moving conduction electrons in a current element related to the retarded positions of the conduction electrons*

Consider the current element of length dl that is at the position $\mathbf{r}_s$ in Figure 5.2. The current element is not an isolated current element, but forms part of a stationary circuit carrying a conduction current I, so that there is no accumulation of electric charge at the ends of the current element. We shall determine the contribution to the electric field at the field point P, at a position $\mathbf{r}$ at a distance $\mathbf{R} = (\mathbf{r} - \mathbf{r}_s)$ from the current element, at the time of observation t, due to the electric charges that were passed and counted by the information collecting sphere inside the current element $I\,\mathbf{dl}$ at the retarded time $t^* = (t - R/c)$. A simplified model of a current element will be used. We shall assume that, when they are counted at the same retarded time t^*, there are N_0 stationary positive ions per metre length, each of charge $+e$, and N_0 moving conduction electrons per metre length each of charge $-e$. In this idealized example there is no resultant electric charge on the current element. It will be assumed that at the retarded time t^* all the conduction electrons have the same velocity $[\mathbf{u}]$ and the same acceleration $[\mathbf{a}]$, which are both in the direction opposite to the direction of current flow in Figure 5.2.

It follows from equation (5.5) that the number of conduction electrons passed and counted by the information collecting sphere, while it is crossing

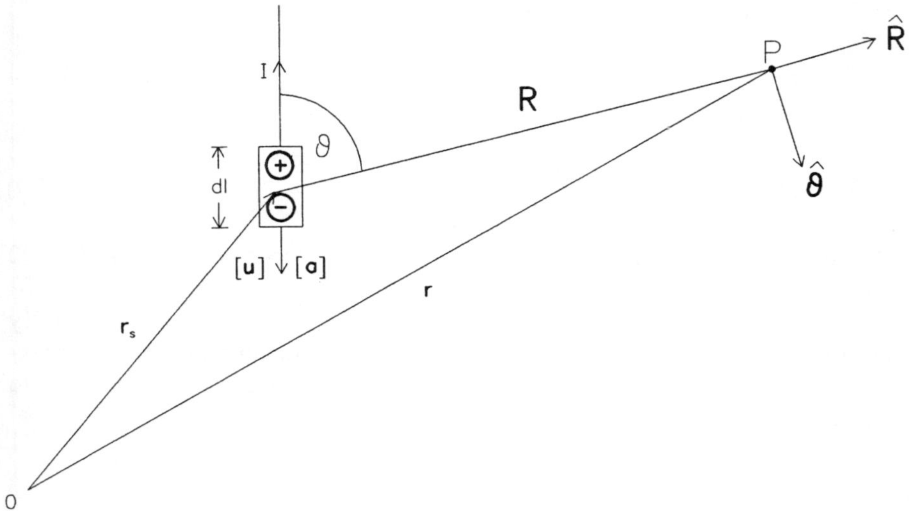

Figure 5.2. Calculation of the electric field at the field point P at the time of observation t due to the charges in a current element, that forms part of a complete circuit. The electric fields of the charges are related to the retarded positions of the charges, that is to the positions of the charges, when the information collecting sphere passed the current element at the retarded time $(t - R/c)$.

the current element $I\,\mathbf{dl}$ at the retarded time t^* is

$$\delta N = N_0\left(\frac{s}{R}\right)\mathrm{d}l. \tag{5.5}$$

The electric field due to each of these conduction electrons is given by equation (3.10). Hence according to the principle of superposition of electric fields, the contribution of the moving conduction electrons, that were counted by the information collecting sphere inside the current element $I\,\mathbf{dl}$ at the retarded time $t^* = (t - R/c)$, to the electric field at the field point P in Figure 5.2, at the time of observation t, is

$$\mathrm{d}\mathbf{E}_- = N_0\left(\frac{s}{R}\right)(\mathbf{E}_V + \mathbf{E}_A)\mathrm{d}l \tag{5.7}$$

where $\mathbf{E}_V$ and $\mathbf{E}_A$ are given by equations (3.11) and (3.12) respectively. In laboratory experiments using DC or the electricity mains, that is in the quasi-stationary approximation, we are invariably in the near zone, where the contribution of the $\mathbf{E}_V$ term is generally much greater than the contribution of the $\mathbf{E}_A$ term which will be neglected for the moment. Using equation (3.11) with $q = -e$ and retaining only the $\mathbf{E}_V$ contribution we have

$$\mathrm{d}\mathbf{E}_- = N_0\left(\frac{s}{R}\right)\frac{(-e)[\mathbf{R} - R\mathbf{u}/c][1 - \beta^2]\mathrm{d}l}{4\pi\varepsilon_0 s^3} \tag{5.8}$$

where $[\beta] = [u]/c$ and where according to equation (5.3)

$$s = \left[R - \frac{\mathbf{u} \cdot \mathbf{R}}{c} \right].$$

Hence

$$\mathrm{d}\mathbf{E}_{-} = -\frac{eN_0 \,\mathrm{d}l[\mathbf{R} - R\mathbf{u}/c][1 - \beta^2]}{4\pi\varepsilon_0 R^3[1 - \mathbf{u} \cdot \mathbf{R}/Rc]^2} . \tag{5.9}$$

The conduction electrons in a metal generally have speeds of the order of $c/200$ corresponding to kinetic energies in the electron volt range. The mean drift velocity of the conduction electrons is very much less than this. To simplify the discussion of the quasi-stationary limit it is reasonable as a first approximation to neglect terms of the order β^2, β^3 etc. Expanding equation (5.9) to the first order in β we find that

$$\begin{aligned} \mathrm{d}\mathbf{E}_{-} &\approx -\frac{eN_0 \,\mathrm{d}l}{4\pi\varepsilon_0 R^3} \left[\mathbf{R} - \frac{R\mathbf{u}}{c} \right] \left[1 + \frac{2\mathbf{u} \cdot \mathbf{R}}{Rc} \right] \\ &\approx \frac{eN_0 \,\mathrm{d}l}{4\pi\varepsilon_0 R^3} \left[-\mathbf{R} - 2\mathbf{R} \left(\frac{\mathbf{u} \cdot \mathbf{R}}{Rc} \right) + \frac{R\mathbf{u}}{c} \right]. \end{aligned} \tag{5.10}$$

Since the positive ions in the current element in Figure 5.2 are at rest, the number of positive ions counted by the information collecting sphere inside I **dl** at the retarded time t^* is equal to $N_0 \,\mathrm{d}l$. The electric field due to each stationary positive ion is given by Coulomb's law, so that the contribution of the positive ions to the electric field at the field point P in Figure 5.2 is

$$\mathrm{d}\mathbf{E}_{+} = N_0 \,\mathrm{d}l \left(\frac{e\mathbf{R}}{4\pi\varepsilon_0 R^3} \right). \tag{5.11}$$

Adding equations (5.10) and (5.11) we find that the electrostatic field due to the stationary positive ions is cancelled by one of the terms in equation (5.10) leaving what we shall call the **induction electric field** denoted $\mathbf{E}_{\text{ind}}$. The value of the induction electric field at the field point P in Figure 5.2 due to the charges counted by the information collecting sphere inside I **dl** at the retarded time t^* is

$$\mathrm{d}\mathbf{E}_{\text{ind}} = \mathrm{d}\mathbf{E}_{+} + \mathrm{d}\mathbf{E}_{-} = \frac{eN_0 \,\mathrm{d}l}{4\pi\varepsilon_0 R^3} \left[-2\mathbf{R} \left(\frac{\mathbf{u} \cdot \mathbf{R}}{Rc} \right) + \frac{R\mathbf{u}}{c} \right]. \tag{5.12}$$

In practice, when a current flows in a conductor, the conduction electrons have a drift velocity **v** superimposed on an isotropic velocity distribution. Averaging equation (5.12) over the velocity distribution of the conduction electrons, we have

$$\mathrm{d}\mathbf{E}_{\text{ind}} = \frac{eN_0 \,\mathrm{d}l}{4\pi\varepsilon_0 R^3} \left[-2\mathbf{R} \frac{\langle \mathbf{u}_i \cdot \mathbf{R} \rangle}{Rc} + \frac{R\langle \mathbf{u}_i \rangle}{c} \right] \tag{5.13}$$

where $\mathbf{u}_i$ is the total velocity of the ith conduction electron. If **v** is the mean drift velocity, then

$$\langle \mathbf{u}_i \rangle = \mathbf{v} \tag{5.14}$$

and, since for moving electrons $\mathbf{v}$ is opposite to $I\mathbf{dl}$,

$$eN_0 \, dl \, \langle \mathbf{u}_i \rangle = eN_0 \, dl\mathbf{v} = -I \, \mathbf{dl}. \tag{5.15}$$

Since $\mathbf{R}$ is a constant vector,

$$\langle \mathbf{u}_i \cdot \mathbf{R} \rangle = \langle \mathbf{u}_i \rangle \cdot \mathbf{R} = \mathbf{v} \cdot \mathbf{R}.$$

Hence, using equation (5.15), we have

$$eN_0 \, dl \, \langle \mathbf{u}_i \cdot \mathbf{R} \rangle = eN_0 \, dl \, \mathbf{v} \cdot \mathbf{R} = -I \, \mathbf{dl} \cdot \mathbf{R}. \tag{5.16}$$

Substituting from equations (5.15) and (5.16) into equation (5.13), we finally obtain

$$\mathrm{d}\mathbf{E}_{\mathrm{ind}} = \frac{[I]}{4\pi\varepsilon_0 c} \left[\frac{2\mathbf{R}(\mathbf{R} \cdot \mathbf{dl})}{R^4} - \frac{\mathbf{dl}}{R^2} \right]. \tag{5.17}$$

If θ is the angle between the direction of the current flow and the vector $\mathbf{R} = (\mathbf{r} - \mathbf{r}_s)$ from the current element at $\mathbf{r}_s$ to the field point at $\mathbf{r}$ in Figure 5.2, then

$$[\mathbf{R} \cdot \mathbf{dl}] = R \cos \theta \, dl.$$

Hence equation (5.17) can be rewritten in the form

$$\mathrm{d}\mathbf{E}(\mathbf{r}, t)_{\mathrm{ind}} = \frac{2[I(\mathbf{r}_s)]\mathbf{R} \, dl \cos \theta}{4\pi\varepsilon_0 c R^3} - \frac{[I(\mathbf{r}_s)]\mathbf{dl}}{4\pi\varepsilon_0 c R^2} \tag{5.18}$$

where $\mathrm{d}\mathbf{E}(\mathbf{r}, t)_{\mathrm{ind}}$ is the contribution to the induction electric field at the field point P in Figure 5.2, at the time of observation t, due to the current $[I]$ flowing in the current element $[I]$ $\mathbf{dl}$ at $\mathbf{r}_s$ at the retarded time $t^* = (t - R/c)$. Equation (5.18) is valid for steady currents and for quasi-stationary conditions when the current is varying at mains frequency and the radiation electric fields are negligible. Equation (5.18) gives the electric field due to the moving electrons and stationary positive ions in a current element that forms part of a complete circuit. In the case of an isolated current element, charge distributions would build up at its ends. (Reference: Section 5.12). For the conditions shown in Figure 5.2, the vector $\mathbf{dl}$ can be expressed in the form

$$\mathbf{dl} = dl \cos \theta \, \hat{\mathbf{R}} - dl \sin \theta \, \hat{\boldsymbol{\theta}} \tag{5.19}$$

where $\hat{\mathbf{R}}$ is a unit vector in the direction from the current element to the field point and $\hat{\boldsymbol{\theta}}$ is a unit vector at the field point P in the direction of increasing θ in Figure 5.2. Substituting for $\mathbf{dl}$ in equation (5.18) we obtain

$$\mathrm{d}\mathbf{E}_{\mathrm{ind}} = \left(\frac{[I] \, dl \cos \theta}{4\pi\varepsilon_0 c R^2} \right) \hat{\mathbf{R}} + \left(\frac{[I] \, dl \sin \theta}{4\pi\varepsilon_0 c R^2} \right) \hat{\boldsymbol{\theta}}. \tag{5.20}$$

The magnitude of the induction electric field given by equation (5.20) is

$$\mathrm{d}E_{\text{ind}} = \frac{[I]\,\mathrm{d}l}{4\pi\varepsilon_0 cR^2} = \beta\left(\frac{N_0 e\,\mathrm{d}l}{4\pi\varepsilon_0 R^2}\right) \tag{5.21}$$

where $\beta = v/c$.

It is straight forward for the reader to show that the direction of $\mathbf{dE}_{\text{ind}}$, given by equation (5.20), is at an angle θ to the line joint the current element to the field point P in Figure 5.2 and at an angle 2θ to the direction of current flow. Comparing equations (5.21) and (5.11), we see that the magnitude of the resultant electric field is $\beta = v/c$ times the magnitude of the electric field $\mathbf{dE}_+$ due to the N_0 dl stationary positive ions inside the current element I $\mathbf{dl}$ in Figure 5.2.

If the conduction electrons in the current element in Figure 5.2 were at rest, the electric fields due to the stationary positive ions and stationary conduction electrons would compensate each other at external field points. The resultant induction electric field $\mathbf{dE}_{\text{ind}}$ given by equations (5.18) and (5.20) arises from

(1) The change in the number of conduction electrons counted by the information collecting sphere inside the current element when the conduction electrons are moving.

(2) The changes in the electric field due to a conduction electron when it is moving, compared with the electrostatic field due to a stationary electron.

5.3.3. *Repeat calculation using the projected positions of the conduction electrons*

Some readers may get a better insight into the origin of the resultant electric field, given by equation (5.20), particularly the origin of the component in the direction of $\hat{\boldsymbol{\theta}}$, if the electric field of a moving conduction electron is related to the projected position of the conduction electron, which is the position

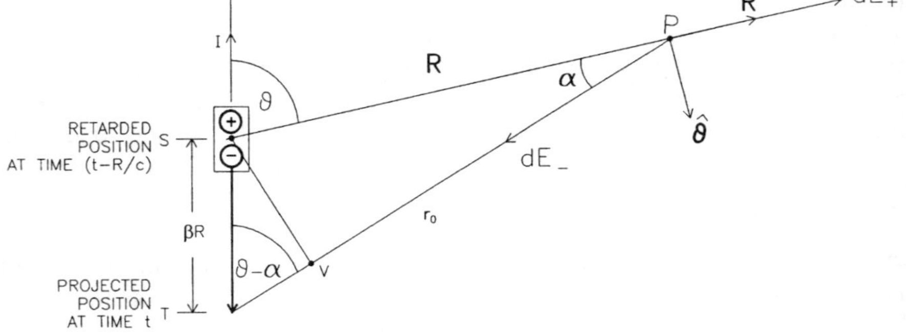

Figure 5.3. The contribution to the electric field due to the charges in a current element that forms part of a complete circuit. In this case the electric fields of the moving conduction electrons are related to their projected positions, which are the positions they would have reached if they had carried on with uniform velocity until the time of observation t.

the electron would have reached by the time of observation t if it had carried on with uniform velocity $[\mathbf{u}]$ in Figure 5.2. Consider again the current element shown in Figure 5.2. The vector $\mathbf{R}$ from the current element to the field point P makes an angle θ with the direction of current flow. A simplified model will be used, in which we shall assume that, at the retarded time t^*, the positive ions are at rest and all the conduction electrons are moving with velocity $[\mathbf{u}]$. Consider one conduction electron and one positive ion, as shown in Figure 5.3. If it had carried on with uniform velocity $[\mathbf{u}]$, in the time R/c it takes the information collecting sphere to go from the current element to the field point P, the conduction electron would have moved downwards a distance $[u]R/c = \beta R$, where $\beta = [u]/c$, from the current element at S to reach the projected position T as shown in Figure 5.3. The $\mathbf{E}_V$ contribution of the conduction electron to the electric field at P at the time of observation t is related to the projected position T by equation (3.38) of Chapter 3. Expanding equation (3.38) to first order of β we have

$$\mathbf{E}_V \approx \frac{q\mathbf{r}_0}{4\pi\varepsilon_0 r_0^3} \tag{5.22}$$

where $\mathbf{r}_0$ is a vector from the projected position T to the field point P in Figure 5.3. Equation (5.22) is similar to Coulomb's law, except that $\mathbf{r}_0$ is measured from the projected position T of the conduction electron in Figure 5.3. It follows from equation (5.4) that the number of conduction electrons counted by the information collecting sphere inside the current element of length dl is

$$\delta N = (1 + \beta \cos \theta)\, N_0 \, \mathrm{d}l \tag{5.4}$$

where N_0 is equal to both the number of conduction electrons and the number of stationary positive ions per metre length counted at the retarded time t^*. If it is related to the projected position of the charge, the electric field due to each of these conduction electron is given by equation (5.22). Hence it follows that the electric field at P at the time of observation t, due to the conduction electrons counted by the information collecting sphere inside the current element at the retarded time t^*, is given to a good approximation by

$$\mathrm{d}E_- = (1 + \beta \cos \theta)\, N_0 \, \mathrm{d}l \left(\frac{e}{4\pi\varepsilon_0 r_0^2} \right) \tag{5.23}$$

in the direction from the field point P to the projected position T, as shown in Figure 5.3. Since they are at rest, the number of positive ions, counted by the information collecting sphere inside the current element is equal to N_0 dl. Since the electric field due to each of these is given by Coulomb's law, the electric field due to the stationary positive ions is

$$\mathrm{d}E_+ = \frac{eN_0 \, \mathrm{d}l}{4\pi\varepsilon_0 R^2} \tag{5.24}$$

in the direction from the current element to the field point P, as shown in

Figure 5.3. Equations (5.23) and (5.24) do not compensate each other. It can be seen from Figure 5.3 that the resultant of $\mathrm{d}\mathbf{E}_+$ and $\mathrm{d}\mathbf{E}_-$ has a component in the direction of $\hat{\boldsymbol{\theta}}$. Resolving in the direction of the vector **R** from the current element to the field point, we have

$$\mathrm{d}E_R = \frac{eN_0\,\mathrm{d}l}{4\pi\varepsilon_0 R^2} - (1 + \beta\cos\theta)\left(\frac{eN_0\,\mathrm{d}l}{4\pi\varepsilon_0 r_0^2}\right)\cos\alpha \tag{5.25}$$

where α is the angle between the vectors **R** and $\mathbf{r}_0$, as shown in Figure 5.3.

Resolving in the direction of $\hat{\boldsymbol{\theta}}$, that is in a direction perpendicular to **R**, we have

$$\mathrm{d}E_\theta = \mathrm{d}E_-\sin\alpha = (1 + \beta\cos\theta)\left(\frac{eN_0\,\mathrm{d}l}{4\pi\varepsilon_0 r_0^2}\right)\sin\alpha \tag{5.26}$$

Since the distance from S to T in Figure 5.3 is $[u]R/c = \beta R$, it follows that the perpendicular distance from S to V is given by

$$SV = \beta R\sin(\theta - \alpha).$$

If $\beta \ll 1$, then βR is very small and $\alpha \ll \theta$, so that to a very good approximation

$$SV = \beta R\sin\theta. \tag{5.27}$$

It follows from Figure 5.3 that

$$\sin\alpha = \frac{\beta R\sin\theta}{R} = \beta\sin\theta \tag{5.28}$$

$$\cos\alpha = (1 - \sin^2\alpha)^{1/2} = (1 - \beta^2\sin^2\theta)^{1/2}$$

$$r_0 = R + \beta R\cos(\theta - \alpha).$$

Since $\alpha \ll \theta$ and $\beta \ll 1$, we have

$$\cos\alpha \approx 1 \tag{5.29}$$

$$r_0 \approx R(1 + \beta\cos\theta). \tag{5.30}$$

Substituting for $\sin\alpha$, $\cos\alpha$ and r_0 from equations (5.28), (5.29) and (5.30) into equations (5.25) and (5.26) we finally obtain

$$\begin{aligned}\mathrm{d}E_R &= \frac{eN_0\,\mathrm{d}l}{4\pi\varepsilon_0 R^2} - \frac{(1 + \beta\cos\theta)eN_0\,\mathrm{d}l}{4\pi\varepsilon_0 R^2(1 + \beta\cos\theta)^2} = \frac{eN_0\,\mathrm{d}l}{4\pi\varepsilon_0 R^2}[1 - (1 + \beta\cos\theta)^{-1}]\\ &\approx \frac{eN_0\,\mathrm{d}l}{4\pi\varepsilon_0 R^2}\beta\cos\theta = \frac{eN_0 u\,\mathrm{d}l\cos\theta}{4\pi\varepsilon_0 cR^2} = \frac{I\,\mathrm{d}l\cos\theta}{4\pi\varepsilon_0 cR^2}\end{aligned} \tag{5.31}$$

$$\mathrm{d}E_\theta = \frac{eN_0\,\mathrm{d}l(1 + \beta\cos\theta)}{4\pi\varepsilon_0 R^2(1 + \beta\cos\theta)^2}\left(\frac{u}{c}\right)\sin\theta \approx \frac{I\,\mathrm{d}l\sin\theta}{4\pi\varepsilon_0 cR^2}. \tag{5.32}$$

Equations (5.31) and (5.32) are in agreement with equation (5.20). It is of interest to note that in the approach used in this section when $\beta \ll 1$, we used the mathematical expression for Coulomb's law for both the positive ions and the conduction electrons, except that in the case of the moving conduction electrons we had to apply Coulomb's law relative to the projected positions of the conduction electrons, that is the positions they would have reached if they had carried on with uniform velocity $[u]$ until the time of observation t. We also had to allow for the fact that the number of moving conduction electrons counted by the information collecting sphere inside the current element in Figure 5.3, which is given by equation (5.5), is not equal to the number of stationary positive ions counted.

5.4. The absence of a resultant induction electric field due to a steady conduction currect flowing in a complete circuit

5.4.1. *General case*

According to equation (5.18), the conduction current in a current element, which forms part of a complete electrical circuit, gives rise to an induction electric field at an external field point. In this section, we shall show that the resultant induction electric field due to the **steady** current in a complete circuit, that has no resultant charge distribution anywhere on the conductors, is zero. This result follows directly from conventional classical electromagnetism, since, if the charge density ρ is zero everywhere, the scalar potential ϕ is zero and the electric field $\mathbf{E}$ is given by $-\dot{\mathbf{A}}$ where $\mathbf{A}$ is the vector potential. When the conditions are steady $\dot{\mathbf{A}}$ is zero showing that $\mathbf{E}$ should be zero.

Consider the electric circuit shown in Figure 5.4. The vector $\mathbf{R}$ is a vector from the current element $I\,\mathrm{d}\mathbf{l}$ at $\mathbf{r}_s$ to the field point P at $\mathbf{r}$. The resultant induction electric field at the field point P will be determined when a steady current I flows in the circuit. Integrating equation (5.18) we have for a steady conduction current I in a complete circuit

$$\mathbf{E}_{\mathrm{ind}} = \frac{I}{4\pi\varepsilon_0 c}\left[2\oint \frac{\mathbf{R}\cos\theta\,\mathrm{d}l}{R^3} - \oint \frac{\mathrm{d}\mathbf{l}}{R^2}\right] \tag{5.33}$$

where θ is the angle between $\mathrm{d}\mathbf{l}$ and $\mathbf{R}$. It is shown in Appendix A3 that

$$2\oint \frac{\mathbf{R}\cos\theta\,\mathrm{d}l}{R^3} = \oint \frac{\mathrm{d}\mathbf{l}}{R^2}\,. \tag{A3.8}$$

Substituting in equation (5.33) we find that, when the current I is steady,

$$\mathbf{E}_{\mathrm{ind}} = 0. \tag{5.34}$$

The currents in the various current elements forming the circuit in Figure 5.4 do give rise to induction electric fields at the field point P, which are given by equation (5.18). It is the resultant induction electric field due to the complete circuit that is zero when the current is steady. This result will now be illustrated using a simple example.

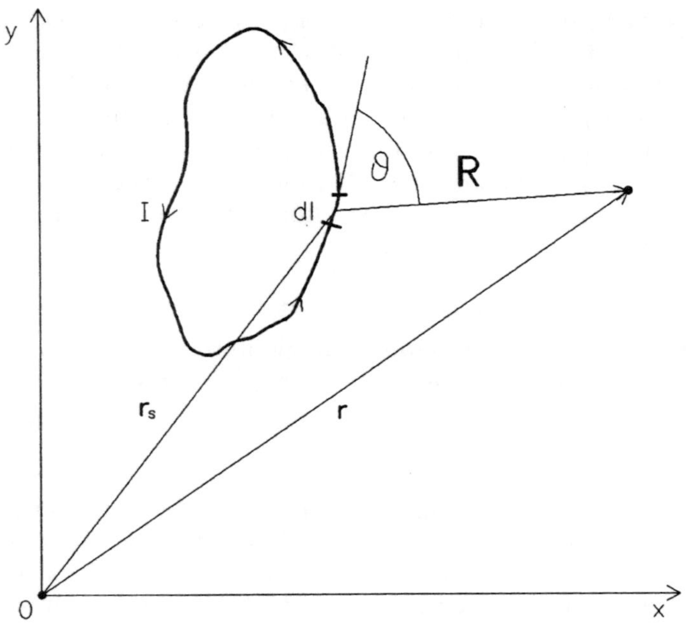

Figure 5.4. Calculation of the electric field at the field point P due to a complete circuit for both steady and varying currents.

5.4.2. *Example to illustrate that* $\mathbf{E}_{\text{ind}} = 0$ *when the conduction current in a circuit is constant*

Consider the small current carrying coil $ABCD$ shown in Figure 5.5. We shall determine the resultant electric field at the field point P, at the time of observation t. A steady current I flows from A to B to C to D to A in Figure 5.5. The sections AD and BC are arcs of circles with centres at the field point P and having radii r and $(r + \delta r)$ respectively, where $r \gg \delta r$. The length

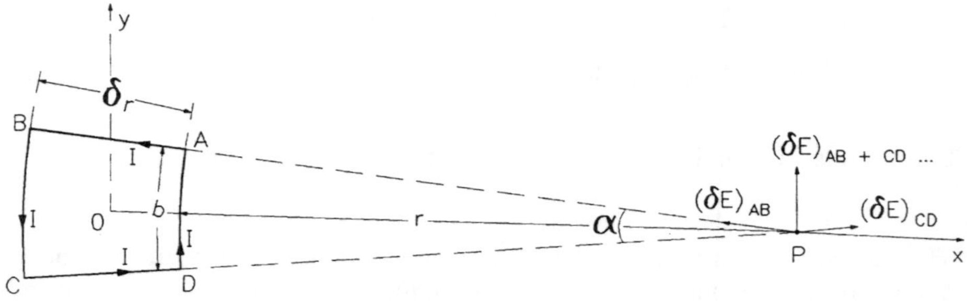

Figure 5.5. The calculation of the electric fields at the field point P due to both steady and varying currents in the coil $ABCD$.

of the sections AD and BC are equal to b and $(b + \delta b)$ respectively, where $b \ll r$. Since AD and BC are arcs of circles with centres at P

$$\frac{b}{r} = \frac{(b + \delta b)}{(r + \delta r)} = \frac{b(1 + \delta b/b)}{r(1 + \delta r/r)} . \tag{5.35}$$

It follows from equation (5.35) that $(1 + \delta b/b)$ is equal to $(1 + \delta r/r)$. Hence the length of the arc BC is equal to $b(1 + \delta r/r)$. Choose the directions of the x and y axes as shown in Figure 5.5.

In the case of the section AB of the coil $ABCD$ in Figure 5.5, the angle θ between $I\,\mathbf{dl}$ and a vector from the current element to the field point P is equal to π, so that $\cos\theta = -1$ and $\sin\theta = 0$ in equation (5.20), which reduces to

$$(\delta E)_{AB} = -\frac{I\,\delta r}{4\pi\varepsilon_0 c(r + \delta r/2)^2} \approx -\frac{I\,\delta r}{4\pi\varepsilon_0 c r^2} . \tag{5.36}$$

This contribution is in the direction from P to A, as shown in Figure 5.5. In the case of the section CD, $\theta = 0$ so that $\cos\theta = 1$ and $\sin\theta = 0$ in equation (5.20), which reduces to

$$(\delta E)_{CD} = +\frac{I\,\delta r}{4\pi\varepsilon_0 c r^2} . \tag{5.37}$$

This contribution is in the direction from D to P as shown in Figure 5.5. The resultant of $(\delta \mathbf{E})_{AB}$ and $(\delta \mathbf{E})_{CD}$ is in the $+y$ direction in Figure 5.5. Since the angle α between the lines AP and DP in Figures 5.5 is very small, $\sin(\alpha/2) \approx \alpha/2 = b/2r$. Hence

$$(\delta \mathbf{E})_{AB + CD} = \frac{2I\,\delta r \sin(\alpha/2)}{4\pi\varepsilon_0 c r^2}\,\hat{\mathbf{j}} = \frac{Ib\,\delta r}{4\pi\varepsilon_0 c r^3}\,\hat{\mathbf{j}} \tag{5.38}$$

where $\hat{\mathbf{j}}$ is a unit vector in the $+y$ direction.

In the case of the section DA of the circuit $ABCD$ in Figure 5.5 $I\,\mathbf{dl}$ is perpendicular to the vector from the current element to the field point P so that $\theta = \pi/2$, $\cos\theta = 0$ and $\sin\theta = 1$ in equation (5.20) which reduces to

$$(\delta \mathbf{E})_{AD} = -\frac{Ib}{4\pi\varepsilon_0 c r^2}\,\hat{\mathbf{j}} \tag{5.39}$$

This contribution to the electric field is in the $-y$ direction in Figure 5.5. In the case of the section BC of the circuit, $\theta = \pi/2$ and equation (5.20) reduces to

$$(\delta \mathbf{E})_{BC} = \frac{Ib(1 + \delta r/r)}{4\pi\varepsilon_0 c r^2(1 + \delta r/r)^2}\,\hat{\mathbf{j}} \approx \frac{Ib(1 - \delta r/r)}{4\pi\varepsilon_0 c r^2}\,\hat{\mathbf{j}}. \tag{5.40}$$

This contribution is in the $+y$ direction in Figure 5.5. Adding equations (5.39) and (5.40) we have

$$(\delta \mathbf{E})_{AD + BC} = -\frac{Ib\,\delta r}{4\pi\varepsilon_0 c r^3}\,\hat{\mathbf{j}}. \tag{5.41}$$

Comparing equations (5.41) and (5.38), we see that $(\delta\mathbf{E})_{AB+CD}$ is equal and opposite to $(\delta\mathbf{E})_{AD+BC}$ so that, when the conduction current in the coil *ABCD* in Figure 5.5 is steady, the resultant induction electric field at the field point *P* is zero. This example illustrates how the conduction currents in the various current elements *AB*, *BC*, *CD* and *DA* of the circuit *ABCD* in Figure 5.5 do individually give rise to induction electric fields at the field point *P*. It is the resultant induction electric field due to the complete circuit that is zero, when the current is steady. We shall go on to consider the case when the current is varying in Section 5.5.1.

Problem. The steady conduction current I flowing in a circular coil of radius b is due to N_0 conduction electrons per metre length each of charge $-e$ and all moving at the same uniform speed u, plus N_0 stationary positive ions per metre length each of charge $+e$. Use equation (5.18) or equation (5.20) to show that, to first order of $\beta = u/c$, the electric field $\mathbf{E}$ at a field point on the axis of the circular coil at a distance x from the centre of the coil is zero. [Hint: For any current element $I\,d\mathbf{l}$, $\theta = 90°$ in equation (5.18) so that the contribution in the direction of $\hat{\mathbf{R}}$ is zero. Show that the contributions of the components in the directions of $-d\mathbf{l}$ add up to zero]. It is straight forward in this example to show that $\mathbf{E} = 0$ to all orders of β as follows. The centripetal accelerations of the electrons is u^2/b. Use equation (5.5) plus the full expressions for $\mathbf{E}_V$ and $\mathbf{E}_A$ given by equations (3.11) and (3.12) respectively to show that the resultant contributions of the $\mathbf{E}_V$ terms due to the speeds of the electrons and the $\mathbf{E}_A$ terms due to their centripetal accelerations are both numerically equal to $\beta^2 N_0 ebx/2\varepsilon_0(b^2 + x^2)^{3/2}$, but, since these contributions are in opposite directions, the resultant electric field is zero to all orders of β. (Comment: Notice that at the higher orders of β we must include the contributions of the acceleration dependent term $\mathbf{E}_A$ to get the result $\mathbf{E} = 0$. The first order theory given by equations (5.18) and (5.20) will however be sufficiently accurate for our interpretation of electromagnetic induction in the quasi-stationary limit.)

5.5. The induction electric field due to a varying conduction current in a coil in the quasi-stationary limit

5.5.1. *Introductory example*

Before going on in Section 5.5.2 to consider the general case, we shall illustrate the role of retardation effects in giving a resultant induction electric field using the example of the coil *ABCD* in Figure 5.5. We shall assume that the conduction current flowing in the coil *ABCD* in Figure 5.5 is varying with time, but at such a slow rate that the quasi-stationary approximations are valid and equation (5.20) can be applied. We shall assume that the value of the conduction current in the coil *ABCD* in Figure 5.5. is equal to I when the information collecting sphere, that reaches the field point P at the time of observation t, passes the section BC at the retarded time $t - (r + \delta r)/c$. According to equation (5.40)

$$(\delta \mathbf{E})_{BC} = \frac{Ib(1 - \delta r/r)}{4\pi\varepsilon_0 cr^2}\hat{\mathbf{j}} \tag{5.42}$$

If the rate of change of the conduction current is $\partial I/\partial t^*$, the value of the current in the coil $ABCD$ when the information collecting sphere passes the section AD at a time $(\delta r/c)$ after it has passed the section BC, is

$$I_{AD} = I + \left(\frac{\partial I}{\partial t^*}\right)\left(\frac{\delta r}{c}\right). \tag{5.43}$$

Hence, according to equation (5.39)

$$(\delta \mathbf{E})_{AD} = -\frac{\left\{I + \left(\frac{\partial I}{\partial t^*}\right)\left(\frac{\delta r}{c}\right)\right\} b}{4\pi\varepsilon_0 cr^2}\hat{\mathbf{j}}. \tag{5.44}$$

Adding equations (5.42) and (5.44) we have

$$(\delta \mathbf{E})_{BC} + (\delta \mathbf{E})_{AD} = -\frac{Ib}{4\pi\varepsilon_0 cr^2}\left(\frac{\delta r}{r}\right)\hat{\mathbf{j}} - \frac{\left(\frac{\partial I}{\partial t^*}\right)\left(\frac{\delta r}{c}\right) b}{4\pi\varepsilon_0 cr^2}\hat{\mathbf{j}}. \tag{5.45}$$

The value of the conduction current in the circuit $ABCD$ is varying continuously from I to $I + (\partial I/\partial t^*)(\delta r/c)$ as the information sphere passes along BA and CD. If δr is infinitesimal, it is sufficiently accurate to use the mean value of current in equation (5.38), which then gives

$$(\delta \mathbf{E})_{AB+CD} = \frac{b\,\delta r\left\{I + \frac{1}{2}\left(\frac{\partial I}{\partial t^*}\right)\left(\frac{\delta r}{c}\right)\right\}}{4\pi\varepsilon_0 cr^3}\hat{\mathbf{j}}.$$

Neglecting terms of order $(\delta r)^2$ we have

$$(\delta \mathbf{E})_{AB+CD} = \frac{b\,\delta r\,I}{4\pi\varepsilon_0 cr^3}\hat{\mathbf{j}}. \tag{5.46}$$

Adding equations (5.45) and (5.46), we find that when the current I in the coil $ABCD$ in Figure 5.5 is varying there is a resultant induction electric field $\mathbf{E}_{\text{ind}}$ at the field point P in Figure 5.5, which at the time of observation t is given by

$$\mathbf{E}_{\text{ind}} = -\frac{b\,\delta r}{4\pi\varepsilon_0 c^2 r^2}\left(\frac{\partial I}{\partial t^*}\right)\hat{\mathbf{j}}. \tag{5.47}$$

The product $b\delta r$ is equal to the area S of the coil $ABCD$. The magnetic moment $\mathbf{m}$ of the plane coil is defined as a vector of magnitude IS pointing in the $+z$ direction in Figure 5.5. Using $\mu_0 = 1/\varepsilon_0 c^2$, equation (5.47) can be rewritten in the form

$$\mathbf{E}_{\text{ind}} = -\frac{S[\dot{I}]}{4\pi\varepsilon_0 c^2 r^2}\,\hat{\mathbf{j}} = -\frac{[\dot{m}]}{4\pi\varepsilon_0 c^2 r^2}\,\hat{\mathbf{j}} = -\left(\frac{\mu_0}{4\pi}\right)\frac{[\dot{m}]}{r^2}\,\hat{\mathbf{j}}. \tag{5.48}$$

We can now see the origin of the induction electric field due to a varying current in a coil in the quasi-stationary limit. When the current in the coil *ABCD* in Figure 5.5 is steady, the contributions of the various current elements making up the circuit to the induction electric field at *P* add up to zero. However, when the current in the coil is varying, the value of current to be used in equation (5.20) is different when the information collecting sphere passes the various current elements making up the circuit and the contributions of the various current elements no longer add up to zero. For example in the special case of the coil *ABCD* in Figure 5.5, the induction electric field, given by equation (5.47), is due to the fact that the conduction current is bigger by an amount $(\partial I/\partial t^*)(\delta r/c)$ when the information collecting sphere passes *AD* compared with when it passes *BC*. This illustrates the essential role of retardation effects in giving a resultant induction electric field, even in the quasi-stationary limit when the overwhelming contribution to the induction electric field comes from the $\mathbf{E}_V$ term in the expression for the electric field due to a moving conduction electron. Notice that there is a factor of $1/\varepsilon_0 c^2$ in equation (5.47) showing that the induction electric field due to the varying current in the coil *ABCD* in Figure 5.5 is very much less than the electrostatic field due to the stationary positive ions or the total electric field due to the moving conduction electrons. One factor of $1/c$ in equation (5.47) comes from equation (5.20) giving the induction electric field due to a current element. The other factor of $1/c$ comes from the fact that the resultant induction electric field at the field point *P* in Figure 5.5 is due to the difference of $(\delta r\dot{I})/c$ in the current when the information collecting sphere passes *AD* compared with *BC*. Putting $1/\varepsilon_0 c^2$ equal to μ_0 we see from equation (5.48) that the induction electric field has a similar mathematical form to the magnetic field $\mu_0 IS/4\pi r^2$ due to a steady current *I* in the coil *ABCD* in Figure 5.5, which for a steady current can be obtained by putting $\dot{m}$ and $\ddot{m}$ equal to zero in equation (6.49) of Chapter 6. (See also the problem at the end of Section 5.6.)

5.5.2. *General formula for the induction electric field due to the varying current in a complete circuit in the quasi-stationary limit*

Consider the current element $I\,d\mathbf{l}$ that forms part of the complete circuit shown previously in Figure 5.4. It will now be assumed that the current in the coil is varying, but at a low enough frequency for the quasi-stationary approximations to be valid. According to equation (5.18), the actual contribution of the charges, that were counted by the information collecting sphere inside the current element $I\,d\mathbf{l}$ at $\mathbf{r}_s$ at the retarded time $t^* = (t - R/c)$, to the electric field at the field point *P* at $\mathbf{r}$ at the time of observation *t* is given in the quasi-stationary limit by

$$\mathrm{d}\mathbf{E}_{\text{ind}} = \frac{2I(t^*)\mathbf{R}\cos\theta\,\mathrm{d}l}{4\pi\varepsilon_0 cR^3} - \frac{I(t^*)\,\mathrm{d}\mathbf{l}}{4\pi\varepsilon_0 cR^2} \tag{5.49}$$

where $\mathbf{R} = (\mathbf{r} - \mathbf{r}_s)$ and θ is the angle between $\mathrm{d}\mathbf{l}$ and $\mathbf{R}$. Since the different current elements, making up the circuit in Figure 5.4, are at different distances R from the field point, the value of $I(t^*)$ is different for the various current elements. If the rate of change of current is $\dot{I}$, then, in the quasi-stationary limit, the change in the value of the current in the circuit in the time R/c it takes the information collecting sphere to go from the current element to the field point is $(R/c)\dot{I}$. Hence, if $I(t)$ is the value of the current in the circuit at the fixed time of observation t, we have to a very good approximation

$$I(t^*) = I(t) - \left(\frac{R}{c}\right)\dot{I}. \tag{5.50}$$

We can ignore the change in $\dot{I}$ in the time R/c in quasi-stationary limit, since for example, for a laboratory experiment of dimensions 3 m, R/c is only of the order of 10^{-8} s. Substituting for $I(t^*)$ in equation (5.49) and integrating around the complete circuit in Figure 5.4, we have

$$\mathbf{E}(\mathbf{r}, t)_{\text{ind}} = \frac{1}{4\pi\varepsilon_0 c}\left(2\oint\frac{[I(t) - (R/c)\dot{I}]\mathbf{R}\cos\theta\,\mathrm{d}l}{R^3} - \oint\frac{[I(t) - (R/c)\dot{I}]\,\mathrm{d}\mathbf{l}}{R^2}\right). \tag{5.51}$$

The reader should note that we are allowing for retardation effects in equation (5.51). The value of the current $I(t)$ at the time of observation t is the same for all the current elements. It follows from equation (A3.8) of Appendix A3 that, for a fixed circuit, we have at the time of observation t

$$2I(t)\oint\frac{\mathbf{R}\cos\theta\,\mathrm{d}l}{R^3} = I(t)\oint\frac{\mathrm{d}\mathbf{l}}{R^2}. \tag{A3.8}$$

Hence equation (5.51) reduces to

$$\mathbf{E}(\mathbf{r}, t)_{\text{ind}} = \frac{\dot{I}}{4\pi\varepsilon_0 c^2}\left(-2\oint\frac{\mathbf{R}\cos\theta\,\mathrm{d}l}{R^2} + \oint\frac{\mathrm{d}\mathbf{l}}{R}\right). \tag{5.52}$$

According to equation (A3.9) of Appendix A3 we have

$$-2\oint\frac{\mathbf{R}\cos\theta\,\mathrm{d}l}{R^2} = -2\oint\frac{\mathrm{d}\mathbf{l}}{R}. \tag{A3.9}$$

Substituting in equation (5.52), we finally obtain for the quasi-stationary limit

$$\mathbf{E}(\mathbf{r}, t)_{\text{ind}} = -\frac{\dot{I}}{4\pi\varepsilon_0 c^2}\oint\frac{\mathrm{d}\mathbf{l}(\mathbf{r}_s)}{R}. \tag{5.53}$$

The finite value of $\mathbf{E}_{\text{ind}}$ arises from equation (5.49) by retardation effects. The integrands $(\mathrm{d}\mathbf{l}/R)$ and $(\mathbf{R}\cos\theta\,\mathrm{d}l/R^2)$ in equation (A3.9) are not equal, so that equation (A3.9) and hence equation (5.53) are only valid when they are integrated around a complete circuit. Equation (5.53) was derived from equation (5.18), which was developed from the $\mathbf{E}_V$ contribution to the electric field due to a moving classical point charge. Hence equation (5.53) is only valid for complete circuits in the quasi-stationary limit, when the radiation electric fields can be neglected.

For a stationary circuit, R and $\mathbf{dl}$ are time independent so that equation (5.53) can be rewritten in the form

$$\mathbf{E}_{\text{ind}} = -\frac{\partial}{\partial t}\left\{\frac{\mu_0 I}{4\pi}\oint\frac{\mathbf{dl}}{R}\right\} = -\frac{\partial \mathbf{A}}{\partial t} \tag{5.54}$$

where $\mathbf{A}$ is given in the present context by the equation

$$\mathbf{A} = \frac{\mu_0 I}{4\pi}\oint\frac{\mathbf{dl}}{R}. \tag{5.55}$$

Comparing equation (5.55) with equation (1.74) of Chapter 1, we see that the quantity $\mathbf{A}$, defined by equation (5.55), is the same as the expression for the vector potential $\mathbf{A}$ at the field point P in Figure 5.4, when the current in the coil is constant. This shows that, when the scalar potential ϕ is zero, the induction electric field due to the varying current in an electrically neutral stationary circuit can be determined, in the quasi-stationary limit, by first determining the vector potential $\mathbf{A}$ assuming that the current in the coil is steady and then determining $-\dot{\mathbf{A}}$. There is no need to use the retarded vector potential in the quasi-stationary limit and allow for retardation effects, even though we showed in Sections 5.5.1 and 5.5.2 that retardation effects play a vital role in giving a resultant induction electric field, when we start with equation (5.18) for the electric field due to a current element, which we had determined using the expression for the electric field due to a moving classical point charge. Equations (5.18), (5.20) and (5.53) must be extended at high frequencies, when the contribution of the acceleration dependent (radiation) term $\mathbf{E}_A$ in the expression for the electric field due to an accelerating classical point charge is important.

5.5.3. *Application of equation (5.53) to the coil* ABCD *in Figure 5.5*

Consider again the coil *ABCD* in Figure 5.5. The current in the coil is varying slowly enough for the quasi-stationary approximations to be valid. The contribution of the current in the section *AB* of the coil *ABCD* in Figure 5.5 to the right hand side of equation (5.53) is

$$(\Delta E)_{AB} = \frac{\dot{i}\,\delta r}{4\pi\varepsilon_0 c^2(r + \delta r/2)} \tag{5.56}$$

in the direction from the point A to the field point P. Notice that, to avoid confusion, we are now using ΔE and not δE, as we did when we applied equation (5.20). This is because ΔE, which is the contribution of the section

AB of the coil *ABCD* to the right hand side of equation (5.53), is not equal to the contribution δE to the resultant induction electric field at the field point *P* actually due to the moving conduction electrons and stationary positive ions in the section *AB*, which is given by equation (5.20).

The contribution of the section *CD* to the right hand side of equation (5.53) is

$$(\Delta E)_{CD} = \frac{\dot{I}\,\delta r}{4\pi\varepsilon_0 c^2(r + \delta r/2)} \tag{5.57}$$

in the direction from the field point *P* to the point *D* in Figure 5.5. The resultant of the contributions of the sections *AB* and *CD* to the right hand side of equation (5.53) is

$$(\Delta \mathbf{E})_{AB\,+\,CD} = -\frac{2\dot{I}(\delta r)\sin(\alpha/2)}{4\pi\varepsilon_0 c^2(r + \delta r/2)}\,\hat{\mathbf{j}}$$

where $\hat{\mathbf{j}}$ is a unit vector in the $+y$ direction and α is the angle between the lines *AP* and *DP* in Figure 5.5. Since α is very small, $\sin(\alpha/2) \approx \alpha/2 = b/2r$. Hence

$$(\Delta \mathbf{E})_{AB\,+\,CD} = -\frac{\dot{I}b(\delta r)}{4\pi\varepsilon_0 c^2 r^2}\,\hat{\mathbf{j}}. \tag{5.58}$$

The contributions of the sections *BC* and *DA* to the right hand side of equation (5.53) are

$$(\Delta \mathbf{E})_{BC} = \frac{\dot{I}b(1 + \delta r/r)}{4\pi\varepsilon_0 c^2 r(1 + \delta r/r)}\,\hat{\mathbf{j}} = \frac{\dot{I}b}{4\pi\varepsilon_0 c^2 r}\,\hat{\mathbf{j}}. \tag{5.59}$$

$$(\Delta \mathbf{E})_{DA} = -\frac{\dot{I}b}{4\pi\varepsilon_0 c^2 r}\,\hat{\mathbf{j}}. \tag{5.60}$$

Adding equations (5.58), (5.59) and (5.60), we find that the resultant electric field at the field point *P* in Figure 5.5 due to the varying conduction current in the coil *ABCD* is

$$\mathbf{E}_{\text{ind}} = -\frac{\dot{I}b(\delta r)}{4\pi\varepsilon_0 c^2 r^2}\,\hat{\mathbf{j}} = -\frac{\dot{I}S}{4\pi\varepsilon_0 c^2 r^2}\,\hat{\mathbf{j}} = -\frac{\mu_0[\dot{m}]}{4\pi r^2}\,\hat{\mathbf{j}} \tag{5.61}$$

where $S = b\,\delta r$ is the area of the coil and $m = IS$ is the magnetic moment of the coil. Equation (5.61) is in agreement with equation (5.48). However, by comparing equations (5.56), (5.57), (5.59) and (5.60) with equations (5.36), (5.37), (5.40) and (5.39) respectively, the reader can see that equations (5.56), (5.57), (5.59) and (5.60) do not give the correct expressions for the induction electric fields due to the moving conduction electrons and positive ions in individual current elements, but they do give the correct resultant induction electric field in the quasi-stationary limit, when summed around a complete circuit.

5.6. The induction electric field due to a varying current in a circular coil in the quasi-stationary limit: an example of detached electric field lines

Consider the plane circular coil of radius b, shown in Figure 5.6. We shall assume that there are no resultant charge distributions anywhere on the coil, and that the conduction current I, that is flowing in the clockwise direction in Figure 5.6, is varying at such a slow rate that equation (5.53) can be applied. Consider a field point P at a distance $r > b$ from the origin O at the centre of the coil and situated in the plane of the coil. Choose the line OP as the x axis with the $+z$ axis pointing upwards from the paper in Figure 5.6. We shall also use the cylindrical coordinates r, ϕ, z with ϕ measured from the x axis. Consider a current element of length $dl = b\,d\phi$ at the position C in Figure 5.6. The angle between the lines CO and OP is equal to ϕ. The distance R from the current element to the field point P is

$$R = (r^2 + b^2 - 2rb\cos\phi)^{1/2}.$$

To simplify the mathematics by avoiding elliptic integrals, we shall assume that $r \gg b$. Expanding using the binomial theorem, we have

$$\frac{1}{R} \approx \frac{1}{r}\left(1 + \frac{b\cos\phi}{r}\right). \tag{5.62}$$

The contribution of the current $I\,\mathbf{dl}$ at C in Figure 5.6 to the right hand side of equation (5.53) is

$$\Delta\mathbf{E}_{\text{ind}} = -\frac{\dot{I}\,\mathbf{dl}}{4\pi\varepsilon_0 c^2 R} = -\frac{\dot{I}\,\mathbf{dl}}{4\pi\varepsilon_0 c^2 r}\left(1 + \frac{b\cos\phi}{r}\right).$$

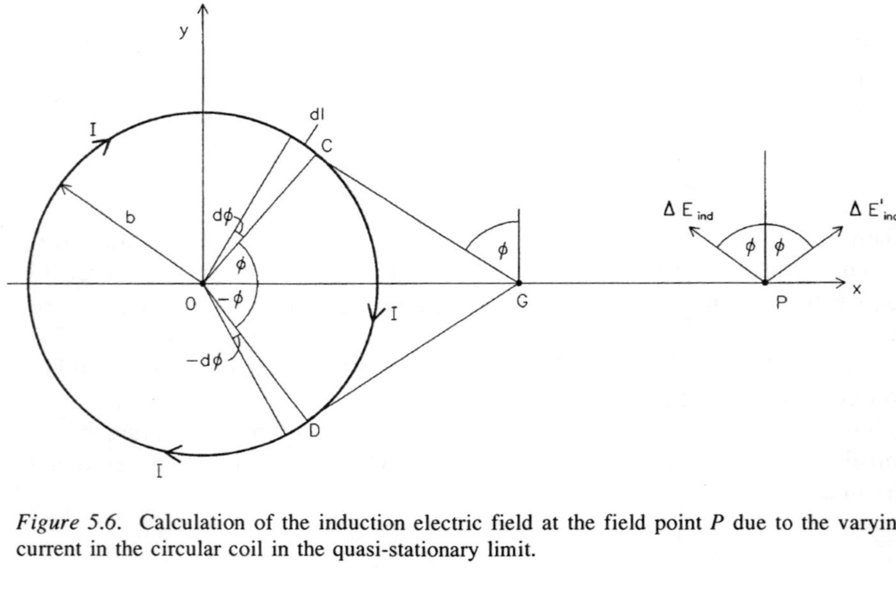

Figure 5.6. Calculation of the induction electric field at the field point P due to the varying current in the circular coil in the quasi-stationary limit.

This contribution is parallel to $-\mathbf{dl}$ and is at an angle $(\pi/2 - \phi)$ to the line OP (the x axis) in Figure 5.6. The component of $\Delta\mathbf{E}_{\text{ind}}$ parallel to the x axis in Figure 5.6 is compensated by the component parallel to the x axis of the contribution $\Delta\mathbf{E}'_{\text{ind}}$ due to a current element of length $b\ \mathrm{d}\phi$ at the point D in Figure 5.6, where the angle between the lines DO and OP is equal to $-\phi$. The resultant of the contributions $\Delta\mathbf{E}_{\text{ind}}$ and $\Delta\mathbf{E}'_{\text{ind}}$ is in the $+y$ direction in Figure 5.6 and is given by

$$\Delta\mathbf{E}_{\text{ind}} + \Delta\mathbf{E}'_{\text{ind}} = \frac{2\dot{I}b\ \mathrm{d}\phi}{4\pi\varepsilon_0 c^2 r}\left(1 + \frac{b\cos\phi}{r}\right)\cos\phi\,\hat{\mathbf{j}}$$

where we have put $\mathrm{d}l = b\ \mathrm{d}\phi$. Summing around the complete circuit, we have

$$\mathbf{E}_{\text{ind}} = -\frac{\dot{I}}{4\pi\varepsilon_0 c^2}\oint\frac{\mathbf{dl}}{R} = \hat{\mathbf{j}}\,\frac{2\dot{I}b}{4\pi\varepsilon_0 c^2 r}\int_0^{\pi}\cos\phi\left(1 + \frac{b\cos\phi}{r}\right)\mathrm{d}\phi. \quad (5.63)$$

Evaluating the integral we find that

$$\mathbf{E}_{\text{ind}} = \frac{\dot{I}\pi b^2}{4\pi\varepsilon_0 c^2 r^2}\,\hat{\mathbf{j}} = \left(\frac{\mu_0}{4\pi}\right)\frac{\dot{m}}{r^2}\,\hat{\mathbf{j}}. \quad (5.64)$$

where $\mathrm{m} = I\pi b^2$ is the magnetic moment of the circular coil. Equation (5.64) is valid at field points in the plane of the coil. Since according to equation (5.64) at distances $r \gg b$ from the centre of the circular coil in Figure 5.6, the electric field is in the plane of the coil and is always perpendicular to the line from the centre of the circular coil to the field point, the electric field lines are circles in the xy plane in Figure 5.6. In cylindrical coordinates we have

$$\mathbf{E}_{\text{ind}} = \left(\frac{\mu_0}{4\pi}\right)\frac{\dot{I}\pi b^2}{r^2}\,\hat{\boldsymbol{\phi}} = \frac{\dot{m}}{4\pi\varepsilon_0 c^2 r^2}\,\hat{\boldsymbol{\phi}} \quad (5.65)$$

where $\hat{\boldsymbol{\phi}}$ is a unit vector in the direction of increasing ϕ. When the current I is flowing in the clockwise direction in the plan view shown in Figure 5.6, and I is increasing in magnitude, the electric field lines are in the anti-clockwise direction, that is in the direction opposite to the direction of the current flow. This example illustrates how we can get detached electric field lines even though, when we derived equation (5.18), which was used to derive equation (5.53), we started with the expression for the electric field due to a moving classical point charge, according to which the electric field lines representing the contributions of individual classical point charges diverge from individual positive ions and converge on individual moving conduction electrons.

It will now be assumed that the field point is a distance z directly above P in Figure 5.6. The distance R from the current element $I\ \mathbf{dl}$ to the new field point is now given by

$$R = (r^2 + b^2 - 2rb\cos\phi + z^2)^{1/2}.$$

Again neglecting b^2 compared with r^2 and using the binomial theorem, we now have

$$\frac{1}{R} = \frac{1}{(r^2 + z^2)^{1/2}} \left(1 + \frac{rb \cos \phi}{(r^2 + z^2)} \right).$$

Since the contribution of I d**l** to the right hand side of equation (5.53) is still parallel to –d**l**, it follows that the analysis given when the field point was in the plane of the coil is still valid, provided we replace the value of $1/R$ given by equation (5.62) by the new value of $1/R$. Equation (5.63) then gives

$$\mathbf{E}_{\mathrm{ind}} = \hat{\mathbf{j}} \frac{2\dot{I}b}{4\pi\varepsilon_0 c^2 (r^2 + z^2)^{1/2}} \int_0^{\pi} \left(1 + \frac{rb \cos \phi}{(r^2 + z^2)} \right) \cos \phi \, d\phi.$$

Evaluating the integral we now find that

$$\mathbf{E}_{\mathrm{ind}} = \left(\frac{\mu_0}{4\pi} \right) \frac{\dot{I}\pi b^2 r}{(r^2 + z^2)^{3/2}} \hat{\boldsymbol{\phi}} = \frac{\dot{m} r}{4\pi\varepsilon_0 c^2 (r^2 + z^2)^{3/2}} \hat{\boldsymbol{\phi}}. \tag{5.66}$$

The electric field lines are again closed circles in the plane containing the field point.

We shall now go on to show that the electric field at the field point P in Figure 5.6 due to the current flowing in the circular current carrying coil obeys Faraday's law of electromagnetic induction. Evaluating the line integral of $\mathbf{E}_{\mathrm{ind}}$ around the circumference of a circular path of radius $r \gg b$, using equation (5.65) we find that in the plane of the coil in Figure 5.6, in the quasi-stationary limit we have

$$\oint \mathbf{E} \cdot d\mathbf{l} = 2\pi r \, E_{\mathrm{ind}} = \frac{\dot{m}}{2\varepsilon_0 c^2 r}. \tag{5.67}$$

According to classical electromagnetism, if we ignore all retardation effects in the quasi-stationary limit, then at a fixed time the magnetic field in the plane of the coil at a distance $r \gg b$ from the magnetic dipole in Figure 5.6 is equal to $\mu_0 m/4\pi r^3$. For example, take the quasi-stationary limit of equation (6.49) of Chapter 6. Since $\nabla \cdot \mathbf{B}$ is zero, the net magnetic flux through a circle of radius r in Figure 5.6 in the direction upwards towards the reader, is equal in magnitude to the magnetic flux in the downwards direction at radial distances $> r$, which is a region where the dipole approximation is valid. Hence the magnitude of the magnetic flux through the circle of radius r in Figure 5.6 is given in the quasi-stationary limit by

$$\Phi = \int_r^{\infty} \frac{\mu_0 m}{4\pi r^3} 2\pi r \, dr = \frac{m}{2\varepsilon_0 c^2 r}.$$

Hence

$$\dot{\Phi} = \frac{\dot{m}}{2\varepsilon_0 c^2 r}.$$

Comparing with equation (5.67) we see that Faraday's law

$$\oint \mathbf{E} \cdot d\mathbf{l} = -\dot{\Phi}$$

is valid for the induction electric field derived using equation (5.18) via equation (5.53).

Problem. Apply equation (A1.37) of Appendix A1.9 to equation (5.65), which gives the induction electric field at the field point P in the plane of the circular current carrying coil in Figure 5.6, to show that at the field point P in Figure 5.6

$$(\nabla \times \mathbf{E})_z = -\frac{S[\dot{I}]}{4\pi\varepsilon_0 c^2 r^3} - \frac{S[\ddot{I}]}{4\pi\varepsilon_0 c^3 r^2} .$$

Comment on the relative magnitudes of the two terms on the right hand side when the current is varying harmonically with angular frequency ω.

According to classical electromagnetism the magnetic field at the field point P in Figure 5.6, when the current is flowing in the direction shown, is given in the dipole approximation by

$$B_z = \frac{IS}{4\pi\varepsilon_0 c^2 r^3} .$$

If we differentiating partially with respect to time, we have

$$\dot{B}_z = \frac{\dot{I}S}{4\pi\varepsilon_0 c^2 r^3} .$$

To satisfy the Maxwell equation (1.117) there must be an extra term on the right hand side. Explain the origin of the missing term. Hint: Have patience until you reach the derivation of equation (6.48) in Chapter 6.

5.7. Induction electric field due to a varying current in a long solenoid in the quasi-stationary limit

Consider the infinitely long solenoid whose axis is along the z axis in Figure 5.7. The solenoid consists of n turns per metre length and carries a varying current I. If we neglect the pitch of the solenoid, we can treat each turn of the solenoid as a circle of radius b. The induction electric field will be determined at the field point P on the x axis at a distance $r \gg b$ from the origin in the plane $z = 0$. According to equation (5.66), the contribution of the $n\,dz$ turns between z and $z + dz$ to the induction electric field at the field point P is given in the quasi-stationary limit by

$$dE_{\text{ind}} = n\,dz \left(\frac{\mu_0}{4\pi}\right) \frac{\dot{I}\pi b^2 r}{(r^2 + z^2)^{3/2}} .$$

By symmetry, the contribution of that part of the solenoid below the $z = 0$ plane to the induction electric field at P is equal to the contribution of that part of the solenoid above the $z = 0$ plane. Hence the total induction electric field at P is given by

$$\begin{aligned}\mathbf{E}_{\text{ind}} &= 2\hat{\boldsymbol{\phi}}\left(\frac{\mu_0}{4\pi}\right) n\dot{I}\pi b^2 r \int_0^\infty \frac{\mathrm{d}z}{(r^2+z^2)^{3/2}} \\ &= \hat{\boldsymbol{\phi}}\left(\frac{\mu_0 n\dot{I}b^2 r}{2}\right)\left[\frac{z}{r^2(r^2+z^2)^{1/2}}\right]_0^\infty \\ &= \frac{\mu_0 n\dot{I}b^2}{2r}\hat{\boldsymbol{\phi}}. \end{aligned} \tag{5.68}$$

The resultant induction electric field lines are again closed circles. When the current I flows in the clockwise direction in Figure 5.7 and is increasing in magnitude, the induction electric field lines are in the anticlockwise direction, that is in the direction opposite to the direction of current flow. The reader can check that we have derived equation (5.68) from the expression for the electric field due to a moving classical point charge. The steps are as follows.

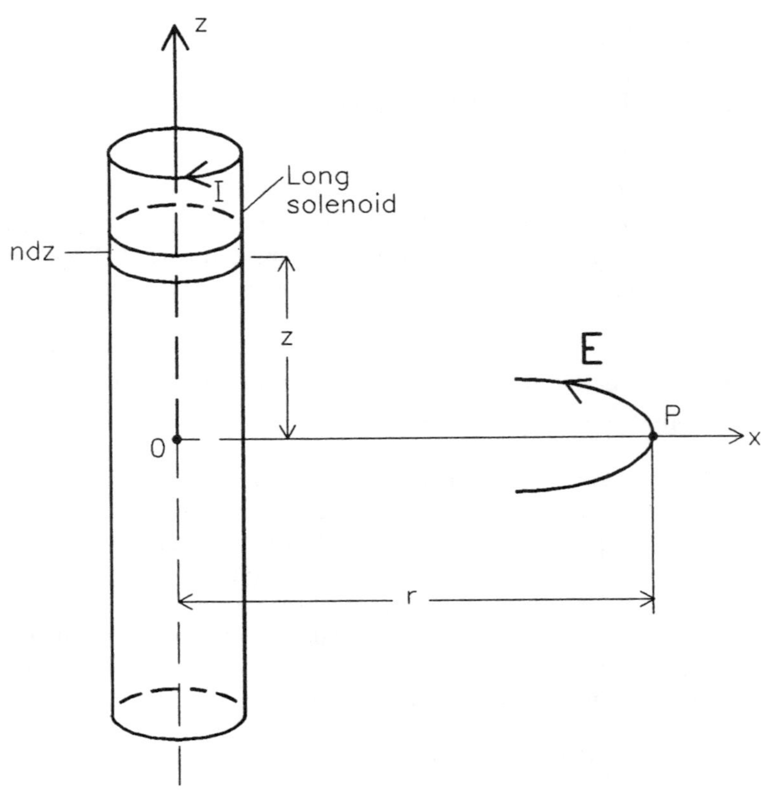

Figure 5.7. Calculation of the induction electric field due to the varying current in a solenoid of infinite length in the quasi-stationary limit.

We derived equation (5.18) in Section 5.3.2 using the expression for the $\mathbf{E}_V$ contribution to the electric field due to a moving classical point charge which is given by equation (3.11). Then in Section 5.5.2 we used equation (5.18) to derive equation (5.53) which we then used to derive equations (5.66) and (5.68). An alternative derivation of equation (5.68), which is valid for all values of $r > b$, will be given at the end of Section 7.2.1 of Chapter 7.

5.8. Detached electric field lines

It was shown in Section 5.7 that the electric field lines that represent the resultant electric field due to the varying conduction current in the infinitely long solenoid, shown in Figure 5.7, are closed circles with centres on the z axis. These resultant electric field lines do not start on positive charges and end on negative charges, but are detached from the moving conduction electrons and stationary positive ions that are in the windings of the solenoid. We derived equation (5.68) using equation (5.53), which had previously been derived from the expression for the electric field due to individual moving classical point charges. The contributions of individual classical point charges to the resultant electric field diverge from the individual stationary positive ions and converge on individual moving conduction electrons. It is the resultant of these individual contributions that gives the detached electric field lines in the example shown in Figure 5.7.

To illustrate how essential it is to include the electric fields due to the stationary positive ions in Figure 5.7 when determining the resultant electric field, we shall now assume that, instead of a conduction current in the solenoid in Figure 5.7, we have a convection current due to the rotation of a negative surface charge distribution, of magnitude $-\sigma$ coulombs per square metre which is on the surface of a hollow dielectric cylinder of outer radius b. The dielectric cylinder rotates with a varying angular velocity ω about its axis. The linear velocity of a point on the surface of the rotating cylinder is equal to $b\omega$. Hence the instantaneous value of the varying convection current is equal to $\sigma b\omega$ amperes per metre length. We shall assume that the convection current is varying slowly enough for the quasi-stationary approximations we used in Section 5.7 to be valid. According to equation (5.68), which was derived from equation (5.53) in Section 5.7, there is an induction electric field outside the solenoid, given by

$$\mathbf{E}_{\text{ind}} = \frac{\mu_0 \sigma \dot{\omega} b^3}{2r}\hat{\boldsymbol{\phi}}. \tag{5.69}$$

where we have replaced nI in equation (5.68) by $\sigma b\omega$. The negative charge distribution on the surface of the rotating cylinder also gives a contribution to the component of the electric field in the radial direction outside the rotating cylinder of charge. Applying Gauss' flux law, we find that, since the total charge per metre length of the cylinder is $\lambda = 2\pi b\sigma$, then outside the rotating negative charge distribution on the cylinder, the radial component of the electric field is given by

$$E_r = -\frac{\lambda}{2\pi\varepsilon_0 r} = -\frac{\sigma b}{\varepsilon_0 r}. \tag{5.70}$$

In this example of a convection current, the resultant electric field is given by the resultant of the contributions given by equations (5.69) and (5.70). Hence the electric field lines due to the rotating negative charge distribution are spirals, the direction of **E** being towards the rotating charge distribution.

If we now add a stationary positive charge distribution of charge $+\sigma$ coulombs per square metre, just outside the rotating negative charge distribution, then the stationary positive charge distribution gives a contribution to the resultant electric field outside the rotating cylinder, given by

$$E_r = +\frac{\sigma b}{\varepsilon_0 r}. \tag{5.71}$$

This contribution compensates the radial component of the electric field due to the rotating negative charge distribution, given by equation (5.70), which leaves only the resultant electric field given by equation (5.69). It is the electric field lines corresponding to the resultant macroscopic electric field, given by equation (5.69), that are detached from the charge distributions.

Problem. The closed circuit shown in Figure 5.8 consists of concentric semicircles *CB* and *DA* of radii *b* and *a* respectively with $b > a$, joined by straight sections *CD* and *AB*. The current in the circuit is equal to αt were α is a constant and *t* is the time. The current varies slowly enough for the quasi-stationary approximations to be valid. Apply equation (5.20) to show that the resultant induction electric field at the origin *O* at the time of observa-

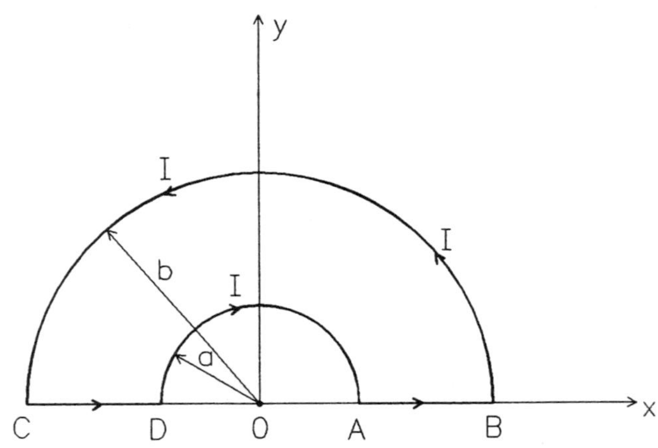

Figure 5.8. Calculation of the induction electric field at the origin *O* due to the varying current in the circuit shown.

tion t is given by

$$\mathbf{E}_{\text{ind}} = -\frac{\alpha}{2\pi\varepsilon_0 c^2}\ln\left(\frac{b}{a}\right)\hat{\mathbf{i}}$$

where $\hat{\mathbf{i}}$ is a unit vector in the $+x$ direction in Figure 5.8. Repeat the calculation using equation (5.53).

(Hints: The appropriate value of current for CB and DA are $\alpha(t - b/c)$ and $\alpha(t - a/c)$ respectively when you apply equation (5.20). The value for the current in AB is $\alpha(t - x/c)$ where x is the distance of the current element from the origin O in Figure 5.8).

5.9. The electric field due to the accelerations of the conduction electrons in a conductor

Consider again the current element that forms part of the stationary electrical circuit, shown previously in Figure 5.2. The conduction current is equal to I. The current element is at the position $\mathbf{r}_s$, the field point P is at the position $\mathbf{r}$ and $\mathbf{R} = (\mathbf{r} - \mathbf{r}_s)$ is a vector from the current element to the field point. We shall again start with a simplified model and assume that at a fixed time there are N_0 stationary positive ions per metre length each of charge $+e$, and N_0 moving conduction electrons per metre length each of charge $-e$. We shall assume that, at the retarded time $t^* = (t - R/c)$, all the conduction electrons have the same velocity $[\mathbf{u}]$ and the same acceleration $[\mathbf{a}]$ in the downwards direction, opposite to the direction of current flow in Figure 5.2.

The stationary positive ions do not contribute to the radiation electric field. It follows from equation (5.5) that the number of moving conduction electrons passed by the information collecting sphere inside the current element at the retarded time t^* is

$$\delta N = N_0\left(\frac{s}{R}\right)\mathrm{d}l \tag{5.72}$$

where s is given by equation (5.3). Multiplying equation (5.72) by the acceleration dependent contribution $\mathbf{E}_A$ to the electric field due to an accelerating conduction electron, which is given by equation (3.12), we find that the contribution of the radiation electric field, denoted $\mathrm{d}\mathbf{E}_{\text{rad}}$, to the resultant electric field at the field point P in Figure 5.2 is

$$\mathrm{d}\mathbf{E}_{\text{rad}} = N_0\left(\frac{s}{R}\right)\mathrm{d}l\left(\frac{q}{4\pi\varepsilon_0 s^3 c^2}\right)\left[\mathbf{R}\times\left\{\left(\mathbf{R} - \frac{R\mathbf{u}}{c}\right)\times\mathbf{a}\right\}\right].$$

We shall neglect the term $(R\mathbf{u}/c)\times\mathbf{a}$ which is of order βa, and we shall neglect the term $\mathbf{u}\cdot\mathbf{R}/c$ in equation (5.3) and put $s = R$. Putting $q = -e$ for a conduction electron and using equation (A1.6) of Appendix A1.1 we then have

$$\mathrm{d}\mathbf{E}_{\mathrm{rad}} = -\frac{eN_0\,\mathrm{d}l}{4\pi\varepsilon_0 c^2 R^3}\,[\mathbf{R}\times(\mathbf{R}\times\mathbf{a})] \tag{5.73}$$

$$= -\frac{eN_0\,\mathrm{d}l}{4\pi\varepsilon_0 c^2 R^3}\,[\mathbf{R}(\mathbf{R}\cdot\mathbf{a}) - \mathbf{a}R^2]. \tag{5.74}$$

A conduction electron can have both an acceleration **a** which leads to a change in the speed of the electron and a centripetal acceleration u^2/ρ where the circuit is curved and has a radius of curvature ρ. The ratio of the magnitudes of the centripetal acceleration u^2/ρ and the acceleration **a** is $u^2/\rho a$. If we assume that the current varies as $I_0 \cos \omega t$, where ω is the angular frequency, then the velocity u must vary as $u_0 \cos \omega t$ so that the acceleration a is equal to $\dot{u} = -\omega u_0 \sin \omega t$. Hence the ratio of the maximum centripetal acceleration to the maximum value of **a** is equal to $u_0/\omega\rho$. In the case of a typical conduction current in a metallic conductor u_0 is of the order of 10^{-4} m s^{-1} and in a typical laboratory experiment ρ may be of the order of 0.1 m. Hence at the mains frequency of 50 Hz the ratio $u_0/\omega\rho$ is of the order of 3×10^{-6}. This ratio is very much smaller at the microwave frequencies, say 10^9 Hz, when the radiation electric fields are important. Hence for conduction currents at mains frequencies it is generally safe to ignore the contributions of the centripetal accelerations of the conduction electrons in regions where the circuit is curved, and to include in equations (5.73) and (5.74) only the acceleration **a** that changes the speeds of the conduction electrons. (The centripetal accelerations of relativistic charged particles moving in helical paths in magnetic fields are important as these centripetal accelerations give rise to synchrotron radiation).

We shall assume that the number of conduction electrons per cubic metre remains constant, and that all the changes in conduction current come from changes in the speeds of the conduction electrons. In these conditions

$$eN_0\,a = \frac{\mathrm{d}}{\mathrm{d}t}(eN_0\,u) = [\dot{I}]. \tag{5.75}$$

Since θ is the angle between the direction of the current I and the vector **R** from the current element to the field point P in Figure 5.2, the angle between **R** and [**a**] is $(\pi - \theta)$. Hence equation (5.74) can be rewritten in the form

$$\mathrm{d}\mathbf{E}_{\mathrm{rad}} = -\frac{eN_0\,\mathrm{d}l}{4\pi\varepsilon_0 c^2 R}\,[\hat{\mathbf{R}}a\cos(\pi-\theta) - \mathbf{a}] \tag{5.76}$$

where $\hat{\mathbf{R}}$ is a unit vector in the direction of **R**. It follows from Figure 5.2 that, if $\hat{\boldsymbol{\theta}}$ is a unit vector at the field point P in Figure 5.2 in the direction of increasing θ, then the acceleration **a** of a conduction electron can be expressed in the form

$$\mathbf{a} = (-a\cos\theta)\hat{\mathbf{R}} + (a\sin\theta)\hat{\boldsymbol{\theta}}. \tag{5.77}$$

Substituting for **a** in equation (5.76) and remembering that $\cos(\pi - \theta) = -\cos\theta$, we find that equation (5.76) can be rewritten in the form

$$\mathrm{d}\mathbf{E}_{\mathrm{rad}} = -\frac{eN_0\,\mathrm{d}l}{4\pi\varepsilon_0 c^2 R}[-a\sin\theta]\hat{\boldsymbol{\theta}}.$$

Using equation (5.75) we finally find that

$$\mathrm{d}\mathbf{E}_{\mathrm{rad}} = \frac{[\dot{I}]\sin\theta\,\mathrm{d}l}{4\pi\varepsilon_0 c^2 R}\hat{\boldsymbol{\theta}}. \tag{5.78}$$

In practice the accelerations of individual conduction electrons are not always in the direction of their drift velocity, as was assumed in Figure 5.2, but the accelerations can be in any direction. If we put $\langle \mathbf{a}_i \rangle$ equal to $\mathbf{a}$, where $\mathbf{a}$ is now the average acceleration of the conduction electrons, then

$$\langle \mathbf{R}\cdot\mathbf{a}_i\rangle = \mathbf{R}\cdot\langle\mathbf{a}_i\rangle = \mathbf{R}\cdot\mathbf{a}. \tag{5.79}$$

Averaging equation (5.75), we have

$$\langle -eN_0\mathbf{a}_i\rangle\,\mathrm{dl} = -eN_0\,\mathrm{d}l\langle\mathbf{a}_i\rangle = -eN_0\mathbf{a}\,\mathrm{d}l = \dot{I}\,\mathbf{dl}. \tag{5.80}$$

It is now straight forward for the reader to show, by averaging equation (5.74) over the distribution of accelerations and using equations (5.79) and (5.80), that equation (5.78) is valid in the general case in the limit when $\beta \ll 1$.

Averaging equations (5.73) and (5.75) over the velocities and accelerations of the conduction electrons, and then using equation (5.80) we have

$$\mathrm{d}\mathbf{E}_{\mathrm{rad}} = \frac{[\dot{I}]}{4\pi\varepsilon_0 c^2 R}[\hat{\mathbf{R}}\times(\hat{\mathbf{R}}\times\mathbf{dl})] \tag{5.81}$$

$$\mathrm{d}\mathbf{E}_{\mathrm{rad}} = \frac{[\dot{I}]}{4\pi\varepsilon_0 c^2}\left[\frac{\hat{\mathbf{R}}\cos\theta\,\mathrm{d}l}{R} - \frac{\mathbf{dl}}{R}\right]. \tag{5.82}$$

Equations (5.81) and (5.82) are alternative forms of equation (5.78). We shall now go on to discuss applications of equations (5.78), (5.81) and (5.82), which are all equivalent to each other.

5.10. Radiation electric field due to a varying conduction current in the coil *ABCD* in Figure 5.5

Consider again the coil *ABCD* shown in Figure 5.5. We shall assume that at any fixed time the conduction electrons all have the same speed $[u]$ and the same acceleration $[a]$, which is in the direction of $[u]$. The radiation electric field at the field point P will be calculated using equation (5.78).

In the case of the section *AB* of the coil *ABCD* in Figure 5.5 $\theta = \pi$ in equation (5.78), so that $\sin\theta = 0$ and the contribution of the varying conduction current in the section *AB* to the radiation electric field at P is zero. For the section *CD*, $\theta = 0$ so that $\sin\theta = 0$ and the contribution of the varying conduction current in the section *CD* to the radiation electric field at P is also zero.

For the sections BC and CD, $\theta = \pi/2$ and $\sin\theta = 1$, so that according to equation (5.78)

$$(\mathrm{d}\mathbf{E}_{\mathrm{rad}})_{BC} = \frac{b(1 + \delta r/r)}{4\pi\varepsilon_0 c^2 r(1 + \delta r/r)}\left[\frac{\partial I}{\partial t^*}\right]_{BC}\hat{\mathbf{j}} = \frac{b}{4\pi\varepsilon_0 c^2 r}\left[\frac{\partial I}{\partial t^*}\right]_{BC}\hat{\mathbf{j}}. \quad (5.83)$$

$$(\mathrm{d}\mathbf{E}_{\mathrm{rad}})_{AD} = -\frac{b}{4\pi\varepsilon_0 c^2 r}\left[\frac{\partial I}{\partial t^*}\right]_{AD}\hat{\mathbf{j}}. \quad (5.84)$$

If $[\dot{I}]$, the rate of change of current is constant then $[\dot{I}]_{BC}$ is equal to $[\dot{I}]_{AD}$. Adding equations (5.83) and (5.84) and remembering that the contributions of the sections AB and CD are both zero we find that the resultant radiation electric field at the field point P in Figure 5.5 is zero. To show that this result is true in the general case, we shall integrate equation (5.82) around any arbitrary closed circuit. If $\dot{I}$ is constant, we have

$$\mathbf{E}_{\mathrm{rad}} = \frac{\dot{I}}{4\pi\varepsilon_0 c^2}\left[\oint\frac{\hat{\mathbf{R}}\,\mathrm{d}l\cos\theta}{R} - \oint\frac{\mathrm{d}\mathbf{l}}{R}\right]. \quad (5.85)$$

It follows from equation (A3.9) of Appendix A3, that the two integrals in equation (5.85) are equal showing that $\mathbf{E}_{\mathrm{rad}}$ is zero when $\dot{I}$ is constant, whatever the shape of the electrical circuit.

To obtain a resultant radiation electric field at the field point P in Figure 5.5, $\dot{I}$ must vary with time. If δr is small, then in the time $(\delta r/c)$ it takes the information collecting sphere to cross the coil $ABCD$

$$\left[\frac{\partial I}{\partial t^*}\right]_{AD} = \left[\frac{\partial I}{\partial t^*}\right]_{BC} + \left[\frac{\partial^2 I}{\partial t^{*2}}\right]\left(\frac{\delta r}{c}\right).$$

Substituting in equation (5.84) and then adding equations (5.83) and (5.84) we obtain

$$\mathbf{E}_{\mathrm{rad}} = -\frac{b\,\delta r}{4\pi\varepsilon_0 c^3 r}\left[\frac{\partial^2 I}{\partial t^{*2}}\right]\hat{\mathbf{j}} = -\frac{\mu_0[\ddot{m}]}{4\pi c r}\hat{\mathbf{j}} \quad (5.86)$$

where $m = I\,b\,\delta r$ is the magnetic moment of the coil $ABCD$ in Figure 5.5. Notice the important role of retardation in giving a resultant radiation electric field at the field point P. Adding equations (5.48) and (5.86), we find that the total electric field at the field point P in Figure 5.5 is given by

$$\mathbf{E} = -\frac{\mu_0}{4\pi}\left[\frac{\dot{m}}{r^2} + \frac{\ddot{m}}{rc}\right]\hat{\mathbf{j}}. \quad (5.87)$$

Equation (5.87) is the standard expression for the electric field due to a magnetic dipole, when the field point is in the plane of the coil.

The ratio of the radiation electric field given by equation (5.86) to the induction electric field given by equation (5.48) is $r\ddot{m}/(\dot{m}c)$. Due to the r term in the numerator of this expression, the ratio of the radiation to induction electric field increases with r and the $\mathbf{E}_{\mathrm{rad}}$ (radiation) contribution to the electric field predominates at very large distances from the coil. The two contributions are equal when $r\ddot{m} = \dot{m}c$. If the current in the coil varies as $I = I_0 \cos \omega t$, the maximum amplitudes of the two contributions are equal when $r = c/\omega$ or $r = \lambda/2\pi$. This is the same condition as for the case of the magnetic field of the oscillating electric dipole discussed in Section 2.4.2 of Chapter 2 and given by equation (2.49). In the quasi-stationary limit, say at 50 Hz, $\omega = 100\ \pi$ and the two contributions are equal when r is about 1000 km. It is clear that the induction electric field, associated with the $\mathbf{E}_V$ term, generally predominates in all laboratory experiments carried out at mains frequency. However, if the frequency were 50 MHz, then $\lambda/2\pi$ would be approximately 1 m showing that the radiation electric fields are extremely important in laboratory experiments carried out at microwave frequencies.

5.11. Expressions for the total electric field due to a current element that forms part of a circuit

So far we have considered the induction electric field and the radiation electric field separately. We shall now add the two contributions to give a single circuit equation. Consider again the current element I **dl** that forms part of the complete electrical circuit shown in Figure 5.2. Adding equations (5.20) and (5.78), we find that the total contribution of the charges counted by the information collecting sphere inside the current element I **dl** at the retarded time $t^* = (t - R/c)$ to the electric field at the field point P in Figure 5.2 at the time of observation t is

$$\begin{aligned} d\mathbf{E} &= d\mathbf{E}_{\text{ind}} + d\mathbf{E}_{\text{rad}} \\ &= \frac{\cos\theta\, dl}{4\pi\varepsilon_0}\left[\frac{I}{cR^2}\right]\hat{\mathbf{R}} + \frac{\sin\theta\, dl}{4\pi\varepsilon_0}\left[\frac{I}{cR^2} + \frac{\dot{I}}{c^2R}\right]\hat{\boldsymbol{\theta}}. \end{aligned} \tag{5.88}$$

It was assumed when equation (5.78) was derived that the average acceleration [**a**] and hence $[\dot{I}]$ were parallel to the wire. We also assumed that there were no resultant charge distributions. Since $\hat{\boldsymbol{\theta}} \sin\theta\, dl$ is equal to $\hat{\mathbf{R}} \times (\hat{\mathbf{R}} \times \mathbf{dl})$ and $\cos\theta\, dl$ is equal to $\hat{\mathbf{R}} \cdot \mathbf{dl}$, we can rewrite equation (5.88) in the vector form

$$d\mathbf{E} = \frac{(\hat{\mathbf{R}} \cdot \mathbf{dl})}{4\pi\varepsilon_0}\left[\frac{I}{cR^2}\right]\hat{\mathbf{R}} + \frac{\hat{\mathbf{R}} \times (\hat{\mathbf{R}} \times \mathbf{dl})}{4\pi\varepsilon_0}\left[\frac{I}{cR^2} + \frac{\dot{I}}{c^2R}\right]. \tag{5.89}$$

An alternative form of equation (5.88) is obtained by adding equations (5.17) and (5.81) to give

$$d\mathbf{E} = \frac{2\hat{\mathbf{R}}(\hat{\mathbf{R}} \cdot \mathbf{dl})}{4\pi\varepsilon_0}\left[\frac{I}{cR^2}\right] - \frac{\mathbf{dl}}{4\pi\varepsilon_0}\left[\frac{I}{cR^2}\right] + \frac{\hat{\mathbf{R}} \times (\hat{\mathbf{R}} \times \mathbf{dl})}{4\pi\varepsilon_0}\left[\frac{\dot{I}}{c^2R}\right]. \tag{5.90}$$

Yet another form of equation (5.88) is obtained by adding equation (5.18) and (5.82) to give

$$\mathrm{d}\mathbf{E} = \frac{\cos\theta\,\mathrm{d}l}{4\pi\varepsilon_0}\left[\frac{2I}{cR^2} + \frac{\dot{I}}{c^2R}\right]\hat{\mathbf{R}} - \frac{\mathrm{d}\mathbf{l}}{4\pi\varepsilon_0}\left[\frac{I}{cR^2} + \frac{\dot{I}}{c^2R}\right]. \tag{5.91}$$

The electric field due to a complete electrical circuit is obtained by integrating any one of the equations (5.88), (5.89), (5.90) or (5.91) around the complete circuit allowing for retardation effects. There is, in addition to this electric field an electric field due to any surface or boundary charge distributions of the type discussed in Section 1.3 of Chapter 1 and in Appendix B, and which are associated with conduction current flow.

If we had an isolated current element, equations (5.88), (5.89), (5.90) or (5.91) would still give the contribution to the electric field of those charges counted by the information collecting sphere inside the current element at the retarded time, but in the case of an isolated current element, the current flowing in the current element would also give rise to varying charge distributions at the ends of the current element, which would also contribute to the total electric field. This would be similar to the example of the oscillating electric dipole, discussed previously in Section 2.4 of Chapter 2, and which we shall now solve using equation (5.88).

5.12. The electric field due to an oscillating electric dipole

Consider again the example of the oscillating electric dipole shown previously in Figure 2.2 of Chapter 2 and shown again in Figure 5.9. The mid-point of the electric dipole is at the origin O in Figure 5.9. The electric field will be determined at the field point P at a distance r from the origin O at the time of observation t. It was assumed, when equation (5.88) was derived, that the current element formed part of a complete electrical circuit, in which case there were no accumulations of charge at the ends of the current element. In the example of the oscillating electric dipole in Figure 5.9 there are accumulations of positive and negative charges, which also contribute to the total electric field at the field point P. In the example of the oscillating electric dipole, equation (5.88) only gives the contribution, denoted $\mathbf{E}_1$, of the moving and accelerating conduction electrons and stationary positive ions actually counted by the information collecting sphere at the retarded time t^* inside the connecting wire of length $\mathrm{d}l$ in Figure 5.9, to the electric field at the field point P at the time of observation t. Since the oscillating electric dipole is at the origin in Figure 5.9, $\mathbf{R} = \mathbf{r}$ in equation (5.88). According to equations (2.32) and (2.39) we have

$$[I]\,\mathrm{d}l = [\dot{p}] \tag{2.32}$$

$$[\dot{I}]\,\mathrm{d}l = [\ddot{p}] \tag{2.39}$$

where $[p] = Q\,\mathrm{d}l$ is the electric dipole moment at the retarded time t^*. Hence

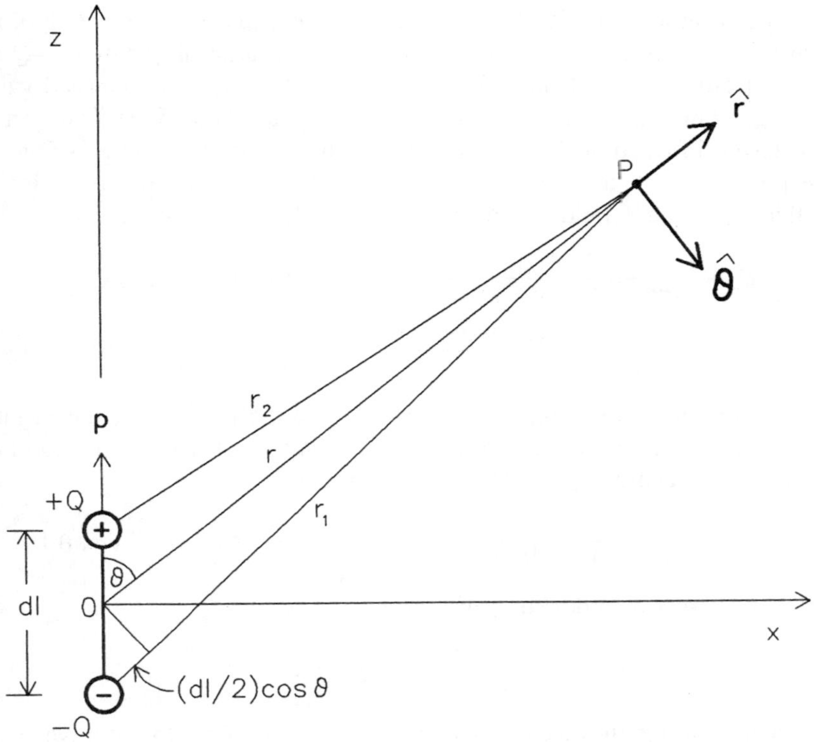

Figure 5.9. Calculation of the electric field due to an oscillating electric dipole.

according to equation (5.88) we have

$$\mathbf{E}_1 = \frac{\cos\theta}{4\pi\varepsilon_0}\left[\frac{\dot{p}}{cr^2}\right]\hat{\mathbf{r}} + \frac{\sin\theta}{4\pi\varepsilon_0}\left[\frac{\dot{p}}{cr^2} + \frac{\ddot{p}}{c^2 r}\right]\hat{\boldsymbol{\theta}}. \tag{5.92}$$

Comparing equation (5.92) with equation (2.54) of Chapter 2, we see that not only are the terms proportional to $1/r^3$ missing from equation (5.92) but there is a factor of 2 missing in the term proportional to $1/r^2$ in the component of $\mathbf{E}_1$ in the direction of $\hat{\mathbf{r}}$. We shall now show that these differences can be accounted for by the contributions of the varying charge distributions that accumulate at the ends of the current element shown in Figure 5.9. We shall assume that, when the information collecting sphere passes the lower end of the electric dipole, the charge on the lower end of the dipole is $-Q$, and when it passes the upper end at a time $(\mathrm{d}l\cos\theta)/c$ later the charge on the upper end is $Q + \mathrm{d}Q$.

We have

$$\mathrm{d}Q = \left[\frac{\partial Q}{\partial t^*}\right]\left(\frac{\mathrm{d}l\cos\theta}{c}\right) = \frac{[I]\,\mathrm{d}l\cos\theta}{c} = \frac{[\dot{p}]\cos\theta}{c}. \tag{5.93}$$

We shall assume that, since the charges at the ends of the electric dipole are at rest, the electric fields due to the varying charge distributions $-Q$ and $(Q + \mathrm{d}Q)$ are given by Coulomb's law, provided we use the retarded values of charge. This assumption is consistent with equation (5.101), which we shall derive in Section 5.13. Since the distance from the charge $-Q$ to the field point P in Figure 5.9 is $r + (\mathrm{d}l/2)\cos\theta$, it follows from Coulomb's law that the electric field due to the charge $[-Q]$ is of magnitude

$$E_- = \frac{[Q]}{4\pi\varepsilon_0 r^2\{1 + (\mathrm{d}l/2r)\cos\theta\}^2} \approx \frac{[Q]}{4\pi\varepsilon_0 r^2}\left(1 - \frac{\mathrm{d}l}{r}\cos\theta\right)$$

$$= +\frac{[Q]}{4\pi\varepsilon_0 r^2} - \frac{[Q]\,\mathrm{d}l\cos\theta}{4\pi\varepsilon_0 r^3} \tag{5.94}$$

and is in the direction from the field point P to the position of the negative charge $[-Q]$ in Figure 5.9. According to Coulomb's law the electric field due to the positive charge $[Q + \mathrm{d}Q]$ is of magnitude

$$E_+ = \frac{[Q + \mathrm{d}Q]}{4\pi\varepsilon_0 r^2\{1 - (\mathrm{d}l/2r)\cos\theta\}^2} \approx \frac{[Q + \mathrm{d}Q]}{4\pi\varepsilon_0 r^2}\left(1 + \frac{\mathrm{d}l}{r}\cos\theta\right).$$

Neglecting terms of order $(\mathrm{d}Q)(\mathrm{d}l)$ we have

$$E_+ = \frac{[Q]}{4\pi\varepsilon_0 r^2} + \frac{[Q]\,\mathrm{d}l\cos\theta}{4\pi\varepsilon_0 r^3} + \frac{[\mathrm{d}Q]}{4\pi\varepsilon_0 r^2}. \tag{5.95}$$

This contribution to the electric field is in the direction from the positive charge $[Q + \mathrm{d}Q]$ to the field point P in Figure 5.9. Adding equations (5.94) and (5.95) we find that the contribution $\mathbf{E}_2$ to the electric field at P due to the charges at the ends of the oscillating electric dipole is given by

$$\mathbf{E}_2 = \mathbf{E}_+ + \mathbf{E}_- = \left\{E_+\cos\left(\frac{\alpha}{2}\right) - E_-\cos\left(\frac{\alpha}{2}\right)\right\}\hat{\mathbf{r}}$$

$$+ \left\{E_+\sin\left(\frac{\alpha}{2}\right) + E_-\sin\left(\frac{\alpha}{2}\right)\right\}\hat{\boldsymbol{\theta}}. \tag{5.96}$$

where α is the angle between $\mathbf{r}_1$ and $\mathbf{r}_2$ in Figure 5.9. Since $\mathrm{d}l$ is infinitesimal, we have $\sin(\alpha/2) \approx (\mathrm{d}l\sin\theta)/2r$ and $\cos(\alpha/2) \approx 1$. Using the value of $\mathrm{d}Q$ given by equation (5.93), putting $[Q]\,\mathrm{d}l = [p]$ in equations (5.94) and (5.95) before substituting in equation (5.96) and then neglecting terms of order $(\mathrm{d}Q)(\mathrm{d}l)$ and $(\mathrm{d}l)^2$ we finally obtain

$$\mathbf{E}_2 = \left(\frac{2[p]\cos\theta}{4\pi\varepsilon_0 r^3} + \frac{[\dot{p}]\cos\theta}{4\pi\varepsilon_0 c r^2}\right)\hat{\mathbf{r}} + \frac{[p]\sin\theta}{4\pi\varepsilon_0 r^3}\hat{\boldsymbol{\theta}}. \tag{5.97}$$

Adding equations (5.92) and (5.97), we find that the total electric field at the field point P in Figure 5.9 at the time of observation t is

$$\mathbf{E} = \left(\frac{\cos\theta}{4\pi\varepsilon_0}\right)\left(\frac{2[p]}{r^3} + \frac{2[\dot{p}]}{cr^2}\right)\hat{\mathbf{r}} + \left(\frac{\sin\theta}{4\pi\varepsilon_0}\right)\left(\frac{[p]}{r^3} + \frac{[\dot{p}]}{cr^2} + \frac{[\ddot{p}]}{c^2 r}\right)\hat{\boldsymbol{\theta}}. \tag{5.98}$$

Equation (5.98) is the same as equation (2.54) of Chapter 2. When we derived equation (5.88) from the expression for the electric field due to an accelerating classical point charge, given by equation (3.10), we ignored higher order terms in β. If we had kept these higher order terms, the predictions we made in Section 2.4 of Chapter 2 using the vector potential **A**, would not be in agreement with our new results to all orders of β, but the differences would be of no practical significance when the theories are applied to conduction currents in metallic conductors. The origins of these differences will be discussed in Section 6.8 of Chapter 6. There are in practice examples when the higher order terms in β are very important, for example when charged particles are moving at relativistic speeds in some astrophysical plasmas such as the Crab Nebula.

Notice we had to allow for retardation effects when we derived equation (5.97) from the charge distributions, but not when we applied equation (5.88). This was because we had already allowed for retardation effects when we used equation (5.5) in the derivations of equations (5.20) and (5.78) which added up to give equation (5.88).

5.13. The electric field due to a varying current distribution

The example of the oscillating electric dipole shown in Figure 5.9 and discussed in Section 5.12 shows that equation (5.88) must be extended when there are resultant charge distributions in the system. Furthermore, when we derived equation (5.88) we assumed that $\dot{I}$, the rate of charge of current, was parallel to the wire. In the general case of a varying charge and current distribution, that is not confined to electric wires, for example in a plasma, there may be resultant charge distributions and $\dot{\mathbf{J}}$ is not necessarily parallel to **J**, as for example in a plasma that is in a strong magnetic field. Consider the macroscopic charge and current distribution that consists of moving and accelerating classical point charges shown in Figure 5.10. We shall assume in this section that all velocities are very much smaller than the speed of light. The electric field will be determined at the field point P at the position **r** in Figure 5.10 at the time of observation t. Let n_i be the number of classical point charges per cubic metre that have charge q_i, velocity $[\mathbf{u}_i]$, acceleration $[\mathbf{a}_i]$ and are inside the volume element $\mathrm{d}V_s$ at the position $\mathbf{r}_s$ in Figure 5.10 at the retarded time $t^* = (t - R/c)$, where **R** is equal to $(\mathbf{r} - \mathbf{r}_s)$. It follows from equation (5.5) that the number of charges of velocity $[\mathbf{u}_i]$ counted inside $\mathrm{d}V_s$ by the information collecting sphere that reaches the field point P at the time of observation t is

$$\delta N_i = n_i \left(\frac{s_i}{R}\right) \mathrm{d}V_s \tag{5.99}$$

where

$$s_i = \left[R - \frac{\mathbf{u}_i \cdot \mathbf{R}}{c}\right]. \tag{5.100}$$

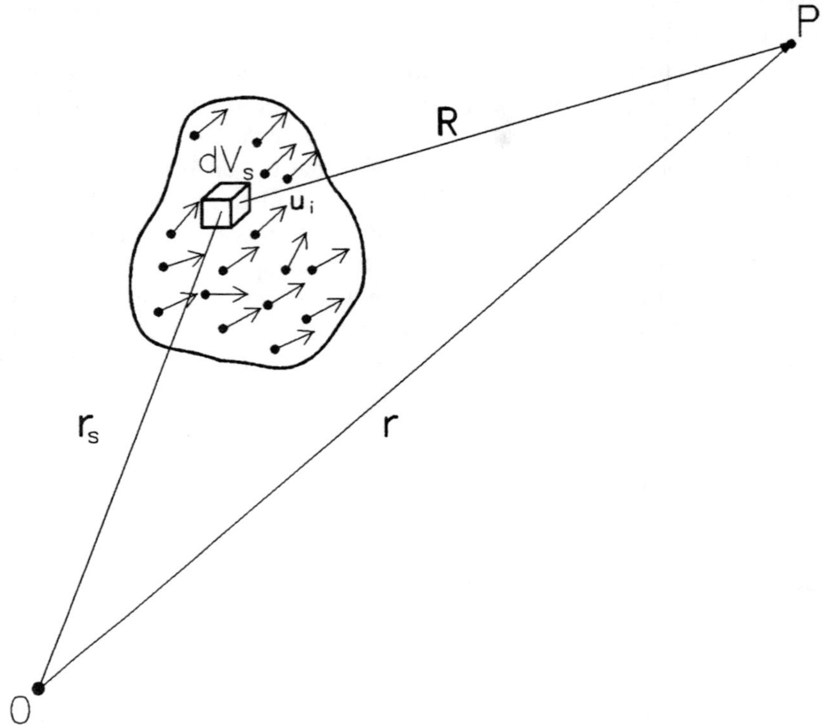

Figure 5.10. Calculation of the total electric field due to a varying macroscopic charge and current distribution made up of moving and accelerating classical point charges.

Multiplying equation (5.99) by the expression for the electric field due to a moving and accelerating classical point charge, which is given by equation (3.10), we have

$$\mathrm{d}\mathbf{E} = n_i \left(\frac{s_i}{R}\right) (\mathbf{E}_V + \mathbf{E}_A)\ \mathrm{d}V_s = \mathrm{d}\mathbf{E}_{\mathrm{ind}} + \mathrm{d}\mathbf{E}_{\mathrm{rad}}.$$

Using equation (3.11) for $\mathbf{E}_V$ and making the same approximations as when we derived equations (5.10), (for example we can put $(-e) = q_i$, $N_0\ \mathrm{d}l = n_i\ \mathrm{d}V_s$ and $u = u_i$ in equation (5.10)), we find that provided $u_i \ll c$

$$\mathrm{d}\mathbf{E}_{\mathrm{ind}} = \frac{q_i n_i\ \mathrm{d}V_s}{4\pi\varepsilon_0 R^3}\left(\mathbf{R} + \frac{2\mathbf{R}([\mathbf{u}_i]\cdot\mathbf{R})}{Rc} - \frac{R[\mathbf{u}_i]}{c}\right).$$

Summing over all possible values of $[\mathbf{u}_i]$ and using the relations

$$\Sigma q_i n_i = [\rho]$$

$$\Sigma q_i n_i [\mathbf{u}_i] = [\mathbf{J}]$$

we have

$$\mathrm{d}\mathbf{E}_{\mathrm{ind}} = \frac{1}{4\pi\varepsilon_0}\left\{\left(\frac{[\rho]}{R^2} + \frac{2\hat{\mathbf{R}}\cdot[\mathbf{J}]}{cR^2}\right)\hat{\mathbf{R}} - \frac{[\mathbf{J}]}{cR^2}\right\}\mathrm{d}V_s. \tag{5.101}$$

Notice that we are now allowing for the possibility of a finite resultant charge density $[\rho]$. Corresponding to equation (5.73), with $(-e) = q_i$, $N_0\,\mathrm{d}l = n_i\,\mathrm{d}V_s$, $u = u_i$ and $a = a_i$, we have

$$\mathrm{d}\mathbf{E}_{\mathrm{rad}} = \frac{q_i n_i\,\mathrm{d}V_s}{4\pi\varepsilon_0 c^2 R^3}\left(\mathbf{R}\times(\mathbf{R}\times[\mathbf{a}_i])\right).$$

Summing over all possible values of $[\mathbf{a}_i]$ and using the equation

$$\Sigma q_i n_i[\mathbf{a}_i] = [\dot{\mathbf{J}}]$$

we find that

$$\mathrm{d}\mathbf{E}_{\mathrm{rad}} = \frac{\{\hat{\mathbf{R}}\times(\hat{\mathbf{R}}\times[\dot{\mathbf{J}}])\}\,\mathrm{d}V_s}{4\pi\varepsilon_0 c^2 R}. \tag{5.102}$$

Adding equations (5.101) and (5.102) gives

$$\mathrm{d}\mathbf{E}(\mathbf{r},t) = \frac{1}{4\pi\varepsilon_0}\left\{\left(\frac{[\rho]}{R^2} + \frac{2\hat{\mathbf{R}}\cdot[\mathbf{J}]}{cR^2}\right)\hat{\mathbf{R}} - \frac{[\mathbf{J}]}{cR^2} + \frac{\hat{\mathbf{R}}\times(\hat{\mathbf{R}}\times[\dot{\mathbf{J}}])}{c^2R}\right\}\mathrm{d}V_s \tag{5.103}$$

where $[\rho]$ and $[\mathbf{J}]$ are the charge and current densities at the source point at $\mathbf{r}_s$ in Figure 5.10 at the retarded time $(t - R/c)$, $\mathbf{R} = (\mathbf{r} - \mathbf{r}_s)$ and $\hat{\mathbf{R}}$ is a unit vector in the direction of $\mathbf{R}$. Expanding the triple vector product using equation (A1.6) of Appendix A1.1 and then rearranging, we obtain

$$\mathrm{d}\mathbf{E}(\mathbf{r},t) = \frac{1}{4\pi\varepsilon_0}\left\{\left(\frac{[\rho]}{R^2} + \frac{2\hat{\mathbf{R}}\cdot[\mathbf{J}]}{cR^2} + \frac{\hat{\mathbf{R}}\cdot[\dot{\mathbf{J}}]}{c^2R}\right)\hat{\mathbf{R}} - \frac{[\mathbf{J}]}{cR^2} - \frac{[\dot{\mathbf{J}}]}{c^2R}\right\}\mathrm{d}V_s. \tag{5.104}$$

Equations (5.103) and (5.104) are valid in differential form in the low velocity limit when $u_i \ll c$. It is straightforward for the reader to develop equations (5.103) and (5.104) for the more general case when there is more than one type of moving classical point charge in the system. The reader can also check that equations (5.103) and (5.104) lead to equations (5.90) and (5.91) respectively in the case of the electrically neutral circuit shown in Figure 5.2.

The **differential** form of Jefimenko's equation (1.136) is not equivalent to our equations (5.103) and (5.104), which are both valid in differential form in the low velocity limit. However, both our approach and that based on Jefimenko's equation (1.136) give the same value for the resultant electric field when integrated over a complete charge and current distribution. To illustrate this we shall now solve the problem of the oscillating electric dipole shown in Figures 2.2 and 5.9 using Jefimenko's equation (1.136), which is

$$\mathbf{E} = \frac{1}{4\pi\varepsilon_0}\int\frac{[\rho]\hat{\mathbf{R}}}{R^2}\,\mathrm{d}V_s + \frac{1}{4\pi\varepsilon_0}\int\frac{[\dot{\rho}]\hat{\mathbf{R}}}{cR}\,\mathrm{d}V_s - \frac{1}{4\pi\varepsilon_0}\int\frac{[\dot{\mathbf{J}}]}{c^2R}\,\mathrm{d}V_s$$

$$= \mathbf{E}_1^0 + \mathbf{E}_2^0 + \mathbf{E}_3^0. \tag{5.105}$$

The contribution $\mathbf{E}_1^0$, which comes from the first integral on the right hand side of equation (5.105), arises from the charges $-Q$ and $(Q + \mathrm{d}Q)$ at the ends of the oscillating electric dipole in Figure 5.9. This contribution is given by equation (5.97) which, with R = r, is

$$\mathbf{E}_1^0 = \frac{\cos\theta}{4\pi\varepsilon_0}\left(\frac{2[p]}{r^3} + \frac{[\dot{p}]}{cr^2}\right)\hat{\mathbf{r}} + \frac{[p]\sin\theta}{4\pi\varepsilon_0 r^3}\hat{\boldsymbol{\theta}}. \tag{5.106}$$

By repeating the method used in the derivation of equation (5.97), it is straightforward for the reader to show that $\mathbf{E}_2^0$ is given by

$$\mathbf{E}_2^0 = \frac{\cos\theta}{4\pi\varepsilon_0}\left(\frac{[\dot{p}]}{cr^2} + \frac{[\ddot{p}]}{c^2r}\right)\hat{\mathbf{r}} + \frac{[\dot{p}]\sin\theta}{4\pi\varepsilon_0 cr^2}\hat{\boldsymbol{\theta}}. \tag{5.107}$$

According to equation (2.39)

$$[\dot{\mathbf{J}}]\mathrm{d}V_s = \dot{I}\,\mathrm{d}\mathbf{l} = [\ddot{\mathbf{p}}].$$

After substituting in the third integral on the right hand side of equation (5.105) we find that

$$\mathbf{E}_3^0 = -\frac{[\ddot{\mathbf{p}}]}{4\pi\varepsilon_0 c^2 r} = -\frac{[\ddot{p}]\cos\theta}{4\pi\varepsilon_0 c^2 r}\hat{\mathbf{r}} + \frac{[\ddot{p}]\sin\theta}{4\pi\varepsilon_0 c^2 r}\hat{\boldsymbol{\theta}}. \tag{5.108}$$

Adding equations (5.106), (5.107) and (5.108) we obtain

$$\mathbf{E} = \frac{\cos\theta}{4\pi\varepsilon_0}\left[\frac{2p}{r^3} + \frac{2\dot{p}}{cr^2}\right]\hat{\mathbf{r}} + \frac{\sin\theta}{4\pi\varepsilon_0}\left[\frac{p}{r^3} + \frac{\dot{p}}{cr^2} + \frac{\ddot{p}}{c^2r}\right]\hat{\boldsymbol{\theta}}. \tag{5.109}$$

This is agreement with equation (5.98).

At first sight equation (5.105) due to Jefimenko might suggest that the radiation electric field due to the accelerating charges inside the wire connecting the two charge distribution in Figure 5.9 was proportional to and in the direction of $[\dot{\mathbf{J}}]$ whereas, according to our equations (5.78) and (5.102) the radiation electric field is proportional to and in the direction of the component of $[\dot{\mathbf{J}}]$ perpendicular to the vector $\hat{\mathbf{r}}$ from the oscillating electric dipole to the field point P, that is in the direction of the unit vector $\hat{\boldsymbol{\theta}}$ in Figure 5.9. The reader can check that the contribution of the first term on the right hand side of equation (5.108), which is in the $-\,\hat{\mathbf{r}}$ direction and which comes from the $[\dot{\mathbf{J}}]$ term in equation (5.105), is compensated by the term proportional to $[\ddot{p}]\cos\theta$ in $+\ \hat{\mathbf{r}}$ direction in equation (5.107) which comes from the $[\dot{\rho}]$ term in equation (5.105). This leaves only a component of the radiation electric field in the $\hat{\boldsymbol{\theta}}$ direction in equation (5.109), in agreement with equation (5.98). This result illustrates that we only get the correct radiation electric field when using the Jefimenko equation (5.105), if we allow for the variations in the charge distributions that accumulate at the ends of the connecting wire in Figure 5.9. According to our equations (5.103) and (5.104) these charge distributions do not contribute to the radiation electric field. When a current element forms part of a complete electrical circuit carrying a varying current, no charge distributions accumulate at the end of the current element, so that unlike our equations (5.103) and (5.104) the differential form of the Jefimenko equation (5.105) would not give the correct contribution to the radiation electric field of those accelerating conduction electrons counted by the information collecting sphere inside the current element at the appropriate retarded time, but Jefimenko's equation (1.105) would give the correct value for the resultant radiation electric field when it is integrated over the complete circuit.

Throughout this chapter, we have been assuming that all velocities are much less than c. We shall discuss examples, where the higher order terms are important, in Section 6.8 of Chapter 6.

CHAPTER 6

Magnetic fields due to electrical circuits

6.1. Magnetic field due to a current element

6.1.1. *Use of the retarded positions of the moving charges*

Consider the stationary current element, that is of length $\mathbf{dl}$ and is at the position $\mathbf{r}_s$ in Figure 6.1. The field point P is a the position $\mathbf{r}$ at a distance $\mathbf{R} = (\mathbf{r} - \mathbf{r}_s)$ from the current element. We shall determine the contribution to the magnetic field at the field point P, at the time of observation t, due to the moving classical point charges that were passed and counted by the information collecting sphere inside the current element at the retarded time $t^* = (t - R/c)$. We shall assume that, when they are counted at the fixed time t^*, there are N_0 moving classical point charges per metre length inside the current element, each of charge q and all moving with the same velocity $[\mathbf{u}]$ and the same acceleration $[\mathbf{a}]$. Since we shall find that there are differences in the higher order terms in β between positive and negative moving charges we shall consider both cases, starting with moving positive charges as shown in Figure 6.1. We shall assume that N_0 is so large that the fluctuations in the number of moving charges inside the current element can be neglected.

According to equation (5.5) of Chapter 5, the number of moving charges counted by the information collecting sphere, while it is crossing the current element at the retarded time t^* is

$$\delta N = \left(\frac{s}{R}\right) N_0 \, \mathrm{d}l \tag{6.1}$$

where

$$s = R\left[1 - \frac{\mathbf{u} \cdot \mathbf{R}}{Rc}\right]. \tag{6.2}$$

According to equation (3.13), the magnetic field $\mathbf{B}$ due to each of these moving charges is given by

$$\mathbf{B} = \mathbf{B}_V + \mathbf{B}_A \tag{6.3}$$

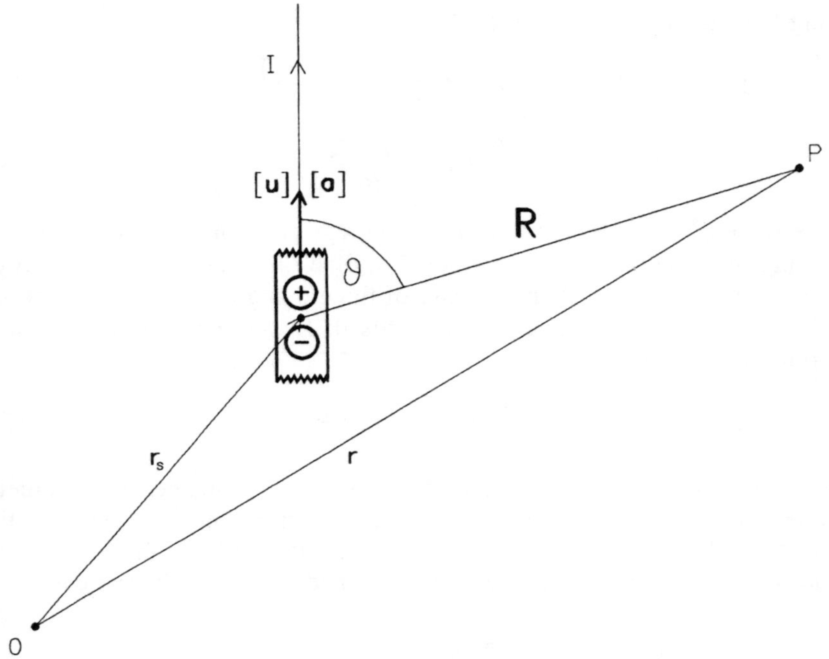

Figure 6.1. The calculation of the magnetic field due to the moving charges in a current element in which the positive charges are moving with velocity **[u]** at the retarded time $(t - R/c)$ and the negative charges are at rest.

where $\mathbf{B}_V$ and $\mathbf{B}_A$ are given by equation (3.14) and (3.15) respectively. According to the principle of superposition of magnetic fields, the resultant magnetic field d**B** due to the δN charges, counted by the information collecting sphere inside d**l**, is the vector sum of the contributions of the individual charges, so that

$$\mathrm{d}\mathbf{B} = \left(\frac{s}{R}\right)(N_0\,\mathrm{d}l)(\mathbf{B}_V + \mathbf{B}_A). \tag{6.4}$$

Using equation (3.14), we find that the contribution of the $\mathbf{B}_V$ term to d**B**, which will be denoted by $\mathrm{d}\mathbf{B}_V$, is, with $[\beta] = [u]/c$, given by

$$\begin{aligned}\mathrm{d}\mathbf{B}_V &= \left(\frac{s}{R}\right)(N_0\,\mathrm{d}l)\left(\frac{\mu_0}{4\pi}\right)\frac{q[\mathbf{u}\times\mathbf{R}][1-\beta^2]}{s^3}\\ &= \left(\frac{\mu_0}{4\pi}\right)\frac{N_0 q\,\mathrm{d}l}{Rs^2}[\mathbf{u}\times\mathbf{R}][1-\beta^2].\end{aligned} \tag{6.5}$$

Using the binomial theorem we have

$$\frac{1}{s^2} = \frac{[1 - \mathbf{u}\cdot\mathbf{R}/Rc]^{-2}}{R^2} = \frac{1}{R^2}\left[1 + \frac{2\mathbf{u}\cdot\mathbf{R}}{Rc} + \ldots\right].$$

Substituting in equation (6.5) and neglecting terms involving β^2 and remembering that ($\mathbf{u} \cdot \mathbf{R}/Rc$) is of order β, we obtain

$$d\mathbf{B}_V = \left(\frac{\mu_0}{4\pi}\right)\frac{N_0 q\, dl}{R^3}[\mathbf{u} \times \mathbf{R}]\left[1 + \frac{2\mathbf{u} \cdot \mathbf{R}}{Rc}\right] \tag{6.6}$$

$$= \left(\frac{\mu_0}{4\pi}\right)\frac{[I]\, d\mathbf{l} \times \mathbf{R}}{R^3}\left[1 + \frac{2\mathbf{u} \cdot \mathbf{R}}{Rc}\right] \tag{6.7}$$

where $[I] = qN[u]$ is the current in the current element at the retarded time t^* due to the moving positive charges. Let θ be the angle between the direction of current flow and $\mathbf{R}$. In the case of the moving positive charges shown in Figure 6.1, $\mathbf{u} \cdot \mathbf{R}/Rc$ is equal to $\beta \cos \theta$. Hence equation (6.7) can be rewritten in the form

$$d\mathbf{B}_V = \left(\frac{\mu_0}{4\pi}\right)\frac{[I]\, d\mathbf{l} \times \mathbf{R}}{R^3}[1 + 2\beta \cos \theta]. \tag{6.8}$$

A more important practical example is that of a conduction current in a metal. In this case negatively charged conduction electrons move in the direction opposite to the direction of current flow. We shall still assume that θ is the angle between the direction of current flow and $\mathbf{R}$, so that in this case

$$\frac{\mathbf{u} \cdot \mathbf{R}}{Rc} = \beta \cos(\pi - \theta) = -\beta \cos \theta.$$

Hence, for negative moving conduction electrons equation (6.7) becomes

$$d\mathbf{B}_V = \left(\frac{\mu_0}{4\pi}\right)\frac{[I]\, d\mathbf{l} \times \mathbf{R}}{R^3}[1 - 2\beta \cos \theta]. \tag{6.9}$$

The positive ions are generally at rest in a stationary metallic conductor and do not contribute to the magnetic field.

In the absence of conduction current flow, the conduction electrons in a metal have velocities of the order of $c/200$ in all directions. When a conduction current flows, the conduction electrons have a drift velocity $\mathbf{v}$ of the order of 10^{-4} m s^{-1}, in the direction opposite to the direction of current flow, which is superimposed on their velocity distribution We shall now average equation (6.6) over the velocity distribution of the conduction electrons. Let

$$\langle \mathbf{u}_i \rangle = \mathbf{v}$$

where $\mathbf{u}_i$ is the velocity of the ith conduction electron. Since $\mathbf{R}$ is a constant vector

$$\langle \mathbf{u}_i \cdot \mathbf{R} \rangle = \langle \mathbf{u}_i \rangle \cdot \mathbf{R} = \mathbf{v} \cdot \mathbf{R} = -vR \cos \theta. \tag{6.10}$$

Similarly,

$$\langle \mathbf{u}_i \times \mathbf{R} \rangle = \mathbf{v} \times \mathbf{R}$$

so that, with $q = -e$, for conduction electrons we have

$$qN_0\, dl\, \langle \mathbf{u}_i \times \mathbf{R} \rangle = -eN_0\, dl(\mathbf{v} \times \mathbf{R}) = I\, d\mathbf{l} \times \mathbf{R} \tag{6.11}$$

where **dl** is in the direction of current flow. Hence, for moving conduction electrons, after averaging equation (6.6) and using equations (6.10) and (6.11), we have

$$d\mathbf{B}_V = \left(\frac{\mu_0}{4\pi}\right)\frac{[I]\ \mathrm{d}\mathbf{l}\times\mathbf{R}}{R^3}\ [1 - 2\beta\cos\theta] \tag{6.12}$$

where β in equation (6.12) is equal to v/c. It was shown in section 1.3.3 of Chapter 1 that, for a typical current in a typical metallic conductor β is of the order of 10^{-12}, so that $2\beta\cos\theta \ll 1$ and the second term inside the brackets in equation (6.12) is generally very much smaller than the first for conduction current flow in a metal.

As we shall need the results when we come to discuss the Biot-Savart law in Section 6.2, we shall now derive the expression for the contributions of the accelerations of the charges, in the current element in Figure 6.1, to the magnetic field. According to equation (3.15), for a charge of magnitude q moving with velocity [**u**] and having an acceleration [**a**] at its retarded position, we have for the conditions shown in Figure 6.1.

$$\mathbf{B}_A = \left(\frac{\mu_0}{4\pi}\right)\frac{q}{cs^3}\left[\mathbf{u}\left(\frac{\mathbf{R}\cdot\mathbf{a}}{c}\right) + s\mathbf{a}\right]\times\mathbf{R}. \tag{6.13}$$

The first and second terms inside the square brackets are of order βRa and Ra respectively, where $\beta = u/c$. To the accuracy we shall generally work to with conduction currents in metals, we can neglect the first term inside the square bracket since it is much smaller than the second term. Using equation (6.4) we then find that the contribution, associated with the $\mathbf{B}_A$ term in equation (6.4), is given, to a very good approximation, by

$$\begin{aligned} d\mathbf{B}_A &= \left(\frac{s}{R}\right)N_0\ \mathrm{d}l\left(\frac{\mu_0}{4\pi}\right)\frac{q}{cs^3}\ [s\mathbf{a}\times\mathbf{R}] \\ &= \left(\frac{\mu_0}{4\pi}\right)\frac{qN_0\ \mathrm{d}l}{csR}\ [\mathbf{a}\times\mathbf{R}]. \end{aligned} \tag{6.14}$$

Since we shall generally only need the first order acceleration dependent term, we shall generally ignore terms of order β and put $s = R$ in equation (6.14), which then reduces to

$$d\mathbf{B}_A = \left(\frac{\mu_0}{4\pi}\right)\frac{qN_0\ \mathrm{d}l}{cR^2}\ [\mathbf{a}\times\mathbf{R}]. \tag{6.15}$$

It is left as an exercise for the reader to show that, if we did retain both terms inside the square brackets in equation (6.13) and expanded $1/s$ to first order of β, we would have to add the terms

$$\left(\frac{\mu_0}{4\pi}\right)\frac{qN_0\ \mathrm{d}l}{cR^2}\left[(\mathbf{a}\times\mathbf{R})\left(\frac{\mathbf{u}\cdot\mathbf{R}}{Rc}\right) + (\mathbf{u}\times\mathbf{R})\left(\frac{\mathbf{a}\cdot\mathbf{R}}{Rc}\right)\right]$$

to the right hand side of equation (6.15). The reader can show that, if we have a distribution of accelerations, equation (6.15) is still valid if **a** is now the average acceleration $\langle \mathbf{a}_i \rangle$ of the conduction electrons in the current element in Figure 6.1.

Adding equations (6.8) and (6.15), we find that the total contribution to the magnetic field due to the moving and accelerating positive classical point charges counted by the information collecting sphere inside the current element, shown in Figure 6.1, is

$$\begin{aligned} \mathrm{d}\mathbf{B} &= \mathrm{d}\mathbf{B}_V + \mathrm{d}\mathbf{B}_A \\ &= \left(\frac{\mu_0}{4\pi}\right) \frac{[I]\,\mathrm{d}\mathbf{l} \times \mathbf{R}}{R^3} [1 + 2\beta \cos\theta] + \left(\frac{\mu_0}{4\pi}\right) \frac{qN_0\,\mathrm{d}l}{cR^2} [\mathbf{a} \times \mathbf{R}]. \end{aligned} \tag{6.16}$$

For negative moving charges of charge $q = -e$, adding equations (6.9) and (6.15) we have

$$\mathrm{d}\mathbf{B} = \left(\frac{\mu_0}{4\pi}\right) \frac{[I]\,\mathrm{d}\mathbf{l} \times \mathbf{R}}{R^3} [1 - 2\beta \cos\theta] - \left(\frac{\mu_0}{4\pi}\right) \frac{eN_0\,\mathrm{d}l}{cR^2} [\mathbf{a} \times \mathbf{R}]. \tag{6.17}$$

where $\mathrm{d}\mathbf{l}$ is in the direction of current flow, which for conduction electrons is in the direction opposite to their mean drift velocity $\mathbf{v}$, $\beta = v/c$ and θ is the angle between the direction of current flow and the vector $\mathbf{R}$ from the current element to the field point.

6.1.2. *Use of the projected positions of the conduction electrons*

Some readers may prefer to derive equation (6.9) using the expression for the magnetic field related to the projected positions of the conduction electrons, for the conditions shown previously in Figure 5.3 of Chapter 5. We shall again assume that the conduction electrons move with the same velocity $[\mathbf{u}]$. In the time R/c it takes the information collecting sphere to go from the current element at S to the field point P in Figure 5.3, the conduction electrons would move downwards a distance $uR/c = \beta R$ to the projected position T, if they moved with velocity $[\mathbf{u}]$. According to equation (6.1) the number of conduction electrons counted by the information collecting sphere inside the current element $I\,\mathrm{d}l$ at the retarded time $(t - R/c)$ is

$$\delta N = (1 + \beta \cos\theta)\, N_0\, \mathrm{d}l \tag{6.18}$$

where N_0 is the number of conduction electrons per metre length and θ is the angle between $\mathbf{R}$ and the direction of current flow. By analogy with the derivation of equation (3.38) relating the electric field of a moving charge to its projected position, we can relate the $\mathbf{B}_V$ term in equation (3.13) to the projected position of a conduction electron using equation (3.28), which is

$$\mathbf{B}_V = \frac{q[\mathbf{u}] \times \mathbf{r}_0[1 - \beta^2]}{4\pi\varepsilon_0 c^2 r_0^3[1 - \beta^2 \sin^2(\theta - \alpha)]^{3/2}} \simeq -\frac{e[u] \sin(\theta - \alpha)}{4\pi\varepsilon_0 c^2 r_0^2}\,\hat{\mathbf{k}} \tag{6.19}$$

where r_0, θ and α are as shown in Figure 5.3. The vector $\mathbf{r}_0$ is measured from the projected position T of the moving charge and $\hat{\mathbf{k}}$ is a unit vector in the direction of the vector product $[\mathbf{u}] \times \mathbf{r}_0$, which is in the upward direction from the paper in Figure 5.3.

Multiplying equations (6.18) and (6.19), we find that the contribution to the magnetic field at the field point P in Figure 5.3, due to the moving conduction electrons counted by the information collecting sphere inside $I\,\mathrm{d}l$ at the retarded time $(t - R/c)$, is

$$\mathrm{d}\mathbf{B} = -\frac{[1 + \beta \cos \theta] N_0 e [u] \mathrm{d}l \sin (\theta - \alpha)}{4\pi\varepsilon_0 c^2 r_0^2} \hat{\mathbf{k}}. \tag{6.20}$$

According to equation (5.30) of Chapter 5,

$$r_0^2 = (1 + \beta \cos \theta)^2 R^2.$$

Using equations (5.28) and (5.29), we have

$$\sin (\theta - \alpha) = \sin \theta \cos \alpha - \cos \theta \sin \alpha = \sin \theta\,(1 - \beta \cos \theta).$$

Substituting in equation (6.20) and putting $Ne[u]$ equal to I we find that the to first order in β

$$\mathrm{d}\mathbf{B} = -\frac{[I]\,\mathrm{d}l \sin \theta}{4\pi\varepsilon_0 c^2 R^2} [1 - 2\beta \cos \theta] \hat{\mathbf{k}} = \frac{[I]\,\mathrm{d}\mathbf{l} \times \mathbf{R}}{4\pi\varepsilon_0 c^2 R^3} [1 - 2\beta \cos \theta]. \tag{6.21}$$

Equation (6.21) is the same as equation (6.9). It is left as an exercise for the reader to use the same method to derive equation (6.8) for positive moving charges.

It is of interest to note that we were able to derive both equations (5.20) and (6.21), which are both valid in the quasi-stationary limit, using Coulomb's law equation (3.33) and the Biot-Savart expression, equation (3.34), for the magnetic field due to a moving charge in the low velocity limit, provided we measured all quantities from the projected positions of the moving charges and used equation (6.1) for the number of moving charges counted by the information collecting sphere inside the current element at the retarded time.

6.2. The Biot-Savart law for steady currents

On the basis of experiments on current balances of different geometrical configurations, we concluded in Section 1.4.1 that the magnetic field due to a steady current I in a complete circuit, of the type shown in Figure 6.2, is given by the Biot-Savart law, according to which the magnetic field at the field point P at the position $\mathbf{r}$ is given by

$$\mathbf{B}(\mathbf{r}) = \frac{\mu_0 I}{4\pi} \oint \frac{\mathrm{d}\mathbf{l} \times \mathbf{R}}{R^3} \tag{6.22}$$

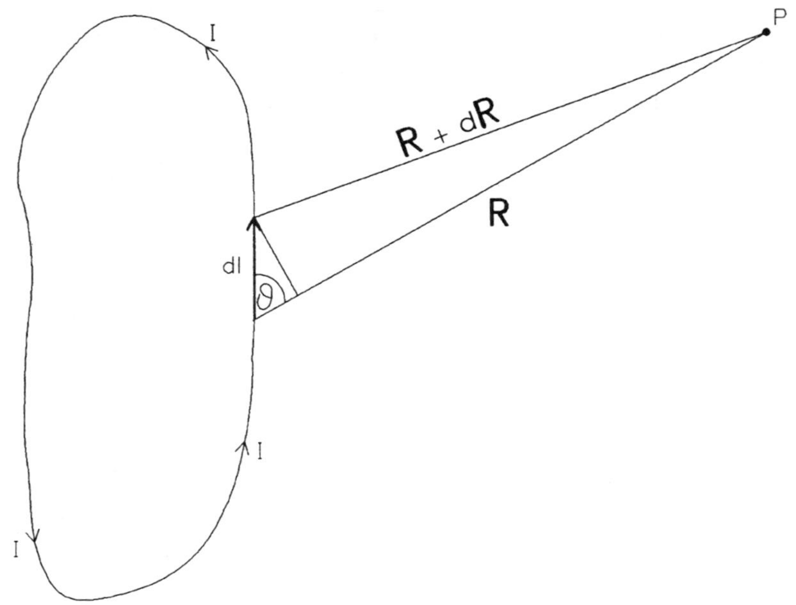

Figure 6.2. Calculation of the magnetic field due to the steady current in a complete circuit.

where **dl** is an element of the circuit at the source position $\mathbf{r}_s$ and $\mathbf{R} = (\mathbf{r} - \mathbf{r}_s)$ is a vector from **dl** to the field point at **r**. In this section we shall start by considering the contribution to the magnetic field of those charges that were counted by the information collecting sphere inside the current element I **dl** in Figure 6.2. We shall use a simplified model in which we shall assume that there are N_0 classical point charges per metre length, each of charge q and moving with speed u in the current element **dl**, such that the magnitude of the current in **dl** is

$$I = qN_0u. \tag{6.23}$$

According to equation (6.16), for moving positive charges we have

$$\mathrm{d}\mathbf{B}(\mathbf{r}) = \frac{\mu_0 I}{4\pi}\left(\frac{\mathrm{d}\mathbf{l} \times \mathbf{R}}{R^3}\right) + \frac{\mu_0 I}{4\pi}\left(\frac{\mathrm{d}\mathbf{l} \times \mathbf{R}}{R^3}\right)\left(\frac{2u \cos \theta}{c}\right) + \frac{\mu_0 N_0 q \,\mathrm{d}l}{4\pi}\left(\frac{\mathbf{a} \times \mathbf{R}}{cR^2}\right). \tag{6.16}$$

(According to equation (6.17), the second and third terms on the right hand side are both negative in the case of negatively charged moving particles such as conduction electrons). The first term on the right hand side of equation (6.16) is the differential form of the Biot-Savart law. Since the second and third terms are finite, it follows that the **differential** form of the Biot-Savart is not exact to all orders of β, though since the second and third terms are much smaller than the first term, the differential form of the Biot-Savart law is a very good approximation for the magnetic field due to the moving charges counted by the information collecting sphere inside the current element $I\,\mathbf{dl}$ when the conduction current I is steady or is varying at mains frequency. We shall now go on to consider the complete circuit shown in Figure 6.2. We shall allow for the possibility that the circuit is made up of wires of different electrical conductivities so that N_0 and u may be different in different parts of the circuit, but the current $I = q\,N_0 u$ is the same in all parts of the circuit. The charges are accelerated or decelerated where wires of different electrical conductivity are joined. Integrating equation (6.16) around the complete circuit, in the case of positive moving charges, when the current I is constant, we have

$$\mathbf{B} = \frac{\mu_0 I}{4\pi}\oint \frac{\mathbf{dl}\times\mathbf{R}}{R^3} + \Delta\mathbf{B} \tag{6.24}$$

where with $I = qN_0 u$

$$\Delta\mathbf{B} = \frac{\mu_0}{4\pi c}\left\{2\oint qN_0 u\left(\frac{\mathbf{dl}\times\mathbf{R}}{R^3}\right) u\cos\theta + \oint qN_0\left(\frac{\mathbf{a}\times\mathbf{R}}{R^2}\right)\mathrm{d}l\right\}. \tag{6.25}$$

The first term on the right hand side of equation (6.24) is the integral form of the Biot-Savart law. It is shown in Appendix A4 that the value of $\Delta\mathbf{B}$ given by equation (6.25) is zero when the current I in the circuit in Figure 6.2 is steady. Hence for steady currents in the complete circuit in Figure 6.2, equation (6.24) reduces to equation (6.22), showing that the **integral** form of the Biot-Savart law, namely equation (6.22), is valid at the next level of approximation in β. To get $\Delta\mathbf{B} = 0$ in equation (6.25) we had to include the accelerations of the charges. Some examples of the deviations from the differential form of the Biot-Savart law will now be considered.

6.3. Examples of deviations from the differential form of the Biot-Savart law

6.3.1. *A straight line section of finite length*

To illustrate the corrections to the differential form of the Biot-Savart law, given by equation (6.25), consider the section AD of the circuit shown in Figure 6.3. The current I in the circuit is constant. The section AD of the circuit lies along the x axis. We shall assume that there are N_0 *positive* classical point charges per metre length each of magnitude $+q$ and all moving with the same uniform speed u in the direction of the current flow from D to A, that is in the $-x$ direction in Figure 6.3. The magnetic field will be determined at the field point P on the y axis in Figure 6.3 at a distance b from the origin. We shall assume that the accelerations of the charges are zero, except possibly at the ends of the section AD, so that the $\mathrm{d}\mathbf{B}_A$ term in equation

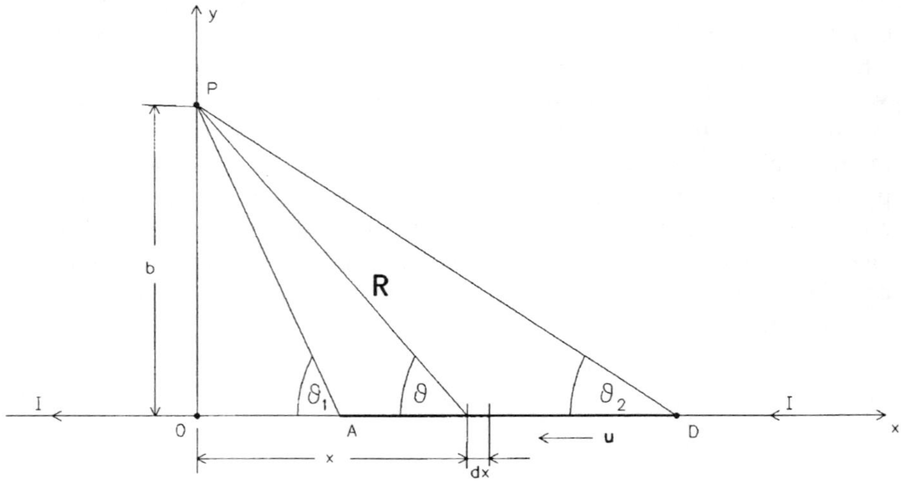

Figure 6.3. Calculation of the contribution to the magnetic field of the current in the section *AD* of a long straight wire carrying a steady current *I*.

(6.16) is zero for the section *AD*. Consider an element of length dx between x and x + dx. Let θ be the angle between **u** and the vector **R** from the element of length dx to the field point. Substituting in equation (6.8), we find that the contribution of the current element of length dx to the magnetic field at the field point P in Figure 6.3 is for positive moving charges

$$dB = \left(\frac{\mu_0 I}{4\pi}\right)\left(\frac{\sin\theta}{R^2}\right)(1 + 2\beta\cos\theta)dx. \tag{6.26}$$

This contribution is in the $-z$ direction in Figure 6.3. Since in Figure 6.3, $x = b\cot\theta$

$$dx = -b\,\mathrm{cosec}^2\,\theta\,d\theta.$$

Substituting in equation (6.26) and using $R = b\,\mathrm{cosec}\,\theta$, we find that

$$dB = -\left(\frac{\mu_0 I}{4\pi b}\right)\sin\theta(1 + 2\beta\cos\theta)d\theta.$$

Integrating from $\theta = \theta_1$ to $\theta = \theta_2$ we obtain

$$\mathbf{B} = -\left(\frac{\mu_0 I}{4\pi b}\right)(\cos\theta_2 - \cos\theta_1)\hat{\mathbf{k}} - \left(\frac{\mu_0 I}{4\pi b}\right)\beta(\cos^2\theta_2 - \cos^2\theta_1)\hat{\mathbf{k}}. \tag{6.27}$$

where $\hat{\mathbf{k}}$ is a unit vector in the $+z$ direction in Figure 6.3. The first term on the right hand side of equation (6.27) is the standard expression obtained using the differential form of the Biot-Savart law. Since

$$\cos^2\theta_2 - \cos^2\theta_1 = (\cos\theta_2 - \cos\theta_1)(\cos\theta_2 + \cos\theta_1)$$

the second term on the right hand side of equation (6.27) is equal to β ($\cos\theta_2 + \cos\theta_1$) times the first term.

If the section AD in Figure 6.3 extended from $x = -\infty$ to $x = +\infty$, θ_1 and θ_2 would be equal to π and 0 respectively and equation (6.27) would reduce to $\mathbf{B} = -\mu_0 I\hat{\mathbf{k}}/2\pi b$ in agreement with the integral form of the Biot-Savart law.

If the section AB extended from the origin O to $x = +\infty$, the limits of θ in equation (6.27) would be from $\theta_1 = \pi/2$ to $\theta_2 = 0$, and equation (6.27) would give

$$\mathbf{B} = -\left(\frac{\mu_0 I}{4\pi b}\right)(1 + \beta)\hat{\mathbf{k}} \tag{6.28}$$

which is not in agreement with the predictions of the differential from of the Biot-Savart law, according to which $\mathbf{B}$ should be equal to $-\mu_0 I\hat{\mathbf{k}}/4\pi b$. If the conductor AB in Figure 6.3 extended from $x = -\infty$ to $x = 0$, θ_1 and θ_2 would be equal to π and $\pi/2$ respectively and equation (6.27) would give

$$\mathbf{B} = -\left(\frac{\mu_0 I}{4\pi b}\right)(1 - \beta)\hat{\mathbf{k}}. \tag{6.29}$$

6.3.2. *A circuit made up of two conductors of different electrical conductivities*

To illustrate equations (6.28) and (6.29), consider the example shown in Figure 6.4. Two very long, straight conductors of different electrical conductivities lie along the x axis. They are joined at the origin O in Figure 6.4. A steady current I flows from right to left in the $-x$ direction along the composite conductor in Figure 6.4. In conductor 1, there are N_1 **positive** classical point charges per metre length, each of magnitude q and each moving with uniform velocity u_1 to the left, whereas in conductor 2 there are N_2 **positive** classical point charges per metre length also of magnitude q, but moving with uniform velocity u_2 to the left giving rise to a steady current flow in the $-x$ direction

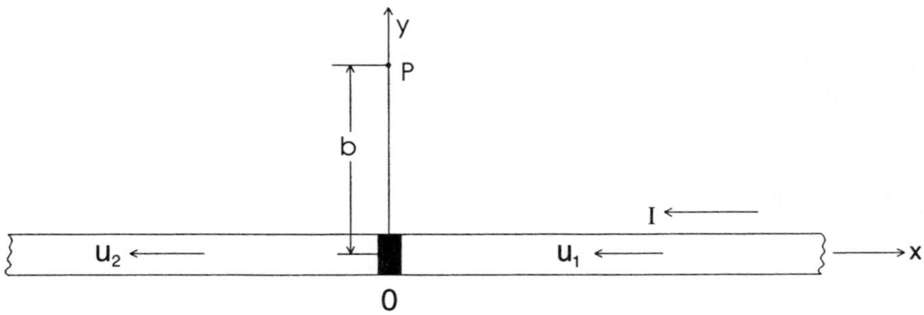

Figure 6.4. The magnetic field due to the current in an infinitely long conductor made up of two conductors of different electrical conductivities. The velocities of the moving positive charges increases fro $\mathbf{u}_1$ to $\mathbf{u}_2$ in a small transition region at the origin.

in Figure 6.4. If we ignored the accelerations of the charges in the transition region in Figure 6.4 and then applied equations (6.28) and (6.29), we would find that the magnetic field at the field point P on the y axis, at a distance b from the origin in Figure 6.4, would be

$$\mathbf{B} = -\left(\frac{\mu_0 I}{4\pi b}\right)\{2 + (\beta_1 - \beta_2)\}\hat{\mathbf{k}}. \tag{6.30}$$

This prediction is not in agreement with the integral form of the Biot-Savart law, according to which the magnetic field at the field point P in Figure 6.4 should be

$$\mathbf{B} = -\left(\frac{\mu_0 I}{2\pi b}\right)\hat{\mathbf{k}}.$$

However, we have neglected the accelerations of the charges when their velocities are changed from u_1 in conductor 1 to u_2 in conductor 2. To simplify the calculation, we shall assume that $u_1 \ll u_2$ and that $u_2 \ll c$. We shall also assume that the charges undergo uniform accelerations from u_1 to u_2 over an extremely small distance L, in a thin transition region where the conductors are joined at the origin O in Figure 6.4. In the non-relativistic limit, since $u_1 \ll u_2$ the acceleration a is given to an excellent approximation by

$$a = \frac{u_2^2}{2L}. \tag{6.31}$$

The velocity u of the charges, when they have travelled a distance l into the transition region, is given by

$$u^2 = u_1^2 + 2al.$$

Neglecting u_1 and substituting for a using equation (6.31) we find that

$$u = (2al)^{1/2} = u_2\left(\frac{l}{L}\right)^{1/2}. \tag{6.32}$$

Since the current I is the same in conductors 1 and 2 and in the transition region, then if N is the number of moving charges per metre length at a distance l into the transition region, we have

$$I = qN_1u_1 = qNu = qN_2u_2.$$

Using equation (6.32), we have

$$N = \frac{N_2u_2}{u} = N_2\left(\frac{L}{l}\right)^{1/2}. \tag{6.33}$$

According to equation (6.15), the acceleration dependent contribution to the magnetic field at the field point P in Figure 6.4 due to a section of length dl of the transition region is, for the conditions shown in Figure 6.4 where $\theta = \pi/2$, given by

$$\mathrm{d}\mathbf{B}_A = -N\,\mathrm{d}l\left(\frac{\mu_0}{4\pi}\right)\left(\frac{qa}{cb}\right)\hat{\mathbf{k}}.$$

Substituting for N and a from equations (6.33) and (6.31) respectively we find that

$$\mathrm{d}\mathbf{B}_A = -N_2\left(\frac{L}{l}\right)^{1/2}\frac{\mu_0 q\,\mathrm{d}l}{4\pi cb}\left(\frac{u_2^2}{2L}\right)\hat{\mathbf{k}} = -(qN_2u_2)\left(\frac{\mu_0}{4\pi b}\right)\beta_2\left(\frac{\mathrm{d}l}{2L^{1/2}l^{1/2}}\right)\hat{\mathbf{k}}.$$

where $\beta_2 = u_2/c$. Integrating from $l = 0$ to $l = L$, we obtain

$$\mathbf{B}_A = -\beta_2\left(\frac{\mu_0 I}{4\pi b}\right)\hat{\mathbf{k}}. \tag{6.34}$$

Since β_1 is negligible compared with β_2, equation (6.30) becomes

$$\mathbf{B} = -\left(\frac{\mu_0 I}{4\pi b}\right)(2 - \beta_2)\hat{\mathbf{k}}. \tag{6.35}$$

Adding equations (6.34) and (6.35) we finally obtain

$$\mathbf{B} = -\left(\frac{\mu_0 I}{2\pi b}\right)\hat{\mathbf{k}}. \tag{6.36}$$

This is in agreement with the predictions of the integral form of the Biot-Savart law. This illustrates how the two terms in equation (6.25) cancel each other so that, for a complete circuit, equation (6.24) reduces to the integral form of the Biot-Savart law, provided the current in the circuit is constant.

6.3.3. *Use of equation (3.28) which gives the magnetic field due to a charge moving with uniform velocity*

According to equation (3.28), the magnetic field due to a classical point charge moving with **uniform** velocity can be related to the position of the charge at the time of observation t. It is tempting to assume, as was done for example by Rosser [1] and Vybiral [2], that the magnetic field due to the moving charges that are actually inside the current element I **d**l in Figure 6.1 at the time of observation t is given by the product of the number of charges N_0 dl counted at the time of observation t and the magnetic field due to each charge which is given by equation (3.28). We would then have

$$\mathrm{d}\mathbf{B} = \frac{\mu_0 N_0 q(\mathbf{u}\times\mathbf{R})(1-\beta^2)\,\mathrm{d}l}{4\pi R^3(1-\beta^2\sin^2\theta)^{3/2}}.$$

Expanding to second order of β we would have

$$\mathrm{d}\mathbf{B} = \frac{\mu_0 I(\mathrm{d}\mathbf{l}\times\mathbf{R})}{4\pi R^3}\left\{1 + \beta^2\left(\frac{3}{2}\sin^2\theta - 1\right)\right\}. \tag{6.37}$$

According to equation (6.37) the correction to the differential form of the Biot-Savart law would be of order β^2 and not of order β as predicted by equation (6.27) for the case of the straight section of wire in Figure 6.3. Equation (6.37) is a satisfactory approximation in the special case of an infinitely long, straight wire, but it breaks down when there is a bend in the wire. As an example, we shall assume that the wire *DA* in Figure 6.3 extends from $x = +\infty$ to the origin *O* in Figure 6.3 and that the wire turns through 90° at the origin to lie along the *y* axis. The moving charges have an acceleration when they turn around the corner, and for the reasons given in Section 3.4.1 of Chapter 3, we must use equation (6.19) which is related to the projected positions of the moving charges, which are the positions the moving charges would have reached if they had carried on with uniform velocity u until the time of observation. It takes the information collecting sphere a time b/c to go from *O* to *P* in Figure 6.3. In this time interval the positive moving charges would have moved a distance $(ub/c) = \beta b$ to the left of the origin, if they had carried on with uniform velocity u. Hence if equation (6.19) is used, we must allow for all the projected positions up to a distance βb on the x axis to the left of the origin in Figure 6.3. Thus, for positive moving charges we must add the contributions of $N_0\beta b$ charges to the left of the origin. Using equation (6.19) we find that the contribution of these $N_0\beta b$ charges is given to first order of β, where $\beta \ll 1$, by

$$\Delta B = N_0\beta b\left(\frac{\mu_0 qu}{4\pi b}\right) = \beta\left(\frac{\mu_0 I}{4\pi b}\right).$$

To first order of β, equation (6.37) is the same as the Biot-Savart law so that the contribution predicted by equation (6.37) for the magnetic field at *P* due to those moving positive charges that were inside the section *AD*, which goes from $x = +\infty$ to the origin in Figure 6.3, at the time of observation t is equal to $\mu_0 I/4\pi b$. Hence the total magnetic field at *P* due to the current in the section *AD* in Figure 6.3 is given, to first order of β, by

$$\mathbf{B} = -\left(\frac{\mu_0 I}{4\pi b}\right)\hat{\mathbf{k}} - \beta\left(\frac{\mu_0 I}{4\pi b}\right)\hat{\mathbf{k}} = -\left(\frac{\mu_0 I}{4\pi b}\right)(1 + \beta)\hat{\mathbf{k}}. \tag{6.38}$$

This is in agreement with equation (6.28). This example illustrates the care that must be exercised if equation (6.37) is used. It is safer in practice to use equation (6.17) in which the magnetic field is related to the charges that were in the current element $I\,\mathbf{dl}$ in Figure 6.2 at the appropriate retarded time.

Problem. The steady conduction current I flowing in a circular coil of radius b is due to N_0 conduction electrons per metre length each of charge $-e$ and all moving at the same uniform speed u, plus N_0 stationary positive ions per metre length each of charge $+e$. Use equations (3.14) and (3.15) to show that to all orders $\beta = u/c$ the magnetic field on the axis of the coil at a distance x from the centre of the coil is given by

$$B = \frac{\mu_0 I b^2}{2(b^2 + x^2)^{3/2}}.$$

Show that the term in β^2 that comes from equation (3.14) is cancelled by the contribution that comes from equation (3.15) due to the centripetal acceleration u^2/b.

6.4. The magnetic field due to an oscillating electric dipole

To illustrate the application of equation (6.17), we shall consider once again the example of the stationary oscillating electric dipole, shown previously in Figure 2.2 of Chapter 2 and in Figure 5.9 of Section 5.12. We shall assume that the current flow in the wire of length dl is due to N_0 conduction electrons per metre length, each of charge $q = -e$ and each moving downwards with velocity [**u**] and acceleration [**a**] at the retarded time $t^* = (t - R/c)$. Let θ be the angle between the direction of current flow and the vector **R** from the oscillating electric dipole to the field point. Since in the present example the electric dipole is at the origin, **R** is equal to **r** in equation (6.17). The accelerations of the conduction electrons are parallel to the wire so that $-eN_0$ dl [**a**] is equal to $[\dot{I}]$ d**l** and equation (6.17) can be written in the form

$$\mathbf{B} = \left(\frac{\mu_0}{4\pi}\right)(\mathrm{d}\mathbf{l}\times\mathbf{r})\left[\frac{I}{r^3}(1 - 2\beta\cos\theta) + \frac{\dot{I}}{cr^2}\right]$$

$$= \left(\frac{\mu_0\,\mathrm{d}l}{4\pi}\right)\left[\frac{I}{r^2}(1 - 2\beta\cos\theta) + \frac{\dot{I}}{cr}\right]\sin\theta\,\hat{\boldsymbol{\phi}} \tag{6.39}$$

where the unit vector $\hat{\boldsymbol{\phi}}$ is in the direction of d**l** × **r**.

From equations (2.32) and (2.39), we have

$$[I]\,\mathrm{d}l = [\dot{p}]; \qquad [\dot{I}]\,\mathrm{d}l = [\ddot{p}].$$

Hence equation (6.39) can be rewritten in the form

$$\mathbf{B} = \left(\frac{\mu_0}{4\pi}\right)\left[\frac{\dot{p}}{r^2}(1 - 2\beta\cos\theta) + \frac{\ddot{p}}{cr}\right]\sin\theta\,\hat{\boldsymbol{\phi}} \tag{6.40}$$

For positive moving and accelerating charges the $(1 - 2\beta\cos\theta)$ term in equation (6.40) must be replaced by $(1 + 2\beta\cos\theta)$.

To obtain agreement with equation (2.46), which we derived using the retarded potentials in Section 2.4.2 of Chapter 2, we must ignore the $2\beta\cos\theta$ term in equation (6.40) as well as all the higher order terms in β which we neglected when we derived equation (6.17). With these approximations, which are reasonable for conduction current flow, equation (6.40) becomes

$$\mathbf{B} = \left(\frac{\mu_0}{4\pi}\right)\left[\frac{\dot{p}}{r^2} + \frac{\ddot{p}}{cr}\right]\sin\theta\,\hat{\boldsymbol{\phi}} \tag{6.41}$$

The origin of the differences between equation (6.40) and equation (6.41) will be discussed in Section 6.8.

The electric field due to the oscillating electric dipole can now be derived from equation (6.41) using Maxwell's equations in the way described in Section 2.6.5.

6.5. The magnetic field due to a varying current distribution

Consider a macroscopic charge and current distribution that consists of moving and accelerating classical point charges as shown previously in Figure 5.10. The magnetic field will be determined at the field point P at the position $\mathbf{r}$ in Figure 5.10 at the time of observation t. Let n_i be the number of classical point charges of type i per cubic metre that have charge q_i, velocity $[\mathbf{u}_i]$, acceleration $[\mathbf{a}_i]$ and are inside the volume element $\mathrm{d}V_s$ at the position $\mathbf{r}_s$ in Figure 5.10 at the retarded time $t^* = (t - R/c)$, where $\mathbf{R}$ is equal to $(\mathbf{r} - \mathbf{r}_s)$. It follows from equation (6.1) that the number of charges of velocity $[\mathbf{u}_i]$ counted inside $\mathrm{d}V_s$ by the information collecting sphere, that reaches the field point P at the time of observation t, is

$$\delta N_i = n_i \left(\frac{s_i}{R}\right) \mathrm{d}V_s \tag{6.42}$$

where

$$s_i = \left[R - \frac{\mathbf{R} \cdot \mathbf{u}_i}{Rc}\right]. \tag{6.43}$$

Multiplying equation (6.42) by the expression for the magnetic field due to a moving and accelerating classical point charge, given by equation (3.13) then making the same approximations as when we derived equation (6.16), for example by putting $(-e) = q_i$, $N_0\,\mathrm{d}l = n_i\,\mathrm{d}V_s$, $[\mathbf{u}] = \mathbf{u}_i$ and $[\mathbf{a}] = \mathbf{a}_i$ in equations (6.6) and (6.15) and ignoring the $\beta \cos \theta$ term we find that the total magnetic field at the field point P in Figure 5.10 due to the charges that have velocity $[\mathbf{u}_i]$ and acceleration $[\mathbf{a}_i]$ and are counted by the information collecting sphere inside $\mathrm{d}V_s$ is

$$\mathrm{d}\mathbf{B}_i = \frac{\mu_0}{4\pi}\left(\frac{q_i n_i [\mathbf{u}_i] \times \mathbf{R}}{R^3} + \frac{q_i n_i [\mathbf{a}_i] \times \mathbf{R}}{cR^2}\right) \mathrm{d}V_s.$$

Summing over all values of $[\mathbf{u}_i]$ and $[\mathbf{a}_i]$ for particles of type i, and using the equations

$$\Sigma q_i n_i [\mathbf{u}_i] = [\mathbf{J}_i]$$

$$\Sigma q_i n_i [\mathbf{a}_i] = [\dot{\mathbf{J}}_i]$$

and then summing over all types of particles we finally obtain

$$\mathrm{d}\mathbf{B}(\mathbf{r}, t) = \frac{\mu_0}{4\pi}\left(\frac{[\mathbf{J}(\mathbf{r}_s)]}{R^2} + \frac{[\dot{\mathbf{J}}(\mathbf{r}_s)]}{cR}\right) \times \hat{\mathbf{R}}\,\mathrm{d}V_s \tag{6.44}$$

where $\hat{\mathbf{R}}$ is a unit vector in the direction of $\mathbf{R}$, $\mathbf{J}$ is the current density and $\dot{\mathbf{J}}$

is the rate of change of current density, which in the example shown in Figure 5.10 is not necessarily in the direction of **J**. Equation (6.44) is the differential form of equation (1.134) which was derived by Jefimenko [3] by solving the differential equation (1.122) for **B**, and which we derived in Section 2.7 of Chapter 2 using the vector potential. This result shows that Jefimenko's equation (1.134) for the magnetic field due to a varying current distribution is valid in differential form, whereas the Jefimenko's equation (1.136) for the electric field due to a varying charge and current distribution is only valid in integral form.

In the case of an electrical circuit, if we ignore the centripetal accelerations of the moving charges where the circuit is curved, then the current flow and the rate of change of current are parallel to the wires making up the circuit and equation (6.44) gives

$$d\mathbf{B}(\mathbf{r}, t) = \frac{\mu_0}{4\pi}\oint\left(\frac{[I]}{R^2} + \frac{[\dot{I}]}{cR}\right) d\mathbf{l} \times \hat{\mathbf{R}}. \tag{6.45}$$

The reader can check back that the term proportional to $[I]$ in equation (6.45) arises from the velocity dependent term $\mathbf{B}_V$ in equation (3.13) and that the $[\dot{I}]$ term in equation (6.45) arises from the acceleration term $\mathbf{B}_A$ in equation (3.13), which is the expression for the magnetic field due to an accelerating classical point charge.

6.6. The magnetic field due to an oscillating magnetic dipole

To illustrate the application of equation (6.45), we shall consider again the small current carrying coil *ABCD* shown previously in Figure 5.5 of Chapter 5. A current I is flowing from A to B to C to D to A, as shown in Figure 5.5. The sections *AD* and *BC* are arcs of circles, with centres at the field point P and have radii r and $r + \delta r$ respectively, where $\delta r \ll r$. It follows from equation (5.35) that if the length of the section *AD* is equal to b, where $b \ll r$, then the length of *BC* is $b(1 + \delta r/r)$. The magnetic field at the field point P in Figure 5.5 will be determined at the time of observation t.

According to equation (6.45) the contributions of the currents in the sections *AB* and *CD* to the magnetic field at P are both zero since $d\mathbf{l} \times \mathbf{R}$ is zero in both cases. If the current and the rate of change of current in the coil *ABCD* are I and $\dot{I}$ respectively, when the information collecting sphere is passing the section *BC*, then according to equation (6.45) the contribution of the charges counted by the information collecting sphere inside *BC* to the magnetic field at the field point P is

$$(\delta\mathbf{B})_{BC} = \left(\frac{\mu_0}{4\pi}\right)\left[\frac{Ib(1 + \delta r/r)}{r^2(1 + \delta r/r)^2} + \frac{\dot{I}b(1 + \delta r/r)}{cr(1 + \delta r/r)}\right]\hat{\mathbf{k}}$$

where $\hat{\mathbf{k}}$ is a unit vector perpendicular to the paper in the direction directly upwards towards the reader in Figure 5.5. Since $\delta r/r \ll 1$, using the binomial theorem we have

$$(\delta \mathbf{B})_{BC} \approx \left(\frac{\mu_0}{4\pi}\right)\left[\frac{Ib}{r^2}\left(1 - \frac{\delta r}{r}\right) + \frac{\dot{I}b}{cr}\right]\hat{\mathbf{k}}. \qquad (6.46)$$

When the information collecting sphere passes AD at a time $\delta r/c$ after it has passed BC, the value of the current in AD is $(I + \dot{I}\,(\delta r/c))$ and the value of the rate of change of current in AD is $(\dot{I} + \ddot{I}\,(\delta r/c))$. Using equation (6.45) we find that the contribution of the current in the section AD to the magnetic field at the field point P in Figure 5.5 is

$$(\delta \mathbf{B})_{AB} = -\left(\frac{\mu_0}{4\pi}\right)\left[\frac{(I + \dot{I}\delta r/c)b}{r^2} + \frac{(\dot{I} + \ddot{I}\delta r/c)b}{cr}\right]\hat{\mathbf{k}}. \qquad (6.47)$$

Adding equations (6.46) and (6.47), we find that the resultant magnetic field at the field point P in Figure 5.5 is

$$\mathbf{B} = -\left(\frac{\mu_0}{4\pi}\right)\left[\frac{Ib\delta r}{r^3} + \frac{\dot{I}b\delta r}{cr^2} + \frac{\ddot{I}b\delta r}{c^2 r}\right]\hat{\mathbf{k}}. \qquad (6.48)$$

The magnetic moment m of a plane coil is defined as the product of the current in the coil and the area of the coil, so that for the coil $ABCD$

$$m = IA = Ib\,\delta r.$$

Equation (6.48) can now be written in the form

$$\mathbf{B} = -\left(\frac{\mu_0}{4\pi}\right)\left\{\frac{[m]}{r^3} + \frac{[\dot{m}]}{cr^2} + \frac{[\ddot{m}]}{c^2 r}\right\}\hat{\mathbf{k}}. \qquad (6.49)$$

Equation (6.49) is the standard expression for the magnetic field due to an oscillating magnetic dipole for the conditions shown in Figure 5.5, where the field point P is in the plane of the coil $ABCD$. If the current in the coil $ABCD$ is steady, then $\dot{m}$ and $\ddot{m}$ are both zero in equation (6.49) and the expression for $\mathbf{B}$ is given by the Biot-Savart law. If the current in the coil $ABCD$ is varying at a constant rate, then $\dot{m}$ is finite, but $[\ddot{m}]$ is zero. The $[\dot{m}]$ term in equation (6.49) arises from the difference in the value of the current $[I]$ when the information collecting sphere passes AD compared with BC and arises from the velocity dependent contribution $\mathbf{B}_V$ to the magnetic field due to an accelerating classical point charge. The $[\ddot{m}]$ term in equation (6.49) arises from the difference in the value of $\dot{I}$, the rate of change of current, when the information collecting sphere passes AD compared with BC and arises from the acceleration dependent term $\mathbf{B}_A$ in the expression for the magnetic field due to an accelerating charge.

6.7. Derivation of the retarded potentials from the Liénard-Wiechert potentials

Consider the varying macroscopic charge and current distribution, consisting of moving and accelerating classical point charges, shown previously in Figure 5.10 of Chapter 5. The resultant scalar potential ϕ and the resultant vector potential $\mathbf{A}$ will be determined at the field point P at the position $\mathbf{r}$ in Figure 5.10 at the time of observation t. It will be assumed in this section that there may be more than one type of moving classical point charge. We shall assume that there are n_i classical point charges of type i per cubic metre, each of charge q_i and each moving with velocity $[\mathbf{u}_i]$ inside the volume element dV_s at the position $\mathbf{r}_s$ in Figure 5.10 at the retarded time $t^* = (t - R/c)$, where $\mathbf{R}$ is equal to $(\mathbf{r} - \mathbf{r}_s)$. It follows from equation (6.1) that the number of charges of type i having velocity $[\mathbf{u}_i]$ counted inside dV_s by the information collecting sphere, that reaches the field point P at the time of observation t, is

$$\delta N_i = n_i \left(\frac{s_i}{R}\right) dV_s \tag{6.50}$$

where

$$s_i = \left[R - \frac{\mathbf{R} \cdot \mathbf{u}_i}{Rc}\right]. \tag{6.51}$$

According to the Liénard-Wiechert potentials, equation (3.4), the contribution of each one of the charges of type i having velocity $[\mathbf{u}_i]$, that are counted inside dV_s, to the scalar potential at the field point P at the time of observation t is

$$\phi_i = \frac{q_i}{4\pi\varepsilon_0 s_i}. \tag{6.52}$$

Multiplying equations (6.50) and (6.52) we find that the resultant contribution to the scalar potential at the field point P in Figure 6.5 of those charges of type i having charge q_i and having velocity $[\mathbf{u}_i]$, that are counted by the information collecting sphere inside dV_s in Figure 5.10, at the retarded time $t^* = (t - R/c)$ is

$$\delta\phi_i = n_i \left(\frac{s_i}{R}\right) dV_s \left(\frac{q_i}{4\pi\varepsilon_0 s_i}\right) = \frac{q_i n_i \, dV_s}{4\pi\varepsilon_0 R}. \tag{6.53}$$

Summing over all directions of $\mathbf{u}_i$ and over all values of $\mathbf{u}_i$ for all the types of classical point charges making up the charge and current distribution, we have

$$d\phi = \frac{(\Sigma q_i n_i)\, dV_s}{4\pi\varepsilon_0 R} = \frac{[\rho]\, dV_s}{4\pi\varepsilon_0 R} \tag{6.54}$$

where

$$[\rho] = \Sigma q_i n_i \tag{6.55}$$

is the resultant macroscopic charge density at $\mathbf{r}_s$ at the retarded time $t^* = (t - R/c)$. Adding the contributions due to all the volume elements in the macroscopic charge distribution shown in Figure 5.10, we obtain

$$\phi(\mathbf{r}, t) = \frac{1}{4\pi\varepsilon_0} \int \frac{[\rho(\mathbf{r}_s)]\, \mathrm{d}V_s}{R}. \tag{6.56}$$

Similarly using equations (6.1) and (3.6), we have for the volume element $\mathrm{d}V_s$ in Figure 5.10

$$\delta\mathbf{A}_i = n_i \left(\frac{s_i}{R}\right) \mathrm{d}V_s \left(\frac{q_i[\mathbf{u}_i]}{4\pi\varepsilon_0 c^2 s_i}\right) = \frac{q_i n_i [\mathbf{u}_i]\, \mathrm{d}V_s}{4\pi\varepsilon_0 c^2 R}.$$

Summing over all directions of $[\mathbf{u}_i]$ for all values of $[\mathbf{u}_i]$ and for all species of moving classical point charge and then using the equation

$$[\mathbf{J}] = \Sigma q_i n_i [\mathbf{u}_i] \tag{6.57}$$

where $[\mathbf{J}]$ is the macroscopic current density at $\mathbf{r}_s$ at the retarded time t^*, we have

$$\mathrm{d}\mathbf{A}(\mathbf{r}, t) = \frac{[\mathbf{J}(\mathbf{r}_s)]}{4\pi\varepsilon_0 c^2 R}\, \mathrm{d}V_s. \tag{6.58}$$

Adding the contributions due to all the volume elements in the macroscopic current distribution shown in Figure 5.10, we obtain for the field point P in Figure 5.10 at the time of observation t

$$\mathbf{A}(\mathbf{r}, t) = \frac{1}{4\pi\varepsilon_0 c^2} \int \frac{[\mathbf{J}(\mathbf{r}_s)]}{R}\, \mathrm{d}V_s. \tag{6.59}$$

Equations (6.56) and (6.59) are the same as equations (2.29) and (2.30) respectively, which are the expressions for the retarded potentials for macroscopic charge and current distributions. These results show that if we start with the Liénard-Wiechert potentials for the potentials due to an accelerating classical point charge then, at external field points the resultant potentials ϕ and $\mathbf{A}$, due to a system of moving and accelerating classical point charges that build up a macroscopic charge and current distribution, are given by the retarded potentials. The most straightforward way of applying equations (6.56) and (6.59) is to use the methods we applied to equations (2.29) and (2.30) in Sections 2.4.2 and 2.4.3 of Chapter 2.

6.8. Discussion of higher order effects

We shall go on now to consider the differences between the microscopic approach of Section 6.4 and the macroscopic approach we used in Section

2.4, when we calculated the magnetic field due to an oscillating electric dipole. In Section 2.4 we used the retarded vector potential, given by equation (2.30) to determine the vector potential due to the current flowing in the wire connecting the charge distributions of the oscillating electric dipole in Figure 2.2. This led to equation (2.46). On the other hand, we derived equation (6.40) from the expression for the magnetic field due to an accelerating classical point charge. Comparing equations (2.46) and (6.40) we can see that they are not exactly the same, but there is an extra term proportional to $\beta \cos \theta$ in equation (6.40) as well as all the higher order terms in β, which we neglected when we derived equation (6.17).

There is an important difference between the methods used when we derived equation (6.40) and the method based on classical electromagnetism which we used in Section 2.4. When we derived the expression for the magnetic field due to an accelerating charge in Appendix C we allowed for the effects of changes in the position of the charge when we determined the spatial variations of the vector potential at the field point at a fixed time to determine $\nabla \times \mathbf{A}$. We then summed the contributions of individual charges using equation (6.1). On the other hand, in Section 2.4 we assumed that the connecting wire of length dl was stationary and we only considered the values of the current and rate of change of current at a fixed point.

When $u/c \ll 1$, which is generally true for conduction currents, the differences between the various predictions for **B** are small and can be neglected. However, if the moving atomic charged particles that make up a charge and current distribution are moving at speeds comparable to the speed of light there could be significant differences between the predictions of equation (2.46) and our method, which is expressed in the form of equation (6.4), which is

$$d\mathbf{B} = qN_0 \, dl \left(\frac{s}{R}\right) (\mathbf{B}_V + \mathbf{B}_A) \tag{6.60}$$

where $\mathbf{B}_V$ and $\mathbf{B}_A$ are given by equations (3.14) and (3.16) respectively. Equation (6.60) should be valid in the relativistic limit. In practice many authors use our microscopic approach when they apply equations (3.10) and (3.13) to the motions of **individual** charged particles moving at relativistic speeds, but without using equation (6.1) to count the particles. This latter method has been applied for example to derive some of the properties of the synchrotron radiation emitted, when charged particles move at relativistic speeds in helical paths in magnetic fields. (References: Panofsky and Phillips [4] and Jackson [5]). The theory has been applied, for example, to the synchrotron radiation emitted by electrons of energies up to 10^{12} eV in the Crab nebula. Another example is the radiation emitted by relativistic electrons trapped in the Van Allen belts of Jupiter. Synchrotron radiation is also emitted by electrons accelerated in electron synchrotrons. Pulsars (neutron stars) are another example of where relativistic effects are important and our equation (6.60) might be the more appropriate method to use.

References

1. Rosser, W. G. V., *Classical Electromagnetism via Relativity* (Butterworths, London, 1968), p. 265.
2. Vybiral, B., *Jour. Phys. D: Appl. Phys.*, Vol. 22, p. 1 (1989).
3. Jefimenko, O. D., *Electricity and Magnetism*, 2nd Edition (Electric Scientific, Star City, W. V., 1989), p. 516, or see the 1st Edition (Appleton Century Crofts, 1966), p. 516.
4. Panofsky, W. K. H. and Phillips, M., *Classical Electricity and Magnetism*, 2nd Edition (Addison-Wesley, Reading, Mass., 1962), p. 363.
5. Jackson, J. D., *Classical Electrodynamics*, 2nd Edition (John Wiley, New York, 1975). p. 672.

CHAPTER 7

Quasi-stationary phenomena and AC theory

7.1. The quasi-stationary approximation

Most household and laboratory equipment works off the electricity mains, in which case the applied emf varies sinusoidally with the mains frequency, which is generally 50 or 60 Hz. It only takes the information collecting sphere 10^{-8} s to cross a piece of laboratory equipment of dimensions 3 m. At mains frequency the change in the emf in 10^{-8} s is less than 3×10^{-6} times the maximum variation in the emf. Since the surface and boundary charge distributions associated with conduction current flow can build up extremely quickly, it is reasonable to assume that at mains frequency the conduction current has the same value in all parts of a series circuit at a given instant. Since the surface and boundary charges are exceedingly small, the current flow associated with changes in their values can be ignored compared with the conduction current flowing in the circuit. Hence Kirchhoff's first law according to which the algebraic sum of the currents into the node of a network is zero is also valid in the quasi-stationary limit, for example at mains frequency.

If the mains frequency is 50 Hz, the wavelength of the electromagnetic variations is 6000 km so that $\lambda/2\pi$ is equal to about 1000 km. Hence for laboratory experiments carried out at mains frequency, we are always in the near zone, so that, when we are working at mains frequency we can generally ignore the radiation fields that arise from the $\mathbf{E}_A$ and $\mathbf{B}_A$ terms in equations (3.10) and (3.13), which give the total electric and magnetic fields due to a moving and accelerating classical point charge. Hence in the quasi-stationary limit, for example, at mains frequency, we need only use equations (3.11) and (3.14) which give the velocity dependent fields $\mathbf{E}_V$ and $\mathbf{B}_V$ and formulae we have derived using them, such as equations (5.18), (5.53) and (6.9).

7.2. The interpretation of transformer induced emf

7.2.1. *A simple air cored transformer*

As a simple introduction to the origin of a transformer induced emf, we shall consider the air cored transformer shown in Figure 7.1. The primary of the air cored transformer is an infinitely long solenoid of the type shown previously in Figure 5.7 of Chapter 5 and shown again in plan view in Figure 7.1, looking downwards from $z = +\infty$. We shall choose the cylindrical coordinate system (r, ϕ, z), whose z axis coincides with the axis of the solenoid and whose origin is in the plane of the paper in Figure 7.1. The solenoid (primary coil) consists of n turns per metre length each of radius b. The secondary of the transformer is a single turn going once around the solenoid, as shown in Figure 7.1. We shall assume that the current I_1 in the primary coil is in the direction $-\hat{\boldsymbol{\phi}}$, that is in the clockwise direction in the plan view in Figure 7.1. According to equation (5.68) of Chapter 5, which was derived from the expression for the electric field due to a moving classical point charge, the moving conduction electrons and stationary positive ions inside the windings of the solenoid, the primary coil, give rise to a resultant induction electric field $\mathbf{E}_i$ outside the solenoid, which at large distances $r \gg b$ from the axis of the solenoid is given in the quasi-stationary limit by

$$\mathbf{E}_i = E_\phi \hat{\boldsymbol{\phi}} = \frac{\mu_0 n b^2}{2r}\left(\frac{\mathrm{d}I_1}{\mathrm{d}t}\right)\hat{\boldsymbol{\phi}}. \tag{7.1}$$

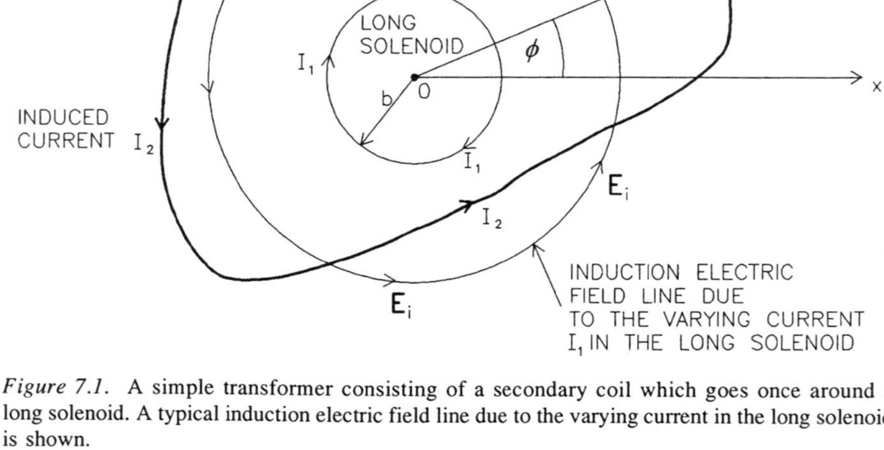

Figure 7.1. A simple transformer consisting of a secondary coil which goes once around a long solenoid. A typical induction electric field line due to the varying current in the long solenoid is shown.

It is left as a problem, given at the end of this section, for the reader to show that equation (7.1) is valid for all values of $r > b$. The induction electric field lines outside the solenoid are closed circles in the (r, ϕ) plane as shown in Figure 7.1. To simplify our initial discussion, we shall assume in this section that the current I_1 in the primary coil varies at a constant rate so that according to equation (7.1) the induction electric field $\mathbf{E}_i$ and hence the induced emf and current flowing in the secondary coil are constant so that, in this special case, there is no back emf due to the self inductance of the secondary coil.

When the varying current I_1 in the primary coil is first switched on, the only electric field inside the stationary wire making up the secondary coil in Figure 7.1 is the induction electric field $\mathbf{E}_i$ given by equation (7.1), which is not, in general, parallel to the wire making up the secondary coil. The initial direction of conduction current flow in the wires making up the stationary secondary coil is in the direction of the induction electric field $\mathbf{E}_i$. This leads to the build up of charge distributions on the surfaces of the wires making up the secondary coil and at boundaries between wires of different electrical conductivities, as explained in Appendix B. The electric field due to these surface and boundary charge distributions will be denoted by $\mathbf{E}_s$. The necessary surface and boundary charge distributions build up very quickly and are of such a magnitude that, once the initial transient state is over, the resultant electric field $\mathbf{E} = (\mathbf{E}_i + \mathbf{E}_s)$ inside the wires making up the secondary coil in Figure 7.1 is always parallel to the wire and is of such a magnitude that, when the conduction current I_1 in the primary coil is varying at a constant rate, the conduction current I_2 in the secondary coil is steady and has the same value in all parts of the secondary circuit, whatever the resistances of the various sections of the secondary coil.

The total emf ε_{sec} in the secondary circuit will be defined by the line integral

$$\varepsilon_{\text{sec}} = \oint \left(\frac{\mathbf{F}}{q}\right) \cdot \mathrm{d}\mathbf{l}_2 \tag{7.2}$$

evaluated around the secondary circuit at a fixed instant of time, where in equation (7.2) $\mathbf{F}$ is the force on a test charge of magnitude q, that is at rest relative to the element of length $\mathrm{d}\mathbf{l}_2$ of the secondary circuit. In the case of a stationary circuit, $\mathbf{F}/q$ is equal to the resultant total electric field $\mathbf{E} = (\mathbf{E}_i + \mathbf{E}_s)$ inside the wire making up the secondary circuit in Figure 7.1. Putting $\mathbf{F}/q$ equal to $\mathbf{E}$ in equation (7.2) and integrating around the secondary circuit, we have

$$\varepsilon_{\text{sec}} = \oint \mathbf{E} \cdot \mathrm{d}\mathbf{l}_2 = \oint (\mathbf{E}_i + \mathbf{E}_s) \cdot \mathrm{d}\mathbf{l}_2 = \oint \mathbf{E}_i \cdot \mathrm{d}\mathbf{l}_2 + \oint \mathbf{E}_s \cdot \mathrm{d}\mathbf{l}_2. \tag{7.3}$$

When the current I_2 in the secondary coil has reached its final constant value, the surface and boundary charge distributions giving rise to $\mathbf{E}_s$ are steady and can be treated as electrostatic. In these conditions the $\mathbf{E}_s$ field is conservative so that $\oint \mathbf{E}_s \cdot \mathrm{d}\mathbf{l}_2$ is zero and equation (7.3) reduces to

$$\varepsilon_{sec} = \oint \mathbf{E}_i \cdot d\mathbf{l}_2. \tag{7.4}$$

If $d\mathbf{l}_2$ is in the direction of the current flow in the secondary coil in Figure 7.1, then using equation (7.1) with $\hat{\mathbf{r}}$, $\hat{\boldsymbol{\phi}}$ and $\hat{\mathbf{k}}$ as unit vectors in the directions of increasing r, ϕ and z respectively, we have

$$\varepsilon_{sec} = \oint \mathbf{E}_i \cdot d\mathbf{l}_2 = \oint (E_\phi \hat{\boldsymbol{\phi}}) \cdot (dr\hat{\mathbf{r}} + r\, d\phi \hat{\boldsymbol{\phi}} + dz\hat{\mathbf{k}}).$$

Since $\hat{\boldsymbol{\phi}} \cdot \hat{\mathbf{r}}$ and $\hat{\boldsymbol{\phi}} \cdot \hat{\mathbf{k}}$ are zero, using equation (7.1) we have

$$\varepsilon_{sec} = \oint E_\phi r\, d\phi = \frac{\mu_0 n b^2}{2}\left(\frac{dI_1}{dt}\right)\oint d\phi. \tag{7.5}$$

For a single turn going around the primary $\oint d\phi = 2\pi$, so that equation (7.5) gives

$$\varepsilon_{sec} = \oint \mathbf{E} \cdot d\mathbf{l}_2 = \oint \mathbf{E}_i \cdot d\mathbf{l}_2 = \mu_0 n \pi b^2 \left(\frac{dI_1}{dt}\right). \tag{7.6}$$

The induction electric field $\mathbf{E}_i$ due to the varying current in the primary coil, which gives the transformer induced emf in the secondary coil is not localized but is distributed all along the secondary coil in Figure 7.1.

In a typical transformer, the secondary consists of more than one turn. If the secondary coil in Figure 7.1 goes around the primary (the long solenoid) N times, $\oint d\phi$ is equal to $2\pi N$ and equation (7.5) gives

$$\varepsilon_{sec} = \oint \mathbf{E}_i \cdot d\mathbf{l}_2 = \mu_0 n N \pi b^2 \left(\frac{dI_1}{dt}\right). \tag{7.7}$$

The emf in the secondary coil of the air cored transformer in Figure 7.1 can be varied either by varying n the number of turns per metre length in the primary coil (the long solenoid) or by varying N the total number of turns in the secondary coil. The emf can also be increased substantially by placing a ferromagnetic core inside the long solenoid.

Problem. It can be shown [1] using Maxwell's equations that the magnetic field due to the solenoid in Figure 7.1 is given by

$$\mathbf{B} = 0 \qquad (r > b), \tag{7.8}$$

$$\mathbf{B} = \mu_0 n I \hat{\mathbf{z}} \qquad (r < b). \tag{7.9}$$

Use the equation

$$\int \mathbf{B} \cdot d\mathbf{S} = \int \boldsymbol{\nabla} \times \mathbf{A} \cdot d\mathbf{S} = \oint \mathbf{A} \cdot d\mathbf{l}$$

to show that

$$\mathbf{A} = -\left(\frac{\mu_0 n I b^2}{2r}\right)\hat{\boldsymbol{\phi}}. \qquad (r > b), \tag{7.10}$$

$$\mathbf{A} = -\left(\frac{\mu_0 n I r}{2}\right)\hat{\boldsymbol{\phi}}. \qquad (r < b), \qquad (7.11)$$

Use equation (2.17) to show that, for quasi-stationary conditions the induction electric field is given by

$$\mathbf{E}_i = \left(\frac{\mu_0 n \dot{I} b^2}{2r}\right)\hat{\boldsymbol{\phi}}. \qquad (r > b), \qquad (7.12)$$

$$\mathbf{E}_i = \left(\frac{\mu_0 n \dot{I} r}{2}\right)\hat{\boldsymbol{\phi}}. \qquad (r < b), \qquad (7.13)$$

Check equations (7.12) and (7.13) using equation (1.105).

7.2.2. *Calculation of the current flow in the secondary coil when $\dot{I}_1$ is constant*

The total electric field $\mathbf{E} = (\mathbf{E}_i + \mathbf{E}_s)$ inside the wire, making up the secondary coil in Figure 7.1, gives rise to a conduction current flow in the way described in Section 1.3 of Chapter 1. According to equation (1.42), the current density $\mathbf{J}_2$ at any point inside the wire making up the secondary coil in Figure 7.1 is related to the total electric field $\mathbf{E} = (\mathbf{E}_i + \mathbf{E}_s)$ inside the wire making up the secondary coil by the constitutive equation

$$\mathbf{J}_2 = \sigma\mathbf{E} = \sigma(\mathbf{E}_i + \mathbf{E}_s) \qquad (7.14)$$

where σ is the local value of the electrical conductivity, which depends on the properties of the conducting wire. When the current I_1 in the primary coil is varying at a constant rate, the emf in the secondary coil, given by equation (7.6), is a constant so that the current I_2 in the secondary coil is a constant. The current density $\mathbf{J}_2$ is then uniform across the wire making up the secondary coil, so that $J_2 = I_2/A_2$, where A_2 is the area of cross section of the wire, and equation (7.14) can be rewritten in the form

$$\mathbf{E} = \mathbf{E}_i + \mathbf{E}_s = \frac{\mathbf{I}_2}{\sigma A}. \qquad (7.15)$$

We now form the scalar product of the total electric field $\mathbf{E}$ and an element of length $\mathbf{dl}_2$ of the wire making up the secondary coil, where the vector $\mathbf{dl}_2$ points in the direction of current flow. Since in the steady state the resultant electric field $\mathbf{E}$ inside the wire is parallel to $\mathbf{dl}_2$, using equation (7.15) we have

$$\mathbf{E}\cdot\mathbf{dl}_2 = E\,dl_2 = I_2\frac{dl_2}{\sigma A} = I_2\,dR$$

where, according to equation (1.43) of Chapter 1, $dl_2/\sigma A_2$ is equal to the resistance dR of the length dl_2 of the secondary coil. Summing $\mathbf{E}\cdot\mathbf{dl}_2$ along the whole length of the secondary coil we have

$$\oint \mathbf{E}\cdot\mathbf{dl}_2 = I_2\Sigma\,dR. \qquad (7.16)$$

Comparing equations (7.4) and (7.16), we conclude that

$$\varepsilon_{sec} = \oint \mathbf{E} \cdot d\mathbf{l}_2 = \oint \mathbf{E}_i \cdot d\mathbf{l}_2 = I_2 \Sigma \, dR. \tag{7.17}$$

According to equation (7.17), the sum of the products I_2 dR taken around the secondary circuit is equal to the emf in the secondary circuit. This is an example of Kirchhoff's second law, which we have derived using the constitutive equation $\mathbf{J}_2 = \sigma\mathbf{E}$. If there is a high resistance R_2 in the secondary circuit, provided the resistances of the wires making up the rest of the secondary coil in Figure 7.1 are negligible compared to R_2, then $I_2\Sigma$ dR can be approximated by I_2R_2 in equation (7.17), which becomes

$$\varepsilon_{sec} = \oint \mathbf{E} \cdot d\mathbf{l}_2 = \oint \mathbf{E}_i \cdot d\mathbf{l}_2 = I_2 R_2. \tag{7.18}$$

Substituting for ε_{sec} using equation (7.6) we find that the current I_2 in the secondary coil in Figure 7.1 is

$$I_2 = \frac{\mu_0 n \pi b^2}{R_2}\left(\frac{dI_1}{dt}\right). \tag{7.19}$$

We have shown by determining the induction electric field directly from the expression for the electric field due to a moving classical point charge using equations (3.10), (5.18), (5.53) and (5.68) that there is no need to mention the magnetic field, when determining and interpreting transformer induced emfs. Still less is it necessary to claim, as some books do, that it is the varying magnetic flux that generates the induction electric field. Our interpretation of Faraday's law of electromagnetic induction is that the same moving and accelerating charges, that give the induction electric field also give a magnetic field whose rate of changed $\dot{\mathbf{B}}$ is equal to $-\nabla \times \mathbf{E}$.

According to equation (7.18) we do not need to know the value of the electric field $\mathbf{E}$ at every point in the secondary circuit in order to determine the current I_2 in the secondary circuit. All that we need is the emf, which is the line integral $\oint \mathbf{E} \cdot d\mathbf{l}_2$ of the electric field around the secondary circuit. It is for this reason that Faraday's law of electromagnetic induction, namely

$$\varepsilon_{sec} = \oint \mathbf{E} \cdot d\mathbf{l} = -\frac{\partial \Phi_{21}}{\partial t} \tag{7.20}$$

works well in practice, if the magnetic field is known. The quantity Φ_{21} in equation (7.20) is the magnetic flux linking coil number 2 due to the current I_1 in coil number 1. As an example, consider the secondary coil that goes once around the long solenoid in Figure 7.1. Since the magnetic field is uniform and equal to $\mu_0 n I_1$ inside the solenoid and is zero outside

$$\Phi_{21} = (\pi b^2)\mu_0 n I. \tag{7.21}$$

Using equations (7.20) and (7.18) we find that

$$\mathcal{E}_{\text{sec}} = -\pi b^2 \mu_0 n \left(\frac{\mathrm{d}I_1}{\mathrm{d}t}\right) = I_2 R_2. \tag{7.22}$$

This is in agreement with equation (7.19). Since it is often easier to calculate magnetic fields using the Biot-Savart law than to calculate the induction electric fields, Faraday's law, equation (7.20), is generally used in practice. It is also often easier to visualize and design an experiment to produce a known varying magnetic field than to design an experiment that produces a known induction electric field. Faraday's law can then be use to determine the transformer induced emf.

7.2.3. *Mutual inductance*

Consider the two stationary coils of arbitrary shape, which are labelled 1 and 2 in Figure 7.2. According to equation (5.53) of Chapter 5, the induction electric field due to the varying current I_1 in coil number 1 in Figure 7.2 at the position $\mathbf{r}_2$ of the element of length $\mathbf{dl}_2$ of coil number 2 is given

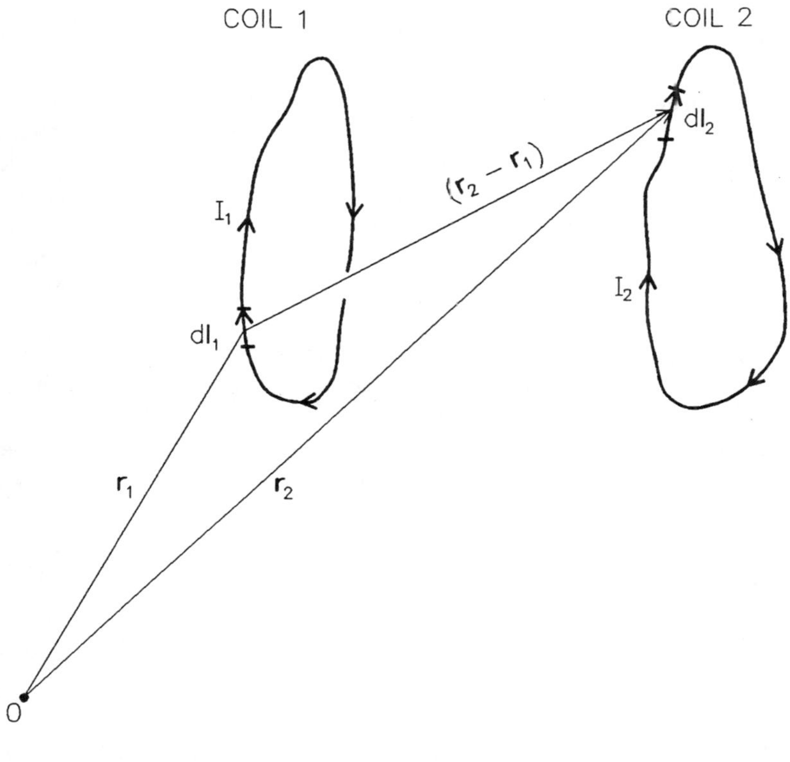

Figure 7.2. The mutual inductance of two spatially separated, rigid, stationary coils.

in the quasi-stationary limit by

$$\mathbf{E}_i(\mathbf{r}_2) = -\frac{\mu_0}{4\pi}\left(\frac{dI_1}{dt}\right)\oint_1 \frac{d\mathbf{l}_1(\mathbf{r}_1)}{|\mathbf{r}_2 - \mathbf{r}_1|}\,. \tag{7.23}$$

In equation (7.23), the element of length $d\mathbf{l}_1$ of coil number 1 is at the position $\mathbf{r}_1$ as shown in Figure 7.2. Equation (7.23) was derived in Section 5.5.2 of Chapter 5 from the expression for the electric field due to a moving classical point charge, in the limit when the radiation electric field could be neglected. Using equations (7.4) and (7.23), we find that the total emf induced in coil number 2 by the varying current I_1 in coil number 1 is

$$\varepsilon_2 = \oint_2 \mathbf{E}_i \cdot d\mathbf{l}_2 = -\frac{\mu_0}{4\pi}\left(\frac{dI_1}{dt}\right)\oint_2\oint_1 \frac{d\mathbf{l}_1 \cdot d\mathbf{l}_2}{r_{12}} \tag{7.24}$$

where $r_{12} = |\mathbf{r}_1 - \mathbf{r}_2|$. Equation (7.24) is generally rewritten in the abbreviated form

$$\varepsilon_2 = \oint \mathbf{E}_i \cdot d\mathbf{l}_2 = -M_{21}\left(\frac{dI_1}{dt}\right) \tag{7.25}$$

where

$$M_{21} = \frac{\mu_0}{4\pi}\oint_2\oint_1 \frac{d\mathbf{l}_1 \cdot d\mathbf{l}_2}{r_{12}} \tag{7.26}$$

is called the **mutual inductance** of the two coils. Equation (7.26) is Neumann's expression for the mutual inductance of two air cored coils. The value of M_{21} depends on the shapes and separation of the two coils in Figure 7.2. According to equation (7.25) the mutual inductance M_{21}, expressed in henries, is numerically equal to the emf induced in coil number 2 when the current in coil number 1 is varying at the rate of one ampere per second. It follows from equation (7.26) that $M_{12} = M_{21}$ for the two air cored coils in Figure 7.2.

The reader should remember that on our approach, when an induced emf is expressed as $-M_{21}\dot{I}_1$, it stands for the line integral $\oint \mathbf{E}_i \cdot d\mathbf{l}_2$ of the induction electric field $\mathbf{E}_i$ around the secondary circuit.

In introductory text books, mutual inductance is generally introduced and defined using Faraday's law of electromagnetic induction, which can be rewritten as follows

$$\varepsilon_2 = -\frac{d\Phi_{21}}{dt} = -\left(\frac{d\Phi_{21}}{dI_1}\right)\left(\frac{dI_1}{dt}\right) = -M_{21}\frac{dI_1}{dt} \tag{7.27}$$

where

$$M_{21} = \frac{d\Phi_{21}}{dI_1} \tag{7.28}$$

is the mutual inductance. For two air cored coils, the magnetic flux Φ_{21} through coil number 2 in Figure 7.2 due to the current I_1 in coil number 1 is proportional to I_1, so that $\Phi_{21} = kI_1$, where k is a constant. Hence

$$M_{21} = \frac{\mathrm{d}}{\mathrm{d}I_1}(kI_1) = k = \frac{\Phi_{21}}{I_1}. \tag{7.29}$$

According to equation (7.29), the mutual inductance of the two air cored coils in Figure 7.2 is equal to the magnetic flux linking coil number 2, when a current of one ampere flows in coil number 1. As an example, consider the case shown in Figure 7.1 where the secondary consists of only one turn. Using equations (7.21) and (7.29) we find that

$$M_{21} = \frac{\Phi_{21}}{I_1} = \mu_0 n\pi b^2. \tag{7.30}$$

The reader can check this result by comparing equations (7.25) and (7.6).

When there are ferromagnetic materials present, the magnetic flux Φ_{21} is not proportional to I_1 and equations (7.29) is not valid. In this case, the mutual inductance is generally defined using equation (7.28).

7.2.4. *Example of Lenz's law*

According to Lenz's law, the direction of the induced emf in the secondary coil of a transformer, due to the varying current in the primary coil, is in such a direction that it tries to give a current flow in the secondary coil which would give a magnetic field in such a direction that it would tend to oppose the change in the magnetic flux through the secondary coil due to the varying current in the primary.

Consider again the example shown in Figure 7.1, when the current I_1 in the primary (the long solenoid) is in the clockwise direction in Figure 7.1 and is increasing in magnitude. Since its magnetic field is downwards away from the reader inside the solenoid and is zero outside the solenoid, the total magnetic flux through the secondary coil due to the varying current in the primary coil is downwards away from the reader in Figure 7.1. According to equation (7.1) the induced field $\mathbf{E}_i$ is in the anticlockwise direction in Figure 7.1 and gives a current flow in the anticlockwise direction in the secondary coil. Such a current flow in the secondary coil gives a magnetic field which is upwards towards the reader in Figure 7.1 and thus opposes the increase in the magnetic flux through the secondary coil in the downward direction due to the increasing current in the primary coil. This direction of current flow in the secondary coil is in agreement with Lenz's law.

7.3. What a voltmeter measures in the presence of an induction electric field in the quasi-stationary limit

7.3.1. *What a voltmeter measures*

A typical laboratory voltmeter consists of a sensitive moving coil galvanometer in series with a high resistance. The moving coil galvanometer responds to the current I_V flowing through the galvanometer. The scale of the moving coil galvanometer is calibrated such that the scale gives the product $I_V R_V$, where R_V is the total resistance of the voltmeter. Another example of a typical voltmeter is a cathode ray oscilloscope. In this case the scale is calibrated to read $I_V R_V$, where R_V is the input resistance of the oscilloscope.

Assume that the voltmeter V in Figure 7.3 is connected by leads of negligible resistance to the terminals A and B of a "black box", that gives rise to a time independent external induction electric field $\mathbf{E}_i$. The total electric field inside the connecting wires and voltmeter outside the "black box" in Figure 7.3 is $\mathbf{E} = (\mathbf{E}_i + \mathbf{E}_s)$, where $\mathbf{E}_s$ is the electrostatic field due to any resultant charge distributions inside the black box and due to the surface and boundary charge distributions on the leads and voltmeter. From the constitutive equation $\mathbf{J} = \sigma\mathbf{E}$ we have

$$E = \frac{J}{\sigma} = \frac{I_V}{\sigma A_0} \tag{7.31}$$

where A_0 is the area of cross section of the conductor and I_V is the current flowing through the voltmeter. We shall now form the scalar product of $\mathbf{E}$ and an element of length $\mathbf{dl}$ of the path from terminal A along the red lead, then through the voltmeter and finally along the black lead to terminal B in Figure 7.3. We have for the element $\mathbf{dl}$

$$\mathbf{E} \cdot \mathbf{dl} = E\,\mathrm{d}l = \left(\frac{I_V}{\sigma A_0}\right)\mathrm{d}l = I_V\,\mathrm{d}R$$

where $\mathrm{d}R = \mathrm{d}l/\sigma A_0$ is the resistance of the section of length $\mathrm{d}l$. Integrate along the path from terminal A along the leads and through the voltmeter to

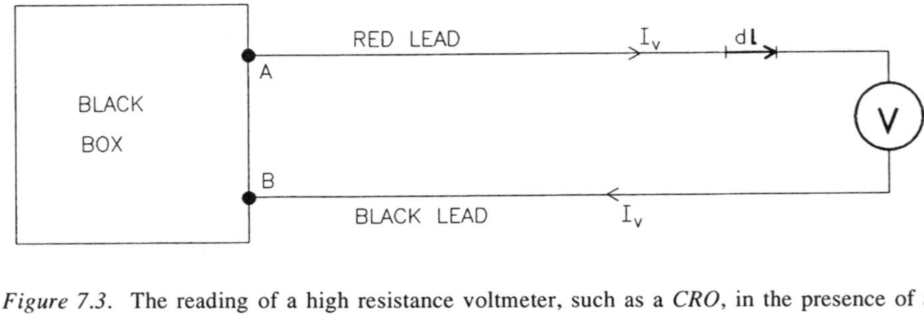

Figure 7.3. The reading of a high resistance voltmeter, such as a *CRO*, in the presence of a steady induction electric field.

terminal B in Figure 7.3. Provided the resistances of the leads are negligible compared with R_V, the resistance of the voltmeter, we have

$$\int_A^B \mathbf{E} \cdot \mathrm{d}\mathbf{l} = I_V \int_A^B \mathrm{d}R = I_V R_V. \tag{7.32}$$

According to equation (7.32), provided the voltmeter is able to respond quickly enough, which an oscilloscope can do in the quasi-stationary limit, the reading of a voltmeter, that is calibrated to measure $I_V R_V$, measures the line integral $\int \mathbf{E} \cdot \mathrm{d}\mathbf{l}$ of the **total** electric field $\mathbf{E} = (\mathbf{E}_i + \mathbf{E}_s)$ from terminal A through the red lead to the voltmeter, then through the voltmeter and finally through the black lead to terminal B in Figure 7.3. Notice that the voltmeter does not measure $\int \mathbf{E}_s \cdot \mathrm{d}\mathbf{l}$ between A and B, so that in the presence of an induction electric field, the voltmeter does not measure an electrostatic potential difference. In the absence of an accepted term to describe $\int \mathbf{E} \cdot \mathrm{d}\mathbf{l}$, the line integral of the total electric field, we shall call it the voltage measured by the voltmeter.

7.3.2. *Example of the effect of the induction electric field on the reading of a voltmeter*

Consider now the example shown in Figures 7.4(a) and 7.4(b). The primary coil is an infinitely long solenoid consisting of n circular turns per metre length, each of radius b. It will be assumed that the current I_1 in the primary coil is in the clockwise direction in Figures 7.4(a) and 7.4(b) and that I_1 is increasing

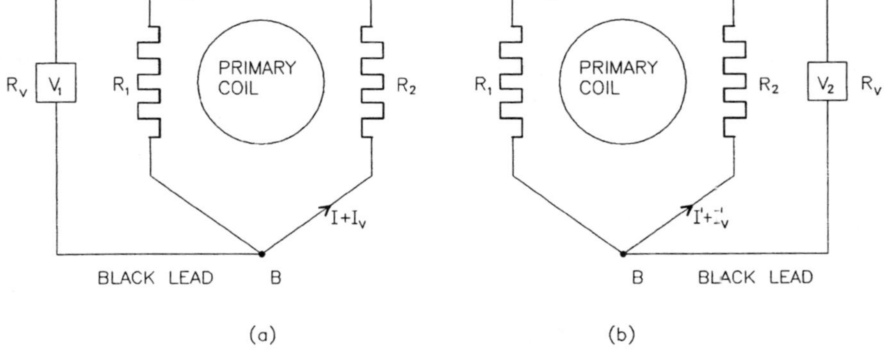

Figure 7.4. Examples to illustrate the influence of an induction electric field on the reading of a voltmeter.

in magnitude at a constant rate $\dot{I}_1$ so that, according to equation (7.1), the induction electric field which is in the anticlockwise direction outside the solenoid does not vary with time. According to equations (7.6) and (7.22) the emf in any coil of any shape going once around the outside of the long solenoid is

$$\varepsilon = \oint \mathbf{E}_i \cdot \mathbf{dl} = \mu_0 n \pi b^2 \dot{I}_1 \tag{7.33}$$

which is constant if $\dot{I}_1$ is constant.

The secondary circuits in Figures 7.4(a) and 7.4(b) both consist of resistors R_1 and R_2 which are connected by leads of negligible resistances to the points A and B. The question we shall discuss is, what will a voltmeter, such as an oscilloscope, read when it is in position 1 in Figure 7.4(a) and position 2 in Figure 7.4(b)? The lead from A to the positive terminal of the voltmeter in both Figures 7.4(a) and 7.4(b) is labelled red and the lead from the negative terminal of the voltmeter to B is labelled black. Experiments, such as those of Romer [2] have shown that, even though the voltmeter, is connected to the same points A and B in Figures 7.4(a) and 7.4(b), the readings of the voltmeter, which will be denoted by V_1 and V_2 respectively, are not the same in the two cases.

Assume that the current through the resistor R_1 in Figure 7.4(a) is I and that the current through the voltmeter, when it is in position 1 in Figure 7.4(a) is I_V. Since the emf given by equation (7.33) is constant, when the current in the primary coil is varying at a constant rate, the current in the secondary circuit is constant and Kirchhoff's first law can be applied so that the current through the resistor R_2 in Figure 7.4(a) is $(I + I_V)$. Since the secondary circuit in Figure 7.4(a) is completely outside the infinitely long solenoid the emf ε, given by equation (7.33), is the same for both the loops $ACBDA$ and $AEBDA$ in Figure 7.4(a). Using the constitutive equation $E = J/\sigma = I/\sigma A_0$ and integrating $\mathbf{E} \cdot \mathbf{dl}$ around the loop $ACBDA$ in Figure 7.4(a) we have

$$\varepsilon = \oint \mathbf{E} \cdot \mathbf{dl} = \sum \frac{I\, dl}{\sigma A_0} = \sum I\, dR = IR_1 + (I + I_V)R_2 \tag{7.34}$$

which is in agreement with Kirchhoff's second law. Similarly for the loop $AEBDA$ in Figure 7.4(b) we have

$$\varepsilon = \oint \mathbf{E} \cdot \mathbf{dl} = I_V R_V + (I + I_V)R_2. \tag{7.35}$$

Solving equations (7.34) and (7.35), we find that the reading V_1 of the voltmeter in position 1 in Figure 7.4(a) is

$$V_1 = I_V R_V = \frac{\varepsilon R_1}{R_1 + R_2 + (R_1 R_2 / R_V)}\,. \tag{7.36}$$

The reading V_1 is positive since the current I_V enters the voltmeter through the red lead, which is connected to the positive end of the voltmeter. If the resistance R_V of the voltmeter is so big that $R_V \gg R_1$ and $R_V \gg R_2$, which is generally true in the case when we use an oscilloscope to measure the voltage, then equation (7.36) reduces to

$$V_1 = I_V R_V = \frac{\varepsilon R_1}{(R_1 + R_2)} \,. \tag{7.37}$$

Consider now the secondary circuit shown in Figure 7.4(b). Let the current through the resistor R_1 now be equal to I'. We shall assume, for purposes of discussion that a current I'_V enters the voltmeter in position 2 in Figure 7.4(b) via the positive terminal. According to Kirchhoff's first law the current through R_2 is $(I' + I'_V)$. The total emf ε in the secondary circuit is still given by equation (7.33) and is still in the anticlockwise direction as shown in Figure 7.4(b). Developing Kirchhoff's second law for the loops *ACBDA* and *ACBFA* in Figure 7.4(b), by using the constitutive equation $E = J/\sigma$ we find that

$$\varepsilon = I'R_1 + (I' + I'_V)R_2, \tag{7.38}$$

$$\varepsilon = I'R_1 - I'_V R_V. \tag{7.39}$$

Solving equations (7.38) and (7.39) we find that I'_V is negative and the reading V_2 of the voltmeter in position 2 in Figure 7.4(b) is

$$V_2 = I'_V R_V = -\frac{\varepsilon R_2}{R_1 + R_2 + (R_1 R_2/R_V)} \,. \tag{7.40}$$

If $R_V \gg R_1$ and $R_V \gg R_2$, we have

$$V_2 = I'_V R_V = -\frac{\varepsilon R_2}{(R_1 + R_2)} \,. \tag{7.41}$$

We can see that, since I'_V is negative, V_2 is negative. This result should not be unexpected since the induced emf ε is trying to drive current around the secondary circuit including the voltmeter in the anticlockwise direction in both Figures 7.4(a) and 7.4(b), so that, when the voltmeter is in position 2 in Figure 7.4(b), the current enters the voltmeter by the black lead. According to equation (7.37) and (7.41), in the presence of the induction electric field outside the long solenoid, not only are V_1 and V_2 not numerically equal but they are of opposite sign.

For example, if $R_1 = 2R_2$, which is the case studied experimentally by Romer [2], then equations (7.37) and (7.41) become

$$V_1 = \frac{2\varepsilon}{3} \tag{7.42}$$

$$V_2 = -\frac{\varepsilon}{3} \,. \tag{7.43}$$

Equations (7.42) and (7.43) were confirmed experimentally by Romer [2].

The current through the voltmeter in Figures 7.4(a) and 7.4(b) is due partly to the induction electric field $\mathbf{E}_i$ due to the varying current in the primary coil (the solenoid) and partly due to the electrostatic field $\mathbf{E}_s$ due to the surface and boundary charge distributions, that give the correct value for the total electric field $\mathbf{E} = (\mathbf{E}_i + \mathbf{E}_s)$ to give the appropriate values of current flow through the voltmeter. For example, if the points A and B in Figure 7.4(a) are diametrically opposite each other, it follows by putting $\int d\phi$ equal to π in equation (7.5) that $\int \mathbf{E}_i \cdot \mathbf{dl}$ from A to B through the voltmeter in Figure 7.4(a) is equal to $\varepsilon/2$. According to equation (7.42) when $R_1 = 2R_2$ and $R_V \gg R_1$, the reading V_1 of the voltmeter in position 1 in Figure 7.4(a) is equal to $2\varepsilon/3$. Hence the contribution of $\int \mathbf{E}_s \cdot \mathbf{dl}$ to $\int \mathbf{E} \cdot \mathbf{dl}$ evaluated from A to B through the voltmeter in Figure 7.4(a), is equal to $+\varepsilon/6$. For the case shown in Figure 7.4(b), the value of $\int \mathbf{E}_i \cdot \mathbf{dl}$ evaluated from A to B through the voltmeter in the direction from A to B is equal to $-\varepsilon/2$. This is negative since the induction electric field $\mathbf{E}_i$ is in the anticlockwise direction in Figure 7.4(b), which is in the direction opposite to $\mathbf{dl}$. Since, according to equation (7.43), when $R_1 = 2R_2$ and $R_V \gg R_1$ we have

$$V_2 = \int \mathbf{E} \cdot \mathbf{dl} = \int \mathbf{E}_i \cdot \mathbf{dl} + \int \mathbf{E}_s \cdot \mathbf{dl} = -\frac{\varepsilon}{3}$$

and since $\int \mathbf{E}_i \cdot \mathbf{dl}$ is equal to $-\varepsilon/2$, then $\int \mathbf{E}_s \cdot \mathbf{dl}$ evaluated from A to B through the voltmeter in Figure 7.4(b) is again equal to $+\varepsilon/6$. Notice that when $R_1 = 2R_2$, the direction of the electrostatic field $\mathbf{E}_s$ due to surface and boundary charge distributions, associated with conduction current flow, is such as to increase the magnitude of the total electric field inside the larger resistance R_1 and to decrease the magnitude of the total electric field inside the smaller resistance R_2, such that in the limit as R_V tends to infinity and I_V tends to zero in Figure 7.4(a), the current is the same in both the resistors R_1 and R_2, despite their different resistances.

If the terminals A and B in Figures 7.4(a) and 7.4(b) were not diametrically opposite each other, the analysis leading to equations (7.37) and (7.41), which can be based on Kirchhoff's laws, would still be valid. When A and B are not diametrically opposite each other the value of $\int \mathbf{E}_i \cdot \mathbf{dl}$ obtained using equation (7.5) for the path from A to B through the voltmeter in Figure 7.4(a) would not be equal to $\varepsilon/2$. However, the surface and boundary charge distributions giving rise to $\mathbf{E}_s$ would be changed in the new situation such that the line integrals $\int \mathbf{E} \cdot \mathbf{dl}$ of the total electric field from A to B through the voltmeters in Figures 7.4(a) and 7.4(b) would still be equal to $2\varepsilon/3$ and $-\varepsilon/3$ respectively and the voltmeters would read the same voltages V_1 and V_2, as was the case when A and B were diametrically opposite each other in Figures 7.4(a) and 7.4(b).

The values of $V_1 = 2\varepsilon/3$ and $V_2 = -\varepsilon/3$, which we determined for the case when $R_1 = 2R_2$ in Figures 7.4(a) and 7.4(b) are the values of the line integrals $\int \mathbf{E} \cdot \mathbf{dl}$ of the total electric field from terminal A along the leads and through the voltmeter to terminal B in Figures 7.4(a) and 7.4(b) respectively. In the special case, when the secondary circuit in Figures 7.4(a) and 7.4(b) is in a region where the magnetic field is zero outside the solenoid, it follows from Faraday's law of electromagnetic induction that $\int \mathbf{E} \cdot d\mathbf{l}$ is zero around the loop *AEBCA* in Figure 7.4(a). Hence for the special case shown in Figure 7.4(a), $\int \mathbf{E} \cdot d\mathbf{l}$ evaluated from A to B through the voltmeter, which is equal to $V_1 = I_V R_V = 2\varepsilon/3$ is also equal to $\int \mathbf{E} \cdot d\mathbf{l}$ evaluated from A to B through the resistance R_1 which is given by

$$\Sigma E \, dl = \Sigma \frac{I \, dl}{\sigma A_0} = I\Sigma \, dR = IR_1.$$

Hence in the special case shown in Figure 7.4(a), V_1 the reading of the voltmeter is equal to the value of IR_1. It should be stressed that IR_1 is not equal to the line integral $\int \mathbf{E}_s \cdot d\mathbf{l}$ of the electrostatic field from A to B through the resistor R, which is only equal to $\varepsilon/6$. The extra contribution of $\varepsilon/2$ comes from the line integral of the induction electric field. In the special case shown in Figure 7.4(b), when $R_V \gg R_2$ the reading $V_2 = -\varepsilon/3$ of the voltmeter is equal to $(I' + I'_V)R_2$, where $(I' + I'_V)$ is the current through the resistor R_2, If the secondary circuit in Figures 7.4(a) were in a varying magnetic field, for example if it were completely inside the solenoid, there would be a varying magnetic flux through the loop *AEBCA* and $\int \mathbf{E} \cdot d\mathbf{l}$ evaluated from A to B through the voltmeter would not be equal to $\int \mathbf{E} \cdot d\mathbf{l}$ evaluated from A to B through the resistor R_1 so that in the more general case V_1 would not be equal to IR_1 in Figure 7.4(a) and V_2 would not be equal to $(I' + I'_V)R_2$ in Figure 7.4(b).

7.4. Self inductance

7.4.1. *Example of self inductance*

Consider the **isolated** long solenoid consisting of n circular turns per metre length each of radius b, as shown in plan view in Figure 7.5. When the current in the solenoid is varying, the moving conduction electrons in the windings of the solenoid give rise to an induction electric field $\mathbf{E}_i$ which, according to equation (7.12), is given in the quasi-stationary limit at the position $r = b$ by

$$E_i = E_\phi = \frac{\mu_0 n b}{2}\left(\frac{dI}{dt}\right). \tag{7.44}$$

If $\dot{I}$ is increasing in magnitude the direction of $\mathbf{E}_i$ is opposite to the direction of current flow in the solenoid, as shown in Figure 7.5. The induction electric field $\mathbf{E}_i$ given by equation (7.44), acts on the conduction electrons inside the windings of the isolated solenoid and tries to give a conduction current flow, which is in such a direction that it opposes the direction of current flow in Figure 7.5. If the conduction current in the isolated solenoid is decreasing in magnitude, $\dot{I}$ is negative and the direction of E_ϕ is reversed. In this case the induction electric field $E_i = E_\phi$ tries to give a conduction current flow that

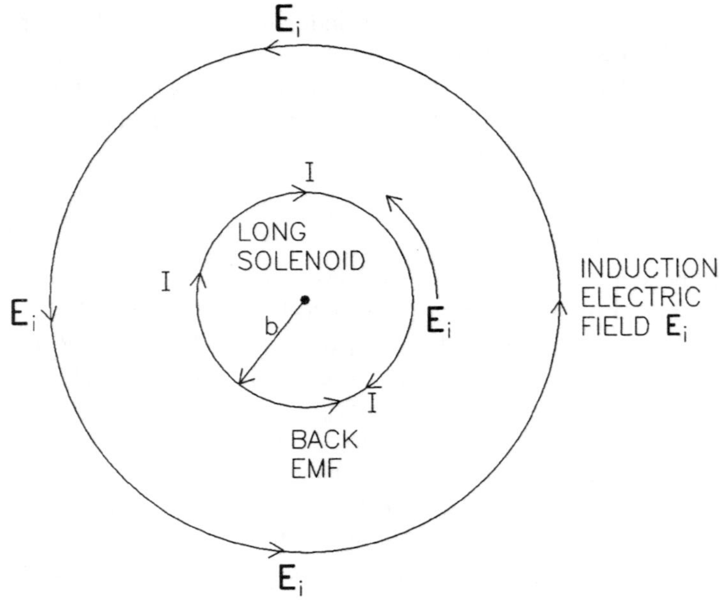

Figure 7.5. Example of self induction. The varying current in the long solenoid gives an induction electric field $\mathbf{E}_i$, which gives a back emf in the long solenoid.

opposes the decrease in current flow. Since the induced emf in the isolated solenoid always opposes the changes in conduction current flow, it is generally called a back emf. This is the phenomenon of self induction. Using equation (7.44) we find that the induced back emf per metre length of the solenoid is given, in the quasi-stationary limit, by

$$\int \mathbf{E}_i \cdot \mathrm{d}\mathbf{l} = 2\pi nb\, E_\phi = \mu_0 \pi n^2 b^2 \left(\frac{\mathrm{d}I}{\mathrm{d}t}\right). \tag{7.45}$$

7.4.2. *General case of the self inductance of an air cored coil*

Consider the general case of an isolated stationary coil of arbitrary shape. The conduction current in the coil is varying at a slow enough rate for the quasi-stationary approximations to be valid. According to equation (5.53) of Chapter 5, which was derived from the expression for the electric field due to a moving classical point charge, the induction electric field $\mathbf{E}_i$ at a position $\mathbf{r}$ due to the varying current in the isolated coil itself is given by

$$\mathbf{E}_i(\mathbf{r}) = -\frac{\mu_0 \dot{I}}{4\pi} \int \frac{\mathrm{d}\mathbf{l}'}{|\mathbf{r} - \mathbf{r}'|} \tag{7.46}$$

where $\mathrm{d}\mathbf{l}'$ is an element of length of the isolated coil at the position $\mathbf{r}'$. The total induced back emf ε in the isolated coil due to the varying current in the coil itself is obtained by evaluating the line integral $\int \mathbf{E}_i \cdot \mathrm{d}\mathbf{l}$ around the coil. We have

$$\varepsilon = \int \mathbf{E}_i \cdot \mathrm{d}\mathbf{l} = -\left\{\frac{\mu_0}{4\pi}\oint\oint \frac{\mathrm{d}\mathbf{l}' \cdot \mathrm{d}\mathbf{l}}{|\mathbf{r} - \mathbf{r}'|}\right\}\left(\frac{\mathrm{d}I}{\mathrm{d}t}\right) \tag{7.47}$$

where $\mathrm{d}\mathbf{l}$ and $\mathrm{d}\mathbf{l}'$ are elements of length of the coil at the positions $\mathbf{r}$ and $\mathbf{r}'$ respectively. Equation (7.47) is generally written in the abbreviated form

$$\varepsilon = \int \mathbf{E}_i \cdot \mathrm{d}\mathbf{l} = -L\left(\frac{\mathrm{d}I}{\mathrm{d}t}\right) \tag{7.48}$$

where

$$L = \frac{\mu_0}{4\pi}\oint\oint \frac{\mathrm{d}\mathbf{l} \cdot \mathrm{d}\mathbf{l}'}{|\mathbf{r} - \mathbf{r}'|} \tag{7.49}$$

is the self inductance of the air cored coil. The direction of the back emf due to self inductance is consistent with Lenz's law, and acts in such a direction as to oppose the changes in the conduction current flowing in the coil. According to equation (7.48), the self inductance of a coil, measured in henries, is numerically equal to the back emf, measured in volts, when the current in the coil is varying at the rate of one ampere per second. In practice the self inductance of a coil can be increased by winding the turns of the coil on a ferromagnetic core.

In introductory courses, self inductance is generally introduced using Faraday's law of electromagnetic induction, according to which the back emf in the coil is

$$\varepsilon = -\frac{\mathrm{d}\Phi}{\mathrm{d}t} \tag{7.50}$$

where Φ is the magnetic flux linking the coil due to the current in the coil itself. Equation (7.50) can be rewritten in the form

$$\varepsilon = -\left(\frac{\mathrm{d}\Phi}{\mathrm{d}I}\right)\left(\frac{\mathrm{d}I}{\mathrm{d}t}\right) = -L\frac{\mathrm{d}I}{\mathrm{d}t} \tag{7.51}$$

where the self inductance L of the coil is defined as

$$L = \frac{\mathrm{d}\Phi}{\mathrm{d}I}\,. \tag{7.52}$$

In the special case of an air cored coil, the magnetic flux Φ is proportional to the current I. If $\Phi = kI$ where k is a constant, equation (7.52) reduces to

$$L = \frac{\mathrm{d}\Phi}{\mathrm{d}I} = \frac{\mathrm{d}}{\mathrm{d}I}(kI) = k = \frac{\Phi}{I}\,. \tag{7.53}$$

According to equation (7.53), the self inductance L of an air cored coil, measured in henries, is numerically equal to the magnetic flux Φ linking the coil, when the current in the coil is one ampere. When there are ferromagnetic materials present Φ is not proportional to I, so that equation (7.53) is no longer valid, but equation (7.52) can still be used to define L. The reader can show either by using equation (7.53) or by comparing equations

(7.48) and (7.45) that the inductance per metre length of a long solenoid is $\mu_0 \pi n^2 b^2$, where b is the radius of the solenoid and n is the number of turns per metre length .

The reader should remember that, in our approach when the back emf due to self induction is written in the form $-L\dot{I}$, the $-L\dot{I}$ term stands for $\int \mathbf{E}_i \cdot \mathrm{d}\mathbf{l}$, the line integral of the induction electric field due to the varying current in the coil making up the inductor evaluated around the inductor coil.

7.5. An air cored transformer working at mains frequency

In Section 7.2, we assumed that the current in the primary coil in Figure 7.1 was varying at a constant rate, so that the current in the secondary coil was constant and the back emf due to the self inductance of the secondary coil was zero. We shall now assume that the current I_1 in the primary coil (the infinitely long solenoid) in Figure 7.1 varies at the frequency of the electricity mains, which is generally 50 or 60 Hz. The sign convention we shall adopt is that the positive direction for current flow, electric fields and emfs in Figure 7.1 is in the direction of increasing angle ϕ, which is in the anticlockwise direction in the plan view shown in Figure 7.1. According to equation (7.25) the emf, now denoted by ε_1, induced in the secondary coil in Figure 7.1 by the varying current I_1 in the primary coil is given by

$$\varepsilon_1 = \oint \mathbf{E}_i \cdot \mathrm{d}\mathbf{l}_2 = -M_{21}\frac{\mathrm{d}I_1}{\mathrm{d}t} \tag{7.54}$$

where $\mathbf{E}_i$ is the induction electric field due to the moving charges in the primary coil. For the special case shown in Figure 7.1, M_{21} is given by equation (7.30). When the current I_1 in the primary coil is in the clockwise direction in Figure 7.1 and is increasing in magnitude $\dot{I}_1$ is negative so that $-M_{21}\dot{I}_1$ and ε_1 are positive, that is ε_1 is in the anticlockwise direction in Figure 7.1. The emf ε_1 gives a current flow in the secondary coil in Figure 7.1 which, after the initial transient state after switching on is over, varies at mains frequency. It follows from the analysis given in Section 7.4 that, due to the inductance L_2 of the secondary coil, the moving conduction electrons, whose motions give the varying current I_2 in the secondary coil, also give rise to an induction electric field, which will be denoted by $\mathbf{E}_L$. This induction electric field $\mathbf{E}_L$ acts on the conduction electrons in the secondary coil. It follows from equation (7.48) that the back emf ε_2 in the secondary coil due to the self inductance L_2 of the secondary coil is

$$\varepsilon_2 = \oint \mathbf{E}_L \cdot \mathrm{d}\mathbf{l}_2 = -L_2\frac{\mathrm{d}I_2}{\mathrm{d}t} \,. \tag{7.55}$$

At mains frequency, there is sufficient time for the necessary quasi-stationary surface and boundary charge distributions to build up such that, at any instant, they give a contribution $\mathbf{E}_s$ to the electric field, that gives a resultant electric field $\mathbf{E} = (\mathbf{E}_i + \mathbf{E}_L + \mathbf{E}_s)$ inside the wire making up the secondary coil in Figure 7.1, which is always parallel to the wire and is of such a magnitude that, at every instant, the conduction current I_2 in the secondary coil has the same value in all parts of the secondary circuit.

The total emf in the stationary secondary circuit is

$$\varepsilon_{\text{sec}} = \oint_2 \mathbf{E} \cdot \mathrm{d}\mathbf{l}_2 = \oint_2 (\mathbf{E}_i + \mathbf{E}_L + \mathbf{E}_s) \cdot \mathrm{d}\mathbf{l}_2. \tag{7.56}$$

The line integrals are evaluated in the positive (anticlockwise) direction in Figure 7.1. In the quasi-stationary limit, the small surface and boundary charge distributions that give rise to $\mathbf{E}_s$ vary at mains frequency, which is slow enough for them to be treated as electrostatic so that, to an excellent approximation, the $\mathbf{E}_s$ field is conservative and

$$\oint_2 \mathbf{E}_s \cdot \mathrm{d}\mathbf{l}_2 = 0.$$

Equation (7.56) for the total emf in the secondary circuit then reduces to

$$\varepsilon_{\text{sec}} = \oint \mathbf{E}_i \cdot \mathrm{d}\mathbf{l}_2 + \oint \mathbf{E}_L \cdot \mathrm{d}\mathbf{l}_2. = -M_{21}\dot{I}_1 - L_2\dot{I}_2 \tag{7.57}$$

where we have used equations (7.25) and (7.48).

The skin depth in copper is 9 mm at 50 Hz. Hence it is a reasonable approximation to assume that the current density J_2 is uniform across the thin wires making up the secondary coil, so that $J_2 = I_2/A_2$, where A_2 is the area of cross section of the wire. The constitutive equation can then be written in the form

$$\mathbf{E} = \mathbf{E}_i + \mathbf{E}_L + \mathbf{E}_s = \frac{\mathbf{J}_2}{\sigma} = \frac{\mathbf{I}_2}{\sigma A_2} \tag{7.58}$$

where σ is the local value of the electrical conductivity. We shall now form the scalar product of $\mathbf{E}$ and the element of length $\mathrm{d}\mathbf{l}_2$ of the secondary coil pointing in the positive (anticlockwise) direction in Figure 7.1. Summing around the secondary circuit, since $\mathbf{E}$, $\mathrm{d}\mathbf{l}_2$ and $\mathbf{I}_2$ are all parallel, using equation (7.58) we have

$$\varepsilon_{\text{sec}} = \Sigma\mathbf{E} \cdot \mathrm{d}\mathbf{l}_2 = \Sigma\frac{I_2\, \mathrm{d}l_2}{\sigma A_2} = I_2\, \Sigma \mathrm{d}R \tag{7.59}$$

where $\mathrm{d}R$ is the resistance of the length $\mathrm{d}\mathbf{l}_2$ of the wire.

If there is a resistor of high resistance R_2 in the secondary circuit, which is very much bigger than the resistances of the wires making up the secondary circuit, then $I_2\, \Sigma\mathrm{d}R$ is given to an excellent approximation by $I_2\, R_2$. Comparing equations (7.57) and (7.59) we see that

$$\varepsilon_{\text{sec}} = \oint \mathbf{E}_i \cdot \mathrm{d}\mathbf{l}_2 + \oint \mathbf{E}_L \cdot \mathrm{d}\mathbf{l}_2. = -M_{21}\dot{I}_1 - L_2\dot{I}_2 = I_2\, R_2 \tag{7.60}$$

For the case shown in Figure 7.1, $\dot{I}_1$ is negative on our sign convention. According to equation (7.60) the sum of the emfs in the secondary circuit in Figure 7.1 is equal to the sum of the products $I_2\,dR$ taken around the circuit. This is another example of Kirchhoff's second law.

The varying current I_2 in the secondary coil in Figure 7.1 gives an emf of magnitude $-M_{12}\dot{I}_2$ in the primary coil of the air cored transformer. It follows from equation (7.26) that $M_{12} = M_{21}$. When the current I_2 in the secondary coil is in the positive (anticlockwise) direction in Figure 7.1 and I_2 is increasing in magnitude the emf $-M_{12}\dot{I}_2$ induced in the primary coil by the varying current in the secondary coil is in the negative (clockwise) direction. The varying current I_1 in the primary coil gives a back emf $-L_1\dot{I}_1$ in the primary coil where L_1 is the self inductance of the primary coil. When $\dot{I}_1$ is negative this back emf is in the anticlockwise direction in Figure 7.1. If there is a resistor of resistance R_1 in the primary circuit and a generator of emf $-\varepsilon_0 \cos \omega t$ to drive the current I_1 in the negative (clockwise) direction around the primary coil, then the reader can show using the methods leading to equation (7.60) that with our sign convention

$$-\varepsilon_0 \cos \omega t - M_{12}\dot{I}_2 - L_1\dot{I}_1 = -I_1R_1. \tag{7.61}$$

The left hand side represents the sum of the emfs in the primary circuit and the right hand side is the sum of the IR drops. Equation (7.61) is in agreement with Kirchhoff's second law. The reader will find that different books use different sign conventions. If $\varepsilon_0 \cos \omega t$, $M_{21} = M_{12}$, L_1, L_2, R_1 and R_2 are given, then equations (7.60) and (7.61) can be used to determine the currents I_1 and I_2 in the primary and secondary coils of the air cored transformer in Figure 7.1.

7.6. AC theory

7.6.1. *Introduction*

In this section we shall develop the interpretation of the LCR circuit shown in Figure 7.6, which consists of an inductor of inductance L, a capacitor of capacitance C and a resistor of resistance R. We shall assume that the applied emf $\varepsilon = \varepsilon_0 \cos \omega t$ comes from the secondary of a transformer working at mains frequency, so that the quasi-stationary approximations discussed in Section 7.1 are valid. We shall assume that the circuit elements can be treated as idealized lumped elements. For example, we shall assume that when the current in the inductor L is varying, the induced electric field due to the varying current in the inductor is confined to the inside of the inductor. We shall assume that the inductance L of the inductor is very much bigger than the self inductances of the other circuit elements. We shall assume that there is no leakage current between the plates of the capacitor and that the capacitance C of the capacitor is very much bigger than any stray capacitances. We shall also assume

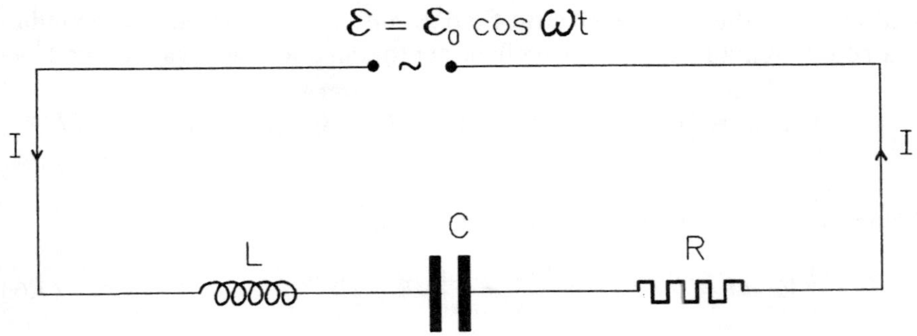

Figure 7.6. The *LCR* circuit.

that the resistance R is very much bigger than the resistances of the leads, inductor and the internal resistance of the source of emf.

7.6.2. *The* LR *circuit*

Consider first the *LR* circuit shown in Figure 7.7. We shall assume that the emf applied to the *LR* circuit comes from a transformer connected to the electricity mains. When the current in the *LR* circuit is varying, according to equation (7.48) there is a back emf $-L\dot{I}$ in the inductor, which affects the value of the current flowing in the circuit. Let $\mathbf{E}_i$ be the induction electric field induced in the secondary coil of the transformer by the varying current in the primary coil of the transformer at the time t. The value of the emf applied to the *LR* circuit at the time t is given by $\varepsilon = \oint \mathbf{E}_i \cdot \mathbf{dl}$ evaluated around the *LR* circuit. We shall ignore the self inductance of the secondary coil of the transformer. Let $\mathbf{E}_L$ be the electric field induced in the windings of the inductor L by the varying current in the inductor. According to equation (7.48), $\oint \mathbf{E}_L \cdot \mathbf{dl}$ evaluated around the circuit is equal to $-L\dot{I}$. Evaluating the line integral of the total electric field around the circuit, remembering that, for quasi-

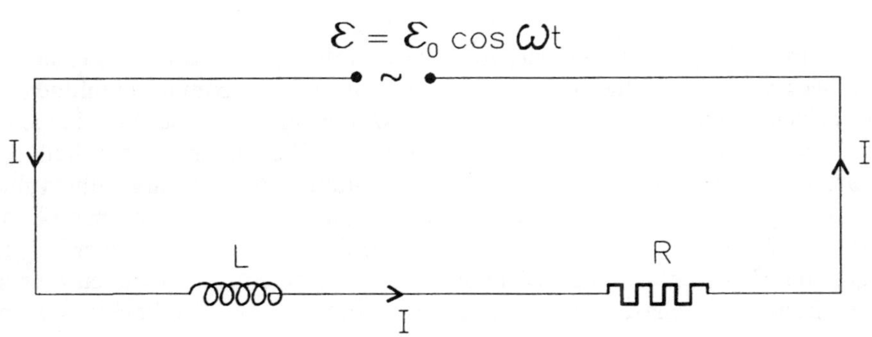

Figure 7.7. The *LR* circuit.

stationary conditions, the electric field $\mathbf{E}_s$ due to the surface and boundary charge distributions is, to an excellent approximation conservative, we have

$$\oint \mathbf{E} \cdot \mathbf{dl} = \oint (\mathbf{E}_i + \mathbf{E}_L + \mathbf{E}_s) \cdot \mathbf{dl}_2 = \oint \mathbf{E}_i \cdot \mathbf{dl}_2 + \oint \mathbf{E}_L \cdot \mathbf{dl}_2 = \varepsilon - L\dot{I}. \tag{7.62}$$

Using the constitutive equation $E = J/\sigma = I/\sigma A$, we have

$$\oint \mathbf{E} \cdot \mathbf{dl} = \oint E\mathrm{d}l = \oint \frac{I\,\mathrm{d}l}{\sigma A} = I \oint \mathrm{d}R = IR \tag{7.63}$$

where I is the current in the circuit, σ is the electrical conductivity, A is the area of the cross section and R is the resistance of the secondary circuit. Comparing equations (7.62) and (7.63), we see that

$$\varepsilon_0 \cos \omega t - L\dot{I} = IR. \tag{7.64}$$

The left hand side of equation (7.64) is the sum of the applied emf $\varepsilon_0 \cos \omega t$ and the back emf $-L\dot{I}$, whereas the right hand side was derived using the constitutive equation $\mathbf{J} = \sigma\mathbf{E}$ to evaluate $\oint \mathbf{E} \cdot \mathbf{dl}$. Equation (7.64) is often rewritten in the form

$$\varepsilon_0 \cos \omega t = L\dot{I} + IR. \tag{7.65}$$

Moving the $L\dot{I}$ term to the right hand side of equation (7.65) does not change the physical interpretation of the role of the $L\dot{I}$ term in equations (7.64) and (7.65).

The steady state solution of equation (7.65), that is valid after the transient state is effectively over, which is after a few cycles after switching on, is

$$I = \frac{\varepsilon_0}{(R^2 + \omega^2 L^2)^{1/2}} \cos(\omega t - \phi) = I_0 \cos(\omega t - \phi). \tag{7.66}$$

where

$$\tan \phi = \frac{\omega L}{R}. \tag{7.67}$$

The current I varies at the same frequency as the applied emf. The effect of the back emf, due to the inductor L, is to reduce the maximum amplitude of the current in the circuit from the value of ε/R it would have had if the inductor were absent to the value of $\varepsilon/(R^2 + \omega^2 L^2)^{1/2}$. The ωL term is called the reactance of the inductor. According to equation (7.66) the larger the value of ωL, then the smaller is the current in the LR circuit. The presence of the inductor L also affects the phase of the current in the circuit. According to equation (7.66), after the transient state is over, the emf leads the current in the circuit by a phase angle ϕ given by equation (7.67). When R is zero, $\phi = \pi/2$.

7.6.3. *The* CR *circuit*

Consider now the *CR* circuit shown in Figure 7.8(a). For quasi-stationary conditions the electric fields due to the surface and boundary charges associated with current flow, and the electric field due to the charges on the plates of the capacitor can be treated, to a good approximation, as electrostatic, so that $\oint \mathbf{E}_s \cdot \mathbf{dl}$, evaluated around the complete circuit, is zero. Hence the integral of the total electric field around the *CR* circuit shown in Figure 7.8(a) reduces to

$$\oint \mathbf{E} \cdot \mathbf{dl} = \oint (\mathbf{E}_i + \mathbf{E}_s) \cdot \mathbf{dl} = \oint \mathbf{E}_i \cdot \mathbf{dl} = \varepsilon \tag{7.68}$$

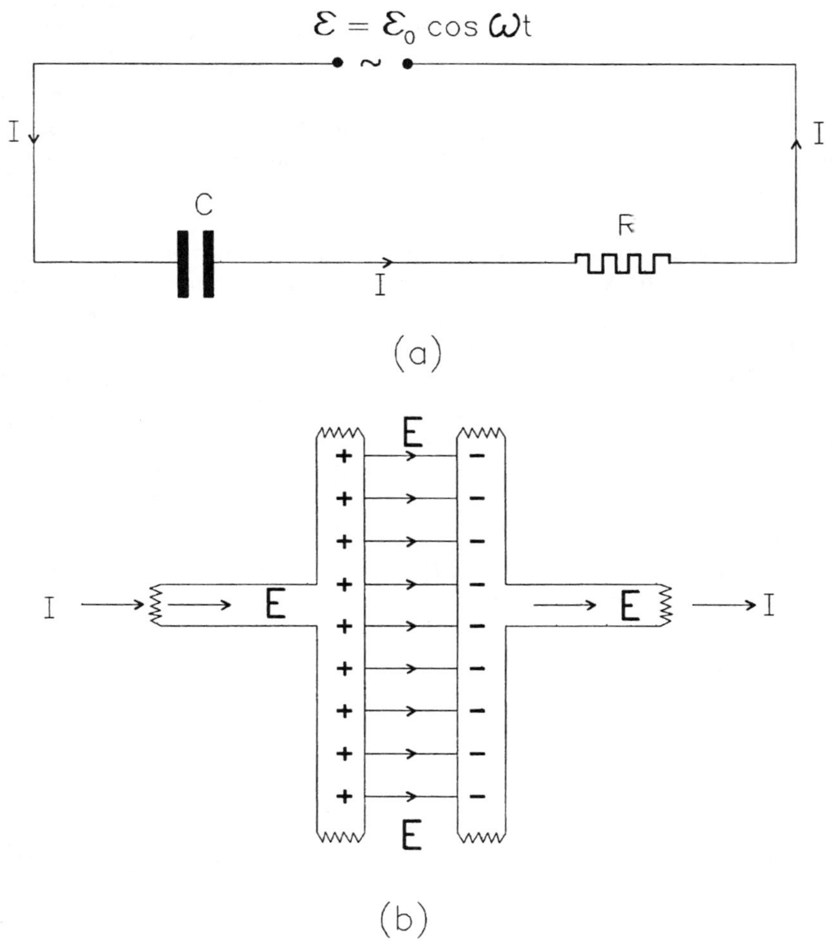

Figure 7.8. (a) The *CR* circuit (b) The electric fields in the leads to and from the capacitor and in the space between the plates of the capacitor.

where ε is the instantaneous value of the applied emf.

To understand how the capacitor acts in the *CR* circuit, assume that, in the way described in Appendix B, the applied emf gives rise to surface and boundary charges which give a resultant electric field inside the wires leading to and from the capacitor, as shown in Figure 7.8(b). The conduction electrons in the wires respond to the local resultant electric fields inside the wires and a conduction current will flow in the leads to and from the capacitor, as long as the emf can maintain electric fields inside the connecting wires in Figure 7.8(b). Even though there is an insulating gap between the plates of the capacitor, which would prevent DC current flow, the current in the connecting leads in Figure 7.8(b) will continue to flow as long as the applied emf is able to maintain an electric field in the connecting leads. For the conditions shown in Figure 7.8(b), the current flow in the connecting leads increases the positive charge on the positive plate of the capacitor and makes the charge on the negative plate more negative. This in turn increases the electric field between the plates of the capacitor and, as we shall see later, this in turn affects the electric fields inside the connecting wires in other parts of the circuit. If at any instant there are charges of $+Q$ and $-Q$ on the plates of the capacitor C in Figure 7.8(a), they will give a contribution $V = Q/C$ to the line integral $\int \mathbf{E} \cdot \mathrm{d}\mathbf{l}$ evaluated between the plates of the capacitor. The electric field in the wire leading to the capacitor is typically of the order of 0.01 V m^{-1}. If there is a potential difference of 10 V across a capacitor whose plates are 1 mm apart, the electric field between the plates of the capacitor due to the charges on the capacitor plates is 10^4 V m^{-1}, which is about 10^6 times the magnitude of the electric field due to the source of emf and the surface and boundary charge distributions on the connecting leads. Hence, to a very good approximation, the line integral $\int \mathbf{E} \cdot \mathrm{d}\mathbf{l}$ of the total electric field $\mathbf{E}$ evaluated from one plate of the capacitor to the other is equal to the line integral of the electric field due to the charges $+Q$ and $-Q$ on the capacitor plates, which is equal to Q/C. Adding this contribution to the contribution of $\int \mathbf{E} \cdot \mathrm{d}\mathbf{l} = \int (I/\sigma A)\, \mathrm{d}l = I \int \mathrm{d}R$ evaluated along the rest of the circuit, excluding the capacitor, we have

$$\oint \mathbf{E} \cdot \mathrm{d}\mathbf{l} = IR + \frac{Q}{C}. \tag{7.69}$$

Comparing equations (7.68) and (7.69) we see that

$$\varepsilon = IR + \frac{Q}{C}.$$

Since, for quasi-stationary conditions, $I = \dot{Q}$, this equation can be rewritten in the form

$$\varepsilon = R\frac{\mathrm{d}Q}{\mathrm{d}t} + \frac{Q}{C}. \tag{7.70}$$

We shall assume initially that the current in the primary of the air cored transformer in Figure 7.8(a) is varying at a constant rate so that the induced emf ε in the secondary of the transformer, which is the emf applied to the *CR* circuit is constant. We shall also assume that there are no charges on the plates of the capacitor at the time $t = 0$. The capacitor cannot charge up instantaneously as the maximum current that can flow through the resistor R is ε/R. Hence it takes a finite time for the capacitor to become fully charged. As the charges on the capacitor build up the contribution of Q/C to $\int \mathbf{E} \cdot \mathbf{dl}$ increases so that $\int \mathbf{E} \cdot \mathbf{dl} = IR$ evaluated around the rest of the circuit must decrease. Consequently I must decrease and the rate at which the capacitor is charging up decreases. The solution of equation (7.70) is

$$Q = \varepsilon C \left\{ 1 - \exp\left(-\frac{t}{CR}\right)\right\}. \tag{7.71}$$

Hence

$$I = \frac{\mathrm{d}Q}{\mathrm{d}t} = \frac{\varepsilon}{R}\exp\left(-\frac{t}{CR}\right). \tag{7.72}$$

According to equations (7.71) and (7.72), as the time t tends to infinity Q tends to εC and the current I tends to zero.

Assume now that the emf ε applied to the *CR* circuit in Figure 7.8(a) is equal to $\varepsilon_0 \cos \omega t$ in equation (7.70). In this case also, the value of the potential difference Q/C across the capacitor in equation (7.69) affects the current flowing in the *CR* circuit. The steady state mathematical solution of equation (7.70), that is valid when the transient state is effectively over after a few cycles, is

$$I = \frac{\varepsilon_0}{(R^2 + 1/\omega^2C^2)^{1/2}} \cos(\omega t + \phi) \tag{7.73}$$

where

$$\tan \phi = \frac{1}{\omega CR}. \tag{7.74}$$

It can be seen from equation (7.73) that the presence of the capacitor in the *CR* circuit in Figure 7.8(a) reduces the current flowing in the circuit from the value of ε/R when there is no capacitor to $\varepsilon/(R^2 + 1/\omega^2C^2)^{1/2}$. The smaller the value of ωC the smaller is the current in the circuit. The quantity $1/\omega C$ is called the reactance of the capacitor. In the case of a *CR* circuit, the current leads the emf by a phase angle ϕ given by equation (7.74).

7.6.4. *The* LCR *circuit*

It is now straight forward to interpret the *LCR* circuit shown in Figure 7.6. Evaluating the line integral of the total electric field around the *LCR* circuit at the time t, remembering that $\oint \mathbf{E}_s \cdot \mathbf{dl}$ is zero for quasi-stationary conditions, we have

$$\oint \mathbf{E} \cdot \mathrm{d}\mathbf{l} = \oint (\mathbf{E}_i + \mathbf{E}_L + \mathbf{E}_s) \cdot \mathrm{d}\mathbf{l} = \oint \mathbf{E}_i \cdot \mathrm{d}\mathbf{l} + \oint \mathbf{E}_L \cdot \mathrm{d}\mathbf{l} = \varepsilon_0 \cos \omega t - L\dot{I}. \tag{7.75}$$

Evaluating $\oint \mathbf{E} \cdot \mathrm{d}\mathbf{l}$ using the method, based on the constitutive equation $E = J/\sigma$ that led to equation (7.69), we again have

$$\oint \mathbf{E} \cdot \mathrm{d}\mathbf{l} = IR + \frac{Q}{C}. \tag{7.76}$$

Comparing equations (7.75) and (7.76) we see that

$$\varepsilon_0 \cos \omega t - L\dot{I} = IR + \frac{Q}{C}. \tag{7.77}$$

Equation (7.77) is another example of Kirchhoff's second law. The left hand side of equation (7.77) is the sum of the emfs in the circuit. The right hand side is the value of $\oint \mathbf{E} \cdot \mathrm{d}\mathbf{l}$ evaluated around the circuit using the equation $E = J/\sigma = I/\sigma A$ for the conductors and adding the contribution of the potential difference Q/C across the capacitor. Since $I = \dot{Q}$, equation (7.77) can be rewritten in the form

$$L\frac{\mathrm{d}^2Q}{\mathrm{d}t^2} + R\frac{\mathrm{d}Q}{\mathrm{d}t} + \frac{Q}{C} = \varepsilon_0 \cos \omega t. \tag{7.78}$$

This is the same differential equation as the equation for forced damped harmonic motion. The steady state solution of equation (7.78), that is valid after the transient stage after switching on is effectively over after the first few cycles, is

$$I = \frac{\mathrm{d}Q}{\mathrm{d}t} = \frac{\varepsilon_0}{Z} \cos (\omega t - \phi) = I_0 \cos (\omega t - \phi) \tag{7.79}$$

where

$$Z = \left[R^2 + \left(\omega L - \frac{1}{\omega C} \right)^2 \right]^{1/2} \tag{7.80}$$

$$\tan \phi = \frac{(\omega L - 1/\omega C)}{R}. \tag{7.81}$$

The current I in the LCR circuit varies at the same angular frequency ω as the applied emf, but there is a phase difference ϕ, given by equation (7.81), between the applied emf and the current I in the LCR circuit. If $\omega L > 1/\omega C$, the applied emf leads the current in the LCR circuit in phase by an angle ϕ given by equation (7.81). According to equation (7.81) if ωL is equal to $1/\omega C$ the current I and the emf ε are in phase and, according to equations (7.79) and (7.80), the current has its maximum value. This condition is known as resonance.

7.6.5. *The voltages across the circuit elements in an* LCR *circuit*

Kirchhoff's second law is sometimes stated in the form that the sum of the voltages across the circuit elements taken in order around a loop or mesh of an AC circuit is zero. An example of this alternative form of Kirchhoff's second law will now be given using the example of the *LCR* circuit shown in Figure 7.9. We shall assume that AC voltmeters, such as cathode ray oscilloscopes, are connected across the various circuit elements of the *LCR* circuit as shown in Figure 7.9. We shall assume that the resistances of the voltmeters are all so large that the presence of the voltmeters has no significant effect on the current I flowing in the *LCR* circuit. The leads joining the circuit elements to the positive terminals of the voltmeters in Figure 7.9 are labelled the red leads. The leads leading from the negative terminals of the voltmeters will be called the black leads. We shall follow standard jargon and call the readings of the voltmeters the voltages across the circuit elements. It was shown in Section 7.3.1 that the reading of the voltmeter in Figure 7.3 is equal to the line integral of the total electric field from the terminal A in Figure 7.3 through the red lead to the voltmeter, then through the voltmeter and finally back through the black lead to terminal B.

Consider first the reading of the voltmeter measuring the voltage V_0 directly across the source of emf in Figure 7.9. We shall assume that at the instant of time under consideration, the induction electric field $\mathbf{E}_i$ inside the source of emf due to the varying current in the primary coil of the transformer is in the direction from B to A in Figure 7.9 and drives a current I around the

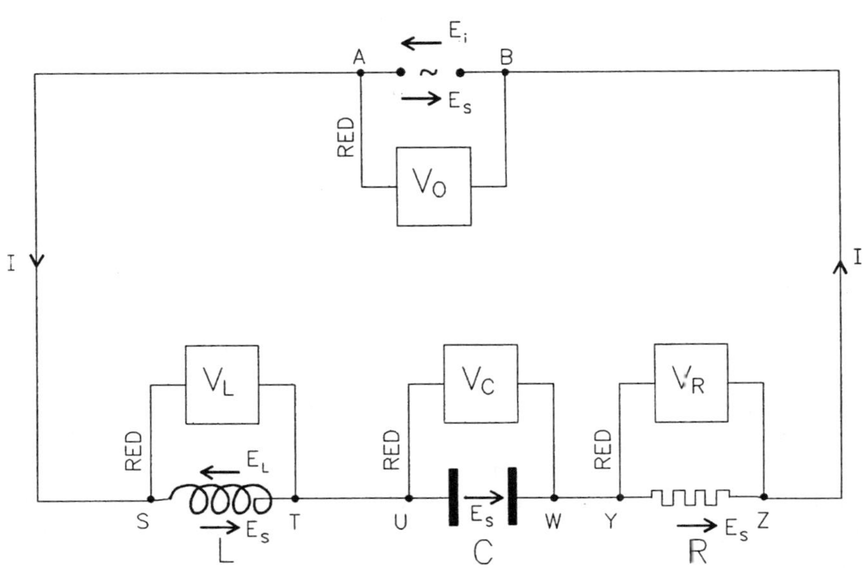

Figure 7.9. Measurement of the voltages across the applied emf a[illegible] the circuit elements in the *LCR* circuit.

LCR circuit in the anticlockwise direction as shown in Figure 7.9. The quasi-static electric field due to the surface and boundary charge distributions associated with current flow and due to the charges on the plates of the capacitor will be denoted by $\mathbf{E}_s$. At mains frequency the $\mathbf{E}_s$ field is to an excellent approximation conservative and so we shall assume that $\oint \mathbf{E}_s \cdot \mathbf{dl}$ is zero for any closed path. If the internal resistance of the source of emf is zero, the total electric field $\mathbf{E} = (\mathbf{E}_i + \mathbf{E}_s)$ inside the source of emf must be zero, otherwise the current I would be infinite. Hence for a path from B to A through the secondary of the transformer (the source of emf) we have

$$\int_B^A \mathbf{E} \cdot \mathbf{dl} = \int_B^A \mathbf{E}_i \cdot \mathbf{dl} + \int_B^A \mathbf{E}_s \cdot \mathbf{dl} = 0. \tag{7.82}$$

If $\mathbf{E}_i$ is in the direction from B to A inside the source of emf, then $\mathbf{E}_s$ must be in the opposite direction, that is from A to B inside the source of emf. (In practice, the source of emf has some internal resistance so that, inside the source of emf, the component of $\mathbf{E}_i$ parallel to the wire must be numerically greater than the component of $\mathbf{E}_s$ in the opposite direction for current to flow). We are assuming in our idealized model of lumped circuit elements that the induction electric field $\mathbf{E}_i$ in the secondary coil of the transformer, which is the source of emf for the *LCR* circuit, is zero outside the source of emf, so that if $\oint \mathbf{E}_i \cdot \mathbf{dl}$ is evaluated around the complete *LCR* circuit,

$$\oint \mathbf{E}_i \cdot \mathbf{dl} = \int_B^A \mathbf{E}_i \cdot \mathbf{dl} = \varepsilon_0 \cos \omega t. \tag{7.83}$$

Substituting in equation (7.82) we find that in our idealized case when the internal resistance and self inductance of the source of emf are both zero

$$\int_B^A \mathbf{E}_s \cdot \mathbf{dl} = -\varepsilon_0 \cos \omega t.$$

Reversing the limits of integration we have

$$\int_A^B \mathbf{E}_s \cdot \mathbf{dl} = \varepsilon_0 \cos \omega t. \tag{7.84}$$

The surface and boundary charge distributions, that give rise to the quasi-static field $\mathbf{E}_s$ of magnitude $(-\mathbf{E}_i)$ inside the source of emf also give rise to a quasi-static electric field outside the source of emf, in a region where $\mathbf{E}_i$ is zero. Since the $\mathbf{E}_s$ field is conservative the line integral $\int \mathbf{E}_s \cdot \mathbf{dl}$ from A to B is the same and according to equation (7.54) is equal to $\varepsilon_0 \cos \omega t$ for any path from A to B outside the source of emf in Figure 7.9, including the path from A along the red lead to the voltmeter measuring V_0, through the voltmeter and back along the black lead to B, which we showed in Section 7.3.1 is equal to the reading of the voltmeter. Hence the reading of the voltmeter placed across the source of emf in Figure 7.9 is

$$V_0 = \varepsilon_0 \cos \omega t. \tag{7.85}$$

which is equal to the instantaneous value of the applied emf.

Consider now the reading V_L of the voltmeter (oscilloscope) connected across the inductor L in Figure 7.9, when the current I in the LCR circuit is in the anticlockwise direction and is increasing in magnitude such that $\dot{I}$ is positive. For positive $\dot{I}$ the direction of the back emf in the inductor is opposite to the direction of current flow. Hence $\int \mathbf{E}_L \cdot \mathbf{dl}$ evaluated from the point S in Figure 7.9 through the windings of the inductor L to the point T is negative. Since in our idealized case $\mathbf{E}_L$ is zero outside the inductor L, $\oint \mathbf{E}_L \cdot \mathbf{dl}$ evaluated around the complete LCR circuit is equal to $\int \mathbf{E}_L \cdot \mathbf{dl}$ evaluated from S to T through the inductor L. Using equation (7.48) we have

$$\int_S^T \mathbf{E}_L \cdot \mathbf{dl} = -L\dot{I}. \tag{7.86}$$

When $\dot{I}$ is positive the line integral on the left hand side of equation (7.86) is negative. If the resistance of the inductor is zero, it follows from the constitutive equation $J = \sigma E$ that the total electric field $\mathbf{E} = (\mathbf{E}_L + \mathbf{E}_s)$ inside the wire making up the inductor must be zero, otherwise the current through the inductor L would be infinite. Hence for a path from S to T through the inductor L in Figure 7.9

$$\int_S^T \mathbf{E} \cdot \mathbf{dl} = \int_S^T \mathbf{E}_L \cdot \mathbf{dl} + \int_S^T \mathbf{E}_s \cdot \mathbf{dl} = 0.$$

$$\int_S^T \mathbf{E}_L \cdot \mathbf{dl} = -\int_S^T \mathbf{E}_s \cdot \mathbf{dl}. \tag{7.87}$$

Substituting for $\int \mathbf{E}_L \cdot \mathbf{dl}$ from equation (7.87) into equation (7.86), we find that the line integral of $\mathbf{E}_s$ from S to T through the inductor L is given by

$$\int_S^T \mathbf{E}_s \cdot \mathbf{dl} = +L\dot{I}.$$

The quasi-static electric field $\mathbf{E}_s$ due to the surface and boundary charges is, to an excellent approximation, conservative. Since $\mathbf{E}_L$ is zero outside our idealized inductor L, for a path from S to T through the voltmeter (oscilloscope) measuring V_L, we have

$$\int_S^T \mathbf{E} \cdot \mathbf{dl} = \int_S^T \mathbf{E}_s \cdot \mathbf{dl} = +L\dot{I}. \tag{7.88}$$

It was shown in Section 7.3.1 that the reading of a voltmeter is equal to the line integral of the total electric field evaluated along the red lead to the voltmeter, through the voltmeter and then back along the black lead. Hence it follows from equation (7.88) that

$$V_L = +L\dot{I}. \tag{7.89}$$

Differentiating equation (7.79) with respect to time to determine $\dot{I}$, and then substituting in equation (7.89) we obtain

$$V_L = L\frac{dI}{dt} = -\omega L\, I_0 \sin(\omega t - \phi) = (\omega L)\, I_0 \cos\left(\omega t - \phi + \frac{\pi}{2}\right). \tag{7.90}$$

Comparing equations (7.90) and (7.79), we see that the voltage V_L across the inductor leads the current I through the inductor by $\pi/2$.

Since in our idealized *LCR* circuit the induction electric fields $\mathbf{E}_i$ and $\mathbf{E}_L$ are zero in the vicinity of the capacitor C in Figure 7.9, the only electric field in the vicinity of the capacitor C is the quasi-static electric field $\mathbf{E}_s$, which, in the region between the plates of the capacitor, is due almost entirely to the charges on the plates of the capacitor so that $\int \mathbf{E} \cdot d\mathbf{l}$ evaluated from U to W through the capacitor is equal to Q/C. Since $\mathbf{E}_s$ is, to a very good approximation, conservative the line integral $\int \mathbf{E} \cdot d\mathbf{l} = \int \mathbf{E}_s \cdot d\mathbf{l}$ from the point U in Figure 7.9 along the red lead to the voltmeter (oscilloscope) measuring V_C, through the voltmeter and back along the black lead to the point W (which we showed in Section 7.3.1 is equal to the voltage V_C measured by the voltmeter), is equal to $\int \mathbf{E}_s \cdot d\mathbf{l}$ evaluated from U to W across the capacitor, which is equal to Q/C, where Q is the charge on the capacitor. Hence, the reading of the voltmeter connected across the capacitor is

$$V_C = \frac{Q}{C}. \tag{7.91}$$

Since $I = \dot{Q}$, integrating equation (7.79) we have

$$Q = \int I\, dt = \left(\frac{I_0}{\omega}\right) \sin(\omega t - \phi).$$

Hence

$$V_C = \frac{Q}{C} = \left(\frac{I_0}{\omega C}\right) \sin(\omega t - \phi) = \left(\frac{I_0}{\omega C}\right) \cos\left(\omega t - \phi - \frac{\pi}{2}\right). \tag{7.92}$$

Comparing equations (7.92) and (7.79), we see that the current I through the capacitor C leads the voltage V_C across the capacitor by $\pi/2$.

Since the induction electric fields $\mathbf{E}_i$ and $\mathbf{E}_L$ due to the applied emf and the back emf in the inductor respectively are both zero in the vicinity of the resistor R, the line integral of the total electric field from the point Y in Figure 7.9 through the resistor R to the point Z reduces to

$$\int_Y^Z \mathbf{E} \cdot d\mathbf{l} = \int_Y^Z (\mathbf{E}_i + \mathbf{E}_L + \mathbf{E}_s) \cdot d\mathbf{l} = \int_Y^Z \mathbf{E}_s \cdot d\mathbf{l}.$$

We can also evaluate the line integral using the constitutive equation $E = J/\sigma = I/\sigma A$. If the resistances of the leads are negligible, we have

$$\int_Y^Z \mathbf{E} \cdot d\mathbf{l} = I \int \frac{dl}{\sigma A} = IR.$$

Hence

$$\int_Y^Z \mathbf{E} \cdot d\mathbf{l} = \int_Y^Z \mathbf{E}_s \cdot d\mathbf{l} = IR. \tag{7.93}$$

Since the $\mathbf{E}_i$ and $\mathbf{E}_L$ fields are zero in the vicinity of both the resistor R and the voltmeter (oscilloscope) measuring V_R and since the $\mathbf{E}_s$ field is conservative, it follows that the value of $\int \mathbf{E}_s \cdot d\mathbf{l}$ evaluated from the point Y in Figure 7.9 to the point Z through the resistor R is equal to the value of the line integral $\int \mathbf{E} \cdot d\mathbf{l}$ of the total electric field evaluated from the point Y in Figure 7.9 along the red lead to the voltmeter, then through the voltmeter measuring V_R and finally along the black lead to the point Z, which we showed in Section 7.3.1 was equal to the reading V_R of the voltmeter. Hence it follows from equation (7.93) that

$$V_R = IR.$$

Substituting for I from equation (7.79), we have

$$V_R = IR = I_0 R \cos(\omega t - \phi). \tag{7.94}$$

The voltage V_R measured across the resistor R is in phase with the current I through the LCR circuit.

Since in our idealized example, the induction electric fields $\mathbf{E}_i$ and $\mathbf{E}_L$ are zero outside the source of emf and the inductor L respectively, there is only the quasi-static electric field $\mathbf{E}_s$ along a continuous closed loop that goes in turn through the voltmeters measuring V_0, V_L, V_C and V_R in Figure 7.9, and avoids the source of emf, the inductor and the capacitor. Since the $\mathbf{E}_s$ field is conservative and since each voltmeter measures $\int \mathbf{E}_s \cdot d\mathbf{l}$, by going around the circuit via the voltmeters in the anticlockwise direction we have

$$V_L + V_C + V_R - V_0 = 0. \tag{7.95}$$

This is the alternative form of Kirchhoff's second law. It is straight forward for the reader to check that the sum of the harmonic variations on the right hand sides of equations (7.90), (7.92) and (7.94) is equal to $\varepsilon_0 \cos \omega t$.

7.6.6. *Discussion*

In this chapter, we have tried to give the reader an insight into the physical principles underlying AC theory by interpreting the roles of electric fields in simple AC circuits. On the way we developed Kirchhoff's first law for quasi-stationary conditions in Section 7.1. Whenever we needed it, we derived Kirchhoff's second law by equating $\int \mathbf{E} \cdot d\mathbf{l}$, the line integral of the total electric field taken around any loop to (a) the sum of the induced emfs in the loop and (b) to the sum of the products IR plus the potential differences across the capacitors, for example leading to equation (7.77). The reader can apply this method in the general case to any loop of an AC circuit to derive the appropriate form of Kirchhoff's second law. On our approach we treated the back emf $-L\dot{I}$ as a source of an induction electric field and hence as an emf. When solving problems it is sometimes mathematically more convenient to take the $-L\dot{I}$ term to the other side, for example in equation (7.77), and treat it as an impedance rather than an emf. In the complex number method of solving AC problems, an inductor of inductance L is treated as a complex impedance $j\omega L$ and capacitor is treated as a complex impedance $1/j\omega C$ where $j = (-1)^{1/2}$, f is the frequency and $\omega = 2\pi f$ is the angular frequency. Kirchhoff's second law is then rewritten in the form

$$\Sigma\varepsilon = \Sigma IZ \tag{7.96}$$

where $\Sigma\varepsilon$ is the sum of the **applied** emfs in the loop and ΣIZ is the sum of the products of current and complex impedance in the loop. Applying equation (7.96) to the LCR circuit shown in Figure 7.8 we have

$$\varepsilon = I(j\omega L) + \frac{I}{j\omega C} + IR = IZ$$

where

$$Z = \left(j\omega L + \frac{1}{j\omega C}\right) + R$$

is the complex impedance. The current in the LCR circuit is then given by the real part of

$$I = \frac{\varepsilon}{Z} = \frac{\varepsilon_0 \exp\,(j\omega t)}{R + j(\omega L - 1/\omega C)} = \frac{\varepsilon_0 \exp\,\{j(\omega t - \phi)\}}{[R^2 + (\omega L - 1/\omega C)^2]^{1/2}}$$

where

$$\tan\phi = \frac{(\omega L - 1/\omega C)}{R}\,.$$

This is in agreement with equations (7.79) and (7.81).

7.7. Motional (dynamo) induced emf

7.7.1. *Introduction*

In the example of the transformer induced emf we discussed in Sections 7.2 and 7.5, the varying current in the primary coil of the transformer gives rise to an induction electric field in the space outside the primary coil. It is this induction electric field that gives rise to a conduction current flow in the secondary of the transformer. There is no direct electrical contact between the primary and secondary coils of the transformer in Figure 7.1. There is another type of induced emf, which was mentioned in Section 1.6 of Chapter 1 and which we called a motional (or dynamo) induced emf. A motional (or dynamo) induced emf is produced when part or all of a circuit moves in a magnetic field. If the magnetic field is steady $\dot{\mathbf{B}}$ is zero and, according to

Maxwell's equations, there is no induction electric field. A simple example of a motional emf is given in the next section. Motional (or dynamo) induced emfs are extremely important in the generation of electricity.

7.7.2. *A simple example of a motional induced emf*

Consider the isolated conductor *RS* that is moving with a velocity $v \ll c$ parallel to the x axis of the laboratory system Σ, as shown in Figure 7.10(a). There is a uniform magnetic field $\mathbf{B}$ in the $-z$ direction, that is away from the reader in Figure 7.10(a). The magnetic field $\mathbf{B}$ does not vary with time so that there is no induction electric field in the laboratory system Σ.

The conduction electrons and positive ions inside the conductor *RS* move with the velocity $\mathbf{v}$ of the moving conductor *RS* in the $+x$ direction in Figure 7.10(a). We shall start with a very simplified model. A charge of magnitude $+q$ that is moving with the velocity $\mathbf{v}$ of the conductor experiences a magnetic force $q\mathbf{v} \times \mathbf{B}$ in the $+y$ direction as shown in Figure 7.10(a). The positive ions are prevented from moving away from their lattice positions by the cohesive forces in the moving conductor, but the conduction electrons are fairly free to move. The magnetic force on a conduction electron of charge $q = -e$ is equal to $-e\mathbf{v} \times \mathbf{B}$, which is in the $-y$ direction, that is towards *R* in Figure 7.10(a). Under the influence of this magnetic force, conduction electrons will drift in the $-y$ direction until charge distributions are built up on the surfaces of the isolated moving conductor *RS*, that are of such a magnitude that they give an electric field $\mathbf{E}_s$ in the $-y$ direction inside the moving conductor *RS*, which gives an electric force eE_s in the $+y$ direction on each of the moving conduction electrons. This force is equal in magnitude, but opposite in direction, to the magnetic force $-e\mathbf{v} \times \mathbf{B}$. The drift of the conduction electrons in the $-y$ direction in the isolated conductor *RS* stops when

$$E_s = vB. \tag{7.97}$$

The charge distributions on the top and bottom surfaces of the moving conductor *RS* also give rise to an electric field $\mathbf{E}_s$ that extends into the space outside the isolated moving conductor *RS* in Figure 7.10(a). When the moving conductor *RS* slides on the conducting rails in Figure 7.10(a) making electrical contact with the rails, this external electric field due to the charges at the ends of the moving conductor *RS* gives rise to conduction current flow in the stationary conducting rails in the clockwise direction in Figure 7.10(a). Surface and boundary charge distributions are built up in the way described in Appendix B, such that the resultant electric field inside the stationary conducting rails in Figure 7.10(a) is parallel to the rails. The current flow in Figure 7.10(a) reduces the charge distributions at the ends of the moving conductor *RS*, which in turn reduces the electric field inside the moving conductor *RS*. The magnetic force of magnitude evB in the $-y$ direction on each of the conduction electrons in the moving conductor then exceeds the reduced electric force of magnitude eE_s in the $+y$ direction, and the conduction electrons in

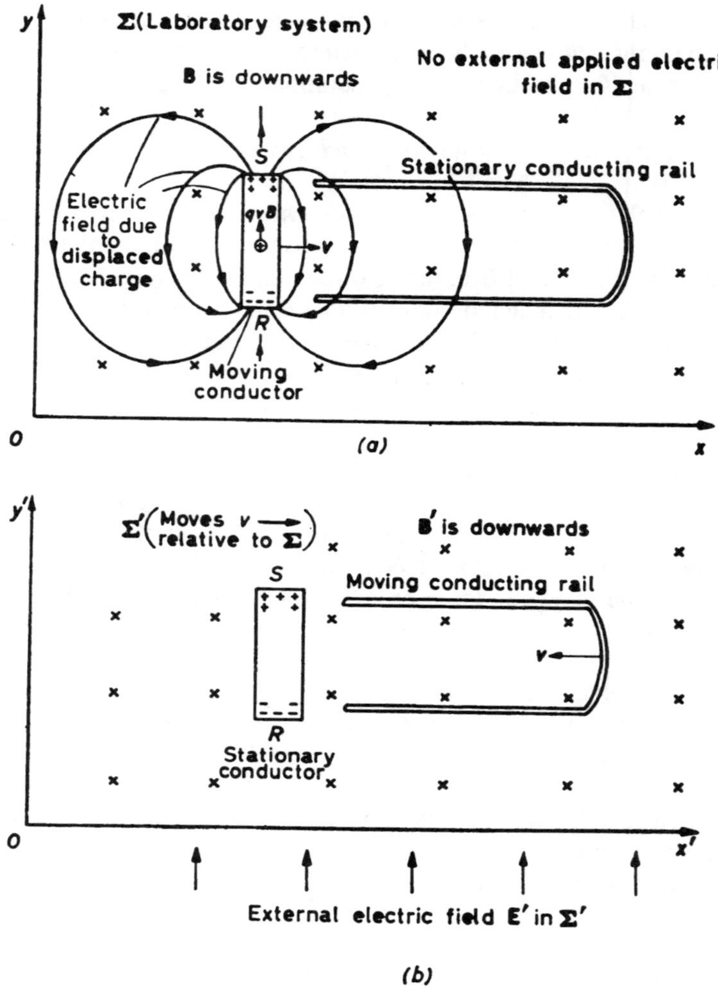

Figure 7.10. Motional emf. There is a magnetic force $q\mathbf{v} \times \mathbf{B}$ on the charges moving with the conductor RS which leads to a displacement of electric charge in the moving conductor RS which in turn gives rise to an electric field. (b) in Σ' the source of the external magnetic field also gives rise to an electric field which gives rise to the displacement of electric charge in the conductor RS, which is 'stationary' in Σ'.

the moving conductor RS in Figure 7.10(a) start to drift again in the $-y$ direction. This replenishes some of the charge removed from the ends of the moving conductor RS by the conduction current flow in the stationary rails. If the conductor RS is kept moving with uniform velocity $\mathbf{v}$ and if the changes in the electrical resistance of the rest of the circuit, associated with the movement of the conductor RS over the stationary conducting rails, is negligible, a state of dynamic equilibrium is reached and the current in the circuit is constant. Before the steady state is reached, the appropriate charge distributions are built up on the surfaces of the conductors and at the boundaries between conductors of different electrical conductivities, in the way described in Appendix B, such that the conduction current has the same numerical value in all parts of the circuit.

The emf ε of the complete circuit in Figure 7.10(a) will be defined as

$$\varepsilon = \int \left(\frac{\mathbf{F}}{q}\right) \cdot \mathbf{dl} \tag{7.98}$$

where $\mathbf{F}$ is the electromagnetic force, measured in the laboratory system Σ, on a test charge of magnitude q, that is at rest relative to the section $\mathbf{dl}$ of the circuit. When the section $\mathbf{dl}$ has a velocity $\mathbf{v}$ relative to the laboratory system Σ the test charge q must have the same velocity $\mathbf{v}$ relative to the laboratory system Σ. According to the Lorentz force

$$\mathbf{F} = q\mathbf{E} + q\mathbf{v} \times \mathbf{B} \tag{7.99}$$

where $\mathbf{E}$ and $\mathbf{B}$ are the local values of the resultant electric and magnetic fields in the laboratory system Σ. Since there is no external applied electric field in the laboratory system Σ, the only electric field is the $\mathbf{E}_s$ field due to the surface and boundary charge distributions on the conductors. After the state of dynamic equilibrium is reached, the surface and boundary charge distributions on the stationary conducting rails are constant and can be treated as electrostatic, in which case their contribution to $\oint \mathbf{E} \cdot \mathbf{dl}$, evaluated around the complete circuit, is zero. If the velocity of the moving conductor RS were 300 m s^{-1}, v/c would only be 10^{-6} and, to an excellent approximation, the electric field due to the charges on the moving conductor RS could be calculated using Coulomb's law, in which case their contribution to $\oint \mathbf{E} \cdot \mathbf{dl}$ would also be zero. Hence $\oint \mathbf{E} \cdot \mathbf{dl}$ evaluated around the complete circuit in Figure 7.10(a) would be zero. Hence using equations (7.98) and (7.99) we have

$$\varepsilon = \oint (\mathbf{E} + \mathbf{v} \times \mathbf{B}) \cdot \mathbf{dl} = \oint \mathbf{v} \times \mathbf{B} \cdot \mathbf{dl}.$$

Applying equations (A1.5) of Appendix A1.1, we have

$$\varepsilon = -\oint \mathbf{B} \cdot (\mathbf{v} \times \mathbf{dl}). \tag{7.100}$$

This is the general expression for the emf generated by a dynamo in a time independent magnetic field.

In the special case shown in Figure 7.10(a), when the moving conductor RS is moving over the stationary conducting rails, the only contribution to the integral in equation (7.100) is along the moving conductor RS, so that the magnitude of the integral $-\int \mathbf{B} \cdot (\mathbf{v} \times \mathbf{dl})$ is equal to Bvl, where l is the length of the conductor RS. The motional emf $\varepsilon = Bvl$ is equal to the magnetic flux cut by the moving conductor RS per second. In the special case of the contracting loop shown in Figure 7.10(a), the magnetic flux cut by the moving conductor *RS* per second is equal to the rate of decrease of the magnetic flux Φ through the loop so that for the special case shown in Figure 7.10(a)

$$\varepsilon = -\frac{d\Phi}{dt}. \tag{7.101}$$

The direction of current flow in the clockwise direction in Figure 7.10(a) is consistent with Lenz's law, since a current flow in this direction gives a magnetic field in the downwards direction in Figure 7.10, which tends to oppose the decrease of the magnetic flux through the loop in the downward direction, as the conductor *RS* moves across the conducting rails. In the general case of a circuit of arbitrary shape $\mathbf{v} \times d\mathbf{l}$ is the area swept out by the section $d\mathbf{l}$ of the moving circuit per second. Hence $\mathbf{B} \cdot (\mathbf{v} \times d\mathbf{l})$ is the rate at which the element of length $d\mathbf{l}$ of the moving circuit is cutting magnetic flux. In some special cases this is equal to rate at which the total magnetic flux through the circuit is varying due to the motion of the element $d\mathbf{l}$. If this is true for all the elements $d\mathbf{l}$, equation (7.100) reduces to equation (7.101). In other cases, equation (7.101) is inappropriate. For example, in the case of the Faraday disk generator the magnetic flux Φ through the circuit is constant. In this case, equation (7.100) must be used and interpreted as a magnetic flux cutting rule. Though equations (7.20) and (7.101) appear to be the same and Lenz's law can be applied, the agreement is largely fortuitous, since the underlying physical principles are different in the laboratory system Σ. The transformer induced emf given by equation (7.20) is due to an induction electric field inside the wire making up the stationary secondary coil in Figure 7.1, that is produced by the varying current in the primary coil. The motional induced emf, given by equation (7.100), is due to the motion of part of the circuit in Figure 7.10(a) through a time independent magnetic field. It is always best to interpret equation (7.100) as a magnetic flux cutting rule associated with the motion of a conductor in a time independent magnetic field. If a conductor is moving through a time dependent magnetic field there is both a motional and a transformer emf. It is straightforward for the reader to apply equation (7.100) to the case of a coil rotating in a uniform magnetic field and then to go on to interpret the action of a dynamo.

7.7.3. *The role of magnetic forces in motional induced emf*

In our simplified discussion of motional induced emf in the last section, we only considered the magnetic force $-e\mathbf{v} \times \mathbf{B}$ on a conduction electron, that is moving with the velocity $\mathbf{v}$ of the conductor *RS* in Figure 7.10(a). This analysis could leave the reader with the impression that it is the magnetic force $-e\mathbf{v} \times \mathbf{B}$ on each conduction electron that drives the conduction current around the circuit and that supplies the energy for the Joule heat generated in the circuit in Figure 7.10(a). Since the magnetic force on a charge moving in a magnetic field is always perpendicular to the velocity $\mathbf{v}$ of the charge the rate $\mathbf{F}_{mag} \cdot \mathbf{v}$ at which the magnetic force is doing work on the moving charge is always zero. What a magnetic field does is to change the direction of motion of the moving charge without changing either its speed, or the magnitudes of its momentum and kinetic energy.

To illustrate the role of magnetic forces in motional induced emf, we shall follow the approach of Mosca [3]. We shall use a simplified model according to which free conduction electrons move inside a uniform potential well inside the moving conductor *RS*.

When the state of dynamic equilibrium is reached in Figure 7.10(a), when the moving conductor *RS* is moving over the conducting rails the conduction electrons drift with a drift velocity, now denoted by $\mathbf{v}_d$, in the $-y$ direction. Due to the motion of the conductor *RS* with velocity $\mathbf{v}$ in the laboratory system Σ in Figure 7.10(a), a conduction electron of mass m, that is at rest inside the conductor *RS*, has a momentum of mv in the $+x$ direction. (In practice this momentum is added to the momentum the conduction electron has due to its random velocity inside the conductor *RS*). The applied magnetic field **B** tries to change the direction of the momentum of the conduction electron. In the absence of collisions and of the Hall potential difference, a conduction electron of total speed u in the laboratory system Σ would go around in a circle of radius $\rho = mu/eB$. For a typical conduction electron $u = c/200$, $e/m = 1.75 \times 10^{11}$ C kg^{-1} and $B = 0.01$T, so that $\rho = 0.85$ mm. At a temperature of 300K the mean free path of a conduction electron in copper is $\lambda = 3 \times 10^{-5}$ mm so that $\lambda/\rho = 3.4 \times 10^{-5}$. Hence collisions prevent the conduction electrons from going around in complete circles.

To simplify the discussion of physical principles, we shall represent the average behaviour of the conduction electrons inside the moving conductor *RS* in Figure 7.10(a) by assuming that when the conductor *RS* is moving over the conducting rails, then on average the conduction electrons have just the velocity $\mathbf{v}$ of the conductor *RS* in the $+x$ direction plus a drift velocity $\mathbf{v}_d$ in the $-y$ direction from *S* to *R* in Figure 7.10(a). A reader familiar with special relativity can see the appropriateness of this approximation by transforming to the inertial reference frame Σ', that is moving with uniform velocity $\mathbf{v}$ relative to the laboratory system Σ, as shown in Figure 7.10(b). The conductor *RS* is at rest in Σ'. It follows from the field transformation, equation (10.80) of Chapter 10 that there is an electric field in the $+y'$ direction of Σ' of magnitude $vB/(1 - v^2/c^2)^{1/2}$. This electric field acts on the conduction electrons in the conductor *RS*, which is at rest in Σ', and a conduction current flows in the stationary conductor *RS* under the influence of this electric field in the direction form *R* to *S*. In Σ' the magnetic deflection of the conduction electrons in the $-x'$ direction in the magnetic field in Σ' is compensated by the Hall electric field in the way described in Section 1.4.3 of Chapter 1, such that, on average, the conduction electrons drift in the $-y'$ direction in Σ' with velocity $\mathbf{v}_d$. Since all velocities are much less than c we can use the Galilean velocity transformations to transform to the laboratory system Σ. Hence in the laboratory system Σ, shown in Figure 7.10(a), each of the conduction electrons inside the moving conductor *RS* has, on average, a velocity $\mathbf{v}_d$ in the $-y$ direction plus a velocity $\mathbf{v}$ in the $+x$ direction giving a total velocity of $(\mathbf{v} + \mathbf{v}_d)$ in the laboratory system Σ, when the conductor RS is moving over the conducting rails. In practice the conduction electrons have speeds of the order of $c/200$ in all directions. It is their average velocity that is equal to $(\mathbf{v} + \mathbf{v}_d)$.

Consider a conduction electron that starts with the velocity $\mathbf{v}$ of the moving conductor RS in Figure 7.10(a). This conduction electron has a momentum $m\mathbf{v}$ in the $+x$ direction of the laboratory system Σ as shown in Figure 7.11. In the laboratory system Σ, the applied magnetic field changes the direction of this momentum a little before the conduction electron undergoes a collision, as shown in Figure 7.11. On our simplified model, after the collision on average, the electron moves again in the $+x$ direction with velocity $\mathbf{v}$ in Figure 7.11. It is again deflected by the applied magnetic field until it collides again and so on, as shown in Figure 7.11. We can see that the effect of the applied magnetic field is to continually turn the momentum of such a conduction electron towards the $-y$ direction in Figure 7.11, such that the conduction electron drifts in the $-y$ direction in the laboratory system Σ, leading to current flow in the way described in Section 7.7.2.

When a conduction electron that starts from S reaches the end of the moving conductor labelled R in Figure 7.10(a), it has gained a potential energy equal to $-e\phi$ compared to a conduction electron that is at the end labelled S where, if $v \ll c$, ϕ is equal to $-\int \mathbf{E}_s \cdot d\mathbf{l}$ evaluated from S to R at a fixed instant of time. Notice that ϕ is negative giving a positive value for $-e\phi$. In a vacuum an electron leaving R would reach S with a kinetic energy equal to $-e\phi$. If a

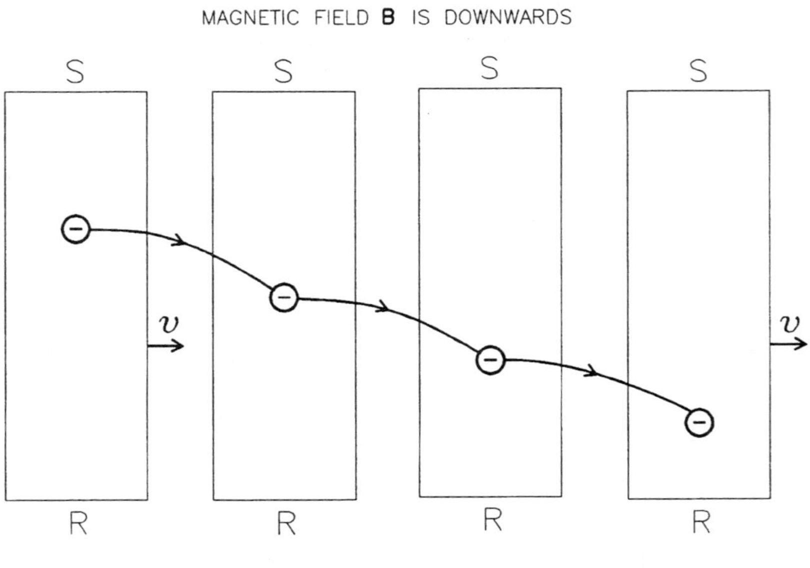

Figure 7.11. The successive positions of the moving conducor RS in Figure 7.10(a), at the instants when the conduction electron undergoes successive collisions.

conduction electron drifts all the way from R to S through the external circuit in Figure 7.10(a), it will still gain a total kinetic energy equal to $-e\phi$, but it will loose this kinetic energy continuously in successive collisions leading to Joule heating, and the conduction electron will finally reach S with only the average drift velocity of about 10^{-4} m s^{-1}.

To discuss the average forces acting on a moving conduction electron, we shall again assume that the conduction electron moves along with the velocity $\mathbf{v}$ of the moving conductor in the $+x$ direction in Figure 7.12 plus its drift velocity $\mathbf{v}_d$ in the $-y$ direction so that its total average velocity is $(\mathbf{v} + \mathbf{v}_d)$. The average magnetic force on such a conduction electron due to the applied magnetic field $\mathbf{B}$ is

$$\mathbf{F} = -e(\mathbf{v} + \mathbf{v}_d) \times \mathbf{B} = \mathbf{F}_1 + \mathbf{F}_2$$

where

$$\mathbf{F}_1 = -e\mathbf{v} \times \mathbf{B}; \qquad \mathbf{F}_2 = -e\mathbf{v}_d \times \mathbf{B}.$$

The contribution $F_1 = -evB$ is in the $-y$ direction in Figure 7.12 and the contribution $F_2 = -ev_dB$ is in the $-x$ direction. Since the rate at which the magnetic field does work is zero, we have

$$\mathbf{F} \cdot (\mathbf{v} + \mathbf{v}_d) = \mathbf{F}_1 \cdot \mathbf{v} + \mathbf{F}_1 \cdot \mathbf{v}_d + \mathbf{F}_2 \cdot \mathbf{v} + \mathbf{F}_2 \cdot \mathbf{v}_d = 0.$$

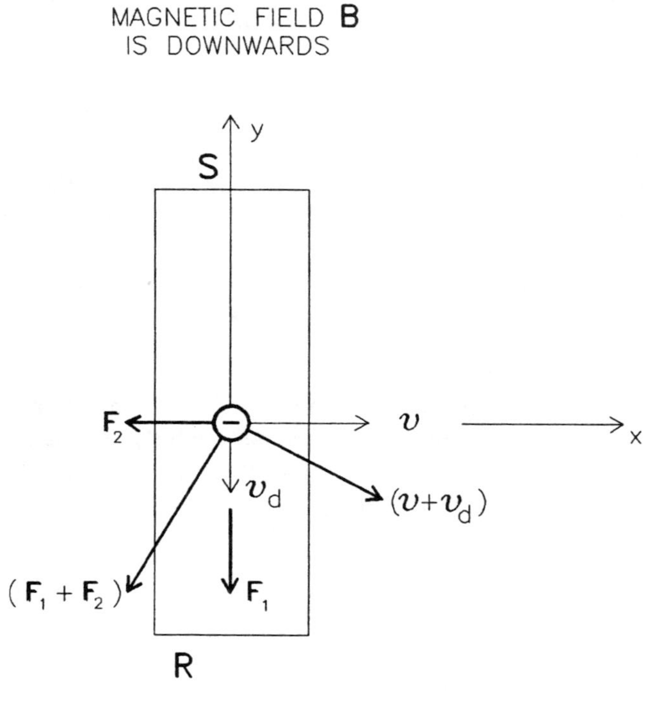

Figure 7.12. The magnetic forces on a conduction electron inside the moving conductor *RS*.

We can see from Figure 7.12 that $\mathbf{v}$ is perpendicular to $\mathbf{F}_1$ so that $\mathbf{F}_1 \cdot \mathbf{v}$ is zero. Since $\mathbf{v}_d$ is perpendicular to $\mathbf{F}_2$ then $\mathbf{F}_2 \cdot \mathbf{v}_d$ is zero. Hence

$$\mathbf{F} \cdot (\mathbf{v} + \mathbf{v}_d) = \mathbf{F}_1 \cdot \mathbf{v}_d + \mathbf{F}_2 \cdot \mathbf{v} = 0. \tag{7.102}$$

Since $\mathbf{F}_1$ and $\mathbf{v}_d$ are both in the $-y$ direction in Figure 7.12, $\mathbf{F}_1 \cdot \mathbf{v}_d$ is a positive and equal to $+evv_dB$, which means that the component $\mathbf{F}_1$ of the magnetic force does work on a conduction electron when it is moved from S to R inside the moving conductor RS against the electric force $-e\mathbf{E}_s$ in the $+y$ direction due to the electric field $\mathbf{E}_s$ which is in the $-y$ direction and is due to the surface and boundary charges on the moving conductor RS and on the conducting rails in Figure 7.10(a). This increases the potential energy $-e\phi$ of the conduction electron. Since $\mathbf{v}$ and $\mathbf{F}_2$ are in opposite directions in Figure 7.12, $\mathbf{F}_2 \cdot \mathbf{v}$ is negative and equal to $-evv_dB$. The negative sign means that the effect of the component $\mathbf{F}_2$ of the magnetic force is to try to reduce the component of the velocity of the conduction electron in the $+x$ direction in Figure 7.12. If there are n conduction electrons per cubic metre and if l is the length and A is the area of cross section of the moving conductor RS in Figure 7.10(a) then the total magnetic force on all the conduction electrons due to the component $\mathbf{F}_2$ is numerically equal to $(nAl)F_2 = (nAl)ev_dB = BIl$ in the $-x$ direction, where $I = nAev_d$ is the current. This force BIl on the conduction electrons is transferred to the moving conductor via the Hall electric field in the way outlined in Section 1.4.3 of Chapter 1 and will reduce the velocity $\mathbf{v}$ of the moving conductor RS unless an equal external force BIl is applied to the moving conductor in the $+x$ direction.

Summarizing, the net result is that the magnetic field in Figure 7.10(a) changes the direction of the momentum of each conduction electron, that is inside and is moving with the moving conductor, towards the $-y$ direction in Figure 7.10(a). This would lead to a loss of momentum in the $+x$ direction unless an external force BIl is applied to the conductor RS to keep the conductor RS moving with uniform velocity $\mathbf{v}$. The rate at which electrical energy is dissipated as Joule heat in the circuit is equal to the product εI of the emf ε and the current I. Since $\varepsilon = Bvl$, it follows that the rate of production of Joule heat εI is equal to $BvlI$, which is equal to the rate at which the external force is doing work on the moving conductor RS in Figure 7.10(a). We have shown that by trying to change the direction of the momentum of the conduction electrons in the moving rod RS, even though the magnetic field itself does no work, the magnetic field helps to convert the mechanical work done on the moving rod into electrical energy.

References

1. Lorrain, P. and Corson, D. R., *Electromagnetic Fields and Waves*, 2nd Edition (W.H. Freeman, San Francisco, 1970), pp. 315–317.
2. Romer, R. H., *Amer. Journ. Phys.*, Vol. 50, p. 1089 (1982).
3. Mosca, E. P., *Amer. Jour. Phys.*, Vol. 42, p. 295 (1974).

CHAPTER 8

Forces, energy and electromagnetic momentum

8.1. Introduction

In this chapter, we shall go on to discuss the concepts of the energy, the linear momentum and the angular momentum of the electromagnetic field. So far in this book, we have generally tried to avoid methods based on the concept of energy, except, for example, when we discussed electrostatic energy and the definition of the electrostatic scalar potential ϕ in Section 1.2.10 of Chapter 1. We have been able to do most of what we have done so far using the Lorentz force law to relate the fields **E** and **B** to experiments. The Lorentz force law will again be the starting point for all of our developments in this chapter. This is similar to the position in Newtonian mechanics, where it is Newton's laws of motion that are the starting point for the development of the concepts of energy, linear momentum and angular momentum and the corresponding conservation laws. To quote French [1]:

> It is an interesting historical sidelight that in pursuing the subject of energy we are temporarily parting company with Newton, although not with what we may properly call Newtonian mechanics. In the whole of the *Principia*, with its awe-inspiring elucidation of the dynamics of the universe, the concept of energy is never once used or even referred to! For Newton, $F = ma$ was enough. But we shall see how the energy concept, although rooted in $F = ma$, has its own special contributions to make.

For example, in Newtonian mechanics it is Newton's laws of motion that are used to determine the expression for the kinetic energy of a particle. The importance of the concept of energy is that energy is conserved. The law of conservation of energy can often be used to solve problems. For example, if we let a particle fall from a height h in the Earth's gravitational field, we can determine the velocity with which it hits the ground, either by applying the law of conservation of energy by equating the gain in the kinetic energy of the particle to its loss of potential energy, or by applying Newton's law of universal gravitation and Newton's second law of motion to determine the acceleration of the particle, and then using the appropriate kinematic relation to determine the velocity with which the particle hits the ground. It is important to realize that the two approaches are alternatives. In this chapter, what we shall be doing is developing alternatives to the direct application of the Lorentz force law.

Since we shall be going on to discuss the law of conservation of linear momentum for a system of moving classical point charges in empty space, it will be instructive to remind ourselves of some of the assumptions made when the law of conservation of linear momentum of Newtonian mechanics is derived from Newton's laws of motion. Consider the collision of two particles labelled 1 and 2 of linear momenta p_1 and p_2 respectively, which are moving in the same straight line before the collision. Let the particles move in this same straight line after the collision with momenta p_1' and p_2' respectively. If, during the collision, the force on particle 1 due to particle 2 is $\mathbf{F}_1$ and the force on particle 2 due to particle 1 is $\mathbf{F}_2$, then, since according to Newton's third law of motion action and reaction are equal and opposite

$$\mathbf{F}_1 = -\mathbf{F}_2$$

Using Newton's second law of motion, we have

$$\frac{\mathrm{d}p_1}{\mathrm{d}t} = -\frac{\mathrm{d}p_2}{\mathrm{d}t}.$$

Integrating over the time of the collision, we find that

$$\int \mathrm{d}p_1 = -\int \mathrm{d}p_2$$

giving

$$p_1' - p_1 = -(p_2' - p_2).$$

Rearranging,

$$p_1' + p_2' = p_1 + p_2.$$

This is an example of the law of conservation of linear momentum. Notice that we used both Newton's second and third laws of motion in its derivation, showing that the law of conservation of linear momentum is an alternative to using Newton's laws of motion directly. We also assumed that the forces $\mathbf{F}_1$ and $\mathbf{F}_2$ were central forces when we assumed that $\mathbf{F}_2 = -\mathbf{F}_1$, which meant that, if the particles were spatially separated, the forces $\mathbf{F}_1$ and $\mathbf{F}_2$ both acted along the line joining the particles. This is true of Newton's law of universal gravitation, but it is not true for the forces between moving charges, as will become clear in Section 8.8.2, where we shall show that Newton's third law is not valid for the forces between moving charges. Whereas it is assumed in Newton's law of universal gravitation that the gravitational force between two masses acts instantaneously, the changes in the electromagnetic forces between two moving charges take time to propagate from one charge to the other at the speed of light c. Hence it will not be surprising to find that, when we consider what happens at one instant of time without allowing for retardation effects, and while there are changes in the electromagnetic interaction on their way between the moving charges we shall have to introduce some of the properties of the electromagnetic field, as for example, when we come to consider the law of conservation of momentum for spatially separated moving charges in Section 8.8.2. We shall show in Section 10.8 of Chapter 10 that for spatially separated charged particles, events that are simultaneous in one inertial reference frame are not simultaneous in any inertial reference frame moving relative to the first, so that we shall have to be particularly careful when we come to discuss some of the properties of a system of spatially separated charges in empty space at a fixed instant of time, such as the conservation laws of energy, momentum and angular momentum. If readers check their mechanics text books, they will also find that it is Newton's laws of motion and the assumption of central forces that are used in the derivation of the law of conservation of angular momentum in Newton mechanics, which is often simpler to apply in practice than using Newton's laws of motion directly.

In Section 8.2 we shall start by showing how the electromagnetic force on a steady charge and current distribution, that is made up of moving classical point charges, can be determined using Maxwell's stress tensor, which we shall derive from the Lorentz force law. We shall then go on to consider varying charge and current distributions leading up to discussions of the energy, momentum and angular momentum of the electromagnetic field and the conservation laws of energy, momentum and angular momentum. We shall always start from the Lorentz force law, or relations derived from it. We shall assume throughout this chapter that the relative permittivity ε_r and the relative permeability μ_r are both equal to unity everywhere.

8.2. The Maxwell stress tensor

8.2.1. *Introduction*

Consider the finite charge and current distribution shown in Figure 8.1, which is due to a system of moving classical point charges in otherwise empty space. There are also external charge and current distributions, not shown in Figure 8.1, that are spatially separated from the charge and current distribution shown in Figure 8.1. The total magnetic field **B** is due partly to the magnetic field due to any external current distributions and due partly to the magnetic field due to the current distribution shown in Figure 8.1. There is also an electric field **E** that is also due partly to external contributions and due partly to the charge distribution shown in Figure 8.1. If the charge and current densities at a point inside the charge and current distribution shown in Figure 8.1, are ρ and **J** respectively, then, according to the Lorentz force

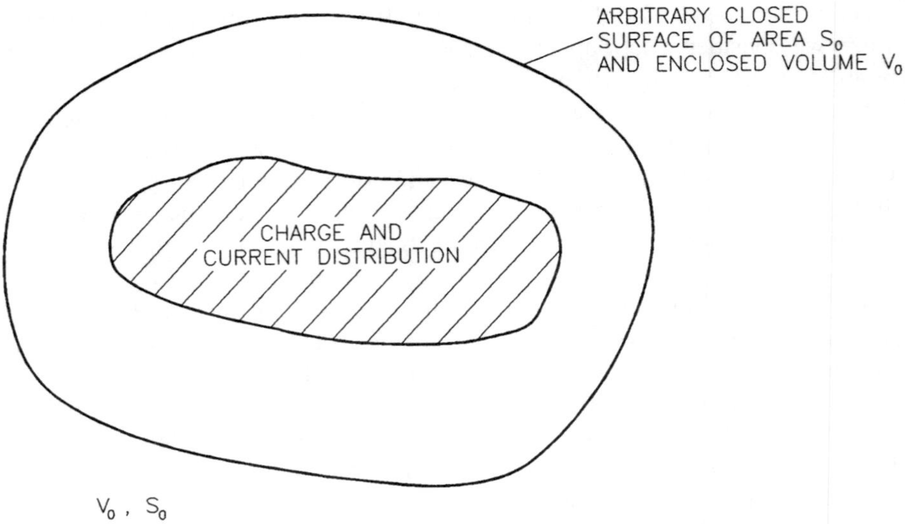

Figure 8.1. A charge and current distribution surrounded by an arbitrary closed surface of area S_0 and enclosed volume V_0. There are other charge and current distributions outside V_0.

law, equation (1.1), the electric and magnetic forces **per unit volume** on the charge and current distribution are

$$\mathbf{f}_{\text{elec}} = \rho\mathbf{E}, \tag{8.1}$$

$$\mathbf{f}_{\text{mag}} = \mathbf{J} \times \mathbf{B}. \tag{8.2}$$

In the case of a plasma there can be a hydrostatic force per unit volume of magnitude

$$\mathbf{f}_{\text{hyd}} = -\nabla p \tag{8.3}$$

where p is the fluid pressure. There may also be a gravitational force per unit volume given by

$$\mathbf{f}_{\text{grav}} = -\rho_m \nabla \phi_g. \tag{8.4}$$

where ρ_m is the mass density and ϕ_g is the gravitational potential. The gravitational force given by equation (8.4) is important, for example, in stellar interiors, but it can be neglected in laboratory experiments on classical electromagnetism.

8.2.2. *The magnetic force on a current distribution*

Consider the charge and current distribution shown in Figure 8.1. The total magnetic force $\mathbf{F}_{\text{mag}}$ on all the currents inside a volume V_0, which may be completely inside the current distribution, partly inside and partly outside the current distribution or completely outside the current distribution, as shown in Figure 8.1, is obtained by integrating equation (8.2) over the volume V_0

to give

$$\mathbf{F}_{\text{mag}} = \int \mathbf{f}_{\text{mag}}\, \mathrm{d}V = \int \mathbf{J} \times \mathbf{B}\, \mathrm{d}V = -\int \mathbf{B} \times \mathbf{J}\, \mathrm{d}V. \tag{8.5}$$

We shall assume in this section that the charge and current distributions both inside and outside the volume V_0 are steady, so that $\dot{\mathbf{B}}$ and $\dot{\mathbf{E}}$ are both zero, so that, from Maxwell's equations we have

$$\boldsymbol{\nabla} \times \mathbf{B} = \mu_0 \mathbf{J}. \tag{8.6}$$

We shall now use equation (8.6) to eliminate $\mathbf{J}$ from equation (8.5) to obtain

$$\mathbf{F}_{\text{mag}} = -\frac{1}{\mu_0} \int \mathbf{B} \times (\boldsymbol{\nabla} \times \mathbf{B})\, \mathrm{d}V. \tag{8.7}$$

To show that for any vector $\mathbf{A}$

$$\mathbf{A} \times (\boldsymbol{\nabla} \times \mathbf{A}) = \boldsymbol{\nabla}\left(\frac{A^2}{2}\right) - (\mathbf{A} \cdot \boldsymbol{\nabla})\mathbf{A} \tag{8.8}$$

consider the x component of the right hand side, namely

$$\begin{aligned}
&\frac{\partial}{\partial x}\left(\frac{1}{2}A_x^2 + \frac{1}{2}A_y^2 + \frac{1}{2}A_z^2\right) - \left(A_x\frac{\partial}{\partial x} + A_y\frac{\partial}{\partial y} + A_z\frac{\partial}{\partial z}\right)A_x \\
&= A_x\frac{\partial A_x}{\partial x} + A_y\frac{\partial A_y}{\partial x} + A_z\frac{\partial A_z}{\partial x} - A_x\frac{\partial A_x}{\partial x} - A_y\frac{\partial A_x}{\partial y} - A_z\frac{\partial A_x}{\partial z} \\
&= A_y\left(\frac{\partial A_y}{\partial x} - \frac{\partial A_x}{\partial y}\right) - A_z\left(\frac{\partial A_x}{\partial z} - \frac{\partial A_z}{\partial x}\right) \\
&= A_y(\boldsymbol{\nabla} \times \mathbf{A})_z - A_z(\boldsymbol{\nabla} \times \mathbf{A})_y = \{\mathbf{A} \times (\boldsymbol{\nabla} \times \mathbf{A})\}_x
\end{aligned}$$

which is the x component of the left hand side of equation (8.8). Since similar results hold for the y and z components of equation (8.8), this proves equation (8.8).

Putting $\mathbf{A} = \mathbf{B}$ in equation (8.8), where $\mathbf{B}$ is now the magnetic field, and then substituting in equation (8.7) we find that

$$\mathbf{F}_{\text{mag}} = \int \boldsymbol{\nabla}\left(-\frac{B^2}{2\mu_0}\right) \mathrm{d}V + \frac{1}{\mu_0}\int (\mathbf{B} \cdot \boldsymbol{\nabla})\mathbf{B}\, \mathrm{d}V. \tag{8.9}$$

According to equation (A1.32) of Appendix A1.7 for any scalar ϕ

$$\int \boldsymbol{\nabla}\phi\, \mathrm{d}V = \int \phi\, \mathrm{d}\mathbf{S}. \tag{8.10}$$

Putting ϕ equal to $(-B^2/2\mu_0)$ we have

$$\int \boldsymbol{\nabla}\left(-\frac{B^2}{2\mu_0}\right) \mathrm{d}V = \int \left(-\frac{B^2}{2\mu_0}\right) \mathrm{d}\mathbf{S}. \tag{8.11}$$

According to equation (A1.31) of Appendix A1.7 for any vector $\mathbf{A}$ and scalar ϕ

$$\int \mathbf{A} \cdot \nabla\phi \, \mathrm{d}V = \int \phi(\mathbf{A} \cdot \mathrm{d}\mathbf{S}) - \int \phi(\nabla \cdot \mathbf{A}) \, \mathrm{d}V.$$

Treating B_x as a scalar and putting $\phi = B_x$ we obtain

$$\int \mathbf{A} \cdot \nabla B_x \, \mathrm{d}V = \int B_x(\mathbf{A} \cdot \mathrm{d}\mathbf{S}) - \int B_x(\nabla \cdot \mathbf{A}) \, \mathrm{d}V$$

with similar results for B_y and B_z. Combining the results, we have

$$\int (\mathbf{A} \cdot \nabla)\mathbf{B} \, \mathrm{d}V = \int \mathbf{B}(\mathbf{A} \cdot \mathrm{d}\mathbf{S}) - \int \mathbf{B}(\nabla \cdot \mathbf{A}) \, \mathrm{d}V \tag{8.12}$$

where so far both **A** and **B** are arbitrary vectors. We shall now put $\mathbf{A} = \mathbf{B}$ in equation (8.12) and assume that **B** is the magnetic field, in which case $\nabla \cdot \mathbf{B}$ is zero. Equation (8.12) then gives

$$\int (\mathbf{B} \cdot \nabla)\mathbf{B} \, \mathrm{d}V = \int \mathbf{B}(\mathbf{B} \cdot \mathrm{d}\mathbf{S}). \tag{8.13}$$

Using equations (8.11) and (8.13) to substitute for the integrals on the right hand side of equation (8.9) we finally obtain

$$\mathbf{F}_{\text{mag}} = \int \left(-\frac{B^2}{2\mu_0}\right) \mathrm{d}\mathbf{S} + \frac{1}{\mu_0}\int \mathbf{B}(\mathbf{B} \cdot \mathrm{d}\mathbf{S}). \tag{8.14}$$

Equation (8.14) gives the total magnetic force on the **steady** current distribution inside the volume V_0 in Figure 8.1. We have only proved equation (8.14) when the integration is over all of the closed surface S_0. We have not derived it in a differential form that can be applied to individual elements of area on the surface S_0. Equation (8.14) is only valid for steady current distributions. The general case of varying charge and current distributions will be considered in Section 8.7. For steady currents (magnetostatics), the forces between the currents satisfy Newton's third law so that the internal magnetic forces inside V_0 in Figure 8.1 add up to zero, and the force given by equation (8.14) is equal to the force due to current distributions outside V_0.

To illustrate the role of the first integral on the right hand side of equation (8.14), we shall now assume that the charge and current distribution is a plasma. In this case, there is a hydrostatic force on the plasma inside a volume V_0 that is completely inside the plasma. Integrating equation (8.3) and then applying equation (8.10) we obtain

$$\mathbf{F}_{\text{hyd}} = \int (-\nabla p) \, \mathrm{d}V = \int (-p) \, \mathrm{d}\mathbf{S}. \tag{8.15}$$

Comparing equation (8.15) with the first integral on the right had side of equation (8.14), we see that we would get the correct value for the first integral, **after integrating** over the surface S_0, if we used the mechanical analogy of a "magnetic pressure" $(B^2/2\mu_0)$ on the surface S_0 acting in the direction of $-\mathrm{d}\mathbf{S}$, that is inwards into the volume V_0.

To interpret the second integral on the right hand side of equation (8.14), let $\hat{\mathbf{b}}$ be a unit vector in the direction of the magnetic field in Figure 8.2. The second integral can then be rewritten in the form

$$\frac{1}{\mu_0}\int \mathbf{B}(\mathbf{B}\cdot \mathrm{d}\mathbf{S}) = \int \hat{\mathbf{b}}\frac{B^2}{\mu_0}\cos\theta\, \mathrm{d}S \tag{8.16}$$

where θ is the angle between $\mathbf{B}$ and $\mathrm{d}\mathbf{S}$ and $(\mathrm{d}S)\cos\theta$ is the magnitude of the projection of the area $\mathrm{d}\mathbf{S}$ on to a plane perpendicular to the direction of the magnetic field line at the position of $\mathrm{d}\mathbf{S}$. It follows from equation (8.16) that, for the conditions shown in Figure 8.2, the contribution to the second integral on the right hand side of equation (8.14), that is associated with the area $\mathrm{d}\mathbf{S}$, is in the direction of the magnetic field $\mathbf{B}$ at the position of $\mathrm{d}\mathbf{S}$. Now consider the element of area $\mathrm{d}\mathbf{S}'$ that is at the point where the magnetic field line, that passes through $\mathrm{d}\mathbf{S}$, enters the volume V_0 in Figure 8.2. In this case, $\mathbf{B}\cdot\mathrm{d}\mathbf{S}'$ is negative in equation (8.16), so that the contribution associated with the area $\mathrm{d}\mathbf{S}'$ to the second integral on the right hand side of equation

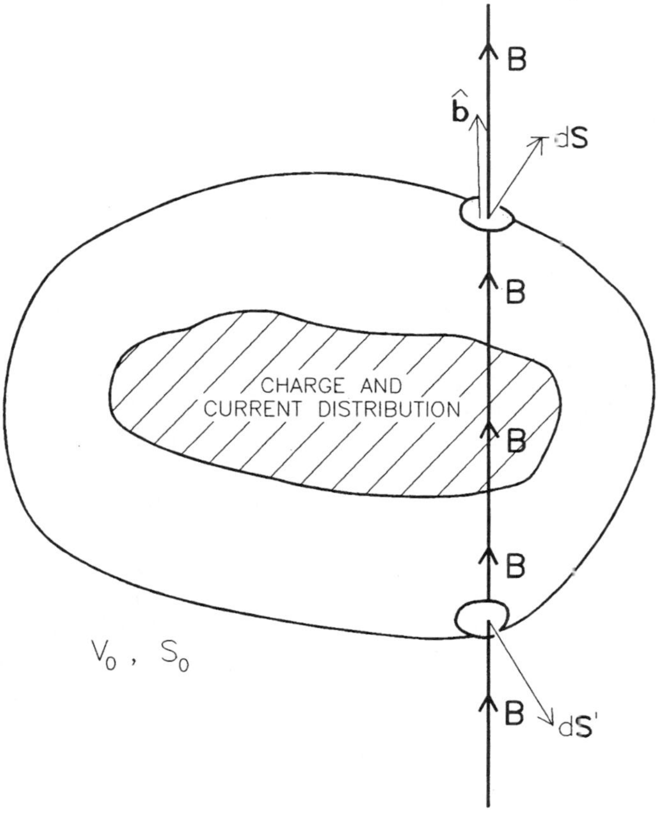

Figure 8.2. The magnetic "tension" along a magnetic field line.

(8.14) is in the direction opposite to the direction of the magnetic field at the position of $d\mathbf{S}'$, that is outwards from the volume V_0. Thus the contributions associated with the elements of area $d\mathbf{S}$ and $d\mathbf{S}'$ to the second integral on the right hand side of equation (8.14) are in opposite directions and are the same as if the magnetic field line were in a state of "tension" equal to (B^2/μ_0) per unit area perpendicular to the magnetic field line. It is straightforward for the reader to consider examples of magnetic field lines to show that the "tension" of (B^2/μ_0) per unit area perpendicular to $\mathbf{B}$ would only give a resultant contribution to the total force on the current distribution inside the volume V_0 in Figure 8.1 when the magnetic field lines are curved.

Our results show that the total magnetic force on the plasma inside the volume V_0 in Figure 8.1 is the same, **after integrating** over the surface S_0, as that given by the mechanical analogy of a magnetic "pressure" $(B^2/2\mu_0)$ on the surface S_0 plus magnetic field lines that are in a state of "tension" of magnitude (B^2/μ_0) per unit area perpendicular to $\mathbf{B}$. We have not shown that there is a force on an individual element of area of the surface S_0 which is the same as that would be given by a magnetic pressure $(B^2/2\mu_0)$ and a magnetic "tension" (B^2/μ_0) per unit area perpendicular to $\mathbf{B}$. For example such a result cannot be true if the surface S_0 is in empty space, since, according to contemporary classical electromagnetism, there are no forces on empty space. To summarize, what we did in this section was to start with the magnetic force on a current distribution, given by equation (8.2) which follows from the Lorentz force law. We then substituted for $\mathbf{J}$ in the volume integral of $\mathbf{J} \times \mathbf{B}$ using Maxwell's equations. We then converted the volume integrals into surface integrals, ending up with equation (8.14). Looking at Figure 8.1, we find that we have expressed the sum of the forces $\mathbf{J} \times \mathbf{B}$ on all the currents inside the volume V_0 in terms of surface integrals evaluated over the surface S_0 of V_0 using only the values of $\mathbf{B}$ on the surface S_0. We have been assuming that $\mathbf{J}$ and the magnetic fields do not vary with time.

If the volume V_0 were completely inside the current distribution in Figure 8.1, equation (8.14) would still give the magnetic force on the plasma inside V_0. However, we should not assume that equation (8.14) gives the stress on the surface S_0 of the volume V_0. The magnetic force on the plasma inside V_0 does not arise in this case from a stress, given by equation (8.14), which is then transmitted to successive surfaces inside V_0 by contiguous action, but the magnetic force on the plasma inside V_0 arises from the action of the resultant magnetic field on all the current distribution inside V_0, which can be evaluated by integrating $\mathbf{J} \times \mathbf{B}$ over the volume V_0 at a fixed time.

It follows from equation (8.14) that, if $\hat{\mathbf{n}}$ is a unit vector in the direction of $d\mathbf{S}$, the contribution associated with the area $d\mathbf{S}$ to the integrals on the right hand side of equation (8.14) can be written in the form

$$d\mathbf{F}_{mag} = -\hat{\mathbf{n}}\left(\frac{B^2}{2\mu_0}\right) dS + \frac{1}{\mu_0}\mathbf{B}(B\cos\theta)\, dS$$

where θ is the angle between $\hat{\mathbf{n}}$ and the magnetic field line crossing $d\mathbf{S}$. If $\hat{\mathbf{t}}$

is a unit vector in the plane containing $\hat{\mathbf{n}}$ and $\mathbf{B}$, in a direction perpendicular to $\hat{\mathbf{n}}$ and pointing towards the direction of the magnetic field line, then we can express $\mathbf{B}$ as follows

$$\mathbf{B} = B \cos\theta\, \hat{\mathbf{n}} + B \sin\theta\, \hat{\mathbf{t}}.$$

Substituting for $\mathbf{B}$ in the expression for $\mathrm{d}\mathbf{F}_{\mathrm{mag}}$ and rearranging we have

$$\begin{aligned}\mathrm{d}\mathbf{F}_{\mathrm{mag}} &= \left(\frac{B^2}{2\mu_0}\right)\left\{(2\cos^2\theta - 1)\hat{\mathbf{n}} + 2\sin\theta\cos\theta\hat{\mathbf{t}}\right\}\mathrm{d}S \\ &= \left(\frac{B^2}{2\mu_0}\right)(\cos 2\theta\hat{\mathbf{n}} + \sin 2\theta\hat{\mathbf{t}})\,\mathrm{d}S. \end{aligned} \tag{8.17}$$

Hence $\mathrm{d}\mathbf{F}_{\mathrm{mag}}$ is of magnitude $(B^2/2\mu_0)\,\mathrm{d}S$ and is in the plane containing $\hat{\mathbf{n}}$ and $\mathbf{B}$ at an angle 2θ to $\hat{\mathbf{n}}$ and an angle θ to $\mathbf{B}$. Equation (8.14) can be rewritten as the integral of a tensor $\mathbf{T}_{\mathrm{mag}}$ as follows

$$\mathbf{F}_{\mathrm{mag}} = \int \mathbf{T}_{\mathrm{mag}} \cdot \mathrm{d}\mathbf{S} \tag{8.18}$$

where we can express $\mathbf{T}_{\mathrm{mag}} \cdot \mathrm{d}\mathbf{S}$ in the matrix form

$$\mathbf{T}_{\mathrm{mag}} \cdot \mathrm{d}\mathbf{S} = \frac{1}{\mu_0}\begin{pmatrix} B_x^2 - \frac{1}{2}B^2 & B_xB_y & B_xB_z \\ B_xB_y & B_y^2 - \frac{1}{2}B^2 & B_yB_z \\ B_xB_z & B_yB_z & B_z^2 - \frac{1}{2}B^2 \end{pmatrix}\begin{pmatrix} \mathrm{d}S_x \\ \mathrm{d}S_y \\ \mathrm{d}S_z \end{pmatrix}. \tag{8.19}$$

The product $\mathbf{T}_{\mathrm{mag}} \cdot \mathrm{d}\mathbf{S}$ is a vector. The tensor $\mathbf{T}_{\mathrm{mag}}$ is the magnetic component of the Maxwell stress tensor. When equation (8.18) is integrated over the surface S_0 in Figure 8.1, it gives the total magnetic force on the current distribution inside V_0. We have only derived equation (8.18) in its integral form.

For the benefit of readers familiar with tensor notation we can express the tensor $\mathbf{T}_{\mathrm{mag}}$ in the form

$$(\mathbf{T}_{\mathrm{mag}})_{ij} = \frac{1}{\mu_0}\left(B_iB_j - \frac{1}{2}\delta_{ij}B^2\right) \tag{8.20}$$

where the indices i and j stand for the x, y and z coordinates respectively and δ_{ij} is defined to be equal to 1 when $i = j$ and equal to zero when i is not equal to j. It is left as an exercise for the reader to show that equation (8.20) gives the correct elements in the matrix in equation (8.19). In tensor notation equation (8.19) can be written in the form

$$(\mathbf{T}_{\mathrm{mag}} \cdot \mathrm{d}\mathbf{S})_i = \sum_j (\mathbf{T}_{\mathrm{mag}})_{ij}\,\mathrm{d}S_j \tag{8.21}$$

where i can stand for x or y or z and the summation is over j equals x, y and z.

8.2.3. *The electric force on an electrostatic charge distribution*

We shall now assume that the charge density of the steady charge and current distribution in Figure 8.1 is ρ. We shall calculate the total electric force on all the charge inside the volume V_0 in Figure 8.1. We shall assume that all the external charge distributions are steady so that $\dot{\mathbf{E}}$ and $\dot{\mathbf{B}}$ are zero. According to equation (8.1), the electrostatic force per unit volume is equal to $\rho\mathbf{E}$, where the electric field $\mathbf{E}$ is due partly to the steady charge distributions outside V_0 and partly to the steady charge distribution inside V_0. Integrating over the volume V_0, we find that the total electric force $\mathbf{F}_{\text{elec}}$ on the charge distribution inside the volume V_0 in Figure 8.1 is

$$\mathbf{F}_{\text{elec}} = \int \rho\mathbf{E}\, \mathrm{d}V. \tag{8.22}$$

Since according to Maxwell's equations ρ is equal to $\varepsilon_0\boldsymbol{\nabla}\cdot\mathbf{E}$, we can rewrite equation (8.22) in the form

$$\mathbf{F}_{\text{elec}} = \varepsilon_0\int \mathbf{E}(\boldsymbol{\nabla}\cdot\mathbf{E})\, \mathrm{d}V. \tag{8.23}$$

Putting $\mathbf{A} = \mathbf{E}$ and $\mathbf{B} = \mathbf{E}$ in equation (8.12) we can substitute for $\mathbf{E}(\boldsymbol{\nabla}\cdot\mathbf{E})$ in equation (8.23) to give

$$\mathbf{F}_{\text{elec}} = -\varepsilon_0\int (\mathbf{E}\cdot\boldsymbol{\nabla})\mathbf{E}\, \mathrm{d}V + \varepsilon_0\int \mathbf{E}(\mathbf{E}\cdot \mathrm{d}\mathbf{S}). \tag{8.24}$$

Putting $\mathbf{A} = \mathbf{E}$ in equation (8.8), where $\mathbf{E}$ is now the electric field, we find that, since $\boldsymbol{\nabla}\times\mathbf{E}$ is zero when $\dot{\mathbf{B}}$ is zero,

$$\varepsilon_0(\mathbf{E}\cdot\boldsymbol{\nabla})\mathbf{E} = \boldsymbol{\nabla}\left(\frac{\varepsilon_0 E^2}{2}\right) - \varepsilon_0\mathbf{E}\times(\boldsymbol{\nabla}\times\mathbf{E}) = \boldsymbol{\nabla}\left(\frac{\varepsilon_0 E^2}{2}\right).$$

Substituting in equation (8.24), we get

$$\mathbf{F}_{\text{elec}} = \int \boldsymbol{\nabla}\left(-\frac{\varepsilon_0 E^2}{2}\right) \mathrm{d}V + \int \varepsilon_0\mathbf{E}(\mathbf{E}\cdot \mathrm{d}\mathbf{S}). \tag{8.25}$$

Putting $\phi = (-\varepsilon_0 E^2/2)$ in equation (8.10) and then substituting in equation (8.25) we can convert the first integral on the right hand side of equation (8.25) into a surface integral, giving finally

$$\mathbf{F}_{\text{elec}} = \int_{S_0} \left(-\frac{\varepsilon_0 E^2}{2}\right) \mathrm{d}\mathbf{S} + \int_{S_0} \varepsilon_0\mathbf{E}(\mathbf{E}\cdot \mathrm{d}\mathbf{S}). \tag{8.26}$$

Equation (8.26) has a similar mathematical form to the magnetostatic case given by equation (8.14). The reader can use similar methods and interpretations, as we used for the magnetostatic case in Section 8.2.2, to show that we get the correct value for the total electrostatic force on the charge distribution inside the volume V_0 in Figure 8.1 **after integrating** over the surface

S_0 of V_0 if we use the mechanical analogy of a "pressure" ($\varepsilon_0 E^2/2$) on the surface S_0 of the volume V_0 plus a "tension" along the electric field lines crossing the surface S_0 of magnitude ($\varepsilon_0 E^2$) per unit area perpendicular to the electric field. Summarizing, we started with equation (8.22) for the force on the steady charge distribution inside the volume V_0 in Figure 8.1. We then used Maxwell's equations to substitute $\varepsilon_0 \nabla \cdot \mathbf{E}$ for ρ, and used the result that $\nabla \times \mathbf{E}$ was zero for steady currents leading up to equation (8.25). We then converted a volume integral into a surface integral ending up with equation (8.26), which gives the total force on all the charge distribution inside the volume V_0 in Figure 8.1 expressed only in terms of the value of the electric field on the surface S_0 of the volume V_0. In electrostatics, the Coulomb forces between the charges inside V_0 add up to zero, so that the force given by equation (8.26) is equal to the force due to the charge distributions outisde V_0.

By analogy with the magnetic case given by equation (8.17), we can express the contribution associated with an element of area $d\mathbf{S}$ to the integrals in equation (8.16) in the form

$$d\mathbf{F}_{\text{elec}} = \left(\frac{\varepsilon_0 E^2}{2}\right)(\hat{\mathbf{n}} \cos 2\theta + \hat{\mathbf{t}} \sin 2\theta) \tag{8.27}$$

where θ is the angle between the direction of $d\mathbf{S}$ and the electric field line passing through $d\mathbf{S}$.

We can rewrite equation (8.26) in the form

$$\mathbf{F}_{\text{elec}} = \int \mathbf{T}_{\text{elec}} \cdot d\mathbf{S} \tag{8.28}$$

where the tensor $\mathbf{T}_{\text{elec}}$ is the electrostatic counterpart of the magnetic tensor $\mathbf{T}_{\text{mag}}$ given by equation (8.20). In tensor notation we have

$$(\mathbf{T}_{\text{elec}} \cdot d\mathbf{S})_i = \sum_j (\mathbf{T}_{\text{elec}})_{ij}\, dS_j \tag{8.29}$$

where

$$(\mathbf{T}_{\text{elec}})_{ij} = \varepsilon_0 \left(E_i E_j - \frac{1}{2}\delta_{ij} E^2 \right). \tag{8.30}$$

It is left as an exercise for the reader to use equation (8.30) to express $\mathbf{T}_{\text{elec}}$ in matrix form.

8.2.4. *The total electromagnetic force on a steady charge and current distribution due to steady electric and magnetic fields*

Adding equation (8.14) and (8.26), we find that the total electromagnetic force on the steady charge and steady current distribution inside the volume V_0 in Figure 8.1 can be expressed in terms of only the values of the steady total electric and magnetic fields on the surface S_0 of the volume V_0 as follows

$$\mathbf{F}_{\text{em}} = \mathbf{F}_{\text{elec}} + \mathbf{F}_{\text{mag}}$$

$$= \int \left\{ \left(-\frac{\varepsilon_0 E^2}{2} \right) \mathrm{d}\mathbf{S} + \varepsilon_0 \mathbf{E}(\mathbf{E} \cdot \mathrm{d}\mathbf{S}) + \left(-\frac{B^2}{2\mu_0} \right) \mathrm{d}\mathbf{S} + \frac{1}{\mu_0} \mathbf{B}(\mathbf{B} \cdot \mathrm{d}\mathbf{S}) \right\} \tag{8.31}$$

$$= \int (\mathbf{T}_{\text{elec}} + \mathbf{T}_{\text{mag}}) \cdot \mathrm{d}\mathbf{S} = \int \mathbf{T} \cdot \mathrm{d}\mathbf{S} \tag{8.32}$$

where

$$\mathbf{T} = \mathbf{T}_{\text{elec}} + \mathbf{T}_{\text{mag}} \tag{8.33}$$

is the Maxwell stress tensor. In tensor notation we have

$$(\mathbf{T} \cdot \mathrm{d}\mathbf{S})_i = \sum_j \mathbf{T}_{ij}\, \mathrm{d}S_j \tag{8.34}$$

where i stands for x or y or z and the summation is over j equal to x, y and z and

$$(\mathbf{T})_{ij} = \varepsilon_0 \left(E_i E_j - \frac{1}{2} \delta_{ij} E^2 \right) + \frac{1}{\mu_0} \left(B_i B_j - \frac{1}{2} \delta_{ij} B^2 \right). \tag{8.35}$$

It is left as an exercise for the reader to express $\mathbf{T} \cdot \mathrm{d}\mathbf{S}$ in a matrix form similar to equation (8.19). The integrations in equations (8.31) and (8.32) are over the closed surface S_0 of the volume V_0 in Figure 8.1. The interpretation of the Maxwell stress tensor is similar to the interpretation of the magnetic component we developed in Section 8.2.2.

8.2.5. *An electrostatic example of the application of Maxwell's stress tensor*

Consider first the example of two point charges of magnitudes $+q$ and $-q$ at a distance $2a$ apart, as shown in Figure 8.3(a). It can be seen from Figure 8.3(a) that the electric field lines diverge from the positive charge and converge on the negative charge. We shall calculate the force on the positive charge using Maxwell's stress tensor. Normally we would say that the negative charge gives an electric field of magnitude

$$E_0 = \frac{q}{4\pi\varepsilon_0(2a)^2} = \frac{q}{16\pi\varepsilon_0 a^2}$$

at the position of the positive charge in Figure 8.3(a) and that the attractive force on the positive charge is given by

$$F = qE_0 = \frac{q^2}{16\pi\varepsilon_0 a^2}. \tag{8.36}$$

When using Maxwell's stress tensor we must apply equation (8.28) to a closed surface S_0 that surrounds the positive charge. As part of the closed surface S_0 we shall use the median plane, which is perpendicular to the line joining the charges and which intersects the line joining the charges at a

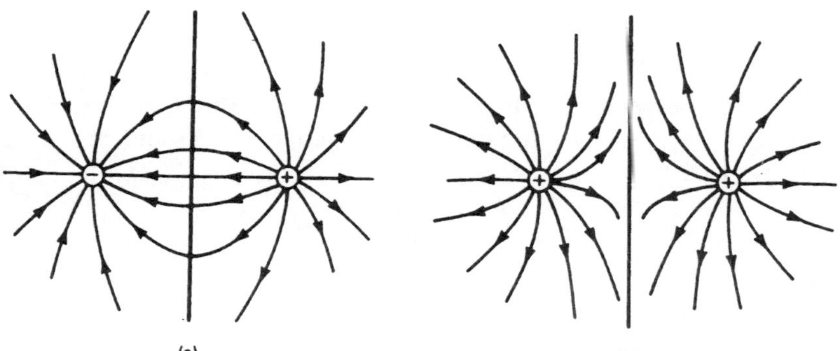

Figure 8.3. Calculation of the electrostatic forces (a) between a positive and negative charge, (b) between two positive charges, using Maxwell's stress tensor. (Reproduced from: *Electromagnetic Fields and Relativistic Particles*, by E. J. Konopinski [6], with the permission of the McGraw-Hill Book Co.)

point halfway between them. The rest of the closed surface S_0, that surrounds the positive charge, is at such a large distance from the charges that the electric field can be taken to be zero on that part of the surface so that, when equation (8.28) is integrated over the surface S_0, it reduces to

$$\mathbf{F}_{\text{elec}} = \int \mathbf{T}_{\text{elec}} \cdot \mathrm{d}\mathbf{S}. \tag{8.37}$$

evaluated over the median plane only.

Since the angle between any element of area $\mathrm{d}\mathbf{S}$ of the median plane and the resultant electric field $\mathbf{E}$ at the median plane is zero, it follows from equation (8.27) that

$$\mathbf{T}_{\text{elec}} \cdot \mathrm{d}\mathbf{S} = \hat{\mathbf{n}} \left(\frac{\varepsilon_0 E^2}{2} \right) \mathrm{d}S. \tag{8.38}$$

In this simple example it is straightforward to use the mechanical analogy. Inspection of Figure 8.3(a) shows that the contribution of the "tension" in the electric field lines is of magnitude $(\varepsilon_0 E^2)$ per unit area of the median plane and is in a direction perpendicular to the median plane pointing to the left. The "pressure" $(\varepsilon_0 E^2/2)$ acts inwards into the median plane, that is to the right in Figure 8.3(a), so that the resultant contribution to the integral in equation (8.37) is the same as if here were a force $(\varepsilon_0 E^2/2)$ per unit area pulling the median plane to the left in Figure 8.3(a). Consider an element of area of the median plane of magnitude $2\pi r\, \mathrm{d}r$ situated between circles of radii r and $r + \mathrm{d}r$, with centres at the mid point of the line joining the two charges. According to Coulomb's law the contribution of the negative charge to the electric field at this element of area is of magnitude $q/4\pi\varepsilon_0(a^2 + r^2)$ in the direction towards the negative charge. The electric field due to the positive charge has the same magnitude but is in the direction radially outwards from the positive charge. The resultant electric field $\mathbf{E}$ at the position of the element of area is of magnitude

$$E = \frac{2qa}{4\pi\varepsilon_0(a^2 + r^2)^{3/2}}$$

in a direction perpendicular to the median plane in the direction to the left in Figure 8.3(a). Substituting for **E** in equation (8.37) and putting $\mathrm{d}S = 2\pi r\ \mathrm{d}r$, we find that

$$\mathbf{F}_{\text{elec}} = \int \mathbf{T}_{\text{elec}} \cdot \mathrm{d}\mathbf{S} = \int \frac{\varepsilon_0 E^2}{2} \mathrm{d}\mathbf{S} = \hat{\mathbf{n}} \int \frac{\varepsilon_0 E^2}{2} 2\pi r\ \mathrm{d}r$$
$$= \hat{\mathbf{n}} \frac{q^2 a^2}{8\pi\varepsilon_0} \int_0^\infty \frac{2r\ \mathrm{d}r}{(a^2 + r^2)^3} = \left(\frac{q^2}{16\pi\varepsilon_0 a^2}\right) \hat{\mathbf{n}} \tag{8.39}$$

This is in agreement with equation (8.36). According to equation (8.28) we would get the same result whatever the shape and position of the surface S_0 that surrounds the positive charge in Figure 8.3(a). These results show that we get the same result if we use Coulomb's law or Maxwell's stress tensor, which is not surprising since, when we derived equation (8.26) we started with equation (8.22) which is equivalent to starting with qE_0 in the present example. This shows that Maxwell's stress tensor adds nothing new to classical electromagnetism and does not imply that there is a force on the empty space in the median plane in Figure 8.3(a), as was assumed in the nineteenth century aether theories.

In the example of two positive charges of magnitude $+q$ shown in Figure 8.3(b), the electric field at the median plane is parallel to the median plane as shown in Figure 8.3(b). If we choose the same surface S_0 as previously, the "tensions" in the electric field lines at the median plane do not give a contribution to the integrals in equation (8.26) and we are left with only the "pressure" ($\varepsilon_0 E^2/2$) which acts into the surface S_0, that is to the right in Figure 8.3(b). The reader can check this result using equation (8.27). It is now straightforward for the reader to show that the magnitude of the repulsive force on the right hand positive charge in Figure 8.3(b) is equal to $q^2/16\pi\varepsilon_0 a^2$ in agreement with Coulomb's law. To determine the repulsive force on the left hand side charge in Figure 8.3(b) consider the mirror image of the surface S_0. In this case the "pressure" acting on the median plane is to the left in Figure 8.3(b).

8.2.6. *Magnetostatic example of the application of Maxwell's stress tensor*

Consider two infinitely long, thin, parallel wires a distance $2a$ apart in empty space as shown in Figures 8.4(a) and 8.4(b). In the example shown in Figure 8.4(a), the two steady parallel currents, each of magnitude I, are upwards towards the reader. The resultant magnetic field lines are shown in Figure 8.4(a). According to the Biot-Savart law the magnetic field B_0 at the position of the right hand side current due to the other current is

$$B_0 = \frac{\mu_0 I}{2\pi(2a)} = \frac{\mu_0 I}{4\pi a}\ .$$

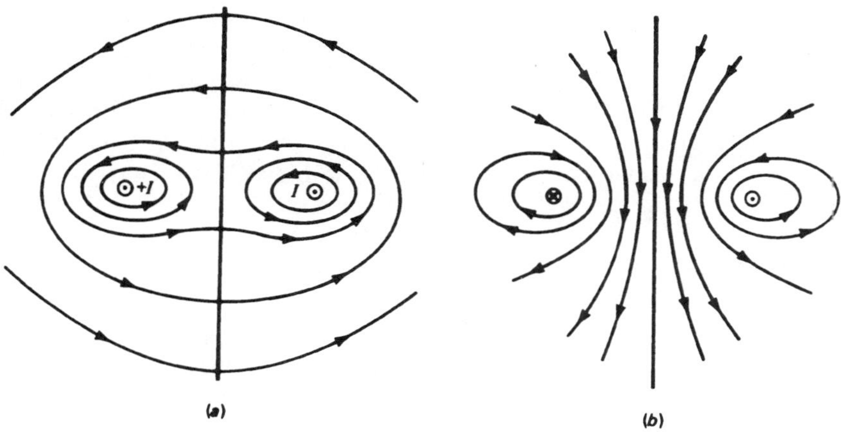

Figure 8.4. Calculation of the magnetic forces between two parallel currents using Maxwell's stress tensor. (Reproduced from: *Electromagnetic Fields and Relativistic Particles*, by E. J. Konopinski [6], with the permission of the McGraw-Hill Book Co.)

The magnetic force on unit length of the right hand side wire is

$$F_{\text{mag}} = BIl = \frac{\mu_0 I^2}{4\pi a}. \tag{8.40}$$

To apply equation (8.14), we shall consider a surface S_0 that surrounds unit length of the right hand side current and is made up of a section of unit width of the median plane in a direction perpendicular to the paper in Figure 8.4(a); S_0 is completed by two planes perpendicular to the median plane which are joined in a region where the magnetic field is negligible to complete the surface S_0 enclosing unit length of the right hand side wire. The contribution associated with the parallel plane surfaces to the integrals in equation (8.14) compensate each other so that we need only evaluate the integrals in equation (8.14) over the strip of unit width of the median plane. At a distance x from the line joining the charges, the resultant magnetic field due to both the currents in Figure 8.4(a) is

$$B_0 = \frac{\mu_0 I x}{\pi(a^2 + x^2)}.$$

According to equation (8.17) for any point on the median plane we have

$$\mathrm{d}\mathbf{F}_{\text{mag}} = \hat{\mathbf{n}}\left(\frac{B_0^2}{2\mu_0}\right)\mathrm{d}S$$

where $\hat{\mathbf{n}}$ is a unit vector normal to the median plane pointing to the left. (The reader can check this result using the mechanical analogy). Substituting for B for a strip of width $\mathrm{d}x$ and integrating we find that

$$\mathbf{F}_{\text{mag}} = \hat{\mathbf{n}}\left(\frac{\mu_0 I^2}{2\pi^2}\right)\int_{-\infty}^{+\infty}\frac{x^2\,\mathrm{d}x}{(a^2 + x^2)^2} = \left(\frac{\mu_0 I^2}{4\pi a}\right)\hat{\mathbf{n}}. \tag{8.41}$$

Equation (8.41) is in agreement with equation (8.40), which shows that Maxwell's stress tensor gives the same result as the standard method using the magnetic field and the Lorentz force law. This is no surprise as we started with equation (8.2) to derive equation (8.14). The use of Maxwell's stress tensor is just an alternative way of deriving a standard result. It is left as an exercise for the reader to interpret and analyse Figure 8.4(b).

8.3. The energy of the electromagnetic field

It is difficult in the general case when there are static, induction and radiation electric and magnetic fields to be explicit about where all the energy of the electromagnetic field resides. As an analogy consider a gravitational example. If we hold a mass m at a height h above the ground, we generally say that it has a potential energy mgh, where g is the acceleration due to gravity. Is this potential energy in the mass m, in the Earth or in the gravitational field? Generally it is only necessary to say that the system has potential energy, though, in the general theory of relativity, energy and momentum are attributed to the gravitational field. In the electromagnetic case, if we assemble an electrostatic charge distribution from its constituent charges, we do work against the electrostatic forces between the charges and the electrostatic charge distribution has a potential energy U_{elec} which can be expressed in the form

$$U_{\text{elec}} = \frac{1}{2}\int \rho\phi \, \mathrm{d}V \tag{8.42}$$

where ρ is the charge density and ϕ is the electrostatic potential. The volume over which the integration in equation (8.42) is evaluated must include all the charge in the system including any test charge. Reference: Griffiths [2].

Since according to Maxwell's equations, $\rho = \varepsilon_0 \nabla \cdot \mathbf{E}$ we can rewrite equation (8.42) in the form

$$U_{\text{elec}} = \frac{\varepsilon_0}{2}\int \phi\nabla \cdot \mathbf{E} \, \mathrm{d}V \tag{8.43}$$

Putting $\mathbf{A} = \mathbf{E}$ in equation (A1.20) of Appendix A1.6, where $\mathbf{E}$ is the electric field, and then rearranging we have

$$\phi\nabla \cdot \mathbf{E} = \nabla \cdot (\phi\mathbf{E}) - \mathbf{E} \cdot \nabla\phi.$$

Since $\nabla\phi = -\mathbf{E}$, after substituting for $\phi\nabla \cdot \mathbf{E}$ in equation (8.43) we find that

$$U_{\text{elec}} = \frac{\varepsilon_0}{2}\int \nabla \cdot (\phi\mathbf{E}) \, \mathrm{d}V + \int \left(\frac{\varepsilon_0 E^2}{2}\right) \mathrm{d}V.$$

After applying Gauss' integral theorem, which is equation (A1.30) of Appendix A1.7, to the first integral on the right hand side we have

$$U_{\text{elec}} = \frac{\varepsilon_0}{2}\int \phi\mathbf{E} \cdot \mathrm{d}\mathbf{S} + \int \left(\frac{\varepsilon_0 E^2}{2}\right) \mathrm{d}V. \tag{8.44}$$

Equation (8.44) is correct for any volume that contains all the electrostatic charge distributions. As we go to larger and larger distances from the electrostatic charge distributions, ϕ and E decrease at least as quickly as $1/r$ and $1/r^2$ respectively so that ϕE goes down at least as quickly as $1/r^3$, whereas the area increases as r^2. Hence **provided we integrate over the whole of space** the first integral on the right hand side of equation (8.44) is zero and equation (8.44) becomes

$$U_{\text{elec}} = \int \left(\frac{\varepsilon_0 E^2}{2} \right) \mathrm{d}V. \tag{8.45}$$

We have only derived equation (8.45) for an electrostatic charge distribution. To illustrate that equations (8.42) and (8.45) are alternative formulae for determining the same quantity, namely the total electrostatic energy, consider an idealized parallel plate capacitor of capacitance $C = \varepsilon_0 A/d$, where A is the area and d the separation of the plates. If Q is the total charge on the positive plate and ϕ is the potential difference between the plates, then corresponding to equation (8.42) we can express the total electrostatic energy in the well known formula

$$U_{\text{elec}} = \frac{Q\phi}{2} = \frac{Q^2}{2C}\,. \tag{8.46}$$

If the fringing field is zero, the electric field, which is then of magnitude $Q/\varepsilon_0 A$, extends only over the volume Ad between the plates of the capacitor, so that, according to equation (8.45),

$$U_{\text{elec}} = \left(\frac{\varepsilon_0 E^2}{2} \right) Ad = \frac{\varepsilon_0 Q^2 Ad}{2\varepsilon_0^2 A^2} = \frac{Q^2 d}{2\varepsilon_0 A} = \frac{Q^2}{2C}\,. \tag{8.47}$$

This shows that both equations (8.46) and (8.47) give the same value for the total electrostatic energy. Is the energy in the charges or is it in the electrostatic field? In practice we only observe the energy when the electromagnetic field interacts with charges.

By considering the work that must be done against the back emfs when we set up a steady current distribution, it can be shown that, when $\mu_r = 1$ everywhere, the total magnetostatic energy can be expressed in the form

$$U_{\text{mag}} = \int \left(\frac{B^2}{2\mu_0} \right) \mathrm{d}V. \tag{8.48}$$

provided we integrate over the whole of space. Reference: Griffiths [3]. Alternatively, we can express the total magnetostatic energy in the form

$$U_{\text{mag}} = \frac{1}{2} \int \mathbf{J} \cdot \mathbf{A}\, \mathrm{d}V \tag{8.49}$$

where $\mathbf{J}$ is the current density and $\mathbf{A}$ is the vector potential. The integration must include all the current distributions and any moving test charge. As an example of the energy stored in an inductor, consider an LR circuit. When the current is I, the total magnetic energy is $LI^2/2$. The current in the LR circuit does not decrease to zero instantaneously when there is no applied emf in the circuit, since, when the magnetic field is decreasing, there is an induction electric field which gives an induced emf which gives a current flow in the resistor leading to the production of observable Joule heating in the resistor. The total Joule heat produced is equal to $LI^2/2$.

Adding equations (8.45) and (8.48) we find that the total electromagnetic energy of a steady charge and current distribution can be expressed in the form

$$U_{\text{em}} = U_{\text{elec}} + U_{\text{mag}} = \int \left(\frac{\varepsilon_0 E^2}{2} + \frac{B^2}{2\mu_0} \right) \mathrm{d}V. \tag{8.50}$$

provided the integration is over the whole of space. Equation (8.50) is often rewritten in the form

$$U_{\text{em}} = \int u_{\text{em}} \, \mathrm{d}V \tag{8.51}$$

where the expression defined by the equation

$$u_{\text{em}} = \left(\frac{\varepsilon_0 E^2}{2} + \frac{B^2}{2\mu_0} \right). \tag{8.52}$$

is generally called the "**energy density**" of the electromagnetic field. When integrated over **all of space** the integral of the "energy density" gives the total electrostatic and magnetostatic energy. In the case of electrostatics and magnetostatics we could just as well define the "energy density" to be

$$u'_{\text{em}} = \frac{1}{2}\rho\phi + \frac{1}{2}\mathbf{J} \cdot \mathbf{A} \tag{8.53}$$

which would be consistent with the alternative view that the energy is stored in the charge and current distributions. Hence in the context of magnetostatics and electrostatics we cannot give a definitive answer to the question: "Where is the energy stored and what is the energy density?" In some cases at least, it is reasonable to attribute energy to the electromagnetic field. For example, radio waves move outwards from a transmitting antenna with the speed of light and exist independently of their source, and will continue outwards after the transmitter is switched off. The electric vector of the radio waves gives an electric force on the conduction electrons in a receiving antenna wherever it is placed, leading to an electrical signal in the receiving antenna. In this way the antenna takes energy from the electromagnetic field. In this example, the capacity to give energy to the receiving antenna must reside in the electromagnetic field and it is sensible to say that there is energy in the radio signal, though we shall find that, since we shall only need to work with integrals of the fields, there will be no need for us, in the context of classical electromagnetism, to be explicit about what the precise expression for the energy density of the electromagnetic field is, and where the energy is located.

So far in this book, we have avoided the use of methods based on energy as the underlying physical principles are often hidden in these methods. For example, if we applied the law of conservation of energy to the *LCR* circuit in Figure 7.6 of Chapter 7, then, by equating the rate εI at which the emf ε supplies energy to the circuit to the rate I^2R at which Joule heat is generated in the resistor R plus the rates of change of the energy stored in the inductor L and in the capacitor C we would have

$$\varepsilon I = I^2R + \frac{\mathrm{d}}{\mathrm{d}t}\left(\frac{1}{2}LI^2 + \frac{1}{2}\frac{Q^2}{C}\right).$$

Carrying out the differentiation and cancelling $I = \mathrm{d}Q/\mathrm{d}t$ we get

$$\varepsilon = IR + L\frac{\mathrm{d}I}{\mathrm{d}t} + \frac{Q}{C} \tag{8.54}$$

$$= L\frac{\mathrm{d}^2Q}{\mathrm{d}t^2} + R\frac{\mathrm{d}Q}{\mathrm{d}t} + \frac{Q}{C}. \tag{8.55}$$

Equations (8.54) and (8.55) are the same as equations (7.77) and (7.78) of Section 7.6.4 of Chapter 7. Equation (8.55) can now be solved mathematically to determine the behaviour of a resistor, an inductor and a capacitor in an AC circuit without having to enquire why the inductor and the capacitor behave in the way that they do. The author would like to stress that the reader should continue to use methods based on energy whenever it is convenient to do so. It is just that in this book our aim is to interpret the underlying physical principles and not to get the results by the quickest method.

8.4. The rate at which the electromagnetic field does work on a charge and current distribution

Consider a system of moving and accelerating classical point charges, each of charge q, that build up a varying macroscopic charge and current distribution. Consider a volume element $\mathrm{d}V$ at a point inside the charge and current distribution, where the total electric and magnetic fields are $\mathbf{E}$ and $\mathbf{B}$ respectively. According to the Lorentz force law, the force on the ith charge q_i is

$$\mathbf{F}_i = q_i\mathbf{E} + q_i\mathbf{u}_i \times \mathbf{B}.$$

The rate at which the electromagnetic field is doing work on the charge q_i is

$$\mathbf{F}_i \cdot \mathbf{u}_i = q_i\mathbf{E} \cdot \mathbf{u}_i + q_i(\mathbf{u}_i \times \mathbf{B}) \cdot \mathbf{u}_i = q_i\mathbf{E} \cdot \mathbf{u}_i.$$

Notice that the magnetic field does no work. Summing over all the charges in the volume element dV we find that

$$\sum\mathbf{F}_i \cdot \mathbf{u}_i = \sum q_i\mathbf{u}_i \cdot \mathbf{E} = \mathbf{E} \cdot \left(\sum q_i\mathbf{u}_i\right). \tag{8.56}$$

If $\langle\mathbf{u}_i\rangle$ is the mean velocity and n is the number of charges per unit volume, then summing over all the charges in the volume element $\mathrm{d}V$, we have

$$\sum q_i \mathbf{u}_i = qn\langle \mathbf{u}_i \rangle \, \mathrm{d}V = \mathbf{J} \, \mathrm{d}V$$

where $\mathbf{J}$ is the current density. Hence equation (8.56) can be rewritten in the form

$$\sum \mathbf{F}_i \cdot \mathbf{u}_i = \mathbf{E} \cdot \mathbf{J} \, \mathrm{d}V. \tag{8.57}$$

Integrating over a finite volume V_0 at the fixed time of observation t, we find that the rate of doing work $\dot{W}$ on the charges inside V_0, which is the power P supplied to the charges in the volume V_0 at the time t is

$$P = \frac{\mathrm{d}W}{\mathrm{d}t} = \int \mathbf{E} \cdot \mathbf{J} \, \mathrm{d}V \tag{8.58}$$

where the symbol W stands for the work done on the charges inside V_0. Equation (8.58) is valid for both static and varying charge and current distributions. Since we derived equation (8.58) using the Lorentz force law, whenever we use equation (8.58) we are using a formula based directly on the Lorentz force law.

8.5. Energy transfer and the Poynting vector in DC circuits

8.5.1. *Theory*

Before going on to discuss the general case of varying charge and current distributions in Section 8.6.1, as an introduction we shall assume in this section that the charge and current distribution in Figure 8.1 is steady, corresponding to DC conditions and that the total electric and magnetic fields due to all the charge and current distributions are also steady so that both $\dot{\mathbf{E}}$ and $\dot{\mathbf{B}}$ are zero, and according to Maxwell's equations

$$\mathbf{J} = \frac{\nabla \times \mathbf{B}}{\mu_0}. \tag{8.59}$$

According to equation (8.58), which we derived using the Lorentz force law, the rate at which the electromagnetic field is doing work on the charges making up the charge and current distribution inside the volume V_0 in Figure 8.1 is

$$\dot{W} = \int \mathbf{E} \cdot \mathbf{J} \, \mathrm{d}V. \tag{8.60}$$

Substituting for $\mathbf{J}$ in the expression $\mathbf{E} \cdot \mathbf{J}$ we obtain

$$\mathbf{E} \cdot \mathbf{J} = \frac{\mathbf{E} \cdot (\nabla \times \mathbf{B})}{\mu_0} \tag{8.61}$$

Putting $\mathbf{A} = \mathbf{E}$ and $\mathbf{B} = \mathbf{B}$ in equation (A1.21) of Appendix A1.6, where $\mathbf{E}$ is now the total electric field and $\mathbf{B}$ is the total magnetic field, and remembering that, according to Maxwell's equations, $\nabla \times \mathbf{E}$ is zero if $\dot{\mathbf{B}}$ is zero, we find that for DC conditions

$$\nabla \cdot (\mathbf{E} \times \mathbf{B}) = -\mathbf{E} \cdot (\nabla \times \mathbf{B})$$

Hence

$$\mathbf{E} \cdot \mathbf{J} = -\nabla \cdot \left(\frac{\mathbf{E} \times \mathbf{B}}{\mu_0}\right) \tag{8.62}$$

Substituting for $\mathbf{E} \cdot \mathbf{J}$ from equation (8.62) into equation (8.61) and applying Gauss' theorem of vector analysis, we find that for DC conditions the rate at which the electromagnetic field is doing work on the charges inside the volume V_0 in Figure 8.1 is also given by

$$\dot{W} = -\frac{1}{\mu_0}\int \nabla \cdot (\mathbf{E} \times \mathbf{B})\, \mathrm{d}V = -\frac{1}{\mu_0}\int (\mathbf{E} \times \mathbf{B}) \cdot \mathrm{d}\mathbf{S}. \tag{8.63}$$

To simplify the equations, we shall introduce a new vector $\mathbf{N}$, called the Poynting vector, which we shall define by the equation

$$\mathbf{N} = \frac{\mathbf{E} \times \mathbf{B}}{\mu_0}. \tag{8.64}$$

In the general case, in the presence of dielectrics and magnetic materials, the Poynting vector is defined as

$$\mathbf{N} = \mathbf{E} \times \mathbf{H}. \tag{8.65}$$

In empty space $\mathbf{H}$ is equal to $\mathbf{B}/\mu_0$ and equation (8.65) reduces to equation (8.64).

Using equation (8.64) we can rewrite equations (8.60) and (8.63) in the forms

$$\dot{W} = \int \mathbf{E} \cdot \mathbf{J}\, \mathrm{d}V = -\int (\nabla \cdot \mathbf{N})\, \mathrm{d}V \tag{8.66}$$

$$\dot{W} = -\int \mathbf{N} \cdot \mathrm{d}\mathbf{S}. \tag{8.67}$$

which are valid for steady (DC) conditions only. By comparing equations (8.60) and (8.67) the reader can see that what we have done in this section is to express $\int \mathbf{E} \cdot \mathbf{J}\, \mathrm{d}V$ in the alternative form $-\int \mathbf{N} \cdot \mathrm{d}\mathbf{S}$. We shall go on to the general case of varying charge and current distributions in Section 8.6.1. It is of interest to note that, according to equation (8.62) for DC conditions $\mathbf{E} \cdot \mathbf{J} = -\nabla \cdot \mathbf{N}$ so that where $\mathbf{E} \cdot \mathbf{J}$ is positive there is a sink for the Poynting vector and where $\mathbf{E} \cdot \mathbf{J}$ is negative there is a source for the Poynting vector. In empty space where $\mathbf{J}$ is zero

$$\nabla \cdot \left(\frac{\mathbf{E} \times \mathbf{B}}{\mu_0}\right) = \nabla \cdot \mathbf{N} = 0. \tag{8.68}$$

8.5.2. *Application to a practical example*

As an example of the application of equation (8.67) to steady (or DC) conditions, we shall consider the example of the Van de Graaff generator, which sustains a steady DC current in the circuit shown in Figure B1 of Appendix B, and which is shown in simplified form in Figure 8.5. After the state of dynamic equilibrium is reached, the belt of the Van de Graaff moves charges against the electric force on them due to the charges on the terminals of the Van de Graaff. The charges on the terminals also give the external electric field that drives current around the circuit in the way described in Section B1 of Appendix B, where it is shown that surface and boundary charge distributions are built up on the conductors, such that the correct value of electric field **E** is sustained inside the conductors to give the same value of current in all parts of the circuit. It is this electric field inside the conductors that acts on the conduction electrons to give the current flow. According to equation (8.58) the rate at which this electric field is doing work on all the conduction electrons inside a volume V_0 is

$$\dot{W} = \int \mathbf{E} \cdot \mathbf{J} \, dV. \tag{8.58}$$

According to equation (8.67), equation (8.58) can be rewritten in the alter-

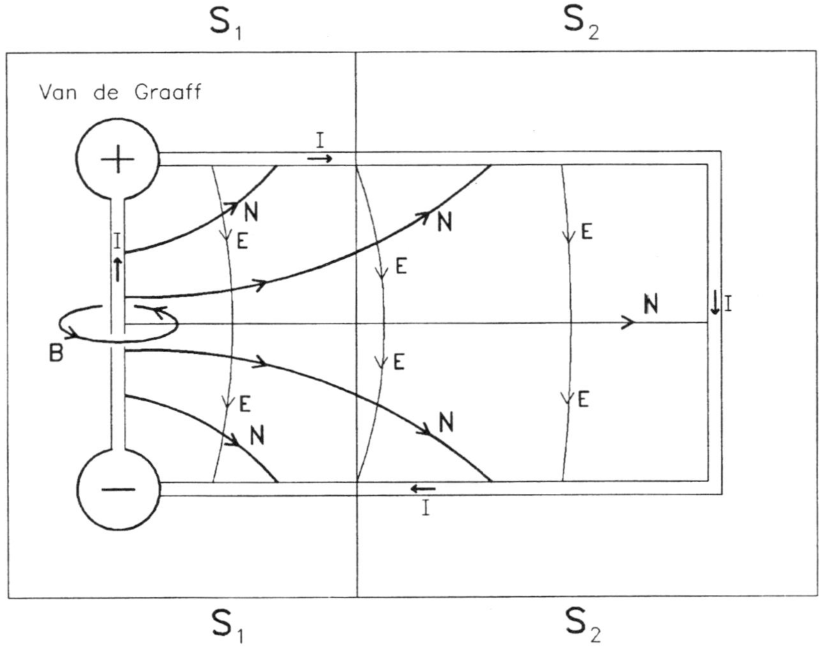

Figure 8.5. The Van de Graaff generator produces a steady current flow in a stationary external circuit. The directions of the resultant electric field **E** and the Poynting vector **N** outside the conductors are sketched in the region inside the circuit.

native form

$$\dot{W} = -\int \mathbf{N} \cdot d\mathbf{S} \tag{8.67}$$

where **N** is the Poynting vector and $\int \mathbf{N} \cdot d\mathbf{S}$ is the Poynting flux out of the surface.

We shall consider first a cylindrical section of the external circuit in Figure 8.5, as shown in Figure 8.6. The wire is of circular cross section, of radius a and area of cross section $A = \pi a^2$. It is of length l and carries a steady conduction current I. According to equation (1.42) when the current is steady the electric field inside the conductor is

$$E = \frac{J}{\sigma} = \frac{I}{\sigma A} \tag{8.69}$$

in a direction parallel to its length. It follows from Ampère's circuital theorem that the magnetic field at a point just inside the surface of the wire, at a distance a from the axis of the long, straight wire is

$$B = \frac{\mu_0 I}{2\pi a} \tag{8.70}$$

in a direction given by the right-handed corkscrew rule, as shown in Figure 8.6. Notice that **E** and **B** are perpendicular to each other. Using equations (8.69) and (8.70) we find that the magnitude of the Poynting vector, just below the surface of the conductor in Figure 8.6, is

$$N = \frac{EB}{\mu_0} = \frac{I^2}{2\pi a \sigma A} \tag{8.71}$$

and its direction is perpendicular to and into the conductor. To apply equation (8.67), consider a cylindrical surface S_0 of height l which is just inside the cylindrical conductor in Figure 8.6. Since the electric field inside the conductor

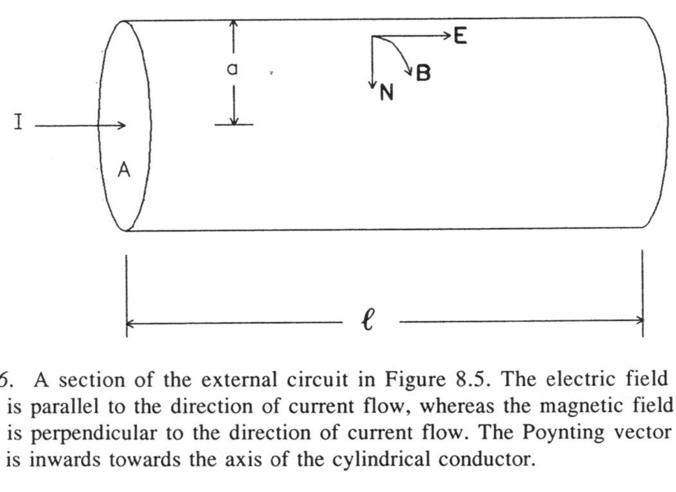

Figure 8.6. A section of the external circuit in Figure 8.5. The electric field **E** inside the conductor is parallel to the direction of current flow, whereas the magnetic field **B** inside the conductor is perpendicular to the direction of current flow. The Poynting vector **N** inside the conductor is inwards towards the axis of the cylindrical conductor.

is parallel to the conductor, the Poynting vector has no component parallel to the conductor and the only contribution to $\int \mathbf{N} \cdot d\mathbf{S}$ comes from the curved surface of S_0. Using equation (8.71) we find that

$$\dot{W} = -\int \mathbf{N} \cdot d\mathbf{S} = 2\pi a l N = I^2 \left(\frac{l}{\sigma A} \right). \tag{8.72}$$

In the conductor this energy is given to the conduction electrons, which subsequently lose this extra kinetic energy in collisions leading to the production of Joule heat. Since $(l/\sigma A)$ is equal to the resistance R, we find that the Joule heat predicted using equation (8.72) is equal to I^2R.

If we used equation (8.58) directly we would find, with $E = I/\sigma A$ and $J = I/A$, that for the cylinder in Figure 8.6

$$\dot{W} = \int \mathbf{E} \cdot \mathbf{J} \, dV = AlEJ = I^2 \left(\frac{l}{\sigma A} \right) = I^2 R.$$

This shows that equations (8.67) and (8.58) make the same predictions, which was only to be expected, since we derived equation (8.67) from equation (8.58).

We return now to consider the complete circuit shown in Figure 8.5. We shall consider the two closed surfaces S_1 and S_2 shown in Figure 8.5. The surface S_1 surrounds the Van de Graaff generator and part of the circuit. The surface S_2 surrounds the rest of the circuit. It can be seen from the practical example given in Figure B2 of Appendix B that, in addition to the electric field inside the conductors, there is an electric field outside the conductors that has components both normal to and tangential to the surfaces of the conductors. The external electric field lines go from regions of high electrical potential to regions of lower potential, as shown in the simplified diagram in Figure 8.5. The conduction current in the external circuit is in a clockwise direction in Figure 8.5. According to the right handed corkscrew rule, the magnetic field **B**, due to the current in the circuit, is downwards into the paper everywhere inside the region bounded by the circuit in Figure 8.5. Inside the Van de Graaff, the electric field is from the positive to the negative terminal, so that just outside the Van de Graaff, the direction of the Poynting vector $(\mathbf{E} \times \mathbf{B})/\mu_0$, in the region bounded by the circuit, is predominantly to the right, away from the source of emf as sketched in Figure 8.5; but, since there is a component of **E** tangential to the surface of the conductors, which is equal to the value of **E** inside the conductors, there is a component of the Poynting vector, perpendicular to and into the surfaces of the conductors, of magnitude given by equation (8.71). The electric and magnetic fields outside the area bounded by the circuit in Figure 8.5 are not shown. It is left as an exercise for the reader to sketch these fields and the direction of the Poynting vector in this region.

Applying equation (8.67) to the surface S_1 that surrounds the Van de Graaff in Figure 8.5 we have

$$\frac{\partial W_1}{\partial t} + \frac{\partial W_2}{\partial t} = -\int \mathbf{N} \cdot d\mathbf{S} \tag{8.73}$$

where $\dot{W}_1$ and $\dot{W}_2$ are respectively the rates at which the electromagnetic field is doing work on the charges on the belt of the Van de Graaff and on the conduction electrons in those parts of the conductors that are inside S_1 in Figure 8.5. It can be seen from Figure 8.5, that the general direction of the Poynting vector is outwards from the surface S_1, so that $\int \mathbf{N} \cdot d\mathbf{S}$ evaluated over the surface S_1 in Figure 8.5 is positive and, according to equation (8.67), $(\dot{W}_1 + \dot{W}_2)$ is negative. The rate $\dot{W}_1$ at which the electromagnetic field is doing work on the charges on the moving Van de Graaff belt is negative since, in this instance, it is the Van de Graaff that is doing work in moving the charges on the belt against the electric forces on them due to the charges on the terminals of the Van de Graaff so that $\mathbf{E} \cdot \mathbf{J}$ is negative in this region. The rate $\dot{W}_2$ at which the electromagnetic field is giving kinetic energy to the conduction electrons in the conductors inside S_1 is positive and this leads to the production of Joule heat in these conductors. Since $(\dot{W}_1 + \dot{W}_2)$ is negative the Van de Graaff does more work in moving charges against the electromagnetic field than is given to the conduction electrons in the conductors inside the surface S_1 in Figure 8.5. According to equation (8.67) the imbalance in the energy equation (8.73) between $|\dot{W}_1|$ and $|\dot{W}_2|$ is equal to the flux of the Poynting vector out of the surface S_1. This shows that it is this Poynting flux out of the surface S_1 that balances the energy balance equation (8.67), when we only consider part of a circuit.

Applying equation (8.67) to the surface S_2 in Figure 8.5, we find that

$$\dot{W}_3 = -\int \mathbf{N} \cdot d\mathbf{S} \tag{8.74}$$

where $\dot{W}_3$ is the rate at which the electromagnetic field is doing work on the conduction electrons in the conductors that are inside the surface S_2 in Figure 8.5. In this case the direction of the Poynting vector is into the surface S_2 so that $\int \mathbf{N} \cdot d\mathbf{S}$ is negative in equation (8.74) and, it follows from equation (8.74), that $\dot{W}_3$ is positive and leads to the generation of Joule heat in the conductors inside S_2.

If we consider a surface, that surrounds the Van de Graaff and external circuit in Figure 8.7, and is so large that the fields **E** and **B** are zero on its surface, then equation (8.67) becomes

$$\dot{W}_1 + \dot{W}_2 + \dot{W}_3 = 0. \tag{8.75}$$

According to equation (8.75) the rate $(-\dot{W}_1)$ at which Van der Graaff is doing work is equal to the rate at which work is done on all the conduction electrons in the external circuit.

All we actually did in going from equation (8.58) to equation (8.67) was to derive an alternative way of determining the energy changes in parts of a steady charge and current distribution. We interpreted current flow in Appendix B in terms of the action of a local electric field on the conduction electrons and we used this approach to derive equation (8.58). Some people like to use the model that there is an actual flow of energy given by the Poynting vector. In this approach they visualise the energy as going from the emf in

Figure 8.5 through empty space following the direction of the Poynting vector **N**, as sketched in Figure 8.5. This approach gives the correct result when the values of the Poynting vector are substituted into equation (8.67) and then integrated over a closed surface. We shall return to discuss this interpretation of the Poynting vector in Section 8.9.

8.6. The propagation of energy and the Poynting vector for the general case of varying charge and current distributions

8.6.1. *Theory*

We shall assume now that the charge and current densities ρ and **J** and the fields **E** and **B** in Figure 8.1 are all varying with time. According to equation (8.58),which we derived using the Lorentz force law, in the general case also, the rate at which the electromagnetic field is doing work on the charge and current distribution inside the volume V_0 in Figure 8.1 is given by

$$\frac{\partial W}{\partial t} = \int \mathbf{E} \cdot \mathbf{J}\, dV. \tag{8.58}$$

We shall now elmiminate **J** from equation (8.58) using the Maxwell equation

$$\mathbf{J} = \frac{1}{\mu_0} \nabla \times \mathbf{B} - \varepsilon_0 \dot{\mathbf{E}}$$

giving

$$\frac{\partial W}{\partial t} = \frac{1}{\mu_0} \int \mathbf{E} \cdot (\nabla \times \mathbf{B})\, dV - \varepsilon_0 \int \mathbf{E} \cdot \dot{\mathbf{E}}\, dV. \tag{8.76}$$

Now

$$\begin{aligned} \mathbf{E} \cdot \dot{\mathbf{E}} &= (E_x \dot{E}_x + E_y \dot{E}_y + E_z \dot{E}_z) \\ &= \frac{1}{2} \frac{\partial}{\partial t} (E_x^2 + E_y^2 + E_z^2) = \frac{\partial}{\partial t} \left(\frac{E^2}{2} \right). \end{aligned} \tag{8.77}$$

According to equation (A1.21) of Appendix A1.6 for any two vectors **A** and **B**

$$\nabla \cdot (\mathbf{A} \times \mathbf{B}) = \mathbf{B} \cdot (\nabla \times \mathbf{A}) - \mathbf{A} \cdot (\nabla \times \mathbf{B}). \tag{8.78}$$

Putting **A** = **E** and **B** = **B** in equation (8.78), where **E** and **B** are now the electric and magnetic fields respectively, and then rearranging we have

$$\mathbf{E} \cdot (\nabla \times \mathbf{B}) = \mathbf{B} \cdot (\nabla \times \mathbf{E}) - \nabla \cdot (\mathbf{E} \times \mathbf{B}). \tag{8.79}$$

Since according to Maxwell's equations, $\nabla \times \mathbf{E} = -\dot{\mathbf{B}}$ we find that

$$\mathbf{B} \cdot \nabla \times \mathbf{E} = -\mathbf{B} \cdot \dot{\mathbf{B}} = -\frac{\partial}{\partial t} \left(\frac{B^2}{2} \right). \tag{8.80}$$

Hence equation (8.79) becomes

$$\mathbf{E} \cdot (\nabla \times \mathbf{B}) = -\frac{\partial}{\partial t}\left(\frac{B^2}{2}\right) - \nabla \cdot (\mathbf{E} \times \mathbf{B}). \tag{8.81}$$

Substituting from equation (8.77) and (8.81) into equation (8.76) we get

$$\frac{\partial W}{\partial t} = -\frac{\partial}{\partial t}\int\left(\frac{\varepsilon_0 E^2}{2} + \frac{B^2}{2\mu_0}\right) \mathrm{d}V - \frac{1}{\mu_0}\int \nabla \cdot (\mathbf{E} \times \mathbf{B})\, \mathrm{d}V. \tag{8.82}$$

Applying Gauss' mathematical theorem, which is equation (A1.30) of Appendix A1.7, to convert the second integral on the right hand side into a surface integral, we finally obtain

$$\frac{\partial W}{\partial t} = -\frac{\partial}{\partial t}\int_{V_0}\left(\frac{\varepsilon_0 E^2}{2} + \frac{B^2}{2\mu_0}\right) \mathrm{d}V - \int_{S_0}\frac{(\mathbf{E} \times \mathbf{B})}{\mu_0} \cdot \mathrm{d}\mathbf{S}. \tag{8.83}$$

The integrations in equation (8.83) are carried out over the volume V_0 in Figure 8.1 at the time of observation t. We should not make any allowance for retardation effects, when we evaluate the integrals at the time t.

It is convenient at this stage to introduce the Poynting vector $\mathbf{N}$, which we defined by equation (8.64), and is given by

$$\mathbf{N} = \frac{\mathbf{E} \times \mathbf{B}}{\mu_0} \tag{8.84}$$

and to introduce the "energy density" u_{em} defined by equation (8.52) which is

$$u_{\mathrm{em}} = \left(\frac{\varepsilon_0 E^2}{2} + \frac{B^2}{2\mu_0}\right). \tag{8.85}$$

We shall denote the integral of the "energy density" over the finite volume V_0 in Figure 8.1 by U_{em} so that

$$U_{\mathrm{em}} = \int u_{\mathrm{em}}\, \mathrm{d}V. \tag{8.86}$$

Equations (8.82) and (8.83) can now be rewritten in the more concise forms

$$\frac{\partial W}{\partial t} = -\int (\nabla \cdot \mathbf{N})\, \mathrm{d}V - \frac{\partial}{\partial t}\int u_{\mathrm{em}}\, \mathrm{d}V \tag{8.87}$$

$$\frac{\partial W}{\partial t} = -\int \mathbf{N} \cdot \mathrm{d}\mathbf{S} - \frac{\partial}{\partial t}\int u_{\mathrm{em}}\, \mathrm{d}V \tag{8.88}$$

$$\frac{\partial W}{\partial t} = -\int \mathbf{N} \cdot \mathrm{d}\mathbf{S} - \frac{\partial U_{\mathrm{em}}}{\partial t}. \tag{8.89}$$

All the integrations in equations (8.87), (8.88) and (8.89) are over the volume V_0 and surface S_0 in Figure 8.1 at the time of observation t. Summarizing, we started with equation (8.58) which gives the rate $\dot{W}$ at which the electromagnetic field is doing work on the moving classical point charges making up the macroscopic charge and current distribution that is inside the finite volume V_0 in Figure 8.1 at the time of observation t. We then used Maxwell's equations and some vector analysis to derive equations (8.87), (8.88) and (8.89) which express $\dot{W} = \int \mathbf{E} \cdot \mathbf{J}\, \mathrm{d}V$ in terms of the values of the fields **E** and **B** on the surface S_0 and their values inside the volume V_0 in Figure 8.1 at the time t. This led up to equations (8.87), (8.88) and (8.89). The observable quantities arise from $\dot{W}$, which can, for example, lead to Joule heating, acceleration of the charge distribution etc. According to equation (8.89), $\dot{W}$ is equal to minus the rate of change of the total electromagnetic energy U_{em} inside V_0 plus the flux of the Poynting vector into the surface S_0 of the volume V_0. When equations (8.87), (8.88) and (8.89) are applied to practical cases, we need only apply them in their integral forms so that at this stage we need only assume that u_{em}, U_{em} and **N** are functions that are defined in terms of **E** and **B** by equations (8.85), (8.86) and (8.84) respectively and whose introduction simplifies the mathematical form of equation (8.83). We shall return to discuss the role of the Poynting vector **N** in Section 8.9.

If equation (8.88) is applied to an infinitesimal volume dV the variations over dV can be neglected. If u_{mech} is the total kinetic energy of the classical point charges per cubic metre, then equation (8.88) becomes

$$\dot{W} = \left(\frac{\partial u_{\mathrm{mech}}}{\partial t}\right) \mathrm{d}V = -(\boldsymbol{\nabla} \cdot \mathbf{N})\, \mathrm{d}V - \left(\frac{\partial u_{\mathrm{em}}}{\partial t}\right) \mathrm{d}V \tag{8.90}$$

where we can interpret $\boldsymbol{\nabla} \cdot \mathbf{N}\, \mathrm{d}V$ as the net Poynting flux coming from dV. If we cancel dV we can rearrange equation (8.90) in the form

$$\boldsymbol{\nabla} \cdot \mathbf{N} + \frac{\partial}{\partial t}(u_{\mathrm{mech}} + u_{\mathrm{em}}) = 0. \tag{8.91}$$

This has the form of an equation of continuity. The reader should remember that when equation (8.91) is applied to a practical case it is equation (8.90) that must be integrated over a finite volume surrounding the detector, which brings us back to equation (8.88).

8.6.2. *A plane electromagnetic wave incident on a perfect absorber*

Consider a plane electromagnetic wave that is propagating in the $+x$ direction in empty space, as shown in Figures 8.7(a) and 8.7(b). The plane wave is polarized with its electric vector **E** in the y direction, and can be represented by the equation

$$E_y = E_0 \cos \omega \left(t - \frac{x}{c}\right) \tag{8.92}$$

where ω is the angular frequency. The magnetic field of the plane wave is in the z direction and can be represented by

$$B_z = B_0 \cos \omega \left(t - \frac{x}{c}\right) \tag{8.93}$$

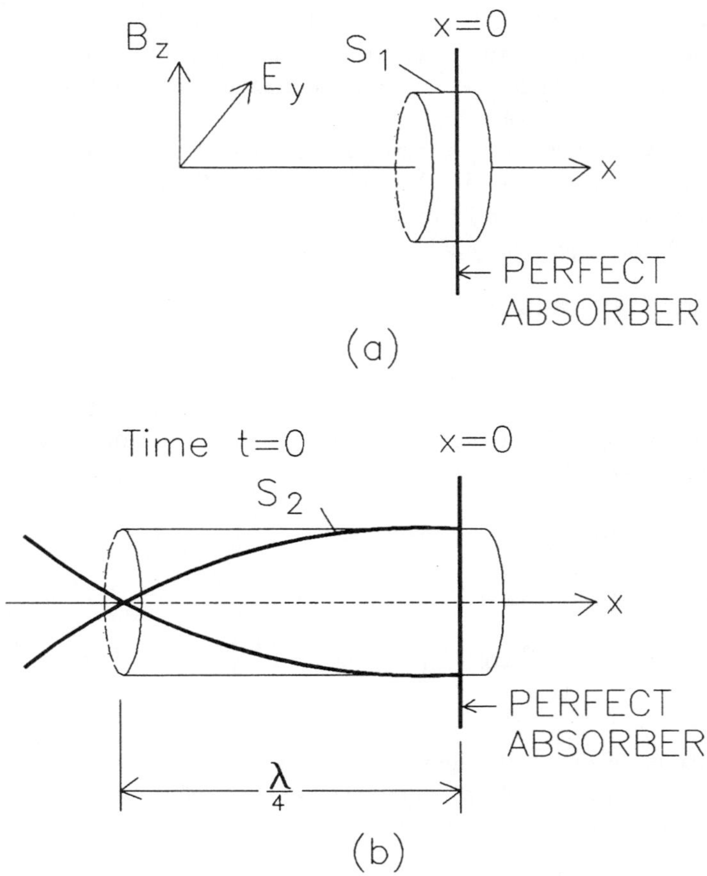

Figure 8.7. Calculation of the radiation pressure due to a plane electromagnetic wave that is incident on a perfect absorber.

where for a plane wave in empty space

$$E_y = cB_z.$$

There is an idealised, very thin, plane, perfect absorber of electromagnetic waves, which is of infinite dimensions in the $x = 0$ plane, as shown in Figure 8.7. We shall consider first the cylindrical surface S_1 shown in Figure 8.7(a) and which has two circular bases each of area 1 m^2, which are just on either side of the perfect absorber. The height of the cylindrical surface S_1 is negligible. Since the volume V_1 enclosed by the surface S_1 is negligible, the second integral on the right hand side of equation (8.88), which is only evaluated over the negligible volume V_1, can be neglected so that for the surface S_1, equation (8.88) reduces to

$$\dot{W} = -\int \mathbf{N} \cdot d\mathbf{S} \tag{8.94}$$

evaluated over the surface S_1. The fields **E** and **B** are zero to the right of the perfect absorber in Figure 8.7(a) so that the Poynting vector **N** is zero on the right hand circular base of S_1. Since the area of the curved surface of S_1 is negligible $\int \mathbf{N} \cdot d\mathbf{S}$ evaluated over the curved surface is zero. The contribution of the left hand circular base of S_1, which is of area 1 m^2, to $\int \mathbf{N} \cdot d\mathbf{S}$ is equal to $-N_i$ where N_i is the Poynting vector of the incident plane wave at the position of the absorber. Hence, $\int \mathbf{N} \cdot d\mathbf{S}$ evaluated over the surface S_1 is equal to $-N_i$ and equation (8.94) reduces to

$$\dot{W} = -\int \mathbf{N} \cdot d\mathbf{S} = N_i. \tag{8.95}$$

This shows that for a plane wave, incident on a perfect absorber at normal incidence, the power given to each square metre of the absorber is equal to the Poynting vector N_i of the incident wave at the position of the absorber. Using equations (8.92) and (8.93) we have

$$\dot{W} = N_i = \frac{E_y B_z}{\mu_0} = \frac{E_0 B_0}{\mu_0} \cos^2 \omega \left(t - \frac{x}{c} \right). \tag{8.96}$$

Using $B_0 = E_0/c$ and $\mu_0 \varepsilon_0 = 1/c^2$ we find that

$$\dot{W} = N_i = c\varepsilon_0 E_0^2 \cos^2 \omega \left(t - \frac{x}{c} \right) = c\left(\frac{\varepsilon_0 E_0^2}{2} + \frac{B_0^2}{2\mu_0} \right) \cos^2 \omega \left(t - \frac{x}{c} \right). \tag{8.97}$$

For the example shown in Figure 8.7(a), when a maximum of the wavefront is reaching the absorber at the time $t = 0$, we find using equations (8.96) and (8.97) that equation (8.95) becomes

$$\dot{W} = N_i = c\left(\frac{\varepsilon_0 E_0^2}{2} + \frac{B_0^2}{2\mu_0} \right) = c\varepsilon_0 E_0^2. \tag{8.98}$$

Consider now the cylindrical surfaces S_2 shown in Figure 8.7(b) at the instant $t = 0$ when a maximum of the wavefront reaches the absorber. The circular bases of S_2 are each of area 1 m^2. One circular surface is just beyond the absorber in Figure 8.7(b) in a region where the radiation electric and magnetic fields are both zero, so that the Poynting vector **N** is zero all over that surface. The other circular surface is at a distance $\lambda/4$ to the left of the absorber in a region where the radiation fields and hence the Poynting vector **N** are zero. The Poynting vector is parallel to the curved surface of S_2 so that $\mathbf{N} \cdot d\mathbf{S}$ is zero all over the curved surface so that integrating over the surface S_2, we have

$$\int \mathbf{N} \cdot d\mathbf{S} = 0.$$

Hence equation (8.88) reduces to

$$\dot{W} = -\frac{\partial}{\partial t} \int \left(\frac{\varepsilon_0 E^2}{2} + \frac{B^2}{2\mu_0} \right) dV. \tag{8.99}$$

In an infinitesimal time dt the wavefront in Figure 8.7(b) moves an infinitesimal distance $c\,dt$ to the right. The change in the integral of "energy density" in the time dt is the same as if we moved the surface S_2 a distance $c\,dt$ to the left at the fixed time $t = 0$. Since the fields **E** and **B** are zero on the left hand side circular base, the change in the integral of the "energy density" due to moving the left hand side circular base of the surface S_2 an infinitesimal distance $c\,dt$ to the left is negligible. The change in the integral of energy density due to moving the right hand side circular base of S_2 an infinitesimal distance $c\,dt$ to the left is a loss of $(\varepsilon_0 E_0^2/2 + B_0^2/2\mu_0)c\,dt$. Dividing by dt and substituting in equation (8.99) we find that, since $E_0 = cB_0$, we have

$$\dot{W} = c\left(\frac{\varepsilon_0 E_0^2}{2} + \frac{B_0^2}{2\mu_0}\right) = c\varepsilon_0 E_0^2 \,. \tag{8.100}$$

This is in agreement with equation (8.98). For the surface S_2 in Figure 8.7(b), the rate at which the electromagnetic field is doing work on the electric charges in the perfect absorber is related to minus the rate of change of the integral of the "energy density" defined by equation (8.85).

If the height of the cylindrical surface S_2 is reduced such that the left hand side circular base of the surface S_2 in Figure 8.7(b) is moved to the right to the position where $E = E_0/2$ and $B = B_0/2$, the integral of the Poynting vector over S_2 is then equal to $-c\varepsilon_0 E_0^2/4$. The change in the integral of the "energy density" in a time $c\,dt$ is obtained by moving S_2 a distance $c\,dt$ to the left losing a contribution of $(\varepsilon_0 E_0^2)c\,dt$ from the right hand side but gaining a contribution of $(\varepsilon_0 E_0^2/4)c\,dt$ from the left hand side giving a net contribution of $-(3\varepsilon_0 E_0^2/4)c$ to the integral of "energy density" which, when added to the contribution of the integral of the Poynting vector in equation (8.88), gives equation (8.100). These examples illustrate that, if we evaluate the right hand side of equation (8.88) at a fixed time, we get the same value for $\dot{W}$ for all the surfaces considered and that in the general case both the surface integral of the Poynting vector and the rate of change of the volume integral of "energy density" contribute to the calculated value of $\dot{W} = \int \mathbf{E} \cdot \mathbf{J}\, dV$ in accordance with equation (8.88). These examples suggest that equation (8.88) is just a book keeping alternative to using equation (8.58) directly.

8.7. The force on a varying charge and current distribution

Equation (8.32) of Section 8.2.4, which expresses the force on a steady charge and current distribution inside a surface S_0 in terms of the resultant fields **E** and **B** on the surface S_0 using Maxwell's stress tensor is only valid for steady (or DC) conditions. We shall now assume that the charge and current distributions both inside and outside the volume V_0 in Figure 8.1 are varying with time. We shall determine the total force **F** on the charge and current distribution inside the volume V_0 in Figure 8.1 at the time of observation t. Let

$\mathbf{P}_{\text{mech}}$ be the total momentum of the moving classical point charges that make up the varying charge and current distribution inside the volume V_0 in Figure 8.1. Integrating equation (8.1) plus equation (8.2), which follow from the Lorentz force law, over the volume V_0 in Figure 8.1 at the time of observation t, we have

$$\mathbf{F} = \frac{d\mathbf{P}_{\text{mech}}}{dt} = \int (\rho\mathbf{E} + \mathbf{J} \times \mathbf{B})\, dV = \int (\rho\mathbf{E} - \mathbf{B} \times \mathbf{J})\, dV. \tag{8.101}$$

We can eliminate ρ and $\mathbf{J}$ using the Maxwell equations:

$$\rho = \varepsilon_0 \nabla \cdot \mathbf{E}; \qquad \mathbf{J} = \frac{1}{\mu_0} \nabla \times \mathbf{B} - \varepsilon_0 \dot{\mathbf{E}}$$

and since $\nabla \cdot \mathbf{B}$ is zero we can add the term $(\nabla \cdot \mathbf{B})\mathbf{B}/\mu_0$ to equation (8.101) which then becomes

$$\mathbf{F} = \int \varepsilon_0 \mathbf{E}(\nabla \cdot \mathbf{E})\, dV - \int \frac{\mathbf{B} \times (\nabla \times \mathbf{B})}{\mu_0}\, dV + \int \varepsilon_0 (\mathbf{B} \times \dot{\mathbf{E}})\, dV + \int \frac{\mathbf{B}(\nabla \cdot \mathbf{B})}{\mu_0}\, dV. \tag{8.102}$$

Differentiating $(\mathbf{E} \times \mathbf{B})$ partially with respect to time gives

$$\frac{\partial}{\partial t}(\mathbf{E} \times \mathbf{B}) = -\frac{\partial}{\partial t}(\mathbf{B} \times \mathbf{E}) = -\dot{\mathbf{B}} \times \mathbf{E} - \mathbf{B} \times \dot{\mathbf{E}}.$$

Since from Maxwell's equation, $-\dot{\mathbf{B}}$ is equal to $\nabla \times \mathbf{E}$, after rearranging we have

$$\mathbf{B} \times \dot{\mathbf{E}} = -\frac{\partial}{\partial t}(\mathbf{E} \times \mathbf{B}) + (\nabla \times \mathbf{E}) \times \mathbf{E} = -\frac{\partial}{\partial t}(\mathbf{E} \times \mathbf{B}) - \mathbf{E} \times (\nabla \times \mathbf{E}).$$

Substituting for $\mathbf{B} \times \dot{\mathbf{E}}$ in equation (8.102) and rearranging we get

$$\mathbf{F} = \int \varepsilon_0[(\nabla \cdot \mathbf{E})\mathbf{E} - \mathbf{E} \times (\nabla \times \mathbf{E})]\, dV + \int \frac{1}{\mu_0}[(\nabla \cdot \mathbf{B})\mathbf{B} - \mathbf{B} \times (\nabla \times \mathbf{B})]\, dV - \frac{\partial}{\partial t} \int \frac{\mathbf{E} \times \mathbf{B}}{\mu_0 c^2}\, dV. \tag{8.103}$$

We shall now apply the methods used in Section 8.2.2 to the second integral on the right side of equation (8.103). Putting $\mathbf{A} = \mathbf{B}$ in equation (8.8) and substituting for $\mathbf{B} \times (\nabla \times \mathbf{B})$ in the second integral, we obtain

$$\int \frac{1}{\mu_0}[(\nabla \cdot \mathbf{B})\mathbf{B} - \mathbf{B} \times (\nabla \times \mathbf{B})]\, dV = \int \frac{1}{\mu_0}\left[(\nabla \cdot \mathbf{B})\mathbf{B} - \nabla\left(\frac{B^2}{2}\right) + (\mathbf{B} \cdot \nabla)\mathbf{B}\right] dV. \tag{8.104}$$

If we put $\mathbf{A} = \mathbf{B}$ in equation (8.12), where $\mathbf{B}$ is now the magnetic field, after rearranging we have

$$\int \frac{1}{\mu_0}[(\nabla \cdot \mathbf{B})\mathbf{B} + (\mathbf{B} \cdot \nabla)\mathbf{B}]\,\mathrm{d}V = \frac{1}{\mu_0}\int \mathbf{B}(\mathbf{B} \cdot \mathrm{d}\mathbf{S}). \tag{8.105}$$

Substituting in equation (8.104) and applying equation (8.11) to the volume integral of $\nabla(-B^2/2\mu_0)$, we have

$$\int \frac{1}{\mu_0}[(\nabla \cdot \mathbf{B})\mathbf{B} - \mathbf{B} \times (\nabla \times \mathbf{B})]\,\mathrm{d}V$$

$$= \int \left(-\frac{B^2}{2\mu_0}\right)\mathrm{d}\mathbf{S} + \frac{1}{\mu_0}\int \mathbf{B}(\mathbf{B} \cdot \mathrm{d}\mathbf{S}) \tag{8.106}$$

$$= \int \mathbf{T}_{\mathrm{mag}} \cdot \mathrm{d}\mathbf{S} \tag{8.107}$$

where $\mathbf{T}_{\mathrm{mag}}$ is the magnetic component of the Maxwell stress tensor, which we derived for the magnetostatic case in Section 8.2.2.

Using the same method we can show that the first integral on the right hand side of equation (8.103) leads to $\int \mathbf{T}_{\mathrm{elec}} \cdot \mathrm{d}\mathbf{S}$. Hence equation (8.103), which gives the total force $\mathbf{F}$ on the varying charge and current distribution inside the volume V_0 in Figure 8.1 at the time of observation t can be written in the forms

$$\mathbf{F} = \int (\rho\mathbf{E} + \mathbf{J} \times \mathbf{B})\,\mathrm{d}V \tag{8.108}$$

$$= \int \left(-\frac{\varepsilon_0 E^2}{2}\right)\mathrm{d}\mathbf{S} + \int \varepsilon_0\mathbf{E}(\mathbf{E} \cdot \mathrm{d}\mathbf{S}) + \int \left(-\frac{B^2}{2\mu_0}\right)\mathrm{d}\mathbf{S}$$
$$+ \frac{1}{\mu_0}\int \mathbf{B}(\mathbf{B} \cdot \mathrm{d}\mathbf{S}) - \frac{\partial}{\partial t}\int \frac{\mathbf{E} \times \mathbf{B}}{\mu_0 c^2}\,\mathrm{d}V \tag{8.109}$$

$$= \int (\mathbf{T}_{\mathrm{elec}} + \mathbf{T}_{\mathrm{mag}}) \cdot \mathrm{d}\mathbf{S} - \frac{\partial}{\partial t}\int \frac{\mathbf{E} \times \mathbf{B}}{\mu_0 c^2}\,\mathrm{d}V$$

$$= \int \mathbf{T} \cdot \mathrm{d}\mathbf{S} - \frac{\partial}{\partial t}\int \frac{\mathbf{E} \times \mathbf{B}}{\mu_0 c^2}\,\mathrm{d}V$$

$$= \int \mathbf{T} \cdot \mathrm{d}\mathbf{S} - \frac{\partial}{\partial t}\int \frac{\mathbf{N}}{c^2}\,\mathrm{d}V \tag{8.110}$$

where $\mathbf{N}$ is the Poynting vector defined by equation (8.64). The surface integral in equation (8.110) is evaluated over the surface S_0 and the volume integral is evaluated over the volume V_0 in Figure 8.1 at the time of observation t. No allowance should be made for any retardation effects when evaluating the integrals at the fixed time t. The tensor $\mathbf{T} = \mathbf{T}_{\mathrm{mag}} + \mathbf{T}_{\mathrm{elec}}$ is the Maxwell stress tensor we derived in Sections 8.2.2 and 8.2.3 for magnetostatic and electrostatic conditions and is given by equation (8.35). The values of the fields $\mathbf{E}$ and $\mathbf{B}$ that must be used in the Maxwell stress tensor are their values on the surface S_0 at the time of observation t. It can be seen from equation (8.110) that, when the charge density ρ and the current density $\mathbf{J}$ are varying, the integral of the Maxwell stress tensor over the surface S_0 is not equal to the integral of the Lorentz force on the moving classical point charges inside the volume V_0 at the time of observation t. For the conditions shown in Figure 8.1, the surface S_0 is outside and spatially separated from the charge and current distribution inside V_0. According to the retarded potentials, the changes in the electromagnetic interactions due to changes in the charge and current distributions outside S_0 take time to travel at the speed of light c from these external sources, so that these changes reach the surface S_0 in Figure 8.1 before they reach the charge and current distribution inside V_0 to effect changes in the Lorentz forces acting on the classical point charges making up the charge and current distribution inside V_0. Changes in the electromagnetic interaction due to the changes in the charge and current distribution inside V_0 take time to travel outwards to reach the surface S_0 in Figure 8.1. Hence, due to the finite time it takes for changes in the electromagnetic interaction to propagate from its sources, it is reasonable to find that, when ρ and $\mathbf{J}$ are varying, the integral of the Maxwell stress tensor over the surface S_0 in Figure 8.1 at the time t is not equal to the volume integral of the Lorentz forces acting on the moving classical point charges inside V_0 evaluated at the same time t. According to equation (8.110) the compensating term is equal to minus the rate of change of the integral of the Poynting vector $\mathbf{N}$ divided by c^2 over the volume V_0 evaluated at the time t. We shall now introduce a new variable $\mathbf{p}_{em}$ defined by the equation

$$\mathbf{p}_{em} = \frac{\mathbf{E} \times \mathbf{B}}{\mu_0 c^2} = \varepsilon_0 \mathbf{E} \times \mathbf{B} = \frac{\mathbf{N}}{c^2} \tag{8.111}$$

where $\mathbf{N}$ is the Poynting vector. The vector $\mathbf{p}_{em}$ is generally called the electromagnetic "momentum density" of the electromagnetic field. The total "electromagnetic momentum" inside the volume V_0 at the time t, which will be denoted by $\mathbf{P}_{em}$, is

$$\mathbf{P}_{em} = \int \mathbf{p}_{em}\, dV = \int \frac{\mathbf{N}}{c^2}\, dV = \int \frac{\mathbf{E} \times \mathbf{B}}{\mu_0 c^2}\, dV. \tag{8.112}$$

Equation (8.110) can now be rewritten in the form

$$\frac{\partial \mathbf{P}_{mech}}{\partial t} = \int \mathbf{T} \cdot d\mathbf{S} - \frac{\partial \mathbf{P}_{em}}{\partial t} \tag{8.113}$$

where $\mathbf{P}_{mech}$ is the total mechanical momentum of the moving classical point charges inside the volume V_0 at the time t, due to the motions of these classical point charges. The integrations are all carried out at the time of observation t.

It can be shown, using the extension of Gauss' theorem of vector analysis to tensors, that the first integral on the right hand side of equation (8.113) can be rewritten as follows:

$$\int \mathbf{T} \cdot d\mathbf{S} = \int (\boldsymbol{\nabla} \cdot \mathbf{T})\, dV$$

where $\nabla \cdot \mathbf{T}$ is the divergence of the Maxwell stress tensor $\mathbf{T}$. The components of $\nabla \cdot \mathbf{T}$ are given in cartesian coordinates by

$$(\nabla \cdot \mathbf{T})_i = \sum_j \nabla_j \mathbf{T}_{ji}.$$

Equation (8.113) can now be rewritten as follows

$$\mathbf{F} = \int \nabla \cdot \mathbf{T}\, \mathrm{d}V - \frac{\partial}{\partial t}\int \mathbf{p}_{\mathrm{em}}\, \mathrm{d}V.$$

If we consider a small volume $\mathrm{d}V$ we can ignore the variations of the variables over the dimensions of $\mathrm{d}V$, so that if $\mathbf{p}_{\mathrm{mech}}$ is the total mechanical momentum of the classical point charges per unit volume, we have

$$\left(\frac{\partial \mathbf{p}_{\mathrm{mech}}}{\partial t}\right) \mathrm{d}V = \nabla \cdot \mathbf{T}\, \mathrm{d}V - \left(\frac{\partial \mathbf{p}_{\mathrm{em}}}{\partial t}\right) \mathrm{d}V. \tag{8.114}$$

Cancelling $\mathrm{d}V$ and rearranging, we have

$$\nabla \cdot (-\mathbf{T}) + \frac{\partial}{\partial t}(\mathbf{p}_{\mathrm{mech}} + \mathbf{p}_{\mathrm{em}}) = 0. \tag{8.115}$$

Equation (8.115) has the mathematical form of a continuity equation. When it is applied to a practical situation, equation (8.115) must be integrated at a fixed time over a finite volume surrounding the detector, which leads us back to equation (8.110).

To illustrate the application of equation (8.110), we shall consider again the example of the plane electromagnetic plane wave, that is incident on a thin, perfect absorber that is perpendicular to the direction of propagation of the plane wave shown previously in Figures 8.7(a) and 8.7(b). The properties of the plane wave are as described in Section 8.6.2. We shall only consider the time $t = 0$ when a maximum of the plane wave reaches the absorber and the electric and magnetic fields have magnitudes E_0 and B_0 respectively at the absorber. In the example shown in Figure 8.7(a), the circular bases of the cylindrical surface S_1, which are both of area 1 m^2, are just on either side of the perfect absorber. Since the height of the cylindrical surface S_1 is negligible, the volume V_1 of the cylindrical surface is negligible and so we can neglect the integral of $\mathbf{N}/c^2$ over the volume V_1, and equation (8.110) reduces to

$$\mathbf{F} = \int \mathbf{T} \cdot \mathrm{d}\mathbf{S} \tag{8.116}$$

where the integration is over the closed surface S_1. Since the angle between $\mathbf{E}$ and any element of area $\mathrm{d}\mathbf{S}$ of the left hand side circular base of the surface S_1 in Figure 8.7(a) is 90° and the angle between $\mathbf{B}$ and $\mathrm{d}\mathbf{S}$ is also 90°, it follows from equations (8.27) and (8.17) that for the left hand side circular base

$$\mathbf{T} \cdot \mathrm{d}\mathbf{S} = -\left(\frac{\varepsilon_0 E_0^2}{2} + \frac{B_0^2}{2\mu_0}\right) \mathrm{d}\mathbf{S} = -(\varepsilon_0 E_0^2)\, \mathrm{d}\mathbf{S} \tag{8.117}$$

where we have used the result that $E_0 = cB_0$ for a plane wave in empty space. Due to the effect of the perfect absorber, the fields **E** and **B** are zero on the right hand side circular base so that $\mathbf{T} \cdot \mathrm{d}\mathbf{S}$ is zero on that surface. It follows from equations (8.27) and (8.17) that there is a net contribution of $(\varepsilon_0 E^2)\mathrm{d}\mathbf{S}$ to the integral in equation (8.116) associated with an element of area $\mathrm{d}\mathbf{S}$ of the curved surface. These contributions from the curved surface add up to zero. Hence we are left with only a contribution from the left hand side circular base of the surface S_1 in Figure 8.7(a). Since the area of the base is 1 m^2, using equations (8.117) and (8.98) we find that equation (8.116) gives

$$F = \varepsilon_0 E^2 = \frac{N_i}{c} \tag{8.118}$$

where F is now the force per unit area, that is the pressure on the perfect absorber. The direction of this force is in the direction opposite to $\mathrm{d}\mathbf{S}$, which is to the right into the perfect absorber. This is an example of radiation pressure.

Consider now the cylindrical surface S_2 in Figure 8.7(b). There is no contribution to $\int \mathbf{T} \cdot \mathrm{d}\mathbf{S}$ from the right hand side circular surface, where the fields **E** and **B** are both zero and from the curved surface, where the resultant contribution is zero. The left hand circular surface of S_2 is at a node of the wavefront, where the fields **E** and **B** are both zero. Hence the integral of $\mathbf{T} \cdot \mathrm{d}\mathbf{S}$ over the surface of S_2 in Figure 8.7(b) is zero, and equation (8.110) reduces to

$$\mathbf{F} = -\frac{\partial}{\partial t}\int \frac{\mathbf{N}}{c^2}\,\mathrm{d}V. \tag{8.119}$$

In this example, the force per unit area on the perfect absorber is related to the rate of change of the integral of the electromagnetic "momentum density" $\mathbf{p}_{\mathrm{em}} = \mathbf{N}/c^2$. The change in the integral of $\mathbf{N}/c^2$ in a time $\mathrm{d}t$ is the same as moving the surface S_2 a distance $c\,\mathrm{d}t$ to the left at a fixed time. Since the area of both the circular bases is 1 m^2, the change in the integral of $\mathbf{N}/c^2$ is equal to $(\mathbf{N}_i/c^2)c\,\mathrm{d}t$, where $\mathbf{N}_i$ is the Poynting vector at the absorber at the time $t = 0$. Dividing by $\mathrm{d}t$ and substituting in equation (8.119), we obtain

$$F = \frac{N_i}{c} = \frac{E_0 B_0}{\mu_0 c} = \frac{E_0^2}{\mu_0 c^2} = \varepsilon_0 E_0^2 \tag{8.120}$$

where F is the force per unit area, which is the radiation pressure on the absorber in Figure 8.7(b). Equation (8.120) is in agreement with equation (8.118). In this example we did not get the correct value for the radiation pressure when we substituted the values of **E** and **B** on the surface S_2 in Figure 8.7(b) into only Maxwell's stress tensor. This result is to be expected, since we used the values of **E** and **B** on the left hand circular surface at the time $t = 0$, which are the values of the fields that will reach the absorber at the time $t = T/4$, where T is the period of the plane wave. We compensated for this by including the rate of change of the integral of $\mathbf{N}/c^2$ over the volume V_2 enclosed by the surface S_2 evaluated at the time $t = 0$.

If we reduce the height of the surface S_2 in Figure 8.7(b), such that the values of the fields on the left hand side circular surface are $E_0/2$ and $B_0/2$, it is straightforward for the reader to show that equation (8.110) becomes

$$\mathbf{F} = \int \mathbf{T} \cdot \mathrm{d}\mathbf{S} - \frac{\partial}{\partial t}\int \frac{\mathbf{N}}{c^2}\,\mathrm{d}V$$

$$F = \frac{\varepsilon_0 E_0^2}{4} + \frac{3\varepsilon_0 E_0^2}{4} = \varepsilon_0 E_0^2 \qquad (8.121)$$

This result is consistent with equations (8.118) and (8.120) and illustrates how we get the same result for the radiation pressure on the perfect absorber in Figure 8.7(b) whatever the shape and dimensions of the closed surface used for the evaluation of the integrals in equation (8.110) at a fixed time. It also illustrates how, in the general case, there are contributions from both the integrals on the right hand side of equation (8.110), and suggests that equation (8.110) is a book keeping alternative to using the Lorentz force law directly.

It will now be assumed that the absorber in Figure 8.7(a) is replaced by a perfect metallic reflector. It is shown in text books on electromagnetism that in this case the plane wave is reflected with a change of phase such that the total electric field at the boundary is zero and the magnetic field is of magnitude $2B_0$. If we apply equation (8.110) to the surface S_1 in Figure 8.7(a) we find that the value of $\mathbf{T} \cdot \mathrm{d}\mathbf{S}$ on the left hand side circular surface is

$$\mathbf{T} \cdot \mathrm{d}\mathbf{S} = -\left(\frac{\varepsilon_0 E^2}{2} + \frac{B^2}{2\mu_0}\right)\mathrm{d}\mathbf{S} = -\frac{(2B_0)^2}{2\mu_0}\,\mathrm{d}\mathbf{S}$$

$$= -\frac{2(B_0)^2}{\mu_0}\,\mathrm{d}\mathbf{S} = -2\varepsilon_0 E_0^2\,\mathrm{d}\mathbf{S}$$

Substituting in equation (8.116) we find that the force per unit area, which is the radiation pressure, is

$$F = 2\varepsilon_0 E_0^2. \qquad (8.122)$$

This is double the value given by equation (8.118). We were able to determine the value of the radiation pressure by substituting the values of $\mathbf{E}$ and $\mathbf{B}$ at the surface of the reflector into equation (8.110) which we derived from the integral of $\mathbf{J} \times \mathbf{B}$ evaluated over the volume V_1 enclosed by S_1. There was no need to enquire about the precise nature of the electromagnetic processes taking place in that portion of the reflector that is inside the volume V_1 when we apply equation (8.110) to determine the radiation pressure. In the case of a metallic conductor the plane wave does penetrate into the conductor but is attenuated. The electric field $\mathbf{E}$ of the plane wave inside the conductor gives a current flow given by $\mathbf{J} = \sigma\mathbf{E}$. This current is then acted upon by the magnetic field of the plane wave giving a resultant force given by $\int \mathbf{J} \times \mathbf{B}\,\mathrm{d}V$ which is directed into the conductor and gives rise to the radiation pressure given by equation (8.122). The full calculation is given by Lorrain and Corson [4], but there is no need to do this calculation as we can derive the result more quickly using equation (8.110).

8.8. Conservation laws for a system of moving charges in empty space

8.8.1. *The law of conservation of linear momentum*

Consider a system of moving classical point charges that build up macroscopic charge and current distributions in otherwise empty space. According to equation (8.113), instead of integrating the Lorentz force to determine the resultant force **F** on the charge and current distribution inside the finite volume V_0 in Figure 8.1 at the time of observation we can, as an alternative use the equation

$$\mathbf{F} = \dot{\mathbf{P}}_{\text{mech}} = \int \mathbf{T} \cdot \mathrm{d}\mathbf{S} - \dot{\mathbf{P}}_{\text{em}} \tag{8.123}$$

where

$$\mathbf{P}_{\text{mech}} = \sum p_i \tag{8.124}$$

is the total linear momentum, which is the sum of the linear momenta of the moving charges inside the volume V_0, and $\mathbf{P}_{\text{em}}$ is the total "electromagnetic momentum", which is defined as the integral of the electromagnetic "momentum density", over the volume V_0 at a fixed time and is given by

$$\mathbf{P}_{\text{em}} = \int \frac{\mathbf{N}}{c^2}\,\mathrm{d}V = \int \frac{\mathbf{E} \times \mathbf{B}}{\mu_0 c^2}\,\mathrm{d}V. \tag{8.125}$$

The Maxwell stress tensor **T** is given by equation (8.35). If we make the volume V_0, over which we are integrating, tend to infinity, the electric and magnetic fields will tend to zero on the surface of V_0, so that the stress tensor will tend to zero and equation (8.123) reduces to

$$\dot{\mathbf{P}}_{\text{mech}} = -\dot{\mathbf{P}}_{\text{em}} \tag{8.126}$$

showing that, instead of using the Lorentz force law to determine $\dot{\mathbf{P}}_{\text{mech}}$ the rate of change of the total mechanical momentum of all the charges in the system at a fixed time, we can determine $\dot{\mathbf{P}}_{\text{mech}}$ from the rate of change of the total electromagnetic "momentum" given by equation (8.125).

Integrating with respect to time, we have

$$\mathbf{P}_{\text{mech}} + \mathbf{P}_{\text{em}} = \text{a constant} \tag{8.127}$$

Equations (8.124) and (8.125) for $\mathbf{P}_{\text{mech}}$ and $\mathbf{P}_{\text{em}}$ are evaluated at the time of observation. The summation in equation (8.124) is now over all the moving charges in the complete system and the integral in equation (8.125) is now over the whole of space. Equation (8.127) is generally called the **law of conservation of linear momentum** for a system of moving charges in empty space. If we only want to consider a part of the system we have to revert to equation (8.123) and the contribution of the Maxwell stress tensor is important. In fact in DC and quasi-stationary conditions, the integral of the Maxwell stress tensor over the surface of the volume V_0 in Figure 8.1 is the main contribution to the right hand side of equation (8.123).

To illustrate the interpretation of equation (8.126) we shall consider the simple, idealized example shown in Figure 8.8. A charge of magnitude q_1 is constrained to move with uniform velocity v along the x axis of the inertial frame Σ. It is at the origin of Σ at the time $t = 0$ in Σ. A second charge of magnitude q_2 is constrained to move with uniform velocity u along the y axis in Figure 8.8. It is at a distance y from the origin at the time $t = 0$, when the charge q_1 is at the origin. According to equation (3.25) the electric field $\mathbf{E}_1$ at the position of q_2 due to the charge q_1 is in the $+y$ direction at the time $t = 0$ and is of magnitude

$$E_1 = \frac{q_1}{4\pi\varepsilon_0 y^2(1 - v^2/c^2)^{1/2}} . \tag{8.128}$$

According to the Lorentz force law, the electric force on the charge q_2 at the time $t = 0$ is

$$(\mathbf{f}_2)_y = q_2 E_1 = \frac{q_1 q_2}{4\pi\varepsilon_0 y^2(1 - v^2/c^2)^{1/2}} . \tag{8.129}$$

According to equation (3.28) the magnetic field at the position of the charge

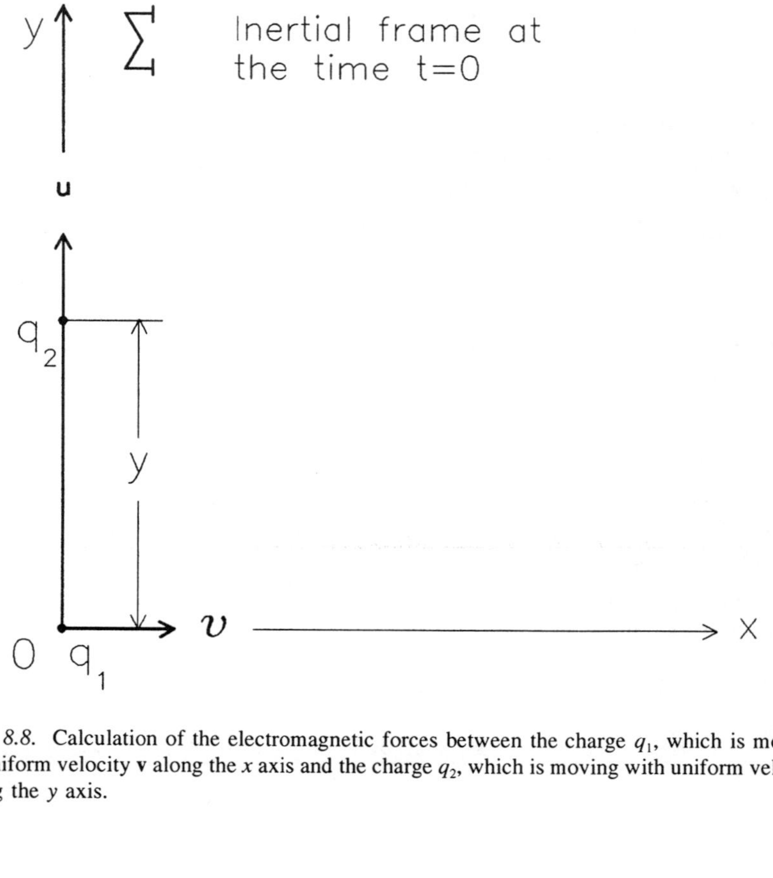

Figure 8.8. Calculation of the electromagnetic forces between the charge q_1, which is moving with uniform velocity **v** along the x axis and the charge q_2, which is moving with uniform velocity **u** along the y axis.

q_2 due to the charge q_1 at the time $t = 0$ is in $+z$ direction and is of magnitude

$$B_1 = \frac{q_1 v}{4\pi\varepsilon_0 c^2 y^2 (1 - v^2/c^2)^{1/2}} . \tag{8.130}$$

It follows from the Lorentz force law that the magnetic force on the charge q_2 is in the $+x$ direction at $t = 0$ and is of magnitude

$$(\mathbf{f}_2)_x = |q_2 \mathbf{u} \times \mathbf{B}_1| = \frac{q_1 q_2 v u}{4\pi\varepsilon_0 c^2 y^2 (1 - v^2/c^2)^{1/2}} . \tag{8.131}$$

According to equation (3.25) the electric field at the position of the charge q_1 due to the charge q_2 in Figure 8.8 is in the $-y$ direction at the time $t = 0$ and is of magnitude

$$E_2 = \frac{q_2(1 - u^2/c^2)}{4\pi\varepsilon_0 y^2} . \tag{8.132}$$

The magnetic field at the position of the charge q_1 due to the charge q_2 is zero at the time $t = 0$. Hence it follows from the Lorentz force law that the total electromagnetic force on the charge q_1 at $t = 0$ is an electric force given by

$$(\mathbf{f}_1)_y = -\frac{q_1 q_2 (1 - u^2/c^2)}{4\pi\varepsilon_0 y^2} . \tag{8.133}$$

Comparing equations (8.129) and (8.133) we see that the components of the forces on q_1 and q_2 in the y direction at the time $t = 0$ in Figure 8.8 are not equal and opposite. Furthermore, there is a component of the total force on the charge q_2 in the $+x$ direction, given by equation (8.131), but there is no component of the force on the charge q_1 in the x direction. These results show that according to classical electromagnetism the force on the charge q_2 due to the charge q_1 is not equal to the force on q_1 due to q_2 at the time $t = 0$, showing that action and reaction are not equal and opposite, though the deviations from equality are only of order u^2/c^2, v^2/c^2 and uv/c^2. These results show that if $\mathbf{p}_1$ and $\mathbf{p}_2$ are the momenta of charges q_1 and q_2 respectively, then

$$\dot{\mathbf{p}}_1 + \dot{\mathbf{p}}_2 = \dot{\mathbf{P}}_{\text{mech}} \neq 0$$

showing that the total linear momentum of the two particles is not conserved. To illustrate equation (8.126), we shall now assume that the charges q_1 and q_2 are not constrained to move with uniform velocities and we shall consider it as a general electromagnetic interaction between the charges. The electromagnetic "momentum density" is

$$\frac{\mathbf{N}}{c^2} = \frac{\mathbf{E} \times \mathbf{B}}{\mu_0 c^2} = \frac{(\mathbf{E}_1 + \mathbf{E}_2) \times (\mathbf{B}_1 + \mathbf{B}_2)}{\mu_0 c^2}$$
$$= \frac{\mathbf{E}_1 \times \mathbf{B}_1 + \mathbf{E}_1 \times \mathbf{B}_2 + \mathbf{E}_2 \times \mathbf{B}_1 + \mathbf{E}_2 \times \mathbf{B}_2}{\mu_0 c^2} .$$

The two terms $\mathbf{E}_1 \times \mathbf{B}_1/\mu_0 c^2$ and $\mathbf{E}_2 \times \mathbf{B}_2/\mu_0 c^2$ are responsible for the electromagnetic masses of the charges when they are accelerated. The other terms vary; for example we can visualize that when q_1 and q_2 move in Figure 8.8, the magnitudes of $\mathbf{E}_1$ and $\mathbf{B}_2$ vary at all points in space such that $\int(\mathbf{E}_1 \times \mathbf{B}_2/\mu_0 c^2)\,\mathrm{d}V$, integrated over the whole of space at a fixed time, varies with time. Similarly, $\int(\mathbf{E}_2 \times \mathbf{B}_1/\mu_0 c^2)\,\mathrm{d}V$ varies with time such that

$$-\frac{\partial \mathbf{P}_{\mathrm{em}}}{\partial t} = -\frac{\partial}{\partial t}\int \frac{(\mathbf{E}_1 \times \mathbf{B}_2 + \mathbf{E}_2 \times \mathbf{B}_1)\,\mathrm{d}V}{\mu_0 c^2}$$

is equal to the difference between the forces between the charges at the time of observation, so that

$$\dot{\mathbf{P}}_{\mathrm{mech}} = -\dot{\mathbf{P}}_{\mathrm{em}}$$

giving after integrating with respect to time

$$\mathbf{P}_{\mathrm{mech}} + \mathbf{P}_{\mathrm{em}} = \text{a constant}$$

according to which it is the sum of the total mechanical momentum and the total "electromagnetic momentum" that is a constant.

When the charges q_1 and q_2 are far apart before and after the collision $\mathbf{E}_1 \times \mathbf{B}_2$ and $\mathbf{E}_2 \times \mathbf{B}_1$ are negligible and it is not necessary to attribute any "momentum" to the electromagnetic field. Hence, the total linear mechanical momentum of the charges, when they are far apart before the collision, should be equal to the total linear mechanical momentum of the charges when they are far apart after the collision. Provided that no radiation is emitted, the collision can be treated as an elastic collision and relativistic mechanics can be applied to the conditions before and after the collision. However during the collision whilst $\mathbf{E}_1 \times \mathbf{B}_2/\mu_0 c^2$ and $\mathbf{E}_2 \times \mathbf{B}_1/\mu_0 c^2$ are not negligible, the total linear momentum of the charges is not conserved. We shall return in Section 10.7 of Chapter 10 to show that equations (8.129), (8.131) and (8.133) are consistent with the transformations of special relativity.

Our discussion of the role of the "electromagnetic momentum" was only qualitative. A quantitative discussion of a similar example was given by Page and Adams [5] who considered the case of two isolated current elements. If we put stationary charges $-q_1$ and $-q_2$ close to the charges $+q_1$ and $+q_2$ respectively in Figure 8.8, it would correspond, to a first rough approximation, to two isolated current elements. It is straightforward for the reader to show that all the electric forces between the charges would cancel out leaving only the magnetic force given by equation (8.131), which is only of order vu/c^2 times the electric forces between the separated charges $+q_1$ and $+q_2$. This illustrates that the magnetic forces between current elements are of second order and do not obey Newton's third law. Page and Adams [5] showed that the unbalanced magnetic forces are equal to minus the rate of change of the "electromagnetic momentum" of the field such that equation (8.127) is satisfied.

8.8.2. *Critique of the law of conservation of momentum*

In our analysis of the example of the two moving charges in Figure 8.8, we found that, at a fixed time, the forces between the charges were not equal and opposite and did not act along the line joining the charges, though the deviations from Newton's third law were only of second order. Generalizing we conclude that the forces between the charges in a system of moving charges do not add up to zero at a fixed instant of time. It is assumed in special relativity, as well as in Newtonian mechanics, that force is equal to the rate of change of momentum. Hence, we conclude that for a system of moving and accelerating classical point charges at a fixed time, we have

$$\dot{\mathbf{P}}_{\text{mech}} = \sum \dot{\mathbf{p}}_i \neq 0$$

According to the law of conservation of linear momentum of Newtonian mechanics, we should have

$$\sum \dot{\mathbf{p}}_i = 0$$

This shows that the finite value of $\dot{\mathbf{P}}_{\text{mech}}$ arises from deviations from Newtonian mechanics, and comparing with equation (8.126) we see that the contribution of these deviations from Newtonian mechanics is given by $-\dot{\mathbf{P}}_{\text{em}}$. We shall show in Chapter 10 that classical electromagnetism is consistent with the theory of special relativity and not with Newtonian mechanics, so that, when we interpret the laws of classical electromagnetism, we must be prepared to replace long held concepts and interpretations developed in Newtonian mechanics, by concepts and interpretations consistent with special relativity.

In the theory of speical relativity, energy and momentum are closely associated. For example, the energy and momentum of a single particle combine to form a 4-vector. Hence, it is reasonable in a relativistic theory, such as classical electromagnetism, to use potential energy as an example to help us to interpret the "momentum" of the electromagnetic field. In Newtonian mechanics, the potential energy of a system is a useful concept, since it leads to the law of conservation of energy, according to which the gain (or loss) of kinetic energy is equal to the loss (or gain) of the potential energy of the system. Everybody accepts, on the basis of their experiences in Newtonian mechanics that the total kinetic energy of a system of particles changes when the positions and velocities of the particles change under the influence of the gravitational forces between the particles. Why then does the total linear momentum of the particles not change also? The answer is that, in Newtonian mechanics, it is assumed that the forces between the particles obey Newton's third law and this leads to the law of conservation of linear momentum, in the way outlined in Section 8.1. We have already illustrated that Newton's third law is not valid for the forces between moving charges. Hence it should not be surprising to find that in classical electromagnetism, the law of conservation of linear momentum is not valid in its Newtonian form for the case of moving classical point charges. It is not unreasonable to find in classical electromagnetism, which is consistent with special relativity where energy and momentum are closely associated, that, if the total kinetic energy of a system of moving charges varies, due to the motions of the charges due to the electromagnetic forces between them, the total linear mechanical momentum of the charges also varies due to deviations from Newtonian mechanics. The rate of change of the total linear mechanical momentum is given by the equation

$$\dot{\mathbf{P}}_{\text{mech}} = \sum \dot{\mathbf{p}}_i = -\dot{\mathbf{P}}_{\text{em}} \tag{8.126}$$

where $\mathbf{P}_{\text{em}}$ has been called the total electromagnetic "momentum" of the field, obtained by integrating equation (8.125) over the whole of space at one instant of time. By analogy with the law of conservation of energy, it is reasonable to interpret equation (8.126) by suggesting that the rate of gain of total mechanical momentum is equal to the rate of loss of a **potential** momentum. In such an interpretation, there is no need to look for a Newtonian mechanical interpretation of electromagnetic field "momentum" any more than we need to look for a mechanical interpretation of, for example, the potential energy of an electrostatic system. Thus we can interpret equation (8.126) by saying that there is a capability in a system of moving classical point charges to increase or decrease their total linear mechanical momentum, when the charges move under the influence of the electromagnetic forces between them; this change in total linear mechanical momentum arises from the contribution of second order relativistic deviations from Newton's laws and from the finite time it takes changes in the electromagnetic interaction to propagate from charge to charge. Examples of the role of second order non-central forces in classical electromagnetism will be given in our discussion of the law of conservation of angular momentum in Section 8.9.5. From now on we shall refer to the electromagnetic field "momentum" obtained by integrating equation (8.125) over the whole of space as **potential momentum**, and treat it in a similar way to the conventional interpretation of potential energy.

Integrating equation (8.126) with respect to time we obtained

$$\mathbf{P}_{\text{mech}} + \mathbf{P}_{\text{em}} = \text{a constant} \tag{8.127}$$

By analogy with potential energy, we can now interpret equation (8.127) as follows. When the (electromagnetic) potential momentum of a system of moving classical point charges is finite, there is a capacity in the electromagnetic system to increase the total linear momentum of the charges when they move under the influence of the electromagnetic forces between them, the (electromagnetic) potential momentum being reduced in the process. Conversely, if the charges have finite linear momentum, they have the capacity to move in such a way as to increase the (electromagnetic) potential momentum losing linear momentum in the process, subject in all cases to the condition that the total linear momentum of the charges plus (electromagnetic) potential momentum is conserved.

Consider the equation

$$\dot{\mathbf{P}}_{\text{mech}} = -\frac{\partial}{\partial t}\int \frac{\mathbf{E}\times\mathbf{B}}{\mu_0 c^2}\,\mathrm{d}V \tag{8.126a}$$

where the integration is over the whole of space at a fixed instant of time. It is assumed nowadays that changes in the electromagnetic interaction between moving charges take a finite time to propagate at the speed of light. It is stretching the imagination to interpret equation (8.126a) by assuming that the rates of change of the fields **E** and **B** at a point well away from any of the moving charges in the system can contribute in a causal way to the rates of change of the linear momenta of distant charges at that precise instant. It is best to interpret equation (8.126a) as a book-keeping exercise carried out at a fixed time rather than as a causal relation between the fields and particles at a particular instant of time. By this we mean that to apply equation (8.126a) we must determine the rate of change of $(\mathbf{E}\times\mathbf{B}/\mu_0 c^2)\,\mathrm{d}V$ at every volume element $\mathrm{d}V$ in the whole of space at the time of observation, and then add up these values to determine $-\dot{\mathbf{P}}_{\text{em}}$, which is then equal to the value of $\dot{\mathbf{P}}_{\text{mech}}$ determined by integrating the Lorentz force law over the whole of space at the time of observation. This interpretation of conservation laws as book-keeping rules will be supported by arguments based on special relativity, which will be given later in Section 10.9 of Chapter 10, where we shall discuss the causal aspects of conservation laws.

8.8.3. *Operational definition of the vector potential*

To develop an operational definition of the vector potential, we shall consider a test charge q that is at rest in a steady magnetic field, at a point where the magnetic field is **B** and the vector potential is **A**. According to equation (8.125) the total "electromagnetic momentum" of the field is

$$\mathbf{P}_{\text{em}} = \int \frac{\mathbf{N}}{c^2}\,\mathrm{d}V = \int \frac{\mathbf{E}\times\mathbf{B}}{\mu_0 c^2}\,\mathrm{d}V \tag{8.134}$$

where **E** is the electric field due to the stationary test charge q and the integration is over the whole of space. The electrostatic field **E** due to the stationary charge q can be expressed in the form $-\nabla\phi$, where ϕ is the electrostatic scalar potential due to the stationary charge q and is given by

$$\phi = \frac{q}{4\pi\varepsilon_0 r} \tag{8.135}$$

where r is the distance from the stationary test charge q to the volume element $\mathrm{d}V$ in equation (8.134). Putting $\mathbf{E} = -\nabla\phi$ in equation (8.134), we get

$$\mathbf{P}_{\text{em}} = \frac{1}{\mu_0 c^2}\int -(\nabla\phi)\times\mathbf{B}\,\mathrm{d}V.$$

It follows from equation (A1.23) of Appendix A1.6 that

$$-\nabla\phi\times\mathbf{B} = -\nabla\times(\phi\mathbf{B}) + \phi\nabla\times\mathbf{B}.$$

Hence,

$$\mathbf{P}_{em} = -\frac{1}{\mu_0 c^2}\int \nabla \times (\phi \mathbf{B})\, dV + \frac{1}{\mu_0 c^2}\int \phi(\nabla \times \mathbf{B})\, dV. \tag{8.136}$$

Using equation (A1.33) of Appendix A1.7, we find that

$$\int \nabla \times (\phi \mathbf{B})\, dV = -\int \phi \mathbf{B} \times d\mathbf{S}. \tag{8.137}$$

Provided the integration is over the whole of space, the product $\phi\mathbf{B}$ tends to zero on the surface of the volume of integration and the integral in equation (8.137) is zero, so that the first integral on the right hand side of equation (8.136) is zero. Since for steady magnetic fields $\nabla \times \mathbf{B} = \mu_0 \mathbf{J}$, equation (8.136) can now be rewritten in the form

$$\mathbf{P}_{em} = \frac{1}{c^2}\int \phi \mathbf{J}\, dV.$$

Using equation (8.135) to substitute for ϕ, we find that the initial total electromagnetic potential momentum of the field can now be expressed in the form

$$\mathbf{P}_{em} = q\left(\frac{\mu_0}{4\pi}\right)\int \frac{\mathbf{J}\, dV}{r}. \tag{8.138}$$

Using equation (1.74) of Chapter 1, which is valid for magnetostatic conditions, we can rewrite equation (8.138) in the form

$$\mathbf{P}_{em} = q\mathbf{A} \tag{8.139}$$

where $\mathbf{P}_{em}$ is the total (electromagnetic) potential momentum of the electromagnetic field, when the test charge q is at rest at a point in a steady magnetic field where the vector potential is $\mathbf{A}$. According to equation (8.127), if the experimental conditions are changed the (electromagnetic) momentum $q\mathbf{A}$ can be associated with an increase of up to $q\mathbf{A}$ in the total momentum of the charges in the system. We can now define the magnetostatic vector potential $\mathbf{A}$ operationally as the ratio of the potential momentum $q\mathbf{A}$ of the system to the magnitude q of the test charge, in the limit as q tends to zero. We can see the reasonableness of interpreting $q\mathbf{A}$ as a potential momentum by considering what would happen if the magnetic field did go to zero quickly enough for the charge q not to move significantly while $\mathbf{B}$ was decreasing and while there was an induction electric field $-\dot{\mathbf{A}}$. The electric force on the test charge would be $-q\dot{\mathbf{A}}$ so that the total impulse given to the charge q, obtained by integrating from $\mathbf{A} = \mathbf{A}$ to $\mathbf{A} = 0$, would be $+q\mathbf{A}$, which would lead to a corresponding change in the linear momentum of the charge q.

In Section 1.2.10 of Chapter 1, we defined the electrostatic scalar potential ϕ operationally in terms of the potential energy of a test charge q using equation (1.35). Thus both $\mathbf{A}$ and ϕ could now, in principle, be defined operationally in terms of the potential momentum and potential energy of a test charge respectively, so that the equations for the potentials ϕ and $\mathbf{A}$ could be treated as relations between operationally defined quantities. The differences between the operational definitions of the field vectors **E** and **B** and the potentials ϕ and **A** were summarized by Konopinski [6] as follows:

> Whereas **E** and **B** describe a field in terms of the forces the field can exert on charged matter, ϕ and **A** describe the same field in terms of energies and momenta that the field makes available to matter.

8.8.4. *The law of conservation of energy*

Consider a system of moving classical point charges that build up macroscopic charge and current distributions in otherwise empty space. If we assume that the volume V_0, to which equation (8.89) is applied, is made large enough for the fields **E** and **B** to be zero on its surface, then equation (8.89) reduces to

$$\frac{\partial W}{\partial t} = -\frac{\partial U_{em}}{\partial t} \tag{8.140}$$

where according to equation (8.86) the total electromagnetic energy of the electromagnetic field is

$$U_{em} = \int \left(\frac{\varepsilon_0 E^2}{2} + \frac{B^2}{2\mu_0} \right) \mathrm{d}V. \tag{8.141}$$

The integral in equation (8.141) is evaluated over the whole of space at one instant of time. If we equate the rate $\dot{W}$ at which the electromagnetic field is doing work on all the charges to the rate at which the total kinetic energy U_{mech} of all the moving classical point charges is increasing, we can rewrite equation (8.140) in the form

$$\frac{\partial U_{mech}}{\partial t} = -\frac{\partial U_{em}}{\partial t}. \tag{8.142}$$

In Section 8.6.1, we derived the $-\dot{U}_{em}$ term in equation (8.142) as an alternative to the $\int \mathbf{E} \cdot \mathbf{J}\, \mathrm{d}V$ term in equation (8.58). Without changing this interpretation of the $-\dot{U}_{em}$, term, we can rewrite equation (8.142) in the form

$$\frac{\partial}{\partial t}(U_{mech} + U_{em}) = 0. \tag{8.143}$$

Integrating with respect to time we get

$$U_{mech} + U_{em} = \text{a constant}. \tag{8.144}$$

This is the law of conservation of energy for a system of moving charges. According to equation (8.144), at any instant of time the sum of the kinetic energies of all the moving charges in the complete system plus the total electromagnetic energy, determined by evaluating the integral in equation (8.141) over the whole of space at that instant, is a constant. There is a capacity in both the electric and magnetic fields to give energy to charges when the field conditions are changed. If we only want to consider the energy changes in a part of the system we have to revert to equation (8.89) and the contribution of the Poynting flux to the energy balance in equation (8.89) must be included. By analogy with our interpretation of equations (8.123) and (8.126) we could also treat equatios (8.89) and (8.140) as book-keeping rules carried out at a fixed time.

We derived a special case of equation (8.144) when we derived equation (1.39) of Chapter 1, which is the law of conservation of energy for a charge moving in an electrostatic field. Consider a charge of magnitude q that is at rest in an applied electrostatic field of magnitude $\mathbf{E}$. Let the electric field due to the charge q be denoted by $\mathbf{E}_q$. According to equation (8.141) the total electromagnetic energy is

$$
\begin{aligned}
U_{\mathrm{em}} &= \frac{\varepsilon_0}{2}\int (\mathbf{E} + \mathbf{E}_q) \cdot (\mathbf{E} + \mathbf{E}_q)\, \mathrm{d}V \\
&= \frac{\varepsilon_0}{2}\int E^2\, \mathrm{d}V + \frac{\varepsilon_0}{2}\int E_q^2\, \mathrm{d}V + \varepsilon_0 \int \mathbf{E}_q \cdot \mathbf{E}\, \mathrm{d}V. \qquad (8.145)
\end{aligned}
$$

The charge q is allowed to go freely to infinity, without any significant changes in the applied electrostatic field $\mathbf{E}$. Assuming that the applied electrostatic field $\mathbf{E}$ is zero at infinity, then, at infinity where $\mathbf{E}_q$ is significant, $\mathbf{E}$ is zero and near the initial position of the charge q, where $\mathbf{E}$ is finite, $\mathbf{E}_q$ is now negligible so that $\mathbf{E}_q \cdot \mathbf{E}$ is zero everywhere when the charge q is at infinity and the electromagnetic energy is then given by just the first two terms on the right hand side of equation (8.145). Hence the decrease in the total electromagnetic energy, when the charge q is at infinity, is given by the third term on the right hand side of equation (8.145). Hence the increase in the total electromagnetic energy when q goes to infinity is given by

$$\delta U_{\mathrm{em}} = -\varepsilon_0 \int \mathbf{E}_q \cdot \mathbf{E}\, \mathrm{d}V \qquad (8.146)$$

where $\mathbf{E}$ is the constant applied electrostatic field and $\mathbf{E}_q$ is the electric field due to the charge q when it is at its initial position. If ϕ is the electrostatic potential of the applied electrostatic field, then since $\mathbf{E}$ is equal to $-\nabla\phi$ we can rewrite equation (8.146) in the form

$$\delta U_{\mathrm{em}} = \varepsilon_0 \int \mathbf{E}_q \cdot \nabla\phi\, \mathrm{d}V. \qquad (8.147)$$

Putting $\mathbf{A} = \mathbf{E}_q$ in equation (A1.20) of Appendix A1.6 we find that

$$\mathbf{E}_q \cdot \nabla\phi = \nabla \cdot (\phi\mathbf{E}_q) - \phi(\nabla \cdot \mathbf{E}_q).$$

Substituting in equation (8.147)

$$\delta U_{\mathrm{em}} = \varepsilon_0 \int \nabla \cdot (\phi\mathbf{E}_q)\, \mathrm{d}V - \varepsilon_0 \int \phi(\nabla \cdot \mathbf{E}_q)\, \mathrm{d}V. \qquad (8.148)$$

If we apply Gauss' theorem of vector analysis, which is equation (A1.30) of

Appendix A1.7, to the first integral, then, since the integration is over the whole of space ϕ is zero on the surface we are integrating over, so that the first integral on the right hand side of equation (8.148) is zero and equation (8.148) reduces to

$$\delta U_{\text{em}} = -\varepsilon_0 \int \phi(\nabla \cdot \mathbf{E}_q)\, \mathrm{d}V.$$

Now $\nabla \cdot \mathbf{E}_q$ is zero except at the charge. Draw a spherical surface of exceedingly small radius to surround the classical point charge q at its initial position. The variation of the scalar potential ϕ of the applied electrostatic field over such a small volume can be neglected and ϕ can be taken outside the integral. Applying first Gauss' theorem of vector analysis, which is equation (A1.30), to convert the volume integral into a surface integral and then applying Gauss' theorem of electrostatics, which is equation (1.13), we obtain

$$\delta U_{\text{em}} = -\varepsilon_0 \phi \int \mathbf{E}_q \cdot \mathrm{d}\mathbf{S} = -\varepsilon_0 \phi \left(\frac{q}{\varepsilon_0} \right) = -q\phi. \tag{8.149}$$

where ϕ is the value of the scalar potential of the applied electrostatic field at the initial position of the charge q. According to equation (8.144),

$$\delta U_{\text{mech}} + \delta U_{\text{em}} = 0 \tag{8.150}$$

where δU_{mech} is the increase in the kinetic energy of the charge q, which, since the charge q starts from rest, is equal to the total kinetic energy T of the charge q when it reaches infinity. Hence equation (8.150) becomes

$$T - q\phi = 0. \tag{8.151}$$

This is in agreement with equation (1.37) and shows that the gain of the kinetic energy of the charge q is equal to the decrease in the total electromagnetic energy of the system determined using equation (8.141).

8.8.5. *The conservation of angular momentum for an electromagnetic system*

It would go beyond the scope of this book to go into a full discussion of the law of conservation of angular momentum. Since the forces between moving charges are non-central forces, the law of conservation of angular momentum of Newtonian mechanics breaks down. We shall only give a brief outline of the general case avoiding the use of tensor analysis. A reader interested in fuller details is referred to Konopinski [7]. Consider a system of moving classical point charges, that build up macroscopic charge and current distributions. The forces on the moving charges, which are given by the Lorentz force law in the form of equations (8.1) and (8.2), have moments about any fixed point. The resultant torque $\mathbf{\Gamma}$ is equal to the rate of change of the angular momentum $\mathbf{L}_{\text{mech}}$ of the particle about that fixed point, that is

$$\mathbf{\Gamma} = \int \mathbf{r} \times (\rho\mathbf{E} + \mathbf{J} \times \mathbf{B})\, \mathrm{d}V = \frac{\mathrm{d}\mathbf{L}_{\text{mech}}}{\mathrm{d}t} \tag{8.152}$$

where $\mathbf{r}$ is the distance of the volume element dV from the fixed point. We now substitute for ($\rho\mathbf{E} + \mathbf{J} \times \mathbf{B}$) d$V$ using equation (8.114) to give

$$\begin{aligned} \frac{\mathrm{d}\mathbf{L}_{\text{mech}}}{\mathrm{d}t} &= \int \mathbf{r} \times \left(\frac{\partial \mathbf{p}_{\text{mech}}}{\partial t} \right) \mathrm{d}V \\ &= \int \mathbf{r} \times (\boldsymbol{\nabla} \cdot \mathbf{T})\, \mathrm{d}V - \int \mathbf{r} \times \left(\frac{\partial \mathbf{p}_{\text{em}}}{\partial t} \right) \mathrm{d}V \end{aligned} \tag{8.153}$$

where $\mathbf{p}_{\text{em}}$ is the (electromagnetic) potential momentum density of the field defined by equation (8.111). Konopinski [7] shows that we can convert the first integral on the right hand side of equation (8.153) into a surface integral that is zero if we integrate over the whole of space. Hence, provided we integrate over the whole of space we can combine equations (8.152) and (8.153) to give

$$\dot{\mathbf{L}}_{\text{mech}} = -\dot{\mathbf{L}}_{\text{em}} \tag{8.154}$$

where

$$\mathbf{L}_{\text{em}} = \int \mathbf{r} \times \mathbf{p}_{\text{em}}\, \mathrm{d}V = \int \frac{\mathbf{r} \times \mathbf{N}}{c^2} \mathrm{d}V \tag{8.155}$$

will be called the total *potential* angular momentum of the electromagnetic field. It is equal to the integral of the moment of the potential momentum density of the field about the fixed point, evaluated over the whole of space at a fixed instant of time. Equation (8.154) is just an alternative way of expressing equation (8.152), and shows that, instead of calculating the total torque using the Lorentz force law, we can, as an alternative, determine the torque from $-\dot{\mathbf{L}}_{\text{em}}$, where $\dot{\mathbf{L}}_{\text{em}}$ is the rate of change of the total potential angular momentum of the electromagnetic field.

Integrating equation (8.154) with respect to time we find that

$$\mathbf{L}_{\text{mech}} + \mathbf{L}_{\text{em}} = \text{a constant.} \tag{8.156}$$

Equation (8.156) is generally called the law of conservation of angular momentum for an electromagnetic system of charge and current distributions. Examples of the applicability of the law were given, among many others, by Romer [8], Pugh and Pugh [9] and Johnson, Cragin and Hodges [10]. We shall return in Section 8.9.5 to discus two simple examples that will illustrate equations (8.154) and (8.156).

8.9. Energy transfer and the Poynting vector

8.9.1. *The Poynting vector hypothesis*

The main results we derived in Sections 8.6 and 8.7 were obtained by applying equations (8.88) and (8.110) in their integral forms. Since we derived equations (8.88) and (8.110) from the equations

$$\dot{W} = \int \mathbf{E} \cdot \mathbf{J} \, \mathrm{d}V \tag{8.58}$$

$$\mathbf{F} = \int (\rho \mathbf{E} + \mathbf{J} \times \mathbf{B}) \, \mathrm{d}V \tag{8.101}$$

respectively by using Maxwell's equations and vector analysis, the use of equations (8.88) and (8.110) is entirely consistent with the approach we adopted in earlier chapters, where we related the fields **E** and **B** to experiments using the Lorentz force law. Since we only needed to apply equations (8.88) and (8.110) in their integral forms, there was no need in the approach we have developed so far to make any precise statements about what may or may not happen at a point in the empty space between moving charges, other than to comment that the electromagnetic field plays a key role in the propagation of the electromagnetic interaction at the speed of light c in empty space. There is also no need in the approach we have used so far, which is based on equations (8.88) and (8.110), to try and interpret what the variables **N** the Poynting vector, u_{em} the energy density and $\mathbf{p}_{\mathrm{em}}$ the electromagnetic potential momentum density mean at a point in empty space other than to say that they are functions of the fields **E** and **B** that enable us to simplify the mathematical forms of equations (8.88) and (8.110). It would be logical in our interpretation of classical electromagnetism to finish at this point. It is however traditional, particularly in the case of electromagnetic waves, to go beyond the application of the Poynting vector in the integral forms we have derived and make the extra assumption, which we shall call the **Poynting vector hypothesis**, that, when the Poynting vector **N** is finite at a point in space, there is an actual local flow of energy that is given exactly by the Poynting vector **N**. We shall show in Section 8.9.2 that, in the case of electromagnetic waves, the Poynting vector hypothesis is a very plausible extension of classical electromagnetism. Then in Section 8.9.3 we shall present further evidence in favour of the Poynting vector hypothesis in the case of electromagnetic waves by considering the quantum theory of radiation, which lies outside the scope of classical electromagnetism. Then in Section 8.9.4 we shall return to discuss the role of the Poynting vector hypothesis in steady (DC) conditions, before returning to discuss the law of conservation of angular momentum in Section 8.9.5.

8.9.2. *The role of the Poynting vector in electromagnetic waves*

To illustrate the plausibility of the Poynting vector hypothesis in the context of electromagnetic waves, we shall consider again the plane electromagnetic wave that is incident at normal incidence on the idealised perfect absorber in Figure 8.7(a). The properties of the plane wave were given in Section 8.6.2. Applying equation (8.88) in its integral form to the closed surface S_1 in Figure 8.7(a) and assuming that **E** and **B** were zero beyond the absorber, we derived equation (8.95) which showed that the rate $\dot{W}$ at which the electromagnetic wave was giving energy to unit area of the perfect absorber in Figure 8.7(a) was equal to the value the Poynting vector of the incident wave had at the surface of the absorber at $x = 0$ at that instant, so that

$$\dot{W} = \int \mathbf{E} \cdot \mathbf{J} \, dV = N_i \tag{8.157}$$

where N_i was given by equation (8.97).

On the basis of equation (8.157), it is not inconsistent with classical electromagnetism in the case of a plane electromagnetic wave to make the Poynting vector hypothesis and assume that the Poynting vector $\mathbf{N}$ gives the local rate at which the energy in the electromagnetic wave is actually flowing across 1 m^2, measured perpendicular to the direction of propagation of the plane electromagnetic wave. Since the electromagnetic wave is travelling at a speed c in empty space, it is then reasonable to go on to postulate that the electromagnetic "energy density" u_{em} in the plane electromagnetic wave is equal to the rate of flow of energy per unit area divided by the speed c, which would give for a plane electromagnetic wave

$$u_{em} = \frac{N}{c} . \tag{8.158}$$

If p_{em} were the electromagnetic momentum per unit volume in the plane wave, then in a time dt a total momentum of $p_{em}(c\, dt)$ would be absorbed by unit area of the perfect absorber in Figure 8.7(a). Hence the force per unit area on the absorber, which is equal to the rate of change of momentum, would be equal to cp_{em}. Comparing with equation (8.118), we could conclude that

$$p_{em} = \frac{N}{c^2} = \frac{u_{em}}{c} . \tag{8.159}$$

The use of the Poynting vector to calculate the rate of energy flow works very well in the case of electromagnetic waves. For example, we used the Poynting vector hypothesis when we determined the angular distribution of the intensity of the radiation emitted by the oscillating electric dipole in Section 2.4.3 of Chapter 2. The use of the Poynting vector hypothesis simplifies the interpretation of Physical Optics. The Poynting vector given by equations (8.64) and (8.65) can, for example, be used to determine the reflection and transmission coefficients, when a plane electromagnetic wave is incident on a plane boundary between air and a refractive medium. There is no doubt that the assumptions made in equations (8.157), (8.158) and (8.159) are extremely helpful in the teaching of the properties of electromagnetic waves. However, when we use a device to measure the energy flow in an electromagnetic wave, the energy detector measures the net power flowing into the detector from all directions, which is related to the integral of the Poynting vector over a surface that surrounds the detector, which in practice leads us back from the Poynting vector hypothesis to the integral form of equation (8.88). Even when we derived equation (8.95), we had to use the information that $\mathbf{E}$ and

B were zero on the right hand side of the perfect absorber in Figure 8.7(a). The use of the Poynting vector hypothesis is, however, far more convenient in the teaching of electromagnetic waves than having to go every time through the type of analysis we gave of Figure 8.7(a).

Since, in all practical cases, we have to integrate over a closed surface surrounding an energy detector when we apply the Poynting vector hypothesis in classical electromagnetism, we could always add a divergence free vector field, such as the curl of a vector, to **N** without changing the predictions of equation (8.88). A discussion of possible alternatives to the Poynting vector was given by Slepian [11]. Evidence in favour of the unique choice of the Poynting vector was given by Romer [12]. The choice of the Poynting vector as a teaching aid is probably the best and is universally chosen in the text books. It leads to correct results when equation (8.88) is applied to electromagnetic waves.

8.9.3. *Evidence in favour of the Poynting vector hypothesis from the quantum theory of radiation*

We can get supplementary evidence in favour of the Poynting vector hypothesis in the case of electromagnetic waves by going outside the scope of classical electromagnetism and using the quantum theory of radiation. We shall assume that there are n photons per unit volume, each of energy $h\nu$, where h is Planck's constant and ν is the frequency. We shall also assume that the photons are all moving with velocity **c**. The energy density u_{em} is

$$u_{em} = n(h\nu). \tag{8.160}$$

The number of photons crossing a unit area, that is measured perpendicular to **c**, per second is equal to nc, so that the energy flow per unit area, which we shall denote by N, is

$$N = nc(h\nu) = u_{em}\, c \tag{8.161}$$

so that

$$u_{em} = \frac{N}{c}\,. \tag{8.162}$$

It has been confirmed by experiments, for example on the Compton effect, that a photon of energy $h\nu$ has a linear momentum of $h\nu/c$. Since there are n photons per unit volume, the momentum density p_{em} in the radiation field is

$$p_{em} = n\left(\frac{h\nu}{c}\right) = \frac{N}{c^2} = \frac{u_{em}}{c}\,. \tag{8.163}$$

Since the photons have linear momenta, they have an angular momentum about any fixed point. Equations (8.162) and (8.163) are in agreement with equations (8.158) and (8.159) respectively. Thus the quantum theory of radiation provides strong evidence in favour of the Poynting vector hypothesis, when the Poynting vector hypothesis is applied to electromagnetic waves.

8.9.4. *The Poynting vector hypothesis in steady (DC) conditions*

The question arises, should we extrapolate from the case of the radiation electromagnetic field and conclude that the Poynting vector hypothesis is valid in steady (DC) conditions and assume that there is an actual local flow of energy through empty space in steady conditions of a magnitude and direction given exactly by the Poynting vector, in conditions where the electromagnetic interaction is not due to real photons moving in one direction? The Poynting vector hypothesis certainly gives the correct results, if the Poynting vector is known and applied to real situations using equation (8.67) in its integral form. The choice of whether to use equation (8.58), which we derived using the Lorentz force law in Section 8.4, or to use the Poynting vector in equation (8.67) is generally settled by choosing whichever of the two alternative approaches that is the more convenient, which in the case of steady (DC) circuits is generally the approach based on circuit theory. In Appendix B, we interpreted the origin of the current flow in Figure 8.5 in terms of the setting up of surface and boundary charge distributions which sustain the steady electric field inside the conductors, which acts on the conduction electrons and gives rise to the rate of increase in the kinetic energies of the conduction electrons, given by equation (8.58). Kirchhoff's laws, which we derived in Chapter 7, can generally be used to determine the current flow if the emf and resistances in the circuit are know. The rate at which Joule heat is generated in a resistor R is then given by I^2R. To use the Poynting vector approach, we could again use circuit theory to determine the current flow and hence the electric field inside the resistors. We could then apply the analysis of Figure 8.6 to derive equation (8.72). In the Poynting vector approach, the surface and boundary charge distributions associated with current flow serve to give the appropriate electric field $\mathbf{E}$ in space, which together with $\mathbf{B}$ guides the direction of the energy flow given by the Poynting vector $\mathbf{E} \times \mathbf{B}/\mu_0$ from the source of emf through space and into the connecting leads, as sketched in Figure 8.5. Since the electric field inside the connecting wires is parallel to the connecting wires in Figure 8.5, the Poynting vector is zero in the direction of current flow inside the connecting wires, so that, according to the Poynting vector hypothesis, there is no energy flow inside the wires. If we wanted to determine the value of the Poynting vector outside the connecting wires in Figure 8.5, we would generally have to solve a complicated boundary value problem to determine the external electric field. An early example was given by Marcus [13]. Due to its complexity, this approach is not used very often to determine quantitative Poynting flow diagrams. We only used qualitative ideas to sketch the Poynting flow from the source of emf in the example of the Van der Graaff generator shown in Figure 8.5, and which we discussed in Section 8.5.2. In Appendix B, we stressed the important role that the electric field has in making a conduction current flow. This shows the importance of field concepts both in the traditional approach of Appendix B and in the Poynting vector method. Whether or not there is, in steady conditions, a local flow of energy in empty space given exactly by the Poynting vector is still a matter of debate, since we only observe changes in the energy of the electromagnetic field when it interacts with charges. Amusingly both sides of the debate have used the example of a stationary charge and a stationary magnetic dipole as evidence both for and against the Poynting vector hypothesis in steady (DC) conditions. We shall now consider this example.

8.9.5. *Examples of the application of the law of conservation of angular momentum*

Consider the stationary electric charge and the stationary magnetic dipole shown in Figure 8.9. A typical example of a magnetic dipole would be a small coil carrying a steady current. Due to the crossed electric and magnetic fields in Figure 8.9, the Poynting vector given by $\mathbf{E} \times \mathbf{B}/\mu_0$ is finite. According to the Poynting vector hypothesis, there should be an energy flow associated with this finite Poynting vector so that, according to the Poynting vector hypothesis energy should be going around in circles in Figure 8.9. If we put a stationary energy detector of some sorts near the charge and magnetic dipole in Figure 8.9, would it be able to perpetually absorb some of this predicted energy flow? The answer is clearly no! Some people treat this as conclusive evidence against the Poynting vector hypothesis. To estimate the energy that would be absorbed by such a detector, we must integrate equation (8.88) over a closed surface S_0 that surrounds the energy detector. Once static conditions prevail, after the introduction of the detector, equation (8.87) reduces to

$$\dot{W} = -\int \nabla \cdot \mathbf{N}\, dV = -\int \mathbf{N} \cdot d\mathbf{S} \tag{8.164}$$

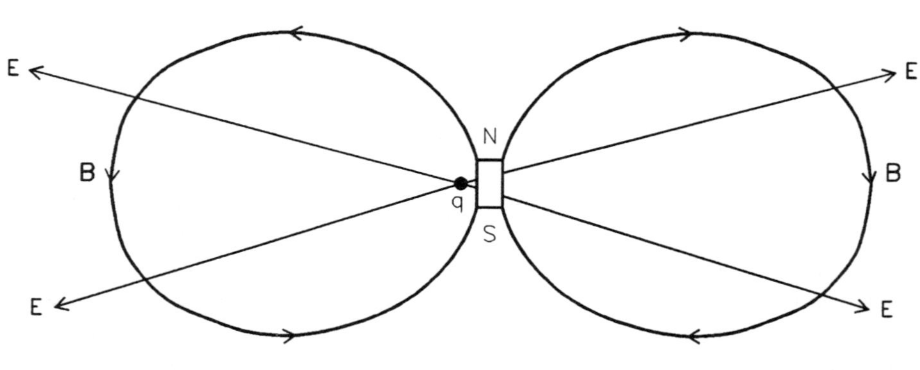

Figure 8.9. Typical electric field lines due to a stationary electric charge and magnetic field lines due to a stationary magnetic dipole, such as a current carrying coil, are shown. The Poynting vector $\mathbf{E} \times \mathbf{B}/\mu_0$ is upwards from the paper to the left of the charge and is downwards to the right of the charge.

where the integrations are over the volume V_0 and area S_0. To have a finite value of $\dot{W}$ inside S_0, we would need a Poynting flux into S_0. To balance the books we would then need an energy source outside S_0 where $\nabla \cdot \mathbf{N}$ would be finite, but according to equation (8.68), in static conditions, $\nabla \cdot \mathbf{N}$ would be zero everywhere outside S_0. This means that the prediction of classical electromagnetism, based on equation (8.88), is that, once the transient state is over, the energy detector should not absorb any energy from the electromagnetic field, which is in agreement with experiment.

Some people claim that when the law of conservation of angular momentum, which, we derived in Section 8.8.5, is applied to the experimental conditions shown in Figure 8.9, it provides definitive evidence in favour of the Poynting vector hypothesis. To quote Feynman, Leighton and Sands [14] when they comment on a similar example:

> This mystic circulating flow of energy, which at first seemed so ridiculous, is absolutely necessary. There is really a momentum flow. It is needed to maintain the conservation of angular momentum in the whole world.

Since, according to the Poynting vector hypothesis there should be a circular flow of energy in Figure 8.9, there should also be a circular flow of "electromagnetic momentum" which would give a finite value for the total "electromagnetic angular momentum" $\mathbf{L}_{\text{em}}$ defined by equation (8.155), which is

$$\mathbf{L}_{\text{em}} = \int \mathbf{r} \times \left(\frac{\mathbf{N}}{c^2}\right) \mathrm{d}V \tag{8.165}$$

We shall prefer to call $\mathbf{L}_{\text{em}}$, which is the integral of the moment of the potential momentum about a fixed axis integrated over the whole of space, the *potential* angular momentum.

We shall now assume that the charge q in Figure 8.9 is moving with a velocity $\mathbf{u}$. According to the Lorentz force law, the charge q is deflected by the magnetic force $q\mathbf{u} \times \mathbf{B}$ due to the magnetic field $\mathbf{B}$ due to the magnetic dipole. If $\mathbf{u}$ is small, the charge q may be trapped by the magnetic field of the magnetic dipole, just as charged particles are trapped in the Van Allen belts by the Earth's magnetic field. If $\mathbf{u}$ is large enough, the charge q will travel outwards in a spiral path to infinity. The moment of the momentum of the charged particle, that is its angular momentum about the original position of the dipole, is then finite. Notice that in this example, the charge q is deflected by a non-central magnetic force, so that the law of conservation of mechanical angular momentum of Newtonian mechanics would not be applicable even in the low velocity limit. However, according to equations (8.152) and (8.154), if $\mathbf{L}_{\text{mech}}$ is the sum of the angular momentum of the charge q and any angular momentum the magnetic dipole may have, then

$$\dot{\mathbf{L}}_{\text{mech}} = \int \mathbf{r} \times (\rho\mathbf{E} + q\mathbf{u} \times \mathbf{B}) \, \mathrm{d}V \tag{8.166}$$

$$\dot{\mathbf{L}}_{\text{mech}} = -\dot{\mathbf{L}}_{\text{em}}. \tag{8.167}$$

Integrating equation (8.167) with respect to time gave

$$\mathbf{L}_{\text{mech}} + \mathbf{L}_{\text{em}} = \text{a constant} \tag{8.168}$$

where $\mathbf{L}_{\text{em}}$, the total potential angular momentum of the electromagnetic field, is given by equation (8.165). Equation (8.168) is the law of conservation of angular momentum for an electromagnetic system made up of moving charges in empty space. The potential angular momentum of the field $\mathbf{L}_{\text{em}}$ is finite initially in Figure 8.9, but it is zero when the charge q reaches infinity. Hence, according to equation (8.168), there should be a corresponding increase in the sum of the mechanical angular momenta of the charge q and the magnetic dipole after the charge q has reached infinity, showing that the changes in the mechanical angular momenta of the charge q and the magnetic dipole do not compensate each other. Just as in Newtonian mechanics, where we can either apply the laws of conservation of angular momentum, or use Newton's laws of motion directly, we can in this case calculate the increase in $\mathbf{L}_{\text{mech}}$ in two ways, either by attributing potential angular momentum to the electromagnetic field and applying the law of conservation of angular momentum in the form of equation (8.168) or by substituting the Lorentz forces (a) on the charge q due to the magnetic field of the magnetic dipole and (b) on the magnetic dipole due to the magnetic field of the moving charge q into equation (8.166), and then integrating with respect to time. When the Lorentz force law is used in equation (8.166), the increase in the total mechanical angular momentum in Figure 8.9 arises from the non-central magnetic forces. It will be illustrated in Section 10.7 of Chapter 10 that these non-central magnetic forces can be interpreted as second order relativistic effects.

As a different type of electromagnetic process, we shall consider a simple example due to Boos [15]. Consider a very long solenoid of radius R of the type shown previously in Figure 5.7 of Chapter 5. The solenoid is of length l, where $l \gg R$, and consists of n turns per metre length. Initially, the current in the solenoid is equal to I_0. There are two cylindrical, dielectric tubes of negligible thicknesses, whose axes coincide with the axis of the solenoid. One of the cylindrical tubes is of radius $a < R$, it is inside the solenoid and has a charge $+Q$ distributed uniformly over its surface. The other cylindrical tube is of radius $b > R$, it is outside the solenoid and has a charge $-Q$ distributed uniformly over its surface. The cylindrical tubes and solenoid are free to rotate without friction about their common axis. Initially, they are all at rest so that initially the total mechanical angular momentum is zero.

The current in the solenoid is reduced to zero slowly enough for the quasi-stationary approximations to be valid. According to equations (7.10) and (7.11) of Chapter 7, provided we can ignore end effects, the magnitudes of the induction electric fields E_i inside the solenoid and E_0 outside the solenoid are given by

$$E_i = \frac{\mu_0 n\dot{I} r}{2}; \qquad E_0 = \frac{\mu_0 n\dot{I} R^2}{2r}$$

where r is the distance from the axis of the solenoid. The electric field lines are circles concentric with the common axis, and, for decreasing current in the solenoid, the electric field lines are in the direction of current flow. According to the Lorentz force law, these induction electric fields give rise to forces on the charges $+Q$ and $-Q$ of magnitudes QE_i and $-QE_0$ respectively in the tangential direction. The torques $QE_i a$ and $-QE_0 b$ about the common axis give rise to rotations of the cylindrical, dielectric tubes. Integrating over the time it takes the current in the solenoid to go to zero and equating the time integral of the torque to the gain L_i in the mechanical angular momentum of the inner cylindrical tube, we have

$$L_i = \int QE_i a \, \mathrm{d}t = Q\left(\frac{\mu_0 n I_0 a^2}{2}\right)$$

The direction of rotation is in the direction of current flow. The gain L_0 in the mechanical angular momentum of the outer cylindrical tube is

$$L_0 = -\int QE_0 b \, \mathrm{d}t = -Q\left(\frac{\mu_0 n I_0 R^2}{2}\right)$$

This rotation is in the direction opposite to the direction of current flow. The resultant total mechanical angular momentum L_{mech} is

$$L_{\mathrm{mech}} = (L_0 - L_i) = -Q\left(\frac{\mu_0 n I_0}{2}\right)(R^2 - a^2) \tag{8.169}$$

The origin of this mechanical angular momentum has been traced to the Lorentz force acting on the charged, dielectric cylindrical tubes due to the induction electric field due to the varying current in the solenoid.

According to equation (8.156), instead of calculating the change in the mechanical angular momentum directly using the Lorentz force law, as we just did, we can calculate the gain in the total mechanical angular momentum from the loss of (electromagnetic) potential angular momentum. Initially the magnetic field due to the current in the solenoid is equal to $\mu_0 n I_0$ inside the solenoid and zero outside the solenoid. It follows from Gauss' flux law of electrostatics that initially the electric field between the charged, dielectric, cylindrical tubes is equal to $Q/2\pi\varepsilon_0 rl$. Initially, the electric and magnetic fields only overlap between the inner tube of charge and the surface of the solenoid. Substituting in equation (8.155) we find that the initial (electromagnetic) potential angular momentum of the system is

$$\begin{aligned} L_{\mathrm{em}} &= \frac{1}{\mu_0 c^2}\int |\mathbf{r} \times (\mathbf{E} \times \mathbf{B})| \, \mathrm{d}V = \frac{1}{\mu_0 c^2}\int r\left(\frac{Q}{2\pi\varepsilon_0 rl}\right)(\mu_0 n I_0) \, \mathrm{d}V \\ &= \frac{\mu_0 Q n I_0}{2\pi l}\int \mathrm{d}V = Q\left(\frac{\mu_0 n I_0}{2}\right)(R^2 - a^2) \end{aligned} \tag{8.170}$$

When the current in the solenoid is zero, the magnetic field is zero so that the final (electromagnetic) potential angular momentum is zero. Hence the loss of potential angular momentum is equal to L_{em}. Comparing equations (8.169) and (8.170) we see that, as an alternative to using the Lorentz force law directly, we can calculate the gain in mechanical angular momentum from the loss of (electromagnetic) potential angular momentum. So far we have only considered the main contributions to L_{mech} and L_{em}. If we did consider all the very much smaller effects, such as the end effects due to the finite lengths of the solenoid and dielectric tubes, it would complicate the analysis, but according to our derivation of equations (8.155) and (8.156), these equations would be valid to all orders in the general treatment at all instants of time. In this example, the non-central force that gives rise to the increase in the total mechanical angular momentum is associated with the induction electric field. We showed in Section 5.7 of Chapter 5 that the induction electric field due to the varying current in a very long solenoid can be derived from the expression for the electric field due to a moving classical point charge and arises partly from the changes in the electric field of the charge when it is moving, which leads up to equation (5.18) and partly from retardation effects, which leads on to the general equation (5.54) which was then applied to derive equation (5.68).

The two examples we have considered have illustrated the role of magnetic fields and induction electric fields respectively in giving the second order non-central forces that give rise to the changes in the total mechanical angular momentum of an electromagnetic system. By analogy with the case of the law of conservation of linear momentum discussed in Section 8.8.2, we can treat the use of the (electromagnetic) potential angular momentum in equations (8.154) and (8.156) as book-keeping rules, since we cannot imagine that the rates of change of the fields **E** and **B** at a distant point of space can influence in a causal way transmitted at infinite speed the rates of change of the momenta and angular momenta of distant charges at that precise instant. We can treat the (electromagnetic) potential angular momentum as a measure of the capability of the system to change the total mechanical angular momentum of the system when the charges move under the influence of the non-central electric and the magnetic forces between them. There is no need to demand a mechanical interpretation of (electromagnetic) potential angular momentum based on Newtonian mechanics and the concept of interactions propagated at infinite speed, anymore than we should demand a mechanical interpretation of potential energy. We can treat potential energy, potential momentum and potential angular momentum as measures of the capability of a system of moving and accelerating charges to change the total kinetic energy, the total mechanical linear momentum and the total mechanical angular momentum of a system of charges when the configuration, velocities and accelerations of the charges change under the influence of the electromagnetic forces between them. It is not necessary to specify precisely where the potential (electromagnetic) angular momentum of a system resides, since it only manifests itself through the actions of second order non-central electric and magnetic forces on charges. Hence our use of the term **potential** angular momentum from the start of our discussions of the conservation of angular momentum.

8.10. Summary

Since the electromagnetic field is only observed when it acts on charges with a force given by the Lorentz force law, it is not necessary in the strict context of classical electromagnetism to try and make any precise conclusions of what may or may not happen in the empty space between moving charges, other than to acknowledge that the electromagnetic field plays a key role in the transmission of the forces between moving charges at the finite speed c. In this chapter, our starting point was always the Lorentz force, or equation (8.58) which was derived from it. In the strict context of classical electromagnetism, equations (8.88) and (8.110) and the conservation laws are as far as we need go, if we are only interested in predicting the results of experiments, in which case we need only have treated the Poynting vector $\mathbf{N}$, the energy density u_{em} and the (electromagnetic) potential momentum density $\mathbf{p}_{\text{em}}$ as abbreviations defined by equations (8.64), (8.85) and (8.111) respectively. It would have been logical to finish this chapter at the end of Section 8.8.5 with the occasional use of equations (8.153) and (8.154). We have suggested that equations (8.88) and (8.110) can be interpreted as book keeping rules evaluated at a fixed time, and which are alternatives to using the Lorentz force directly. Our approach does not preclude the Poynting vector hypothesis or any other model that goes beyond classical electromagnetism, such as the photon model, and which leads to equations (8.88) and (8.110). The Poynting vector hypothesis works extremely well when it is applied to electromagnetic waves. If in addition it is assumed that the electromagnetic field has an energy density given by equation (8.85), then the use of the Poynting vector hypothesis leads to equation (8.88). In steady (DC) conditions, we can rewrite $\mathbf{E} \cdot \mathbf{J}$ as $(-\nabla \cdot \mathbf{N})$ using equations (1.89), (A1.21) and (1.117), so that the Poynting vector hypothesis can be used as an alternative to circuit theory. If a momentum density $\mathbf{N}/c^2$ is attributed to the electromagnetic field, the Poynting vector can be used to set up equation (8.125) for the (electromagnetic) potential momentum which can then be substituted in equation (8.127), which is the law of conservation of momentum for a system of moving charges. However, care must be exercised if the Poynting vector hypothesis is applied to only that part of the system of moving charges inside V_0 in Figure 8.1, when it is equation (8.110) that must be used and the Maxwell stress tensor term is dominant. As an illustrative example from Newtonian mechanics, consider an isolated system of particles that move under the influence of the gravitational forces between them. If we only consider a part, of volume V_0, of the system then, according to Newtonian mechanics, the rate of change of the total momentum of the particles inside V_0 is equal to the resultant gravitational force on them due to the particles outside V_0. According to Newtonian mechanics, there is an equal and opposite gravitational force on the particles outside V_0 due to the particles inside V_0. These forces cancel if we combine the two systems to form an isolated system, and momentum is then conserved. In the electromagnetic case the main contribution to the resultant force on the charges inside V_0 in Figure 8.1 is due to the moving charges outside V_0, though there is a small internal contribution due to the fact that Newton's third law is not valid for the forces between the moving charges inside V_0. There is also a force on the charges outside V_0 due to the charges inside V_0. What we have is that both parts of the system of moving charges in Figure 8.1 transfer momentum to each other. (Such forces have been interpreted in terms of the exchange of virtual photons.) Such a result is predicted by equation (8.110), since the surface integral of the Maxwell stress tensor is a vector that can point in different directions on either side of the surface S_0 in Figure 8.1. An example was given in our discussion of Figure 8.3(a) in Section 8.2.5. If we combine the two systems of moving charges inside and outside V_0 in Figure 8.1 to form a composite isolated system, the forces between the two systems do not add up to zero and we are left with the second order electric and magnetic forces that arise from deviations from Newton's third law. Subsequent changes in the total mechanical momentum of the charges are consistent with equation (8.127).

References

1. French, A.P., *Newtonian Mechanics* (W. W. Norton, New York 1965), p. 368.
2. Griffiths, D. J., *Introduction to Electrodynamics* (Prentice-Hall, Englewood Cliffs, New Jersey, 1981), pp. 80–85.
3. Griffiths, D. J., see ref. 2, pp. 268–270.
4. Lorrain, P. and Corson, D. R., *Electromagnetic Fields and Waves*, 2nd Edition (W. H. Freeman, San Francisco, 1970), pp. 543–545.
5. Page, L. and Adams, N. I., *Amer. Journ. Phys.*, Vol. 13, p. 141 (1945).
6. Konopinski, E. J., *Electromagnetic Fields and Relativistic Particles* (McGraw-Hill, New York, 1981), p. 159.
7. Konopinski, E. J., see ref. 6, pp. 168–170.
8. Romer, R. H., *Amer. Journ. Phys.*, Vol. 34, p. 772 (1966) and Vol. 35, p. 445 (1967).
9. Pugh, E. M. and Pugh, G. E., *Amer. Journ. Phys.*, Vol. 35, p. 153, (1967).
10. Johnson, F. S., Cragin, B. L. and Hedges, R. R., *Amer. Journ. Phys.*, Vol. 62, p. 33 (1994).
11. Slepian, J., *Journ. App. Phys.*, Vol. 13, p. 512 (1942).
12. Romer, R. H., *Amer. Journ. Phys.*, Vol. 50, p. 512 (1982).
13. Marcus, A., *Amer. Journ. Phys.*, Vol. 9, p. 225 (1941).
14. Feynman, R. P., Leighton, R. B. and Sands, M., *The Feynman Lectures on Physics, Vol. 2, The Electromagnetic Field* (Addison-Wesley, Reading, Mass., 1964), pp. 27–28.
15. Boos, F. L., *Amer. Journ. Phys.*, Vol. 52, p. 756 (1984).

CHAPTER 9

Stationary dielectrics and stationary magnetic materials

9.1. Introduction

So far in this book we have always assumed that the relative permittivity ε_r and the relative permeability μ_r were both equal to unity everywhere. We shall now go on to discuss the form Maxwell's equations take at points inside stationary dielectrics and stationary magnetic materials. Dielectrics are polarized in the presence of an applied electric field and make contributions to the total electric field. Magnetic materials are magnetized in the presence of an applied magnetic field and make contributions to the total magnetic field. The development of Maxwell's equations for field points inside dielectrics and magnetic materials from the Maxwell-Lorentz equations, namely equations (1.137), (1.138), (1.139) and (1.140) of Chapter 1 is covered comprehensively in the literature, for example Russakoff [1] and in the text books such as Robinson [2]. Consequently we shall confine our remarks to a few headlines and comments to indicate how Maxwell's equations for field points inside dielectrics and magnetic materials can be interpreted in a way consistent with the approach we developed in Chapter 4. We shall not go through the full processes of averaging the microscopic variables using equation (1.147) of Chapter 1 to determine the corresponding macroscopic variables, but we shall just use simplistic arguments to illustrate the plausibility of the final results. We shall find it convenient to introduce two new macroscopic variables to describe the dielectric and magnetic properties of materials, namely the polarization vector **P** and the magnetization vector **M**. We shall illustrate how a knowledge of the vectors **P** and **M** is sufficient for us to determine the contributions of polarized dielectrics and magnetized bodies to the macroscopic electric and magnetic fields. We shall also develop the form Maxwell's equations take in a material medium that has both dielectric and magnetic properties.

It is important for the reader to realize that, in the methods used by Russakoff [1] and Robinson [2], the macroscopic fields inside dielectrics and magnetic materials are derived from the Maxwell-Lorentz equations, so that, on the microscopic scale, the electromagnetic interaction inside dielectrics and magnetic materials is propagated at the speed c of light in empty space, even though we shall find in Section 9.5.2 that the resultant macroscopic fields in a medium are transmitted at a speed c/n where n is the refractive index.

9.2. A stationary polarized dielectric

9.2.1. *The types of atomic electric dipoles*

Consider a stationary dielectric, which is polarized under the influence of an applied electric field. The electric forces on the positive and negative charges in a molecule inside the dielectric are in opposite directions. It goes beyond the limits imposed by the uncertainty principle to give a precise classical model for an atomic electric dipole. According to quantum theory, the wave function of a molecule can be used to determine the average charge distribution in the molecule. For a non-polar molecule, in the absence of an applied electric field, the "centre of the time averaged positive charge distribution", due to the positive nuclei in the molecules coincides with the "centre of the time averaged negative charge distribution" due to the electrons in the molecule. Under the influence of an applied electric field the two "centres of charge" are displaced in opposite directions giving rise to a resultant electric dipole moment.

Polar molecules have a permanent electric dipole moment. For example, in a water molecule the nuclei of the hydrogen and oxygen atoms are not in a straight line, but the molecule is bent at an angle of 108°. The electrons tend to cluster around the oxygen nucleus, leaving, on average, net positive charge distributions in the vicinities of the hydrogen nuclei and a net negative charge distribution in the vicinity of the oxygen nucleus. This gives the water molecule a permanent electric dipole moment. Just as in the case of non-polar molecules there is also a displacement of charge in polar molecules, when an electric field is applied. This gives a contribution to the total electric dipole moment of the polar molecule, which is generally much smaller than the permanent electric dipole moment. In the absence of an applied electric field the dipole moments of individual polar molecules point in random directions, and the resultant polarization of the dielectric is zero. When an external electric field $\mathbf{E}$ is applied there is a couple $\mathbf{p} \times \mathbf{E}$ on each electric dipole, of dipole moment $\mathbf{p}$, which tries to make the dipoles point in the direction of the electric field. Complete alignment is prevented by the thermal motions of the molecules, but the electric dipole moments of the polar molecules do point preferentially in the direction of the applied electric field giving a resultant polarization.

The positive ions in crystals, such as sodium chloride, tend to be displaced in the direction of the applied electric field, whilst the negative ions tend to be displaced in the opposite direction giving a contribution to the polarization of the dielectric.

9.2.2. *The electric dipole moment of a molecule*

The electric dipole moment **p** of a polarized molecule is given by

$$\mathbf{p} = \int \mathbf{r}_s \, \rho^{\mathrm{mic}} \, \mathrm{d}V_s \qquad (9.1)$$

where ρ^{mic} is the average microscopic charge density at a point at the position $\mathbf{r}_s$ inside the molecule. To simplify our discussions we shall use a simple classical model in which we shall assume that each electric dipole consists of two classical point charges of magnitudes $+q$ and $-q$ a small distance l apart, which gives an electric dipole moment p equal to ql. The direction of the vector **p** is from the negative to the positive charge. When an electric field is applied to a non-polar molecule, both the positive and negative charge distributions are displaced. Similarly, when polar molecules rotate towards the direction of the applied electric field both positive and negative charge distributions move. However, to simplify the discussions we shall assume that, when a dielectric is polarized, the positive charge $+q$ in a molecule moves a distance l in the direction of the electric field, but the negative charge $-q$ does not move. It is a straightforward for the reader to show that the results that we shall derive are equally valid if both positive and negative charges inside a molecule moved, or if only the negative charges moved.

9.2.3. *The polarization vector* **P**

We shall find it useful to introduce a vector quantity to specify the macroscopic properties of a polarized dielectric. Due to the influence of the applied electric field a large number of atomic electric dipole moments are induced or aligned in the direction of the applied field (for isotropic dielectrics). Consider a small element of volume ΔV. Let

$$\sum_{i=1}^{N} \mathbf{p}_i = \mathbf{P}\Delta V; \qquad \text{or} \qquad \mathbf{P} = \frac{1}{\Delta V} \sum_{i=1}^{N} \mathbf{p}_i \qquad (9.2)$$

where the vector summation is carried out over all the N atomic electric dipoles in the volume element ΔV. Equation (9.2) is a simplistic definition of the polarization vector **P**, in terms of the microscopic atomic electric dipole moments. The volume element ΔV can be made small on the laboratory scale (say a 10^{-7} m cube). In a macroscopic theory **P** can be treated as a continuous function of position, and defined as the average resultant electric dipole moment per unit volume.

9.2.4. *The electric field due to a stationary polarized dielectric*

It can be shown that the electrostatic scalar potential at a distance **R** from an electric dipole consisting of point charges $+q$ and $-q$ a distance l apart (such that $p = ql$) is

$$\phi = \frac{\mathbf{p} \cdot \mathbf{R}}{4\pi\varepsilon_0 R^3} \tag{9.3}$$

provided $R \gg l$. This is the dipole approximation. The direction of $\mathbf{p}$ is from the negative to the positive charge. Equation (9.3) follows from equation (2.51) of Chapter 2, if we assume that the dipole moment does not vary with time.

The electric field due to the polarized dielectric shown in Figure 9.1 will be determined at the field point T at the position $\mathbf{r}$. In this chapter we shall denote the position of the field point by T to avoid confusion with the polarization $\mathbf{P}$. Consider the volume element ΔV_s that is at the position $\mathbf{r}_s$ inside the dielectric, where the polarization vector is $\mathbf{P}$. The distance from ΔV_s to T is $\mathbf{R} = (\mathbf{r} - \mathbf{r}_s)$ as shown in Figure 9.1.

Consider one atomic electric dipole, of dipole moment $\mathbf{p}_i$, inside the volume element ΔV_s. According to equation (9.3), its contribution to the electrostatic scalar potential at the field point T is equal to $\mathbf{p}_i \cdot \mathbf{R}/4\pi\varepsilon_0 R^3$. The electrostatic scalar potential at T due to all the atomic electric dipoles inside ΔV_s, which are all at distance $\mathbf{R}$ from T is equal to

$$\Delta\phi = \sum \frac{\mathbf{p}_i \cdot \mathbf{R}}{4\pi\varepsilon_0 R^3}. \tag{9.4}$$

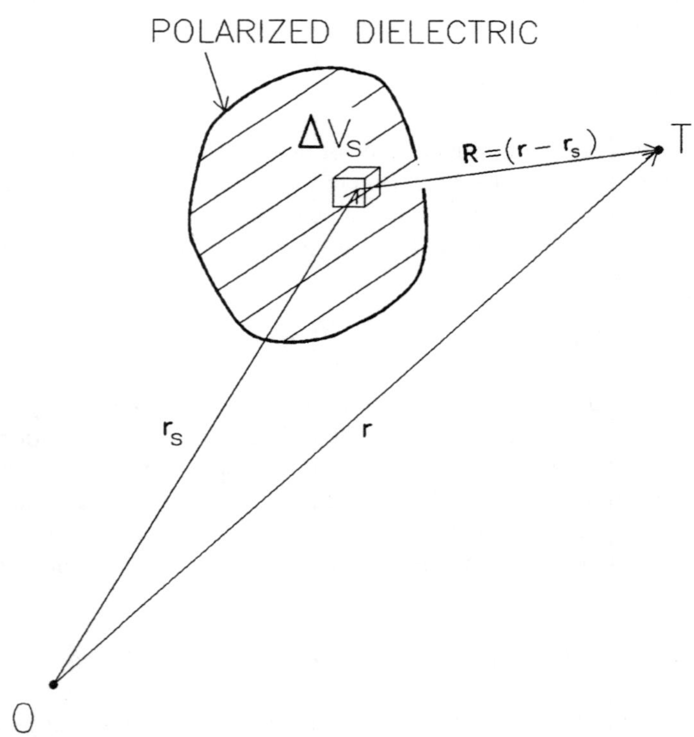

Figure 9.1. Calculation of the electric field due to a polarized dielectric.

The summation in equation (9.4) is over all N atomic electric dipoles inside ΔV_s. It follows from equation (A1.2) of appendix A1.1 that

$$\mathbf{p}_1 \cdot \mathbf{R} + \mathbf{p}_2 \cdot \mathbf{R} + \ldots + \mathbf{p}_N \cdot \mathbf{R} = \left(\sum_{i=1}^{N} \mathbf{p}_i \right) \cdot \mathbf{R}.$$

But from equation (9.2)

$$\sum_{i=1}^{N} \mathbf{p}_i = \mathbf{P}\Delta V_s.$$

Hence, equation (9.4) becomes

$$\Delta\phi = \frac{\mathbf{P} \cdot \mathbf{R}\Delta V_s}{4\pi\varepsilon_0 R^3}.$$

Integrating, we have

$$\phi = \frac{1}{4\pi\varepsilon_0} \int_{V_0} \frac{\mathbf{P} \cdot \mathbf{R}}{R^3} \, \mathrm{d}V_s. \tag{9.5}$$

Since $\mathbf{R}$ is equal to $\{(x - x_s)^2 + (y - y_s)^2 + (z - z_s)^2\}^{1/2}$, it follows that, if the position of ΔV_s is changed when the field point T is fixed, then

$$\boldsymbol{\nabla}_s \left(\frac{1}{R} \right) = \frac{\mathbf{R}}{R^3} \tag{9.6}$$

where

$$\boldsymbol{\nabla}_s = \hat{\mathbf{i}} \frac{\partial}{\partial x_s} + \hat{\mathbf{j}} \frac{\partial}{\partial y_s} + \hat{\mathbf{k}} \frac{\partial}{\partial z_s}. \tag{9.7}$$

Hence equation (9.5) becomes

$$\phi = \frac{1}{4\pi\varepsilon_0} \int_{V_0} \mathbf{P} \cdot \boldsymbol{\nabla}_s \left(\frac{1}{R} \right) \mathrm{d}V_s. \tag{9.8}$$

According to equation (A1.20) of Appendix A1.6 for any scalar ψ and vector $\mathbf{A}$,

$$\boldsymbol{\nabla} \cdot (\psi\mathbf{A}) = \psi\boldsymbol{\nabla} \cdot \mathbf{A} + \mathbf{A} \cdot \boldsymbol{\nabla}\psi.$$

Substituting $\mathbf{A} = \mathbf{P}$ and $\psi = 1/R$, and rearranging

$$\mathbf{P} \cdot \boldsymbol{\nabla}_s \left(\frac{1}{R} \right) = \boldsymbol{\nabla}_s \cdot \left(\frac{\mathbf{P}}{R} \right) - \frac{1}{R} \boldsymbol{\nabla}_s \cdot \mathbf{P}.$$

Substituting in equation (9.8), we obtain

$$\phi = \frac{1}{4\pi\varepsilon_0} \int_{V_0} \boldsymbol{\nabla}_s \cdot \left(\frac{\mathbf{P}}{R} \right) \mathrm{d}V_s + \frac{1}{4\pi\varepsilon_0} \int_{V_0} \frac{(-\boldsymbol{\nabla}_s \cdot \mathbf{P})}{R} \, \mathrm{d}V_s.$$

Applying Gauss' theorem of vector analysis, which is equation (A1.30) of Appendix A1.7, to the first integral on the right hand side, we finally find that

$$\phi = \frac{1}{4\pi\varepsilon_0}\int_{S_0} \frac{\mathbf{P}\cdot\hat{\mathbf{n}}}{R}\,\mathrm{d}S_s + \frac{1}{4\pi\varepsilon_0}\int_{V_0} \frac{(-\boldsymbol{\nabla}_s\cdot\mathbf{P})}{R}\,\mathrm{d}V_s. \tag{9.9}$$

when $\hat{\mathbf{n}}$ is a unit vector in the direction of the outward normal to the surface S_0. Comparing with equations (1.25) and (1.26) of Chapter 1, we see that, if we had a surface charge distribution $\mathbf{P}\cdot\hat{\mathbf{n}}$ and a volume charge distribution $(-\boldsymbol{\nabla}_s\cdot\mathbf{P})$ *in vacuo*, the geometrical configuration of the surface and volume charge distributions being the same as that of the dielectric, then the scalar potential at the field point T would also be given by equation (9.9). Thus we have shown that for purposes of electric field calculation, the dielectric can be replaced by a surface charge distribution

$$\sigma_p = \mathbf{P}\cdot\hat{\mathbf{n}} \tag{9.10}$$

and a volume charge distribution

$$\rho_p = -\boldsymbol{\nabla}_s\cdot\mathbf{P}. \tag{9.11}$$

Thus the problem of the polarized dielectric can be replaced, for purposes of field calculation, by a distribution of charges, given by equations (9.10) and (9.11), placed *in vacuo*. Coulomb's law, and equations derived from it can be applied to this charge distribution to calculate the electric field and potential. This method is valid provided the dipole approximation is valid for *all* the dipoles in the dielectric. It is therefore valid for a point outside a dielectric, or for a point in the middle of a cavity, large on the atomic scale, cut inside the dielectric. If one wishes to calculate the microscopic field near or inside a molecule in the dielectric, in a region of space where the dipole approximation breaks down, we must use a microscopic theory based on a particular atomic model.

The microscopic electric field varies enormously on the atomic scale inside a polarized dielectric. One can have electric fields exceeding 10^8 V m^{-1} just outside atoms, and very much stronger fields inside atoms. It can be shown that inside a dielectric, the electric field calculated using equations (9.10) and (9.11) is equal to the local space average of the microscopic field, so that equation (9.9) gives the macroscopic electric field inside the dielectric. A rigorous proof is given by Lorrain and Corson [3]. As an example, the macroscopic and microscopic electric fields inside a uniformly polarized, cylindrical dielectric are sketched in Figures 9.2(a) and 9.2(b) respectively.

So far, all we have had to say when interpreting equation (9.9) was that, for purposes of calculating the macroscopic electric field due to a polarized dielectric, we can replace the polarized dielectric by charge distributions given by equation (9.11) and then we can use Coulomb's law or equations derived using Coulomb's law such as Gauss' flux law. When the macroscopic theory of dielectrics is derived from the Maxwell-Lorentz equations, for example in the way given by Russakoff [1] using equation (1.147) of Chapter 1, the authors conclude that there are actual macroscopic charge distributions, of charge density given by equation (9.11), inside a polarized dielectric. We shall now

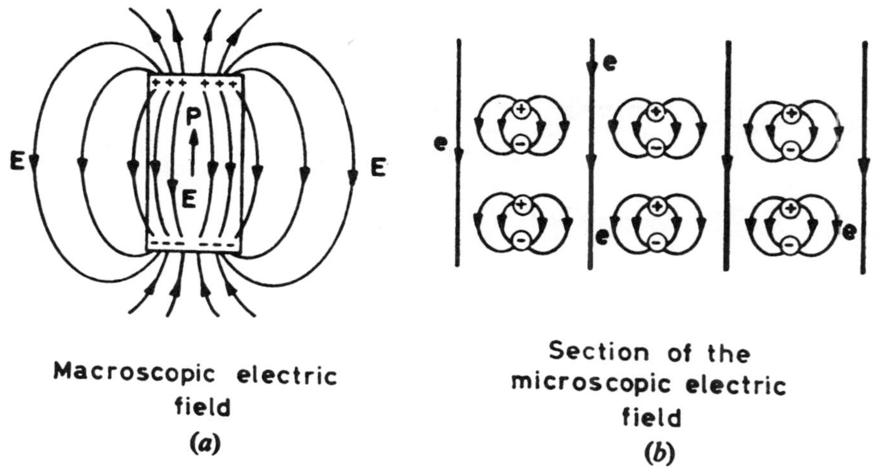

Figure 9.2. (a) The macroscopic electric field **E** due to a polarized cylindrical dielectric. The uniform polarization **P** is parallel to the axis of the cylinder. The macroscopic electric field **E** can be calculated by replacing the dielectric by charge distributions $\mathbf{P} \cdot \hat{\mathbf{n}}$ at each end of the dielectric. (b) A magnified section of the dielectric showing a rough sketch of the electric field on the atomic scale. If these microscopic electric fields **e** are averaged over volumes large on the atomic scale, but small on the laboratory scale, we obtain the smoothly varying macroscopic (or local space average) electric field **E** shown in (a).

use our simplified model of an atomic electric dipole to illustrate the plausibility of this conclusion.

9.2.5. *The macroscopic charge density inside a polarized dielectric*

Consider the plane circular disk shaped surface ΔS that is inside a polarized dielectric as shown in Figure 9.3. The direction of the unit vector $\hat{\mathbf{n}}$ normal to the surface ΔS is in the direction of the polarization vector **P**, which points to the right in the $+x$ direction in Figure 9.3. On our simplified model of atomic electric dipoles, each positive charge $+q$ in each atomic electric dipole moves a distance l when the dielectric is polarized whereas the negative charges do not move. The number of positive charges, that cross the surface ΔS when the dielectric is polarized, is equal to the number of positive charges that were within a distance l of the surface ΔS before the dielectric was polarized. Hence if n is the number of atomic electric dipoles per cubic metre, the total charge ΔQ that crosses the surface ΔS when the dielectric is polarized is

$$\Delta Q = (nl\Delta S)q = np\Delta S = P\Delta S \tag{9.12}$$

where $p = ql$ is the electric dipole moment of each atomic electric dipole and **P** is the polarization vector.

If the polarization vector **P** is at an angle θ to the unit vector $\hat{\mathbf{n}}$ that is

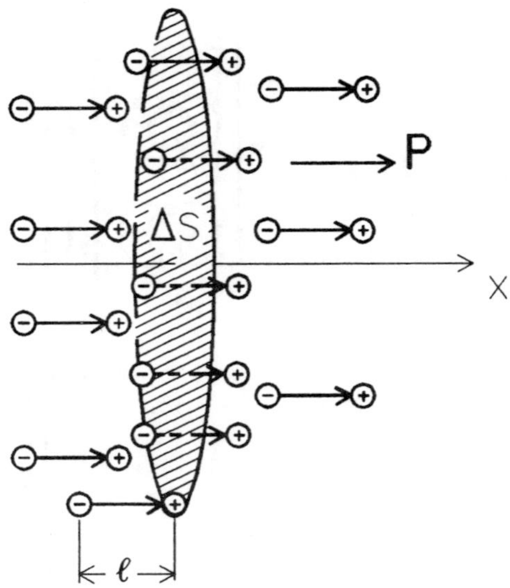

Figure 9.3. A simplified model of a polarized dielectric. It is assumed that each positive charge is displaced a distance l when the external electric field is applied, giving rise to a displacement of electric charge across the surface ΔS, which is perpendicular to the direction in which the positive charges are displaced.

normal to the surface ΔS, it is only those positive charges that were within a distance $l \cos \theta$ of the surface ΔS before the dielectric was polarized that cross the surface ΔS when the dielectric is polarized. Hence, in the general case, the charge that crosses the surface ΔS inside the dielectric when it is polarized is

$$\Delta Q = (nl \cos \theta \, \Delta S)q = P\Delta S \cos \theta = \mathbf{P} \cdot \Delta \mathbf{S}. \tag{9.13}$$

We shall now assume that there is a second circular disk shaped surface of area ΔS at a small distance Δx to the right of and parallel to the surface ΔS in Figure 9.3. The circular disk shaped surfaces are joined by a curved surface to form a small cylindrical surface of volume ΔV equal to $(\Delta S)(\Delta x)$ inside the dielectric. We shall assume, for the moment, that the polarization vector $\mathbf{P}$ is in the $+x$ direction, which is parallel to the axis of the cylinder. We shall assume that P increases with increasing x due to an increase in the separation of the charges in the atomic electric dipoles from l at the position of the first surface of area ΔS to $(l + \Delta l)$ at the position of the second surface of area ΔS. In the case of the second surface of area ΔS, all the positive charges within a distance $(l + \Delta l)$ of that surface cross it when the dielectric is polarized, so that according to equation (9.12) the total charge ΔQ_2 that crosses the second surface ΔS is

$$\Delta Q_2 = nq(l + \Delta l)\Delta S = (P + \Delta P)\Delta S. \tag{9.14}$$

Since the polarization vector $\mathbf{P}$ is parallel to the x axis, it follows from equation (9.13) that no charge crosses the curved surface of the small cylinder of dielectric when the dielectric is polarized. Comparing equations (9.12) and (9.14) we see that more positive charges cross the second surface of area ΔS and leave the cylinder of dielectric than enter across the first surface ΔS. This leaves the cylinder of dielectric with a net charge Q given by $Q = -(\Delta Q_2 - \Delta Q)$.

If ρ_p is the polarization charge density, then Q is equal to $\rho_p\,\Delta V$. After substituting for ΔQ and ΔQ_2 using equations (9.12) and (9.14) respectively we find that

$$\rho_p \Delta V = -(\Delta P)(\Delta S) = -\frac{\partial P}{\partial x}(\Delta x)(\Delta S) = -\frac{\partial P}{\partial x}\Delta V.$$

Hence

$$\rho_p = -\frac{\partial P}{\partial x}. \tag{9.15}$$

This result shows that according to our simplistic model there is a net macroscopic charge density given by equation (9.15) when the dielectric is polarized.

We shall now go on to the general case by considering an arbitrary volume V_0 which is of surface area S_0 and which is completely inside a polarized dielectric. We shall assume that the polarization vector **P** varies with position inside the dielectric. Integrating equation (9.13) over the surface S_0 of the volume V_0 that is inside the dielectric, we find that, on average, the total charge Q that crosses the surface S_0 and leaves the volume V_0 when the dielectric is polarized is

$$Q = \int_{S_0} \mathbf{P} \cdot \mathrm{d}\mathbf{S}. \tag{9.16}$$

Using Gauss' mathematical theorem, which is equation (A1.30) of Appendix A1.7, we can rewrite equation (9.16) in the form

$$Q = \int \boldsymbol{\nabla} \cdot \mathbf{P}\, \mathrm{d}V.$$

If the dielectric inside the volume V_0 was electrically neutral before it was polarised, and if total charge is conserved, then the net charge left inside the volume V_0 after the dielectric is polarized is equal to $-Q$. If the polarization charge density is ρ_p then

$$\int_{V_0} \rho_p\, \mathrm{d}V = -Q = -\int_{V_0} \boldsymbol{\nabla} \cdot \mathbf{P}\, \mathrm{d}V. \tag{9.17}$$

If we make the volume V_0 infinitesimal, we can ignore the variations of ρ_p and $\boldsymbol{\nabla} \cdot \mathbf{P}$ inside V_0 and equation (9.17) reduces to

$$\rho_p = -\boldsymbol{\nabla} \cdot \mathbf{P}. \tag{9.18}$$

The quantity ρ_p is the macroscopic polarization (or bound) charge density. If the polarized dielectric also had a free (or true) charge density ρ, for example due to the removal of some electrons, the total macroscopic charge density at a field point inside the polarized dielectric would be $(\rho + \rho_p)$, where ρ_p is given by equation (9.18).

If we assumed that the surface ΔS in Figure 9.3 was on the surface of the dielectric, the charge that would cross the surface ΔS when the dielectric was polarized would appear as a surface charge distribution. Hence when the dielectric was polarized, in addition to the polarization charge density given by equation (9.18), we would also have a surface charge per unit area, denoted σ_p, which, according to equation (9.13), would be given by

$$\sigma_p = \mathbf{P} \cdot \hat{\mathbf{n}} \tag{9.19}$$

where $\hat{\mathbf{n}}$ is a unit vector in the direction of the outward normal from the surface of the dielectric.

For electrostatic conditions, the contributions of the macroscopic charge distributions ρ, ρ_p and σ_p to the macroscopic electrostatic field can be calculated using Coulomb's law or laws derived from it. Thus a knowledge of $\mathbf{P}$ and its spatial variations is sufficient to determine ρ_p and σ_p and hence the total macroscopic electric field, provided the free (true) charge density ρ is also given.

9.2.6. *The Maxwell equation* $\nabla \cdot \mathbf{D} = \rho$

According to the Maxwell equation (4.14) of Chapter 4, at a field point inside a macroscopic charge distribution made up of classical point charges, we have

$$\nabla \cdot \mathbf{E} = \frac{\rho}{\varepsilon_0} \tag{9.20}$$

where ρ is the free (or true) macroscopic charge density. Since the total macroscopic charge density inside a polarized dielectric is $(\rho + \rho_p)$, it follows that, at a field point inside the dielectric, equation (9.20) becomes

$$\nabla \cdot \mathbf{E} = \frac{(\rho + \rho_p)}{\varepsilon_0}. \tag{9.21}$$

Using equation (9.18) to substitute for ρ_p we obtain

$$\nabla \cdot \mathbf{E} = \frac{(\rho - \nabla \cdot \mathbf{P})}{\varepsilon_0}. \tag{9.22}$$

According to equation (9.22), $\nabla \cdot \mathbf{E}$ is related to both the free charge density ρ and the polarization vector $\mathbf{P}$. It is assumed in classical electromagnetism that equation (9.22) is valid when both ρ and $\mathbf{P}$ are varying with time, whatever the frequency of the variations.

The ratio $\mathbf{P}/\varepsilon_0\mathbf{E}$ is called the electric susceptibility and is denoted by χ_e, so that

$$\mathbf{P} = \chi_e \varepsilon_0 \mathbf{E}. \tag{9.23}$$

Equation (9.22) can be rearranged in the form

$$\nabla \cdot (\varepsilon_0 \mathbf{E} + \mathbf{P}) = \rho. \tag{9.24}$$

The quantity ($\varepsilon_0\mathbf{E} + \mathbf{P}$) appears so frequently in the theory of classical electromagnetism that it is worth giving it a special name, namely the **electric displacement vector** and the special symbol **D**, so that, by definition,

$$\mathbf{D} = \varepsilon_0 \mathbf{E} + \mathbf{P}. \tag{9.25}$$

Equation (9.24) can now be rewritten in the more concise form

$$\nabla \cdot \mathbf{D} = \rho \tag{9.26}$$

where ρ is the macroscopic free (or true) charge density at a point inside a stationary dielectric where the displacement vector is **D**. Equation (9.26) is one of Maxwell's equations for the macroscopic fields inside a polarized dielectric. It is assumed that equation (9.26) is valid when both **D** and ρ vary with time. One could just as well use equation (9.22) and never introduce the auxiliary vector **D**. However the introduction of the vector **D** is often convenient and can help in the solution of problems. For example it follows from equation (9.26) that the flux of **D** from any surface is equal to the total free (true) charge inside that surface.

Equation (9.25) is often rewritten in the form

$$\mathbf{D} = \varepsilon \mathbf{E} = \varepsilon_r \varepsilon_0 \mathbf{E} \tag{9.27}$$

where ε is the permittivity. The ratio $\varepsilon_r = \varepsilon/\varepsilon_0$ is called the relative permittivity. Substituting for **D** and for **P** from equations (9.27) and (9.23) into equation (9.25), we find that

$$\varepsilon_r = 1 + \chi_e \tag{9.28}$$

where χ_e is the electric susceptibility.

For a large class of dielectric, for electrostatic conditions, the relative permittivity ε_r and the electric susceptibility χ_e are independent of the value of the electric field. Such dielectrics are called linear dielectrics. In these cases ε_r is sometimes called the dielectric constant. For single crystals it is sometimes easier to displace the atomic charges inside the molecules and ions in the crystal in some directions of space than others. In these cases ε_r and χ_e are tensors. Some dielectrics, for example electrets, exhibit remanence in which case ε_r and χ_e depend on the past history of the dielectrics. Electrets can retain a permanent polarization after the external field has been removed. In general, ε_r and χ_e depend on the temperature, on the applied electric field and, if the conditions are varying, they depend on the frequency. Equation (9.27) is one of the constitutive equations; the values of ε_r and χ_e depend on the properties of the material at the field point inside the dielectric and on the experimental conditions.

9.3. Stationary magnetic materials

9.3.1. *The Amperian model*

In our interpretation of the origin of the magnetic fields due to magnetized bodies, we shall use the Amperian model according to which we can treat each atomic magnetic dipole as a small electric current system. The magnetic moment **m** of a current distribution is given by

$$\mathbf{m} = \frac{1}{2}\int \mathbf{r} \times \mathbf{J}(\mathbf{r})\, dV \tag{9.29}$$

where $\mathbf{J}(\mathbf{r})$ is the current density at the position **r**. It follows from equation (9.29) that, in the special case of a plane current carrying coil, the magnetic dipole moment **m** is numerically equal to the product of the area of the coil and the current in the coil. The direction of the magnetic dipole moment **m** is in the direction a right-handed corkscrew would advance if it were rotated in the direction of current flow.

The Amperian model is a reasonable model for the contributions of the orbital motions of the electrons in a molecule to the resultant magnetic moment of the molecule. However, in practice, the magnetic moments associated with the spins of the particles can make a significant contribution to the resultant magnetic moment, particularly in the case of ferromagnetism. It would go beyond the uncertainty principle to attempt to give a classical model for electron spin. If one did treat an electron as a slowly rotating, uniformly charged sphere, the calculated gyromagnetic ratio, which is the ratio of the magnetic moment to the angular momentum of the electron, would only be one Bohr magneton whereas the value predicted by Dirac's relativistic wave equation is equal to two Bohr magnetons. Hence we shall consider our use of the Amperian model, particularly in the case of electron spin, as illustrative rather than explanatory.

9.3.2. *Types of magnetic materials*

Diamagnetism: In the case of diamagnetic materials, when a magnetic field is applied to them, there is a weak magnetization in the direction opposite to the direction of the applied magnetic field, It follows from Faraday's law of electromagnetic induction, equation (4.46) of Chapter 4, that, when the applied magnetic field is switched on, there is also and induction electric field while the applied magnetic field is building up to its final value. This induction electric field changes the motions of the electrons in the molecules in such a way that, according to Lenz's law, the consequential changes in the magnetic field are in the direction opposite to the direction of the increasing applied magnetic field, thereby reducing the total magnetic field and giving a relative permeability μ_r slightly less than unity. In the absence of any ohmic resistance these changes in the current distributions inside the molecules persist until the applied magnetic field is reduced to zero, when there is again a transient induction electric field, associated with the decreasing applied magnetic field, which reverses the changes in the electronic motions inside the molecules which brings the magnetic material back to its original state.

Paramagnetism: In paramagnetic materials some of the molecules have a permanent, resultant, magnetic dipole moment associated with the orbital motions of the electrons and with electron spin. In an applied magnetic field, more of these permanent dipole moments point in the direction of the applied magnetic field than in the opposite direction. Complete alignment is prevented by the thermal motions of the atomic magnetic dipole moments. The resultant magnetization due to paramagnetism can be calculated using the Boltzmann distribution. Reference: Rosser [4]. The resultant weak magnetization in the direction of the applied magnetic field due to paramagnetism generally predominates over the weaker diamagnetic effect and gives an extra contribution to the magnetic field in the direction of the applied magnetic field so that the value for the relative permeability μ_r is slightly greater than unity.

Ferromagnetism: In ferromagnetic materials the atomic magnetic moments within a macroscopic region, called a domain are parallel to each other. Before a ferromagnetic material is magnetized the resultant magnetic moments of the domains point in all directions so that the resultant magnetic moment of the ferromagnetic material is zero. When a weak external magnetic field is applied, the sizes of those domains whose resultant magnetic moments point in the direction of the applied magnetic field, grow in size at the expense of the other domains. When the external magnetic field is increased further, the directions of the magnetic moments of the domains are rotated such that they point preferentially in the direction of the crystal axis nearest to the direction of the applied magnetic field. This is the Barkhausen effect. At high magnetic fields of the order of 2T, the magnetization of a ferromagnetic material generally tends to its saturation limit. The relative permeabilities of ferromagnetic materials are typically of the order of 1500. Some ferromagnetic materials exhibit hysteresis effects, and some remanent magnetization can persist after the external applied magnetic field is removed; the material is then a permanent magnet.

9.3.3. *The magnetization vector* **M**

We shall find it useful to introduce a vector quantity to specify the macroscopic properties of a magnetized magnetic material. Due to the influence of the applied magnetic field a large number of atomic magnetic moments are aligned in the direction of the applied magnetic field. Consider a small volume element ΔV. Let

$$\sum_{i=1}^{N} \mathbf{m}_i = \mathbf{M}\Delta V; \qquad \text{or} \qquad \mathbf{M} = \frac{\sum_{i=1}^{N} \mathbf{m}_i}{\Delta V} \tag{9.30}$$

where the vector summation is carried out over all the N atomic magnetic dipoles in the volume element ΔV. Equation (9.30) is a simplistic definition of the magnetization vector $\mathbf{M}$ in terms of the microscopic magnetic dipole moments. In a macroscopic theory $\mathbf{M}$ can be treated as a continuous function of position and defined as the average resultant magnetic dipole moment per unit volume.

9.3.4. *The magnetic field due to a stationary magnetized body*

Consider a volume element ΔV_s at a source point at a position $\mathbf{r}_s$ inside a stationary magnetized body. The vector potential will be determined at a field point T at the position $\mathbf{r}$ which is at a distance $\mathbf{R} = (\mathbf{r} - \mathbf{r}_s)$ from ΔV_s. These are the same geometrical conditions as those shown in Figure 9.1 for the case of a polarized dielectric. It is shown in text books on electromagnetism that the vector potential $\mathbf{A}_i$ at the field point T due to a magnetic dipole of dipole moment $\mathbf{m}_i$ that is inside ΔV_s is

$$\mathbf{A}_i = \left(\frac{\mu_0}{4\pi}\right) \frac{\mathbf{m}_i \times \mathbf{R}}{R^3} . \tag{9.31}$$

Summing over all the magnetic dipole moments inside the volume element ΔV_s we have

$$\Delta\mathbf{A} = \left(\frac{\mu_0}{4\pi}\right) \frac{\Sigma(\mathbf{m}_i \times \mathbf{R})}{R^3} = \left(\frac{\mu_0}{4\pi}\right) \frac{(\Sigma\mathbf{m}_i) \times \mathbf{R}}{R^3} . \tag{9.32}$$

According to equation (9.30), $\Sigma\mathbf{m}_i$ is equal to $\mathbf{M}\Delta V_s$. After substituting in equation (9.32), then integrating over the volume of the magnetized body and using equation (9.6) we find that the vector potential at the field point T due to the magnetized body is

$$\mathbf{A} = \frac{\mu_0}{4\pi} \int \frac{(\mathbf{M} \times \mathbf{R})}{R^3}\, \mathrm{d}V_s = +\frac{\mu_0}{4\pi} \int \mathbf{M} \times \boldsymbol{\nabla}_s \left(\frac{1}{R}\right) \mathrm{d}V_s \tag{9.33}$$

where $\boldsymbol{\nabla}_s$ is given by equation (9.7).

According to equation (A1.23) of Appendix A1.6

$$\mathbf{M} \times \boldsymbol{\nabla}_s \left(\frac{1}{R}\right) = \frac{1}{R} \boldsymbol{\nabla}_s \times \mathbf{M} - \boldsymbol{\nabla}_s \times \left(\frac{\mathbf{M}}{R}\right) .$$

Substituting in equation (9.33) we have

$$\mathbf{A} = \frac{\mu_0}{4\pi} \int \frac{\boldsymbol{\nabla}_s \times \mathbf{M}}{R}\, \mathrm{d}V_s - \frac{\mu_0}{4\pi} \int \boldsymbol{\nabla}_s \times \left(\frac{\mathbf{M}}{R}\right) \mathrm{d}V_s .$$

After applying equation (A1.33) of Appendix A1.7 to the second integral on the right hand side we finally obtain

$$\mathbf{A} = \frac{\mu_0}{4\pi}\int \frac{\nabla_s \times \mathbf{M}}{R}\, dV_s + \frac{\mu_0}{4\pi}\int \frac{\mathbf{M} \times \hat{\mathbf{n}}}{R}\, dS_s. \tag{9.34}$$

Comparing equation (9.34) with equation (1.74) of Chapter 1, we see that, for the purposes of magnetic field calculations we can replace the magnetized body by a surface current

$$\mathbf{K} = \mathbf{M} \times \hat{\mathbf{n}} \tag{9.35}$$

per metre length, where $\hat{\mathbf{n}}$ is a unit vector in the direction of the outward normal, plus a volume current distribution of current density

$$\mathbf{J}_m = \nabla_s \times \mathbf{M} \tag{9.36}$$

distributed throughout the volume of the magnetized body. The current density given by equation (9.36) will be called the magnetization (or bound) current density. The magnetic field due to the magnetized body can be obtained by taking the curl of the vector potential given by equation (9.34) or by applying the Biot-Savart law to the current distributions given by equations (9.35) and (9.36). So far we have only shown that this approach is correct when the dipole approximation leading up to equation (9.31) is valid for every one of the atomic magnetic dipoles. This is generally true at an external field point. It can be shown that, at a field point inside the magnetized body, the magnetic field calculated using equation (9.34) is equal to the macroscopic magnetic field **B**, which is the space average of the microscopic magnetic field **b**. Reference: Lorrain and Corson [5]. There is no need to use a precise atomic model when calculating the macroscopic magnetic field. A knowledge of **M**, the average resultant magnetic moment per unit volume is sufficient. We shall now interpret equations (9.35) and (9.36) using a simplistic Amperian model for the atomic magnetic dipoles.

9.3.5. *Example of the origin of a macroscopic magnetization current*

To illustrate equations (9.35) and (9.36) we shall consider a magnetostatic example using a simplistic Amperian model, in which we shall assume that the magnetic dipoles can be treated as square coils, each of side d and of area $A = d^2$ carrying a steady current. We shall assume that, in the presence of an applied magnetic field, the coils are aligned, such that more of their magnetic moments point in the $+z$ direction than in the $-z$ direction giving a resultant magnetization M_z in the $+z$ direction. The centres of the coils are at a distance d apart as shown in Figure 9.4. We shall assume that there are identical parallel rows of coils in the $z = 0$ plane, consisting of $(N + 1)$ coils per row. The central lines of successive rows of coils are a distance d apart, as shown in Figure 9.4. We shall assume that there are identical layers of coils parallel to the $z = 0$ plane, the layers of coils being at a distance h apart, so that by analogy with crystal structure there is one square coil per

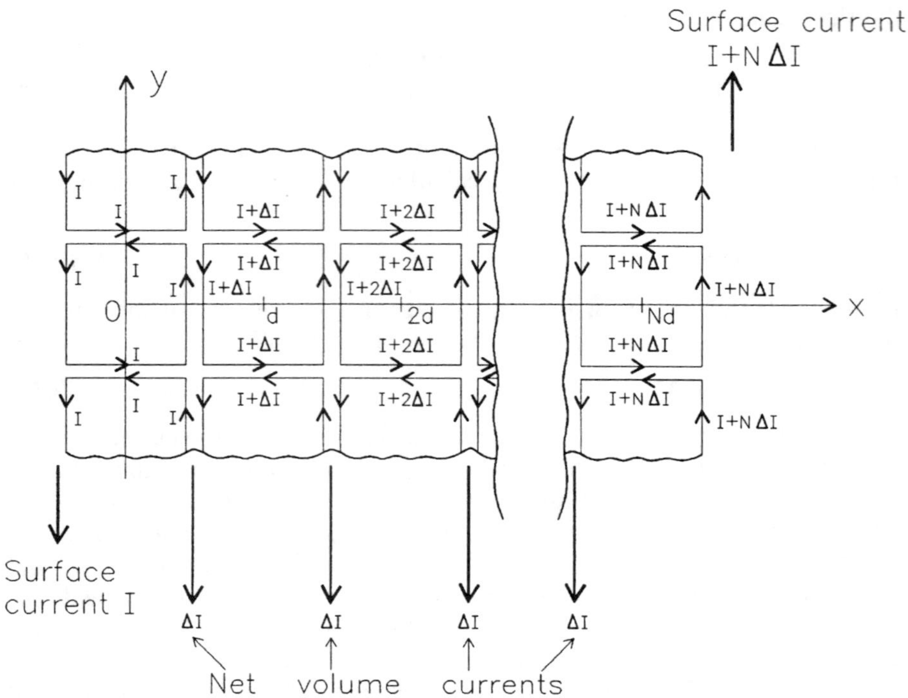

Figure 9.4. A simplistic Amperian model is applied to illustrate the origin of the macroscopic current density inside a magnetized body in which the magnetization **M** varies with position within the magnetized body.

volume of hd^2. We shall assume that there is a uniform rate of increase of M_z with increasing x, which in the real case would be due to the alignment of more and more atomic magnetic moments in the $+z$ direction with increasing x. In our simplistic model, we shall represent this increase in the number of aligned magnetic moments by increasing the currents in the square coils in Figure 9.4 from I at the coil at the origin to $(I + \Delta I)$ in the next etc. . . . up to $(I + N\Delta I)$ in the $(N + 1)$th coil where

$$\Delta I = \left(\frac{\partial I}{\partial x}\right) d. \tag{9.37}$$

Thus each square coil represents the resultant of the atomic magnetic moments in a volume of hd^2. We shall assume that the magnetization does not vary with y for fixed x and z and with z for fixed x and y.

Consider first the currents in the y direction in Figure 9.4. On the left hand side of the row of coils along the x axis, that is at $x = -d/2$, the current I in the first coil is in the $-y$ direction in Figure 9.4. This is true for the extreme left hand edge of all the rows of coils that are in the $z = 0$ plane, so that, for purposes of magnetic field calculation, we can assume that there is a continuous line current I in the $-y$ direction at $x = -d/2$ in Figure 9.4. Moving a distance d to the right to $x = d/2$ on the x axis in Figure 9.4, we have a current I in the $+y$ direction in the first coil and a current of $(I + \Delta I)$ in the $-y$ direction in the second coil so that, when we average to get the macroscopic current density, these two currents give a resultant contribution of ΔI in the $-y$ direction. This is true at $x = +d/2$ for each of the rows of coils in the $z = 0$ plane. These sections of current of magnitude ΔI add up on the macroscopic scale to give the same magnetic field as a continuous line current of magnitude ΔI in the $-y$ direction at $x = +d/2$. Then at $x = +3d/2$ there is another current ΔI in the $-y$ direction. This pattern is repeated for every increment of d in x until we reach the right hand surface at $x = (N + 1/2)d$, where there is a current of magnitude $(I + N\Delta I)$ on the surface in the $+y$ direction. The above analysis is true for every layer of square coils. Hence, for purposes of magnetic field calculation there is inside the magnetized body, an average current of ΔI per area of cross section equal to hd. Using equation (9.37) to substitute for ΔI and then averaging, we find that the y component of the macroscopic magnetization current density inside the magnetized body shown in Figure 9.4 is given by

$$J_y = -\frac{\Delta I}{hd} = -\frac{1}{hd}\left(\frac{\partial I}{\partial x}\right)d = -\frac{1}{h}\left(\frac{\partial I}{\partial x}\right). \tag{9.38}$$

We can see from Figure 9.4 that the currents in the x direction in successive rows of square coils compensate each other so that the x component J_x of the macroscopic magnetization current density is zero. Since on our model we are assuming that there are no atomic currents in the z direction, J_z is also zero. Now M_z is the average resultant magnetic moment per unit volume. The resultant magnetic moments of the atomic magnetic moments represented by the square coil at the origin in Figure 9.4 is equal to Id^2. By analogy with crystal structure, this is the resultant magnetic moment per volume of hd^2 in the vicinity of the origin. Hence the magnetization in the vicinity of the origin in Figure 9.4, which is the average magnetic dipole moment per unit volume at that position, is

$$M_z = Id^2 \div hd^2 = \frac{I}{h}. \tag{9.39}$$

Differentiating partially with respect to x, we have

$$\frac{\partial M_z}{\partial x} = \frac{1}{h}\frac{\partial I}{\partial x}. \tag{9.40}$$

Comparing equations (9.38) and (9.40) we conclude that

$$J_x = 0; \qquad J_y = -\frac{\partial M_z}{\partial x}; \qquad J_z = 0. \tag{9.41}$$

This result is in agreement with equation (9.36).

We have shown that there is a line current of magnitude I on the left hand edges of the rows of coils in the $z = 0$ plane in Figure 9.4. This is true for each successive layer of coils, which are a distance h apart. After using equation (9.39), we find that, at the surface at $x = -d/2$ in Figure 9.4, the surface current per metre is given by

$$K_y = -\frac{I}{h} = -M_z. \tag{9.42}$$

This result is in agreement with equation (9.35).

These results illustrate how, according to the simplistic Amperian model we have used, when the magnetization varies with position the currents in the square coils do not average out to zero, but add up to give macroscopic current distributions given by equations (9.35) and (9.36). Since the Amperian model is a reasonable model when the atomic magnetic moments are due only to the orbital motions of electrons inside molecules, it is reasonable to find in these cases that there are macroscopic current distributions inside magnetized bodies, which arise from the non-cancellation of the currents in molecules, and whose magnitudes are given by equations (9.35) and (9.36). However, we pointed out in Section 9.3.1 that we cannot give an appropriate classical model for electron spin and so we cannot be sure how valid the Amperian model is for electron spin. Hence in the case of electron spin it is probably best to refer to the current density given by equation (9.36) as an equivalent macroscopic magnetization current distribution that would give the same magnetic field as the magnetic dipoles in the magnetized body.

9.3.6. *The magnetizing force (or magnetic field intensity)* **H**

Since in magnetostatic conditions we can use the Biot-Savart law to calculate the magnetic field due to a magnetized body using the equivalent magnetization current $\mathbf{J}_m$ given by equation (9.36), it is straightforward for the reader to show, by following the steps we used in Section 1.4.7 of Chapter 1, that, if $\mathbf{B}_m$ is the macroscopic magnetic field due to the magnetized body then, corresponding to equation (1.89), at a field point inside a magnetized body we have

$$\nabla \times \mathbf{B}_m = \mu_0 \mathbf{J}_m. \tag{9.43}$$

If there is also a free (true) current density $\mathbf{J}$ at the field point due, for example, to a conduction current, then, if $\mathbf{B}_c$ is the macroscopic magnetic field due to all the free (true) currents, we have

$$\nabla \times \mathbf{B}_c = \mu_0 \mathbf{J}. \tag{9.44}$$

Adding equations (9.43) and (9.44), we have

$$\nabla \times \mathbf{B} = \mu_0(\mathbf{J} + \mathbf{J}_m)$$

where $\mathbf{B} = (\mathbf{B}_c + \mathbf{B}_m)$ is the total macroscopic magnetic field at the field point. Using equation (9.36) to substitute for $\mathbf{J}_m$, we find that

$$\nabla \times \mathbf{B} = \mu_0(\mathbf{J} + \nabla \times \mathbf{M}). \tag{9.45}$$

Equation (9.45) can be rearranged in the form

$$\nabla \times \left(\frac{\mathbf{B}}{\mu_0} - \mathbf{M}\right) = \mathbf{J}. \tag{9.46}$$

The quantity $(\mathbf{B}/\mu_0 - \mathbf{M})$ appears so frequently in the theory of classical electromagnetism that it is worth giving it a special name and a special symbol. We shall call it the **magnetizing force** and denote it by the symbol **H**. It is sometimes called the **magnetic field intensity**. Hence by definition

$$\mathbf{H} = \frac{\mathbf{B}}{\mu_0} - \mathbf{M}. \tag{9.47}$$

Rearranging equation (9.47), we have

$$\mathbf{B} = \mu_0(\mathbf{H} + \mathbf{M}). \tag{9.48}$$

Using equation (9.47) to substitute for **B** in equation (9.46) we find that

$$\nabla \times \mathbf{H} = \mathbf{J} \tag{9.49}$$

where **J** is the free (true) current density at a point where the magnetizing force (magnetic field intensity) is **H**. Since the vectors **B** and **M** suffice in magnetostatics, we could always use equation (9.45) and leave **H** out of the theory of classical electromagnetism.

Integrating equation (9.49) over a finite surface S and applying Stokes' theorem of vector analysis, which is equation (A1.34) of Appendix A1.8, we obtain

$$\oint \mathbf{H} \cdot \mathrm{d}\mathbf{l} = \int \mathbf{J} \cdot \mathrm{d}\mathbf{S} = I \tag{9.50}$$

where I is the total free (true) current, such as a conduction current, that crosses the surface S. The line integral $\oint \mathbf{H} \cdot \mathrm{d}\mathbf{l}$ is sometimes called the magnetomotive force. Equation (9.50) is often useful in solving problems, such as the magnetic circuit, details of which will be found in elementary text books.

9.3.7. *The relative permeability μ_r and the magnetic susceptibility χ_m*

Equation (9.48) is often rewritten in the form

$$\mathbf{B} = \mu\mathbf{H} = \mu_r\mu_0\mathbf{H} \tag{9.51}$$

where μ is the **permeability** of the medium and the ratio μ/μ_0 which is denoted by μ_r, is called the **relative permeability** of the medium. Equation (9.51) is one of the constitutive equations. The value of the relative permeability μ_r depends on the material present at the field point and the experimental conditions. (Reference: Section 9.3.2)

As an alternative to equation (9.51) we can introduce the magnetic susceptibility χ_m which is generally defined in terms of the magnetizing force (or magnetic field intensity) **H** by the equation

$$\mathbf{M} = \chi_m \mathbf{H}. \tag{9.52}$$

Putting $\mathbf{B} = \mu_r\mu_0\mathbf{H}$ and $\mathbf{M} = \chi_m\mathbf{H}$ in equation (9.48) and cancelling $\mu_0\mathbf{H}$ we find that

$$\mu_r = 1 + \chi_m. \tag{9.53}$$

Putting $\mathbf{H} = \mathbf{B}/\mu_r\mu_0$ in equation (9.48) and rearranging, we find that

$$\mathbf{M} = \frac{(\mu_r - 1)}{\mu_r\mu_0}\mathbf{B} = \frac{\chi_m}{\mu_0(1 + \chi_m)}\mathbf{B}. \tag{9.54}$$

Equation (9.54) is an alternative constitutive equation expressing **M** in terms of **B**, where **B** is the resultant macroscopic magnetic field.

9.4. Maxwell's equations at a field point inside a magnetic dielectric

9.4.1. *The equation* $\nabla \times \mathbf{H} = \mathbf{J} + \dot{\mathbf{D}}$

We only derived the equation

$$\nabla \times \mathbf{B} = \mu_0(\mathbf{J} + \mathbf{J}_m) = \mu_0(\mathbf{J} + \nabla \times \mathbf{M}) \tag{9.45}$$

for magnetostatic conditions. We shall now go on to consider how equation (9.45) must be modified when the conditions are varying. We shall assume that the magnetization current density $\mathbf{J}_m$ is still given by equation (9.36). However when the conditions are varying, we must include the vacuum displacement current density $\varepsilon_0\dot{\mathbf{E}}$ in equation (9.44) so that equation (9.45) then becomes

$$\nabla \times \mathbf{B} = \mu_0(\varepsilon_0\dot{\mathbf{E}} + \mathbf{J} + \nabla \times \mathbf{M}) \tag{9.55}$$

for a field point inside a magnetic material, for which the relative permittivity $\varepsilon_r = 1$. If the material also has dielectric properties, such that $\varepsilon_r > 1$, there is also a contribution to the magnetic field due to the motions of charges inside molecules that lead to a varying polarization **P**. We assumed in Section 9.2. that, when a dielectric is polarized the negative charges in the molecules do not move but the positive charges are displaced a distance l. According to equation (9.12) the total charge that crosses the surface ΔS, which is normal to the polarization vector **P** in Figure 9.3, when the dielectric is polarized, is

$$\Delta Q = P\Delta S. \tag{9.56}$$

On our simplistic model, when the polarization **P** is varying positive charges go back and forth across the surface ΔS giving rise to a macroscopic current. Differentiating equation (9.56) with respect to time we have

$$J_x\Delta S = \frac{\partial(\Delta Q)}{\partial t} = \frac{\partial P}{\partial t}\Delta S.$$

Hence the polarization current density $\mathbf{J}_p$ is given by

$$\mathbf{J}_p = \frac{\partial \mathbf{P}}{\partial t}. \tag{9.57}$$

Since the macroscopic current density, given by equation (9.57), is due to the motion of charges it gives a contribution to the magnetic field. Hence at a field point inside a stationary body that has both dielectric and magnetic properties we must add the varying polarization current density $\mathbf{J}_p$ to the right hand side of equation (9.55) which finally becomes

$$\nabla \times \mathbf{B} = \mu_0(\varepsilon_0\dot{\mathbf{E}} + \mathbf{J} + \mathbf{J}_m + \mathbf{J}_p) \tag{9.58}$$

$$= \mu_0(\varepsilon_0\dot{\mathbf{E}} + \mathbf{J} + \nabla \times \mathbf{M} + \dot{\mathbf{P}}) \tag{9.59}$$

where $\mathbf{B}$ is the macroscopic magnetic field due to the free (true) current density $\mathbf{J}$, the magnetization current density $\nabla \times \mathbf{M}$ and the varying polarization current density $\dot{\mathbf{P}}$. Equation (9.59) can be rearranged in the form

$$\nabla \times \left(\frac{\mathbf{B}}{\mu_0} - \mathbf{M}\right) = \mathbf{J} + \frac{\partial}{\partial t}(\varepsilon_0\mathbf{E} + \mathbf{P}). \tag{9.60}$$

If we introduce the auxiliary vectors $\mathbf{D}$ and $\mathbf{H}$ defined by the equations

$$\mathbf{D} = \varepsilon_0\mathbf{E} + \mathbf{P} \tag{9.25}$$

$$\mathbf{H} = \frac{\mathbf{B}}{\mu_0} - \mathbf{M}. \tag{9.47}$$

we can rewrite equation (9.60) in the more concise form

$$\nabla \times \mathbf{H} = \mathbf{J} + \dot{\mathbf{D}}. \tag{9.61}$$

It is of interest to note that $\dot{\mathbf{D}}$ is equal to the sum of the vacuum displacement current density $\varepsilon_0\dot{\mathbf{E}}$ and the varying polarization current density $\dot{\mathbf{P}}$, which have completely different origins. It was the addition of the displacement current term $\varepsilon_0\dot{\mathbf{E}}$ to Maxwell's equations that converted the Laplacians of electrostatics and magnetostatics into D'Alembertians and led to the prediction that the electromagnetic interaction is propagated at the speed of light c in empty space. The $\dot{\mathbf{P}}$ term arises from the motions of charges within molecules. We shall comment on its role in the propagation of the macroscopic field vectors $\mathbf{E}$ and $\mathbf{B}$ in a dielectric in Section 9.5.3.

9.4.2. *The equations* $\nabla \cdot \mathbf{B} = 0$ *and* $\nabla \times \mathbf{E} = -\dot{\mathbf{B}}$

It was shown in Section 4.3.2 and 4.5.4 of Chapter 4 that in the absence of magnetic monopoles the equations

$$\nabla \cdot \mathbf{B} = 0 \tag{9.62}$$

$$\nabla \times \mathbf{E} = -\dot{\mathbf{B}} \tag{9.63}$$

which relate to the macroscopic field vectors **E** and **B**, are valid at a field point inside a system of moving classical point charges that build up a macroscopic charge and current distribution. Since the polarization current density $\mathbf{J}_p$ and according to our Amperian model the magnetization current density $\mathbf{J}_m$ are macroscopic currents arising from moving charge distributions, they should behave in the same way as a current distribution due to classical point charges. This shows that equations (9.62) and (9.63) need not be modified in the presence of the contributions due to the $\mathbf{J}_p$ and $\mathbf{J}_m$ terms. The macroscopic fields **E** and **B** are then the resultant fields due to the contributions of all the ρ, ρ_p, $\mathbf{J}$, $\mathbf{J}_p$ and $\mathbf{J}_m$ terms. One cannot be sure how valid the Amperian model is for the contributions due to electron spin. The magnetic field associated with electron spin is a dipole field for which $\nabla \cdot \mathbf{B}$ is zero, so that equation (9.62) should be valid. We can appeal directly to experiments with transformers having ferromagnetic cores to show that equation (9.63) is also valid. Such experiments have confirmed that Faraday's law of electromagnetic induction, which is an integral form of equation (9.63), is valid.

9.4.3. *Summary of Maxwell's equations*

In this chapter we have only given a very simplistic insight into the roles and origins of the various terms in the form Maxwell's equations take at a field point inside a material medium, where the relative permittivity ε_r and the relative permeability μ_r are both greater than unity. For example, we did not average the microscopic variables properly using equation (1.147) and the weighting function shown in Figure 1.12 of Chapter 1, to determine the corresponding macroscopic variables. A reader interested in a more rigorous development is referred to Russakoff [1] or Robinson [2]. The basic field vectors in our approach are the macroscopic field vectors **E** and **B**, which are the averages of the corresponding microscopic variables **e** and **b**. We found it convenient to introduce the properties of the material medium using the polarization vector **P** and the magnetization vector **M**, which we defined as the average electric dipole moment per unit volume and the average magnetic dipole moment per unit volume respectively. It follows from equations (9.22), (9.62), (9.63) and (9.59) that at a field point inside a material medium where $\varepsilon_r > 1$ and $\mu_r > 1$ Maxwell's equations take the form

$$\nabla \cdot \mathbf{E} = \frac{(\rho + \rho_p)}{\varepsilon_0} = \frac{(\rho - \nabla \cdot \mathbf{P})}{\varepsilon_0} \tag{9.22}$$

$$\nabla \cdot \mathbf{B} = 0 \tag{9.62}$$

$$\nabla \times \mathbf{E} = -\dot{\mathbf{B}} \tag{9.63}$$

$$\nabla \times \mathbf{B} = \mu_0(\varepsilon_0\dot{\mathbf{E}} + \mathbf{J} + \mathbf{J}_p + \mathbf{J}_m) \tag{9.58}$$

$$= \mu_0(\varepsilon_0\dot{\mathbf{E}} + \mathbf{J} + \dot{\mathbf{P}} + \nabla \times \mathbf{M}) \tag{9.59}$$

The above form of Maxwell's equations is the form that fits in best with the approach we developed in Chapter 4. In equations (9.22) both the free (true) charge density and the polarization charge density ($-\nabla \cdot \mathbf{P}$) contribute to the flux of the macroscopic electric field **E**. In equation (9.59), in addition to the free (true) current density **J**, the varying polarization current density $\dot{\mathbf{P}}$ and the magnetization current density $\nabla \times \mathbf{M}$ are also treated as sources of the macroscopic magnetic field **B**. Equations (9.22) and (9.59) show explicity the roles of all the contributing terms. The form of the constitutive equations that would fit in best with the above approach to Maxwell's equations would be

$$\mathbf{P} = \chi_e \varepsilon_0 \mathbf{E} \tag{9.23}$$

$$\mathbf{M} = \frac{(\mu_r - 1)}{\mu_r \mu_0}\mathbf{B} = \frac{\chi_m}{\mu_0(1 + \chi_m)}\mathbf{B} \tag{9.54}$$

$$\mathbf{J} = \sigma \mathbf{E}. \tag{9.64}$$

There is no need in our approach to classical electromagnetism to introduce the auxiliary vectors **D** and **H**, defined by the equations

$$\mathbf{D} = \varepsilon_0 \mathbf{E} + \mathbf{P} \tag{9.25}$$

$$\mathbf{H} = \left(\frac{\mathbf{B}}{\mu_0} - \mathbf{M}\right) \tag{9.47}$$

where **D** is the electric displacement and **H** is the magnetizing force (or the magnetic field intensity). It is, however, both conventional and traditional to introduce **D** and **H** in which case Maxwell's equations can be expressed in the more concise form

$$\nabla \cdot \mathbf{D} = \rho \tag{9.65}$$

$$\nabla \cdot \mathbf{B} = 0 \tag{9.66}$$

$$\nabla \times \mathbf{E} = -\dot{\mathbf{B}} \tag{9.67}$$

$$\nabla \times \mathbf{H} = \mathbf{J} + \dot{\mathbf{D}} \tag{9.68}$$

where ρ is the free (true) charge density and **J** is the free(true) current density. The contributions of the polarization charge density in equation (9.65) and the varying polarization and magnetization current densities in equation (9.68) are hidden in the variables **D** and **H**. When the auxiliary vectors **D** and **H** are introduced, it is more convenient to express the constitutive equations in the form

$$\mathbf{D} = \varepsilon \mathbf{E} = \varepsilon_r \varepsilon_0 \mathbf{E} \tag{9.27}$$

$$\mathbf{B} = \mu \mathbf{H} = \mu_r \mu_0 \mathbf{H} \tag{9.51}$$

$$\mathbf{J} = \sigma \mathbf{E} \tag{9.64}$$

where ε is the permittivity, ε_r is the relative permittivity, μ is the permeability, μ_r is the relative permeability and σ is the electrical conductivity. It is assumed in classical electromagnetism that the Maxwell equations (9.65), (9.66), (9.67) and (9.68) are valid for varying charge and current distributions whatever the frequency of the variations.

By repeating the method used in Section 8.5.1 of Chapter 8 but using equations (9.65)–(9.68) the reader can show that the Poynting vector **N** inside a material medium is given by

$$\mathbf{N} = \mathbf{E} \times \mathbf{H} \tag{9.69}$$

9.5. Electromagnetic waves in a linear isotropic homogenous medium

9.5.1. *The wave equations*

Consider a linear, isotropic and homogenous medium, denoted by LIH, for which ε, μ and σ are constants independent of position, of direction and of the magnitudes of the electric and magnetic fields.

Taking the curl of both sides of equation (9.67) and using the relation $\mathbf{B} = \mu\mathbf{H}$ where μ is now a constant, we find that at a field point inside an LIH medium

$$\nabla \times (\nabla \times \mathbf{E}) = -\nabla \times \frac{\partial \mathbf{B}}{\partial t} = -\mu \frac{\partial}{\partial t}(\nabla \times \mathbf{H}).$$

Using equation (9.68) to substitute for $\nabla \times \mathbf{H}$ and then putting $\mathbf{D} = \varepsilon\mathbf{E}$, where ε is now constant, we find that

$$\nabla \times (\nabla \times \mathbf{E}) = -\mu \frac{\partial \mathbf{J}}{\partial t} - \mu\varepsilon \frac{\partial^2 \mathbf{E}}{\partial t^2}.$$

Using equation (A1.27) of Appendix A1.6 to substitute for $\nabla \times (\nabla \times \mathbf{E})$ and putting $\nabla \cdot \mathbf{E} = \rho/\varepsilon = \rho/\varepsilon_r\varepsilon_0$, where ρ is the free(true) charge density, after rearranging we obtain

$$\nabla^2\mathbf{E} - \mu\varepsilon \frac{\partial^2 \mathbf{E}}{\partial t^2} = \nabla\left(\frac{\rho}{\varepsilon}\right) + \mu \frac{\partial \mathbf{J}}{\partial t}. \tag{9.70}$$

This is the wave equation inside an LIH medium.

Taking the curl of equation (9.68) and putting $\dot{\mathbf{D}} = \varepsilon\dot{\mathbf{E}}$ we have

$$\nabla \times (\nabla \times \mathbf{H}) = \nabla \times \mathbf{J} + \varepsilon \frac{\partial}{\partial t}(\nabla \times \mathbf{E}). \tag{9.71}$$

Using equation (A1.27) of Appendix A1.6 and using the relations

$$\nabla \times \mathbf{E} = -\dot{\mathbf{B}} = -\mu\dot{\mathbf{H}}$$

$$\nabla \cdot \mathbf{H} = \frac{\nabla \cdot \mathbf{B}}{\mu} = 0$$

which are valid for an LIH medium, we find that equation (9.71) becomes

$$\nabla^2 \mathbf{H} - \mu\varepsilon \frac{\partial^2 \mathbf{H}}{\partial t^2} = -\nabla \times \mathbf{J}. \tag{9.72}$$

If we put $\mathbf{H} = \mathbf{B}/\mu$ in equation (9.72), after multiplying by the constant μ we obtain

$$\nabla^2 \mathbf{B} - \mu\varepsilon \frac{\partial^2 \mathbf{B}}{\partial t^2} = -\mu\nabla \times \mathbf{J}. \tag{9.73}$$

If the LIH medium has a conductivity σ and $\mathbf{J}$ is the conduction current we can put $\mathbf{J} = \sigma\mathbf{E}$ in equation (9.70). Also since the relaxation time is exceedingly short we can generally assume that ρ is zero inside a good conductor. Equation (9.70) then becomes for a good conductor

$$\nabla^2 \mathbf{E} - \mu\sigma \frac{\partial \mathbf{E}}{\partial t} - \mu\varepsilon \frac{\partial^2 \mathbf{E}}{\partial t^2} = 0. \tag{9.74}$$

The reader can show that in these conditions equation (9.73) becomes

$$\nabla^2 \mathbf{B} - \mu\sigma \frac{\partial \mathbf{B}}{\partial t} - \mu\varepsilon \frac{\partial^2 \mathbf{B}}{\partial t^2} = 0. \tag{9.75}$$

The wave equations are more complicated in a non LIH medium.

9.5.2. *Plane waves in an LIH medium*

Consider a non-conducting medium for which $\sigma = 0$ and in which the charge density ρ is zero. Equations (9.74) and (9.72) become

$$\nabla^2 \mathbf{E} = \mu\varepsilon \frac{\partial^2 \mathbf{E}}{\partial t^2} \tag{9.76}$$

$$\nabla^2 \mathbf{H} = \mu\varepsilon \frac{\partial^2 \mathbf{H}}{\partial t^2}. \tag{9.77}$$

In an LIH medium of infinite extent the wave equations (9.76) and (9.77) for $\mathbf{E}$ and $\mathbf{H}$ have plane wave solutions corresponding to waves propagating with a velocity v given by

$$v = \frac{1}{(\mu\varepsilon)^{1/2}} = \frac{1}{(\mu_r\mu_0\varepsilon_r\varepsilon_0)^{1/2}} = \frac{c}{(\mu_r\varepsilon_r)^{1/2}}. \tag{9.78}$$

These waves propagate at a velocity less than c, the speed of light in empty space. If n is the refractive index, then

$$v = \frac{c}{n}. \tag{9.79}$$

Comparing equations (9.78) and (9.79) we see that

$$n = (\mu_r\varepsilon_r)^{1/2}. \tag{9.80}$$

For most materials, other than ferromagnetic materials, μ_r is always very close to unity. In these cases we have, to an excellent approximation,

$$n = (\varepsilon_r)^{1/2}.$$

The relative permittivity and refractive index of water are frequency dependent. For example the static relative permittivity of water is about 80, in which case $n = (\varepsilon_r)^{1/2}$ is about 9. At radio frequencies the refractive index of water is about 8 corresponding to a value of ε_r of about 64. At optical frequencies the experimental value for the refractive index of water is 1.33 corresponding to a value of 1.77 for the relative permittivity ε_r.

Full details of the properties of electromagnetic waves in a refractive medium are given in all the standard text books on electromagnetism. It is straightforward to apply Maxwell's equations (9.65), (9.66), (9.67) and (9.68) to derive the appropriate boundary conditions before going on to consider topics such as Fresnel's equations for the reflection and refraction of electromagnetic waves etc. As full details of these topics are given in all the standard text books, we shall not repeat them here. A reader interested in a discussion of topics such as Maxwell's stress tensor, and energy and momentum in a material medium is referred to Stratton [6], where an account will also be found of the equations for the potentials in a medium that has finite values of ε_r, μ_r and σ.

9.5.3. *The method of propagation of electromagnetic waves in a dielectric*

It was shown in Section 9.5.2 that, according to Maxwell's equations (9.65)–(9.68) the velocity of electromagnetic waves in a dielectric is equal to c/n where the refractive index n is equal to $(\varepsilon_r)^{1/2}$ and ε_r is the relative permittivity of the dielectric. This change in velocity from c to c/n arises from the addition of the $\dot{\mathbf{P}}$ term to the vacuum displacement current term $\varepsilon_0\dot{\mathbf{E}}$ to give the $\dot{\mathbf{D}}$ term in equation (9.68). We shall illustrate qualitatively in this section how the varying polarization term $\dot{\mathbf{P}}$ influences the speed of electromagnetic waves, even though on the microscopic scale the electromagnetic interaction is still propagated at the speed c.

Consider a linearly polarized plane electromagnetic wave that is propagating in a direction parallel to the x axis and is incident on an LIH dielectric for which $\mu_r = 1$ and which fills the whole of that portion of space for which $x > 0$. We shall consider the problem using classical electromagnetism. The electric vector of the incident electromagnetic wave acts on the electrons and positive charges in the molecules in the dielectric which undergo forced harmonic motions and give rise to varying electric dipole moments which are represented on the macroscopic scale by the $\dot{\mathbf{P}}$ term in equation (9.59). The varying oscillating electric dipoles emit electromagnetic radiation, whose electric vector is given by equation (2.58) of Chapter 2. This radiation is emitted in all directions both in the forward and backward directions and it acts on all the other oscillating electric dipoles in the dielectric. It can be shown that the superposition of all the radiation fields of all the oscillating electric dipoles in the dielectric gives a wave moving in the backward direction with a velocity c in the dielectric which, in a typical dielectric leads to the extinction of the incident electromagnetic wave in a distance of about 10^{-6} m into the dielectric, Fox [7], plus a wave in the forward direction having a velocity of c/n. The full theory is given by Born and Wolf [8]. We shall only make a few qualitative comments. Consider a point T on the x axis inside the dielectric. The various off axis oscillating electric dipoles are at different distances from T and their contributions to the resultant radiation electric field at T take different times to go at the speed c to T, so that their contributions to the resultant electric field at T have different phases corresponding to the different path lengths. Due to these differing phases the harmonic variations in the radiation electric field add up to give a resultant electromagnetic wave moving at the speed c/n in a direction, which by symmetry arguments is parallel to the x axis. We can imagine radiation going from one oscillator to another along a zig-zag path at the speed c, before reaching T. Due to this increased path length it is reasonable to find that the speed of the resultant electromagnetic wave in the dielectric is less than c. A point we wish to stress is that on the microscopic scale the electromagnetic interaction between oscillators goes at the speed c. An introductory account of the theory is given by Rossi [9].

9.5.4. *An example of the choice between the use of Maxwell's equations for the macroscopic and for the microscopic fields*

The choice of whether to use Maxwell's equations for the macroscopic fields or Maxwell's equations for the microscopic fields depends on the scale of the phenomenon under investigation. A typical case in which the use of Maxwell's equations is useful is the theory of the reflection of electromagnetic waves at the surface of a dielectric. In this case, one does not want to relate the electric field **E** and the magnetic field **B**, associated with the radiation, back to the source of radiation. One merely wants to find out what happens, at the boundary of the dielectric, to the electric and magnetic field vectors associated with the radiation. Maxwell's equations (9.65)–(9.68) are relations between the macroscopic electric field **E** and the macroscopic magnetic field **B** at the boundary, valid irrespective of the position of the source of the radiation. For visible radiation the wavelength is ~500 nm which is much greater than atomic dimensions, which are ~0.1 nm. There is no need to allow for the discrete structure of matter and for the microscopic fluctuations in the electric and magnetic fields when discussing the reflection of visible radiation, since, over distances of ~500 nm the irregularities in the fields and dielectric constants associated with atomic structure will average out. Maxwell's equations for the macroscopic fields give the appropriate boundary conditions, and in these boundary conditions the macroscopic (or local space average) values for **E** and **B** and the macroscopic values for dielectric constant and electrical conductivity can be used, treating the dielectric as a continuous medium. However, if radiation in the x-ray region is used, the wavelength is ~0.1 nm, which is comparable with atomic dimensions. The use of the macroscopic dielectric constant and electrical conductivity is no longer appropriate. To calculate the scattering of x-rays, one must use the microscopic theory taking into account the crystal structure of the scattering material.

References

1. Russakoff, G., *Amer. Journ. Phys.*, Vol. 38, p. 1188 (1970).
2. Robinson, F. N. H., *Macroscopic Electromagnetism* (Pergamon Press, Oxford, 1973).
3. Lorrain, P. and Corson, D., *Electromagnetic Fields and Waves*, 2nd Edition (W. H. Freeman, San Francisco, 1970), pp. 97–102.
4. Rosser, W. G. V., *An Introduction to Statistical Physics* (Ellis Horwood, Chichester, 1982), pp. 114–117 and pp. 144–148.
5. Lorrain, P and Corson, D., see reference 3, pp. 387–393.
6. Stratton, J. A., *Electromagnetic Theory* (McGraw-Hill, New York, 1941).
7. Fox, J. G., *Amer. Journ. Phys.*, Vol. 30, p. 297 (1962).
8. Born, M. and Wolf, E., *Principles of Optics* (Pergamon Press, Oxford, 1959), pp. 98–104.
9. Rossi, B., *Optics* (Addison-Wesley, Reading, Mass., 1957), pp. 392–402.

CHAPTER 10

Special relativity and classical electromagnetism

10.1. Introduction

Until now we have deliberately avoided the use of ideas based on special relativity so as to show that all our interpretations can be based on classical electromagnetism itself without any recourse to special relativity. In this chapter we shall give a brief survey of the intimate relationships between classical electromagnetism and special relativity, which will show that the interpretations we have developed are consistent with special relativity. We shall assume that the reader has done an introductory course on special relativity leading up to the Lorentz transformations and the other transformations of special relativity. A summary of the transformations of special relativity is given in Appendix E. Readers interested in a recent book which fits in with the approach we are adopting is referred to Rosser [1].

It was mentioned in Section 4.13.3 of Chapter 4 that Maxwell's equations do not obey the principle of relativity if the coordinates and time are transformed from one inertial reference to another moving with uniform velocity relative to the first using the Galilean transformations, which were based on the assumption of an absolute time. In the nineteenth century it was assumed that Maxwell's equations could only hold in one absolute reference frame. It was pointed out in Section 4.13.3 that, if the Earth were moving with velocity v relative to the hypothetical absolute system, then, if the Galilean transformations were correct, it should have been possible to determine v by experiments such as the Michelson-Morley experiment. All such attempts failed. What Einstein did in his 1905 paper [2], which was called "On the electrodynamics of moving bodies", was to assume that the laws of classical electromagnetism obeyed the principle of relativity. This necessitated the abandonment of the concept of an absolute time. According to Einstein, it was the reinterpretation of the measurement of the time of distant events that was the key new idea in Einstein's 1905 paper. For example, in an account of some conversations he had with Einstein, Shankland [3] wrote:

> I asked Professor Einstein how long he had worked on the Special Theory of Relativity before 1905. He told me that he had started at age 16 and

> worked for 10 years: first as a student when, of course, he could only spend part-time on it, but the problem was always with him. He abandoned many fruitless attempts, "until at last it came to me that time was suspect!" Only then, after all his earlier efforts to obtain a theory consistent with the experimental facts had failed, was the development of the Special Theory of Relativity possible.

Einstein could have shown that the transformations which transform Maxwell's equations into Maxwell's equations were the Lorentz transformations. This was the approach tried by Lorentz, but Lorentz did not reinterpret the measurement of time. For example in 1927, Lorentz [4] said:

> A transformation of the time was necessary so I introduced the conception of local time which is different for different systems of reference which are in motion relative to each other. But I never thought that this had anything to do with real time. This real time for me was still represented by the older classical notion of an absolute time, which is independent of any reference to special frames of coordinates. There existed for me only one true time. I considered my time-transformation only as a heuristic working hypothesis so the theory of relativity is really solely Einstein's work. And there can be no doubt that he would have conceived it even if the work of all his predecessors in the theory of this field had not been done at all. His work in this respect is independent of the previous theories.

In his 1905 paper, in addition to the principle of relativity, Einstein chose the principle of the constancy of the speed of light as his second postulate. To quote from Einstein's 1905 paper:

> Examples of this sort, together with the unsuccessful attempts to discover any motion of the earth relatively to the 'light medium', suggest that the phenomena of electrodynamics as well as of mechanics possess no properties corresponding to the idea of absolute rest. They suggest rather that, as has already been shown to the first order of small quantities, the same laws of electrodynamics and optics will be valid for all frames of reference for which the equations of mechanics hold good. We will raise this conjecture (the purport of which will hereafter be called the 'Principle of Relativity') to the status of a postulate, and also introduce another postulate, which is only apparently irreconcilable with the former, namely, that light is always propagated in empty space with a definite velocity c which is independent of the state of motion of the emitting body. These two postulates suffice for the attainment of a simple and consistent theory of the electrodynamics of moving bodies based on Maxwell's theory for stationary bodies.

The choice of the principle of the constancy of the speed of light enabled Einstein to define how to measure the times of distant events before he derived the Lorentz transformations.

10.2. The postulates of special relativity

10.2.1. *The principle of relativity*

According to the principle of relativity all the laws of physics including the laws of classical electromagnetism are the same in all inertial frames. What this means is that, if an isolated system is observed from two different inertial frames, one moving with uniform velocity relative to the other, though the observations on the system carried out in the two inertial systems yield different numerical values for some quantities, the laws developed on the basis of these observations should have the same mathematical form in both inertial frames. The laws of physics should not contain any terms referring to an absolute system.

The definition of an inertial frame is the same in the theory of special relativity as in Newtonian mechanics. If in a reference frame a particle under the influence of no forces (e.g. a particle far away from any other particles capable of exerting forces) travels in a straight line with constant speed then Newton's first law (the principle of inertia) is valid in that frame, which is then considered suitable for the application of Newton's law of motion. Such a reference frame is called an **inertial reference frame**, or **inertial frame** or sometimes a **Galilean reference frame**. An experimenter at rest in an inertial reference frame is called an **inertial observer**. The same definition of an inertial frame is used in the theory of special relativity. Owing to the rotation of the Earth, the laboratory frame, that is a reference frame fixed to the Earth, is strictly not an inertial frame, and effects associated with the Earth's rotation are sometime important. For example, consider a spaceship coasting with uniform velocity relative to the fixed stars. There are no applied forces acting on the spaceship. If its position is plotted on a coordinate system fixed to the Earth then, owing to the Earth's rotation, the spaceship will appear to travel in a spiral path, going around through 360° every day. According to Newton's first law, since there are no forces acting on the spaceship, it should travel in a straight line in all inertial reference frames. This shows that the laboratory frame, i.e. a reference frame fixed to the Earth, is not an inertial frame, and effects associated with the Earth's rotation are sometimes important even in terrestrial experiments, such as long-range naval gunnery and Foucault's pendulum experiment. In these cases we get a better approximation to an inertial frame by taking a frame at rest relative to the solar system or the fixed stars. However, the angular velocity of rotation of the Earth is only 7.3×10^{-5} rad s^{-1} (the Earth turns through 360° in 24 h), so that the effects associated with the Earth's rotation about its axis are generally very small and play no significant role in classical electromagnetism. Hence the laboratory system is a satisfactory approximation to an inertial frame for the application of the laws of classical electromagnetism.

10.2.2. *The principle of the constancy of the speed of light*

According to the principle of the constancy of the speed of light, the speed of light in empty space has the same numerical value in all inertial frames, so that light should travel in straight lines and have the same speed in all directions of empty space in all inertial frames. We shall now show that, if we assume that Maxwell's equations obey the principle of relativity, then the principle of the constancy of the speed of light follows.

It was shown in Section 1.9.2 of Chapter 1 that Maxwell's equations have wave solutions. In empty space the velocity of these electromagnetic waves, measured in the laboratory frame Σ, is

$$c = (\mu_0\varepsilon_0)^{-1/2},$$

where μ_0 is the magnetic constant (permeability of free space) and ε_0 is the electric constant (permittivity of free space). This velocity is identified with the speed of light in empty space. It is independent of the velocities of the accelerating electric charges giving rise to the electromagnetic waves, illustrating how, according to Maxwell's equations, the velocity of light in empty space is independent of the velocity of the source of light.

If we assume that Maxwell's equations obey the principle of relativity then in an inertial reference frame Σ' that is moving with uniform velocity v relative to Σ, according to the laws of electricity and magnetism, the velocity of electromagnetic waves measured in Σ' should be

$$c' = (\mu_0'\varepsilon_0')^{-1/2},$$

where μ_0' and ε_0' are the values of the magnetic and electric constants in Σ'.

It can be shown, using the Biot-Savart law, that the force of attraction per unit length on each of two infinitely long, thin, parallel wires a distance r apart in empty space and carrying currents I_1 and I_2 is $\mu_0 I_1 I_2/2\pi r$. The ampere is defined as

> that unvarying current which, if present in each of two infinitely thin parallel conductors of infinite length and 1 m apart in empty space, causes each conductor to experience a force of exactly 2×10^{-7} newton per metre of length.

Putting $I_1 = I_2 = 1$ and $r = 1$ in the expression $\mu_0 I_1 I_2/2\pi r$, we find that it follows from the definition of the ampere that, in SI units, $\mu_0 = 4\pi \times 10^{-7}$ H m^{-1}.

If Maxwell's equations are valid in both Σ and Σ' then the force between two parallel currents should be given by the Biot-Savart law in both Σ and Σ'. If the ampere is defined in the same way in Σ and Σ', this implies that $\mu_0 = \mu_0' = 4\pi \times 10^{-7}$ by definition. For the velocity of light to have the same numerical value in Σ and Σ' one must also have $\varepsilon_0 = \varepsilon_0'$. In principle, according to classical physics, one could place two protons a known distance, say 1 m, apart in empty space and measure the force between them when they are at rest in the inertial frame Σ. Similarly, one could place two protons 1 m apart and measure the force between them when they are at rest relative to the inertial frame Σ'. If Maxwell's equations are valid in both Σ and Σ' then Coulomb's law should be applicable in both cases. For ε_0 to be equal to ε_0' it would require that the force between the protons at rest in Σ (measured in Σ) would have the same numerical value as the force between the two protons at rest in Σ' (measured in Σ'). Though it is not always stated specifically, it is generally assumed in the theory of special relativity, that, if two experiments are carried out under *identical* conditions in two *inertial* frames Σ and Σ' that are equivalent in every way then they give the same numerical results, within experimental error. If this assumption is made then ε_0 should equal ε_0'. Since μ_0 is equal to μ_0' by definition, then $(\mu_0\varepsilon_0)^{-1/2}$ should be equal to $(\mu_0'\varepsilon_0')^{-1/2}$. According to Maxwell's theory, these latter expressions are equal to the velocities of electromagnetic waves in free space in Σ and Σ' respectively so that $c = c'$. Thus if it is assumed that Maxwell's equations are correct and obey the principle of relativity then the principle of the constancy of the velocity of light follows. The simultaneous measurement of the velocity of light in Σ and Σ' is not carried out under identical conditions in Σ and Σ', since the velocity of the source of light is not the same relative to inertial observers at rest in Σ and Σ' respectively, so that *a priori* there is no reason why the same numerical value for the velocity of light should be obtained; but according to Maxwell's equations, it should.

At the time the theory was introduced, there was no direct experimental verification of the principle of the constancy of the speed of light. Since then, several direct experimental checks of the postulate have been performed. For example in 1964 Alväger, Farley, Kjellman and Wallin [5] determined the velocities of photons arising from the decay of neutral pions (π°) into two photons each. The velocity of the π° estimated using the equations of the theory of special relativity, was 0.99975 c. The measured velocity of the photons arising from π° decays was $(2.9977 \pm 0.0004) \times 10^8$ m s^{-1}. This value is consistent with the accepted value of 2.9979×10^8 m s^{-1} for the velocity of light emitted by a stationary source. This result shows that the velocity of light does not add on to the velocity of the source, which according to the Galilean velocity transformations it should do. Thus there is now direct experimental evidence for the principle of the constancy of the speed of light.

10.2.3. *The principle of constant charge*

We shall now show that, if we assume that Maxwell's equations hold in all inertial frames, the principle of constant charge follows. Taking the divergence of the Maxwell equation

$$\boldsymbol{\nabla} \times \mathbf{H} = \mathbf{J} + \dot{\mathbf{D}}$$

we find that, since the divergence of the curl of any vector is zero and, since according to Maxwell's equations, the divergence of **D** is equal to ρ, then

$$\nabla \cdot \mathbf{J} + \frac{\partial \rho}{\partial t} = 0 \tag{10.1}$$

which is the continuity equation for charge and current densities. Integrating equation (10.1) over a finite volume and applying Gauss' mathematical theorem of vector analysis, we have

$$-\frac{\partial}{\partial t}\int \rho \, dV = \int \mathbf{J} \cdot d\mathbf{S}. \tag{10.2}$$

If the surface enclosing the system of moving charges is far away from all the charges, such that no charges cross the surface, then equation (10.2) reduces to

$$-\frac{\partial}{\partial t}\int \rho \, dV = 0. \tag{10.3}$$

Thus if Maxwell's equations are correct, the rate of change of the total charge of a system of charges is zero and this holds independently of how the velocities of individual charges may vary. If the surface encloses one accelerating charge only, then the value of this charge should be invariant and independent of both the velocity and acceleration of the charge, and of course it should be equal to the value of the charge when it is at rest in the inertial frame Σ.

If Maxwell's equations are also valid in an inertial frame Σ′ moving with uniform velocity v relative to Σ, then proceeding as for Σ, it can be shown that in Σ′

$$\nabla' \cdot \mathbf{J}' + \frac{\partial \rho'}{\partial t'} = 0$$

and the total charge of a system of charges in Σ′ is also an invariant independent of how the velocity of the charges may vary. For a single charge the charge is again an invariant, but it is now equal to the value of the charge when it is at rest in Σ′. Consider a single proton accelerating relative to the laboratory (Σ) and relative to a spaceship (Σ′) that is moving with uniform velocity relative to the laboratory. In this case, for the principle of constant electric charge to hold, all that need now be assumed is that the proton has the same value of charge, when it is at rest relative to Σ, as it has when it is at rest relative to Σ′, since if Maxwell's equations are valid both in Σ and Σ′ the charge is independent of its velocity in both Σ and Σ′. In the theory of special relativity it is normal to make the assumption that fundamental particles at rest relative to an inertial frame Σ have the same properties and numerical values of mass, charge, lifetime etc. as the same fundamental particles would have when they are at rest, under the same conditions, relative to another inertial frame Σ′, provided the units are defined in the same way in Σ and Σ′. If one makes this assumption, then the principle of constancy of the electric charge of fundamental particles follows from Maxwell's equations, if Maxwell's equations obey the principle of relativity.

It was pointed out in Section 1.2.4 of Chapter 1 that there is now experimental evidence that shows that the total charge on a particle does not vary with its velocity. If the charge on a particle did vary with velocity, then, since on average electrons move faster than the protons inside a hydrogen molecule, the hydrogen molecule would have a resultant electric charge and would be deflected by electric fields. In 1960 King [6] showed that the charges on the electrons and protons inside hydrogen molecules were equal in magnitude, but opposite in sign to within 1 part in 10^{20}, showing that the total charge on a particle is independent of its velocity.

10.2.4. *The other postulates of special relativity*

It must be pointed out that there are additional assumptions implicit in the theory of special relativity. For example, it is assumed that inertial frames exist and that, in such a reference frame, the motion of a body is uniform and rectilinear provided no forces act on the body. This definition is taken over from Newtonian mechanics. It is also assumed that in such a reference frame light is propagated rectilinearly and isotropiclly in free space. This assumes that all regions of space and all directions in space are equivalent. It is also assumed in the the theory of special relativity that Euclidean geometry can be used to calculate the relationships between geometrical quantities. It is assumed that all time intervals are equivalent. The validity of these extra postulates is not known *a priori* but has to be tested experimentally. In fact, some of these postulates must be modified within the context of the theory of general relativity. The refinements due to the general theory are generally very small, and need only be introduced when the accuracy of the measurements make it necessary; they can be ignored in the context of classical electromagnetism.

10.3. Measurement of the times of distant events

Before going on to discuss the derivation of the Lorentz transformations, we shall follow Einstein's approach in his 1905 paper [2] and discuss the measurement of the times of distant events. Einstein assumed that, in every inertial frame, there is an array of rulers and clocks at rest relative to each inertial frame such that the positions and times of events can be determined when and where they occur. The question that then arises is, how do we synchronize the spatially separated clocks in a given inertial frame? The definition of the synchronization of spatially separated clocks chosen by Einstein [2] was:

> If at the point A of space there is a clock, an observer at A can determine the time values of events in the immediate proximity of A by finding the

positions of the hands which are simultaneous with these events. If there is at the point B of space another clock in all respects resembling the one at A, it is possible for an observer at B to determine the time values of events in the immediate neighbourhood of B. But it is not possible without further assumption to compare, in respect of time, an event at A with an event at B. We have so far only defined an 'A' time' and a 'B' time'. We have not defined a common time for A and B. The latter time can now be defined in establishing by definition that the 'time' required by light to travel from A to B equals the time it requires to travel from B to A. Let a ray of light start at the 'A time' t_A from A towards B, let it at the 'B time' t_B be reflected at B in the direction of A, and arrive again at A at the 'A' time 't'_A'.

In accordance with definition the two clocks synchronize if

$$t_B - t_A = t'_A - t_B. \tag{10.4}$$

Thus Einstein required that the time recorded by the clocks for light to pass from A to B in vacuo should be equal to the time for light to pass from B to A. This definition of synchronization is consistent with the principle of the constancy of the velocity of light, according to which the velocity of light *in vacuo* should have the same numerical value in all directions of space. Hence, Einstein's definition of synchronization is in accord with the postulates of the theory of special relativity and classical electromagnetism.

Rearranging equation (10.4), we have

$$t_B = \frac{1}{2}(t_A + t'_A). \tag{10.5}$$

According to equation (10.5), radar techniques can be used to measure the time of an event. If a radio pulse, emitted from a radar base at A at time t_A, is reflected by the event at a point B, and returns to base at time t'_A, then according to equation (10.5) the time of the event at B is equal to $(1/2)(t_A + t'_A)$. In this way it is not necessary to have a second clock at B. One clock at the radar base A is sufficient. The distance of the event from the radar base is equal to $(c/2)(t'_A - t_A)$. If directional antennae are used, the position of the event in space can be determined. This shows that the use of radar methods to determine the positions and times of events is consistent with special relativity, provided we assume that the speed of electromagnetic waves in empty space is the same in both directions. No radar operator ever queries this assumption.

The author's preferred way of deriving the Lorentz transformations is to use radar methods, as this approach makes clear the importance of the finite propagation time of signals. If we did have signals that could be transmitted at infinite speed, then an experimenter near an event could "press a button" and the information would arrive instantaneously at one master clock. This is not possible in practice. In the context of classical electromagnetism the fastest signals that can be transmitted travel at the speed of light in empty space. Since the principle of the constancy of the speed of light follows, if we assume that Maxwell.s equations obey the principle of relativity, the use of radar methods is fully consistent with our approach to classical electromagnetism. A full account of the application of radar methods to derive the basic results of special relativity is given by Rosser [1].

10.4. The Lorentz transformations

In this section we shall derive the Lorentz transformations which transform the coordinates and times of events from one inertial frame Σ to another inertial frame Σ′, that is moving with uniform velocity v relative to Σ along their common x axis as shown in Figure 10.1(a). Let the coordinates of an event relative to Σ be represented by x, y and z, and the coordinates of the same event relative to Σ′ by x', y' and z'. The times of the event relative to Σ and Σ′ will be represented by t and t' respectively. It will be assumed that the origins of Σ and Σ′ coincide at a time $t = t' = 0$, and that the directions of the y' and z' axes of Σ′ coincide with y and z axes of Σ at $t = 0$ as shown in Figure 10.1(a).

The Lorentz transformations will be developed for a simplified case. Consider a pulse of light emitted from the origins of Σ and Σ′ at the instant $t = t' = 0$ when they coincide, as shown in Figure 10.1(a). Let the light reach a light detector at the point P. Let this event be measured, to be at the position x, y, z at a time t relative to Σ, using rulers and synchronized clocks at rest relative to Σ. An observer at rest in Σ would say that the light travelled along the path OP covering a distance $(x^2 + y^2 + z^2)^{1/2}$ in a time t at a speed c such that

$$\frac{(x^2 + y^2 + z^2)^{1/2}}{c} = t.$$

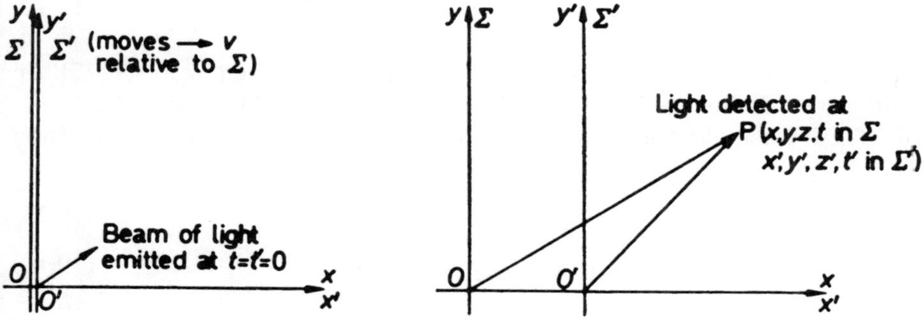

Figure 10.1. The inertial reference frame Σ′ moves with uniform velocity v relative to Σ along the common x axis. The origins O and O' coincide at $t = t' = 0$. The y and y' and the z and z' axes coincide at $t = t' = 0$. Light emitted from the origins O and O' at $t = t' = 0$ is detected at P.

Hence

$$x^2 + y^2 + z^2 - c^2t^2 = 0 \tag{10.6}$$

where c is the velocity of light.

An observer at rest in Σ' would agree that the light did reach the light detector at P (which may be moving relative to both Σ and Σ'). Using rulers and synchronized clocks stationary in Σ', let an observer at rest in Σ' record the same event at a position x', y', z' at a time t' as shown in Figure 10.1(b). Relative to Σ' the light would appear to have travelled the path $O'P$, such that

$$x'^2 + y'^2 + z'^2 - c^2t'^2 = 0 \tag{10.7}$$

where c is the velocity of light.

Notice the same value is used for the speed of light in both Σ and Σ', so as to be in accord with the principle of the constancy of the speed of light. The coordinates x, y, z and t in Σ and x', y', z' and t' in Σ' refer to the same event, namely the detection of the light at P in Figure 10.1(b). The required transformations must transform x', y', z' and t' such that equation (10.7) is transformed into equation (10.6), since both equations refer to the same event. Direct substitution into equation (10.7) verifies that the appropriate transformations are

$$x' = \gamma(x - vt) \tag{10.8}$$

$$y' = y \tag{10.9}$$

$$z' = z \tag{10.10}$$

$$t' = \gamma\left(t - \frac{vx}{c^2}\right) \tag{10.11}$$

where

$$\gamma = \frac{1}{(1 - v^2/c^2)^{1/2}}. \tag{10.12}$$

These are the Lorentz transformations. A derivation is given by Rosser [7]. An alternative derivation based on radar methods is given by Rosser [1].

The inverse transformations which transform equation (10.6) into equation (10.7) are

$$x = \gamma(x' + vt') \tag{10.13}$$

$$y = y' \tag{10.14}$$

$$z = z' \tag{10.15}$$

$$t = \gamma\left(t' + \frac{vx'}{c^2}\right). \tag{10.16}$$

Notice the inverse transformations can be obtained from equations (10.8),

(10.9), (10.10) and (10.11) by interchanging primed and unprimed quantities and replacing v by $-v$. The inverses of all relativistic transformations can be obtained by this procedure.

10.5. Applications of the Lorentz transformations

10.5.1. *Relativity of the simultaneity of events*

Let the two events occur at two separated points x_1 and x_2 in the inertial frame Σ. Let them be measured to occur at the same time t in Σ. According to the Lorentz transformations these events would be recorded at times t_1' and t_2' by clocks at rest in Σ' where t_1' and t_2' are given by

$$t_1' = \gamma\left(t - \frac{vx_1}{c^2}\right); \qquad t_2' = \gamma\left(t - \frac{vx_2}{c^2}\right). \tag{10.17}$$

Since x_1 is not equal to x_2, t_1' cannot be equal to t_2', so that, according to the Lorentz transformations, two spatially separated events which are simultaneous in Σ, would not be measured to be simultaneous in Σ'. Similarly, if two events occur simultaneously at two spatially separated points x_1' and x_2' in Σ', according to the Lorentz transformations, they would not be measured to be simultaneous in Σ. Thus, according to the theory of special relativity the simultaneity of spatially separated events is not an absolute property, as it was assumed to be in Newtonian mechanics.

10.5.2. *Time dilation*

Consider a clock at rest at the point (x', $y' = 0$, $z' = 0$) in Σ'. Let it emit ticks at times t_1' and t_2' in Σ'. In Σ, the events, associated with the successive ticks emitted by the clock at rest in Σ', are recorded at

$$x_1 = \gamma(x' + vt_1'); \qquad t_1 = \gamma\left(t_1' + \frac{vx'}{c^2}\right) \tag{10.18}$$

and

$$x_2 = \gamma(x' + vt_2'); \qquad t_2 = \gamma\left(t_2' + \frac{vx'}{c^2}\right). \tag{10.19}$$

Hence,

$$(t_2 - t_1) = \frac{(t_2' - t_1')}{(1 - v^2/c^2)^{1/2}}. \tag{10.20}$$

The time interval between two events, taking place at the same point in an inertial reference frame and measured by one clock stationary at that point, is called the proper time interval between the events. In equation (10.20) $t_2' - t_1'$ is the proper time interval between the events. The time interval

$t_2 - t_1$ in any other inertial frame is longer than the proper time interval. In equations (10.18) and (10.19), x_1 is not equal to x_2 so that the two events are not at the same point in Σ and so t_1 and t_2 must be measured by spatially separated synchronized clocks in Σ. Time dilation has been confirmed by experiments on the decay of pions and muons (Rosser [8]).

10.5.3. *Length contraction*

Consider a body moving parallel to the x axis with velocity v relative to Σ. In Σ', which moves with velocity v relative to Σ, the body is at rest. Let the length of the body be measured relative to Σ, by recording the positions of its extremities at x_1 and x_2 at the same time t in Σ. In Σ' the corresponding events are

$$x_1' = \gamma(x_1 - vt); \qquad t_1' = \gamma\left(t - \frac{vx_1}{c^2}\right)$$

$$x_2' = \gamma(x_2 - vt); \qquad t_2' = \gamma\left(t - \frac{vx_2}{c^2}\right)$$

Subtracting

$$(x_1' - x_2') = \gamma(x_1 - x_2).$$

Though t_1' is not equal to t_2', since the body is at rest in Σ', $x_1' - x_2'$ is equal to l_0, the proper length of the body measured when it is at rest in Σ'. If $l = (x_1 - x_2)$ is the length of the moving body measured in Σ

$$l = l_0\left(1 - \frac{v^2}{c^2}\right)^{1/2}. \tag{10.21}$$

Thus a body moving with velocity v relative to an observer is measured to be shorter by a factor of $(1 - v^2/c^2)^{1/2}$ in its direction of motion relative to the observer.

A body of proper volume V_0, moving with velocity v relative to an observer, can be divided into thin rods parallel to $\mathbf{v}$. Each one of these rods is reduced in length by a factor $(1 - v^2/c^2)^{1/2}$. Since $y = y'$ and $z = z'$, the area of cross-section of each rod is unchanged. Hence the measured volume of the moving body is

$$V = V_0\left(1 - \frac{v^2}{c^2}\right)^{1/2}. \tag{10.22}$$

10.5.4. *Discussion*

The results derived in Sections 10.5.1, 10.5.2 and 10.5.3, which predict non-absolute simultaneity, time dilation and length contraction, illustrate how the theory of special relativity, which arose from the application of the principle of relativity to the laws of classical electromagnetism, has necessitated a revision of the Newtonian ideas of absolute space and absolute time. These changes are entirely consistent with classical electromagnetism. A reader interested in a discussion of the physical implications and the experimental confirmations of these predictions of the theory of special relativity is referred to Rosser [1].

The Lorentz transformations can be used to develop the transformations for velocity, momentum and force (Reference: Rosser [1]). A summary of these transformations is given in Appendix E.

10.6. Forces between two parallel convection currents derived using the theory of special relativity

The example illustrated in Figure 1.9, and discussed in Section 1.5 using the laws of classical electromagnetism, will now be considered from the viewpoint of the theory of special relativity. It will be assumed that λ, the charge per unit length on the moving wires, measured in the laboratory frame Σ, is made up of n discrete charges per unit length of magnitude q each, as shown in Figure 10.2(b). In Σ,

$$\lambda = nq. \tag{10.23}$$

According to the theory of special relativity, the laws of electromagnetism are the same in all inertial reference systems. Hence the problem can be considered in the reference frame Σ' moving with uniform velocity v relative to Σ along the common x axis. In Σ' the wires are at rest, as shown in Figure 10.2(a). Let λ', the charge per unit length measured in Σ', be made up n' charges per unit length of magnitude q each, so that

$$\lambda' = n'q. \tag{10.24}$$

The same value q is used for the total charge on a particle in Σ and Σ' so as to be in accord with the principle of constant electric charge (see Section 10.2.3). The charge distributions are at rest in Σ'. According to the Lorentz length contraction, equation (10.21), a length l_0 of the wire at rest in Σ' is measured to be $l_0(1 - v^2/c^2)^{1/2}$ in Σ, since the wire is moving with velocity v relative to Σ. The number of charges in a length l_0 of wire in Σ' which is equal to $n'l_0$, is measured to be in length $l_0(1 - v^2/c^2)^{1/2}$ relative to Σ, as illustrated in Figure 10.2(b). Hence the number of charges per unit length in Σ is $n'l_0/l_0(1 - v^2/c^2)^{1/2}$, so that

$$n = \frac{n'}{(1 - v^2/c^2)^{1/2}}$$

showing that the charge per unit length is greater in Σ than Σ', as illustrated in Figure 10.2(a) and 10.2(b). Hence, using equation (10.23)

$$\lambda = nq = \frac{n'q}{(1 - v^2/c^2)^{1/2}} = \frac{\lambda'}{(1 - v^2/c^2)^{1/2}}\,. \tag{10.25}$$

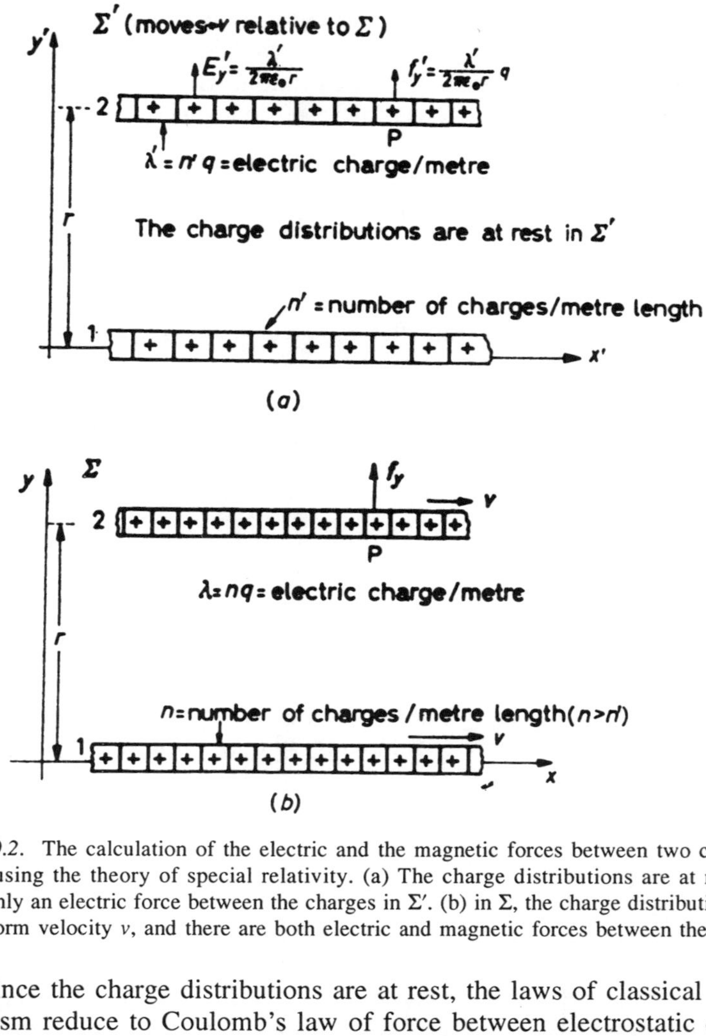

Figure 10.2. The calculation of the electric and the magnetic forces between two convection currents using the theory of special relativity. (a) The charge distributions are at rest in Σ′; there is only an electric force between the charges in Σ′. (b) in Σ, the charge distributions move with uniform velocity v, and there are both electric and magnetic forces between the charges.

In Σ′, since the charge distributions are at rest, the laws of classical electromagnetism reduce to Coulomb's law of force between electrostatic charges. Using Gauss' flux law we find that the electric field at the position of wire 2 due to wire 1 is

$$E'_y = \frac{\lambda'}{2\pi\varepsilon_0 r'} \,. \tag{10.26}$$

According to the Lorentz transformations, the separation of the wires is the same in Σ and Σ′, as it is measured in the y direction, so that $r = r'$. The force on one of the charges (labelled P and of magnitude q) of wire 2 in Figure 10.2(a) has components

$$f'_x = 0; \qquad f'_y = \frac{\lambda' q}{2\pi\varepsilon_0 r} ; \qquad f'_z = 0 \tag{10.27}$$

in Σ′. The force transformations quoted in Appendix E, will now be applied to determine the force acting on the single charge P, measured in the inertial reference frame Σ shown in Figure 10.2(b). Since the velocity $\mathbf{u}'$ of the charged particle P is zero in Σ', the inverses of equations (E14), (E15) and (E16) reduce to

$$f_x = f'_x \tag{10.28}$$

$$f_y = \frac{f'_y}{\gamma} = f'_y\left(1 - \frac{v^2}{c^2}\right)^{1/2} \tag{10.29}$$

$$f_z = \frac{f'_z}{\gamma} = f'_z\left(1 - \frac{v^2}{c^2}\right)^{1/2}. \tag{10.30}$$

Since, according to equations (10.27)

$$f'_x = f'_z = 0$$

it follows from equations (10.29) and (10.30) that in Σ

$$f_x = f_z = 0.$$

Substituting from equation (10.27) into equation (10.29), we have for the force on the charge P measured in Σ

$$f_y = \frac{\lambda' q}{2\pi\varepsilon_0 r}\left(1 - \frac{v^2}{c^2}\right)^{1/2}. \tag{10.31}$$

The force per unit length on wire 2, measured in Σ, is equal to the number of charges per unit length, measured in Σ, times the force on each charge, given by equation (10.31). Hence in Σ

$$\text{force/unit length} = nf_y = n\frac{\lambda' q}{2\pi\varepsilon_0 r}\left(1 - \frac{v^2}{c^2}\right)^{1/2}.$$

But, from equation (10.25), $\lambda' = \lambda(1 - v^2/c^2)^{1/2}$, and, from equation (10.23), $nq = \lambda$, Hence in Σ

$$\text{force/unit length} = \frac{\lambda^2}{2\pi\varepsilon_0 r}\left(1 - \frac{v^2}{c^2}\right). \tag{10.32}$$

This is in agreement with equation (1.98) derived in Section 1.5 by applying the laws of classical electromagnetism in the inertial frame Σ. This example illustrates how the magnetic forces produced by electric currents can be calculated, in some cases, from Coulomb's law for the forces between electrostatic charges, if the principle of constant electric charge and the force transformations of the theory of special relativity are taken as axiomatic. One has to include all second-order relativistic effects, since the magnetic forces between moving charges are themselves of second order. For example, in the present case the effects associated with the Lorentz contraction had to be included, illustrating the intimate connections and consistency between special relativity and classical electromagnetism.

10.7. The inequality of action and reaction in classical electromagnetism

We shall return now to consider the example shown previously in Figure 8.8 of Chapter 8 and shown again in Figure 10.3(a). The charge q_1 is moving along the x axis of the inertial frame Σ with uniform velocity v. It is at the origin of Σ at the time $t = 0$. The charge q_2 is moving along the $+y$ axis of Σ with uniform velocity u, and is at a distance y from the origin of Σ at the time $t = 0$. The inertial frame Σ' shown in Figure 10.3(b) is moving with uniform velocity v relative to Σ along their common x axis. The origins of Σ and Σ' coincide at the time $t = t' = 0$. The charge q_1 remains at rest at the origin of Σ'. Corresponding to the event when the charge q_2 has coordinates $(0, y, 0)$ at the time $t = 0$ in Σ, according to the Lorentz transformations in Σ' we have $(0, y' = y, 0)$ at the time $t' = 0$. According to the principle of constant charge, q_1 and q_2 have the same values in Σ' as in Σ.

Since the charge q_1 is at rest in Σ', it follows from Coulomb's law that the electric field at the position of the charge q_2 at $t' = 0$ in Σ' is with $y' = y$

$$E'_y = \frac{q_1}{4\pi\varepsilon_0 y'^2} = \frac{q_1}{4\pi\varepsilon_0 y^2} \tag{10.33}$$

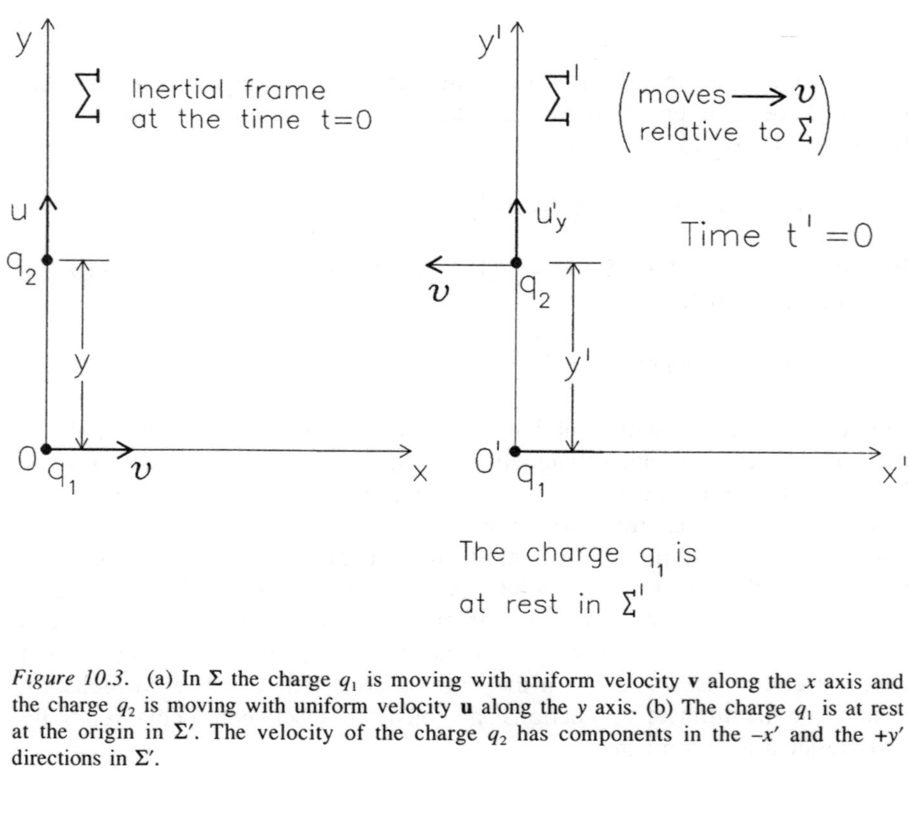

Figure 10.3. (a) In Σ the charge q_1 is moving with uniform velocity $\mathbf{v}$ along the x axis and the charge q_2 is moving with uniform velocity $\mathbf{u}$ along the y axis. (b) The charge q_1 is at rest at the origin in Σ'. The velocity of the charge q_2 has components in the $-x'$ and the $+y'$ directions in Σ'.

According to the Lorentz force law, the force on q_2 at the time $t' = 0$ in Σ' has the components

$$(f_2')_x = 0; \qquad (f_2')_y = \frac{q_1 q_2}{4\pi\varepsilon_0 y^2}; \qquad (f_2')_z = 0. \tag{10.34}$$

We shall now use the force transformations of special relativity to determine the components of the force acting on q_2 in Σ. Using equation (E16) of Appendix E we find that

$$(f_2)_z = 0. \tag{10.35}$$

Rearranging equation (E15), we have for the charge q_2

$$f_y = \gamma\left(1 - \frac{vu_x}{c^2}\right)f_y'. \tag{10.36}$$

Since the charge q_2 is moving along the y axis of Σ, $u_x = 0$ in Σ. Substituting for $(f_2')_y$ from equation (10.34) we find that

$$(f_2)_y = \gamma(f_2')_y = \frac{q_1 q_2}{4\pi\varepsilon_0 y^2(1 - v^2/c^2)^{1/2}}. \tag{10.37}$$

This is in agreement with equation (8.129) of Chapter 8. According to equation (E14)

$$f_x' = f_x - \frac{vu_y}{c^2(1 - vu_x/c^2)}f_y - \frac{vu_z}{c^2(1 - vu_x/c^2)}f_z.$$

For the charge q_2 in Σ, we have $u_x = 0$, $u_y = u$. Using equations (10.37) and (10.35) for $(f_2)_y$ and $(f_2)_z$, and using equation (10.34) to put $(f_2')_x = 0$, we find that

$$(f_2)_x = \frac{vuq_1q_2}{4\pi\varepsilon_0 c^2 y^2(1 - v^2/c^2)^{1/2}}. \tag{10.38}$$

This is in agreement with equation (8.131) of Chapter 8.

It is left as an exercise for the reader to show that the force on q_1 due to q_2 is given by

$$(f_1)_x = 0; \qquad (f_1)_y = -\frac{q_1 q_2(1 - v^2/c^2)}{4\pi\varepsilon_0 y^2}; \qquad (f_1)_z = 0. \tag{10.39}$$

Consider the inertial frame that is moving with velocity u relative to Σ along the y axis of Σ, and in which it is the charge q_2 that is at rest. Apply Coulomb's law in the rest frame of q_2 and then transform the force on the charge q_1 in the rest frame of the charge q_2 to Σ to obtain equation (10.39). Hint. Remember that the separation of q_1 and q_2 is measured in this case in the direction in which one inertial reference frame is moving relative to the other, so that the reader must allow for length contraction.

These results show that the fact that the forces between q_1 and q_2 are not equal and opposite is entirely consistent with the force transformations of special relativity. It is left as an exercise for the reader to derive equations (10.37), (10.38) and (10.39) using the transformations for **E** and **B** which will be derived in Section 10.10.3.

10.8. Electromagnetism via special relativity

Consider again the two moving charges q_1 and q_2 shown in Figures 10.3(a) and 10.3(b). By applying Coulomb's law in the inertial frame Σ', shown in Figure 10.3(b) to determine the force on the charge q_2 due to q_1 in Σ' and then using the principle of constant charge and the transformations of special relativity to transform the force on q_2 from Σ' to Σ, we derived the components of the force $\mathbf{f}_2$ on the charge q_2 at the time $t = 0$ in Σ, which are given by equations (10.38), (10.37) and (10.35). These components are consistent with the Lorentz force

$$\mathbf{F}_2 = q_2\mathbf{E}_1 + q_2\mathbf{u}_2 \times \mathbf{B}_1$$

on the charge q_2, where

$$\mathbf{E}_1 = \frac{q_1}{4\pi\varepsilon_0 y^2(1 - v^2/c^2)^{1/2}}\,\hat{\mathbf{j}} \tag{10.40}$$

$$\mathbf{B}_1 = \frac{q_1 v}{4\pi\varepsilon_0 c^2 y^2(1 - v^2/c^2)^{1/2}}\,\hat{\mathbf{k}} \tag{10.41}$$

are the electric field $\mathbf{E}_1$ and the magnetic field $\mathbf{B}_1$ due to the charge q_1 at the position of the charge q_2 at the time $t = 0$ in Σ and $\hat{\mathbf{j}}$ and $\hat{\mathbf{k}}$ are unit vectors in the $+y$ and $+z$ directions respectively. Equations (10.40) and (10.41) are in agreement with equations (8.128) and (8.130) of Chapter 8, which are a special case of equations (4.2) and (4.4) of Chapter 4, which are the same as equations (3.25) and (3.28) of Chapter 3, and which are given by

$$\mathbf{E}_1 = \frac{q_1\mathbf{r}(1 - \beta^2)}{4\pi\varepsilon_0 r^3(1 - \beta^2 \sin^2 \theta)^{3/2}} \tag{10.42}$$

$$\mathbf{B}_1 = \frac{q_1\mathbf{v} \times \mathbf{r}(1 - \beta^2)}{4\pi\varepsilon_0 c^2 r^3(1 - \beta^2 \sin^2 \theta)^{3/2}} \tag{10.43}$$

where $\beta = v/c$ and $\mathbf{r}$ is the distance from the charge q_1 to the charge q_2 at the time $t = 0$ in Σ and θ is the angle between $\mathbf{v}$ and $\mathbf{r}$. We could derive equations (10.42) and (10.43) by assuming that the charge q_2 in Figure 10.3(a) is at the position (x, y, z) having a velocity $\mathbf{u}$ at the time $t = 0$ in Σ. By applying Coulomb's law in the inertial frame Σ' in which the charge q_1 is at rest and transforming the force on q_2 to Σ allowing for length contraction etc., it is straightforward to derive equations (10.42) and (10.43). References: Rosser [9] and [10]. After deriving equations (10.42) and (10.43) from Coulomb's law and the transformations of special relativity, we could use them to replace equations (4.2) and (4.4) of Chapter 4 and use them to develop Maxwell's equations for the special case of a classical point charge moving with uniform velocity in the ways described in Chapter 4. This approach, starting from Coulomb's law, the principle of constant charge and the transformations of special relativity illustrates the essential unity of classical electromagnetism and shows that classical electromagnetism and special relativity are entirely consistent with each other.

10.9. Causality in classical electromagnetism

After deriving the retarded potentials in Section 2.3 of Chapter 2, we have always stressed that it takes time for the changes in the electromagnetic interaction to propagate in empty space at the finite speed of light, which is also the maximum speed at which electromagnetic signals can propagate. We showed in Section 10.2.2 that the principle of the constancy of the speed of light follows if we assume that Maxwell's equations obey the principle of relativity, and it was the application of the principle of the constancy of the speed of light that led to the prediction of the non-absolute simultaneity of spatially separated events, expressed for example in equation (10.17). In this section we shall show that, despite non-absolute simultaneity, it is still the same events that contribute to the retarded potentials and fields **E** and **B** at any field point in empty space in both Σ and Σ'. We shall consider the determination of the retarded potentials and fields **E** and **B** at the origins of Σ and Σ' at the time $t = t' = 0$ when the origins of Σ and Σ' coincide.

Let the information collecting sphere, that collapses with the speed c in Σ to reach the origin of Σ at $t = 0$, pass an accelerating charge at the position (x, y, z) at a radial distance r from the origin of Σ at the time $t = -r/c$ in Σ. According to the Lorentz transformations, the coordinates and time of this event in the inertial frame Σ', that moves with uniform velocity v along the x axis of Σ, are given by

$$x' = \gamma(x - vt) = \gamma(x + \beta r) \tag{10.44}$$

$$y' = y \tag{10.45}$$

$$z' = z \tag{10.46}$$

$$t' = \gamma\left(t - \frac{vx}{c^2}\right) = -\frac{\gamma}{c}(r + \beta x) \tag{10.47}$$

where $\beta = v/c$. Using equations (10.44), (10.45) and (10.46) we find that the distance r' of this event from the origin of Σ' is given by

$$r'^2 = x'^2 + y'^2 + z'^2 = \gamma^2(x + \beta r)^2 + y^2 + z^2.$$

Using the relation

$$r^2 = x^2 + y^2 + z^2.$$

we find that

$$r'^2 = \gamma^2[x^2 + 2\beta xr + \beta^2 r^2 + (1 - \beta^2)(y^2 + z^2)]$$

$$r'^2 = \gamma^2(r^2 + 2r\beta x + \beta^2 x^2) = \gamma^2(r + \beta x)^2. \tag{10.48}$$

It follows from equation (10.47) that in Σ'

$$c^2 t'^2 = \gamma^2(r + \beta x)^2. \tag{10.49}$$

Comparing equations (10.48) and (10.49) we conclude that in Σ'

$$r' = -ct'. \tag{10.50}$$

This shows that the event when the information collecting sphere, that reaches the origin of Σ at $t = 0$, passes the accelerating charge is also recorded by the information collecting sphere that reaches the origin of Σ' at $t' = 0$, when the origins of Σ and Σ' coincide. These events, when the information collecting sphere passes individual charges at their retarded positions at the appropriate retarded times recording their charges, velocities and accelerations so that the retarded potentials can be calculated, will be called the antecedent events or the antecedents. It follows from equation (10.50) that it is the same antecedent events that contribute to the retarded potentials and the fields $\mathbf{E}'$ and $\mathbf{B}'$ and hence to the force on a moving test charge q at the origin of Σ' at $t' = 0$ as contribute to the retarded potentials, and fields $\mathbf{E}$ and $\mathbf{B}$ and the force on the moving test charge q when it is at the origin of Σ at $t = 0$ when the origins of Σ and Σ' coincide. This shows that, in the general case, provided we allow for retardation, the force on the test charge q at the time t in Σ has the same antecedents as the force on the moving test charge q at the time t' in Σ', where t and t' are related by the Lorentz transformations.

Another type of electromagnetic law is developed by integrating over a fixed volume at a fixed time without allowing for retardation. In this category we should, for example, place equation (8.113) of Section 8.7, which was derived by integrating the Lorentz force over the volume V_0 in Figure 8.1 at a fixed time and then using Maxwell's equations to eliminate ρ and $\mathbf{J}$. Equation (8.113) was then used to derive equations (8.126) and (8.127), which are two ways of expressing the law of conservation of linear momentum. In Section 8.8.2 of Chapter 8 we suggested that we could treat both equation (8.126) and (8.127) as book-keeping rules carried out at a fixed time, which in addition to the momenta of the charges, bring in the values of the fields $\mathbf{E}$ and $\mathbf{B}$ at field points where there are no charges. This latter contribution was expressed in the form of the (electromagnetic) potential momentum. If Maxwell's equations and the Lorentz force law obey the principle of relativity then the analysis of Section 8.7 should be valid in all inertial frames so that, for example, the law of conservation of linear momentum should be valid at a fixed time such as $t = 0$ in Σ and at a fixed time such as $t' = 0$ in Σ'. We shall now consider how such laws, applied over a finite volume at a fixed time, behave under a Lorentz transformation.

As an illustrative example, consider just two of the moving charges, namely

q_1 and q_2, of a system of moving and accelerating classical point charges. Let q_1 be at the origin of Σ at $t = 0$ and let q_2 be on the x axis of Σ at x_2 at $t = 0$. What we showed earlier in this section was that the force on q_1 in Σ and the force on q_1 in Σ', when q_1 is at the origins of Σ and Σ' when the origins coincide, have the same antecedents in both Σ and Σ'. However, according to the Lorentz transformations, the force on q_2 in Σ', that is the same event and has the same antecedents as the force on q_2 when it is at x_2 at $t = 0$ in Σ, is at $x_2' = \gamma x_2$ at $t_2' = -\gamma v x_2/c^2$ in Σ'. Hence the force on q_2 at $t' = 0$ in Σ' is not the same event and does not have the same antecedents as the force on q_2 at $t = 0$ in Σ. Hence, if we sum at $t' = 0$ in Σ', due to non-absolute simultaneity, the values of the forces and hence the rates of change of the momenta of the charges in Σ' are not the same events and do not have the same antecedents as the forces on and the rates of change of the momenta of the charges at the time $t = 0$ in Σ. In the general case the equation

$$\dot{\mathbf{P}}_{\text{mech}} = -\dot{\mathbf{P}}_{\text{em}}$$

at $t = 0$ in Σ and the equation

$$\dot{\mathbf{P}}'_{\text{mech}} = -\dot{\mathbf{P}}'_{\text{em}}$$

at $t' = 0$ in Σ' do not refer to the same events and do not have the same antecedents in Σ and Σ', and due to non-absolute simulaneity, do not transform into each other under a Lorentz transformation. This result is consistent with our suggestion in Section 8.8.2 of Chapter 8 that we can treat the law of conservation of linear momentum as a book-keeping rule involving both fields and charges. Similar analyses apply to the laws of conservation of energy and of angular momentum and other laws obtained by integrating at a fixed time. It was shown earlier in this section that, however complicated these conservation laws become and, even if we treat them as book-keeping rules, the force on any individual moving and accelerating classical point charge has the same antecedents in all inertial frames provided the position and time of the event in Σ and Σ' are related by the Lorentz transformations. The other properties of the charge, such as velocity and momentum and the force on the charge, must be transformed using the appropriate transformations of special relativity.

10.10. Transformation of Maxwell's equations

10.10.1. *Introduction*

Experiments have confirmed that the *macroscopic* electric and magnetic fields, both in empty space and inside *stationary* material bodies, are adequately described by Maxwell's equations, which in an inertial frame Σ', in which the materials are at rest, take the form

$$\nabla' \cdot \mathbf{D}' = \rho' \tag{10.51}$$

$$\nabla' \cdot \mathbf{B}' = 0 \tag{10.52}$$

$$\nabla' \times \mathbf{E}' = -\frac{\partial \mathbf{B}'}{\partial t'} \tag{10.53}$$

$$\nabla' \times \mathbf{H}' = \frac{\partial \mathbf{D}'}{\partial t'} + \mathbf{J}' \tag{10.54}$$

where $\mathbf{E}'$ is the electric field, $\mathbf{D}'$ the electric displacement, $\mathbf{B}'$ the magnetic field, $\mathbf{H}'$ the magnetizing force, ρ' the (free) macroscopic charge density and $\mathbf{J}'$ the (true) macroscopic conduction current density. All the quantities appearing in equations (10.51), (10.52), (10.53) and (10.54) apply to the same point x', y', z' at a time t' in the inertial frame Σ'. All the differential coefficients in the equations are with respect to x', y', z' or t', for example, equation (10.52) is

$$\frac{\partial B'_x}{\partial x'} + \frac{\partial B'_y}{\partial y'} + \frac{\partial B'_z}{\partial z'} = 0.$$

In order to solve problems these equations have to be supplemented by the constitutive equations relating $\mathbf{D}'$ and $\mathbf{E}'$, $\mathbf{B}'$ and $\mathbf{H}'$, and $\mathbf{E}'$ and the conduction current density $\mathbf{J}'$. These constitutive equations depend on the properties of the materials present in the system.

One of the postulates of the theory of special relativity is that the laws of physics have the same mathematical form in all inertial frames of reference. If Maxwell's equations are correct and obey the principle of relativity, then in the inertial frame Σ moving with velocity v in the negative Ox' direction relative to Σ', one should have

$$\nabla \cdot \mathbf{D} = \rho \tag{10.55}$$

$$\nabla \cdot \mathbf{B} = 0 \tag{10.56}$$

$$\nabla \times \mathbf{E} = -\dot{\mathbf{B}} \tag{10.57}$$

$$\nabla \times \mathbf{H} = \dot{\mathbf{D}} + \mathbf{J}. \tag{10.58}$$

These equations should hold for a point x, y, z at a time t in Σ. If all the material bodies are stationary in Σ', then they are moving with uniform velocity v relative to Σ. The constitutive equations may take a different mathematical form for moving matter. The current density $\mathbf{J}$ in Σ includes the convection current density as well as the conduction current density.

If Maxwell's equations do obey the principle of relativity, when the coordinates and time in equations (10.55), (10.56), (10.57) and (10.58) are transformed according to the Lorentz transformations, then equations (10.55)–(10.58) should be changed into equations (10.51)–(10.54). It will now be shown that this is so. Equations which have the same mathematical form in both Σ and Σ', when the coordinates and time are transformed according to the Lorentz transformations, are said to be *Lorentz covariant*.

10.10.2. *Transformation of $\partial/\partial x$, $\partial/\partial y$, $\partial/\partial z$ and $\partial/\partial t$*

Consider a function F of x', y', z' and t' in Σ'. The total differential of F is

$$dF = \frac{\partial F}{\partial x'}dx' + \frac{\partial F}{\partial y'}dy' + \frac{\partial F}{\partial z'}dz' + \frac{\partial F}{\partial t'}dt'. \tag{10.59}$$

Now for a given event, x', y', z' and t' are all functions of x, y, z and t. The total differential of x' can expressed as

$$dx' = \frac{\partial x'}{\partial x}dx + \frac{\partial x'}{\partial y}dy + \frac{\partial x'}{\partial z}dz + \frac{\partial x'}{\partial t}dt. \tag{10.60}$$

From the Lorentz transformations,

$$x' = \gamma(x - vt)$$

$$y' = y; \qquad z' = z$$

$$t' = \gamma\left(t - \frac{vx}{c^2}\right)$$

where $\gamma = (1 - v^2/c^2)^{-1/2}$. Remembering that v and γ are constants, we have

$$\frac{\partial x'}{\partial x} = \gamma; \qquad \frac{\partial x'}{\partial y} = 0; \qquad \frac{\partial x'}{\partial z} = 0; \qquad \frac{\partial x'}{\partial t} = -\gamma v$$

and, substituting in equation (10.60) we obtain

$$dx' = \gamma\, dx - \gamma v\, dt. \tag{10.61}$$

Similarly,

$$dy' = dy \tag{10.62}$$

$$dz' = dz \tag{10.63}$$

$$dt' = \frac{\partial t'}{\partial x}dx + \frac{\partial t'}{\partial y}dy + \frac{\partial t'}{\partial z}dz + \frac{\partial t'}{\partial t}dt = -\frac{\gamma v}{c^2}dx + \gamma\, dt. \tag{10.64}$$

Substituting for dx', dy', dz' and dt' from equation (10.61), (10.62), (10.63) and (10.64) respectively into equation (10.59), we have

$$dF = \frac{\partial F}{\partial x'}(\gamma\, dx - \gamma v\, dt) + \frac{\partial F}{\partial y'}dy + \frac{\partial F}{\partial z'}dz + \frac{\partial F}{\partial t'}\left(\gamma\, dt - \frac{\gamma v}{c^2}dx\right)$$

Rearranging

$$dF = \gamma\left(\frac{\partial F}{\partial x'} - \frac{v}{c^2}\frac{\partial F}{\partial t'}\right)dx + \frac{\partial F}{\partial y'}dy + \frac{\partial F}{\partial z'}dz + \gamma\left(\frac{\partial F}{\partial t'} - v\frac{\partial F}{\partial x'}\right)dt. \tag{10.65}$$

But, if F is a function of x, y, z and t, the total differential of F can be written as

$$dF = \frac{\partial F}{\partial x}dx + \frac{\partial F}{\partial y}dy + \frac{\partial F}{\partial z}dz + \frac{\partial F}{\partial t}dt. \qquad (10.66)$$

Comparing the coefficients of dx, dy, dz and dt, in equations (10.65) and (10.66) we conclude that

$$\frac{\partial}{\partial x} = \gamma\left(\frac{\partial}{\partial x'} - \frac{v}{c^2}\frac{\partial}{\partial t'}\right) \qquad (10.67)$$

$$\frac{\partial}{\partial y} = \frac{\partial}{\partial y'} \qquad (10.68)$$

$$\frac{\partial}{\partial z} = \frac{\partial}{\partial z'} \qquad (10.69)$$

$$\frac{\partial}{\partial t} = \gamma\left(\frac{\partial}{\partial t'} - v\frac{\partial}{\partial x'}\right). \qquad (10.70)$$

10.10.3. *Transformation of* **E** *and* **B**

The y component of equation (10.57) is

$$\frac{\partial E_x}{\partial z} - \frac{\partial E_z}{\partial x} = -\frac{\partial B_y}{\partial t}. \qquad (10.71)$$

Substituting for $\partial/\partial z$, $\partial/\partial x$ and $\partial/\partial t$ from equations (10.69), (10.67) and (10.70) respectively, we have

$$\frac{\partial E_x}{\partial z'} - \gamma\left(\frac{\partial E_z}{\partial x'} - \frac{v}{c^2}\frac{\partial E_z}{\partial t'}\right) = -\gamma\left(\frac{\partial B_y}{\partial t'} - v\frac{\partial B_y}{\partial x'}\right)$$

and rearranging

$$\frac{\partial E_x}{\partial z'} - \frac{\partial}{\partial x'}\gamma(E_z + vB_y) = -\frac{\partial}{\partial t'}\gamma\left(B_y + \frac{v}{c^2}E_z\right). \qquad (10.72)$$

If Maxwell's equations are to be Lorentz covariant, that is invariant in mathematical form in all inertial frames of reference, then in Σ' we must have

$$\frac{\partial E'_x}{\partial z'} - \frac{\partial E'_z}{\partial x'} = -\frac{\partial B'_y}{\partial t'}. \qquad (10.73)$$

Equations (10.72) and (10.73) have the same mathematical form showing that the y-component of equation (10.57) is Lorentz covariant. In fact, if one puts

$$E'_x = E_x$$

$$E'_z = \gamma(E_z + vB_y)$$

$$B'_y = \gamma\left(B_y + \frac{v}{c^2}E_z\right)$$

then equations (10.72) and (10.73) are exactly the same.

It is left as an exercise for the reader to show that the x and z components of equations (10.57) transform into the x' and z' components of equation (10.53) respectively, and to show that equation (10.56) transforms into equation (10.52), if **E** and **B** transform according to the following equations. (Reference: Rosser [11]):

$$E'_x = E_x \qquad E_x = E'_x \tag{10.74}$$

$$E'_y = \gamma(E_y - vB_z) \qquad E_y = \gamma(E'_y + vB'_z) \tag{10.75}$$

$$E'_z = \gamma(E_z + vB_y) \qquad E_z = \gamma(E'_z - vB'_y) \tag{10.76}$$

$$B'_x = B_x \qquad B_x = B'_x \tag{10.77}$$

$$B'_y = \gamma\left(B_y + \frac{v}{c^2}E_z\right) \qquad B_y = \gamma\left(B'_y - \frac{v}{c^2}E'_z\right) \tag{10.78}$$

$$B'_z = \gamma\left(B_z - \frac{v}{c^2}E_y\right) \qquad B_z = \gamma\left(B'_z + \frac{v}{c^2}E'_y\right). \tag{10.79}$$

10.10.4. *Transformation of charge and current densities*

Equations (10.55) and (10.58) include the charge density ρ and the current density **J**. Before showing that equations (10.55) and (10.58) obey the principle of relativity when the Lorentz transformations are used, we shall derive the transformations for ρ and **J**.

We shall start with a simplified model and consider a “uniform” electric charge distribution of volume V in the inertial frame Σ, consisting of n discrete charges per unit volume of magnitude q each. It will be assumed that all the charges have the same velocity **u** relative to the inertial frame Σ. For this simplified case the charge density in Σ is

$$\rho = nq. \tag{10.80}$$

The current density, which is the current crossing unit area normal to the direction of current flow, is given in Σ by

$$\mathbf{J} = nq\mathbf{u}, \tag{10.81}$$

or in components form $J_x = nqu_x$, $J_y = nqu_y$, $J_z = nqu_z$. Let the same charge distribution have a volume V' when measured relative to an inertial frame Σ' that is moving with uniform velocity v relative to Σ, and let it consist of n' discrete charges per unit volume moving with velocity $\mathbf{u}'$ relative to Σ'. We then have in Σ'

$$\rho' = n'q, \tag{10.82}$$

$$\mathbf{J}' = n'q\mathbf{u}'. \tag{10.83}$$

The principle of the invariance of total electric charge will be taken as axiomatic in this section. According to this principle, the total charge on a body is independent of the velocity of the body, and the total charge has the same numerical value in all inertial reference frames.

Consider an inertial frame Σ^0 in which the charge distribution is at rest; Σ^0 moves with velocity $\mathbf{u}$ relative to Σ. Let V_0 be the proper volume of the charge distribution in Σ^0, and let n_0 be the number of charges per unit volume measured in Σ^0, such that the total number of charges is n_0V_0. According to equation (10.22) owing to length contraction the volume of the charge distribution in Σ should be measured to be

$$V = V_0\left(1 - \frac{u^2}{c^2}\right)^{1/2}.$$

The total number of charges measured in Σ is nV, which is equal to $nV_0(1 - u^2/c^2)^{1/2}$. But the total *number* of charges is an invariant, since it is a pure number. Hence

$$nV = nV_0\left(1 - \frac{u^2}{c^2}\right)^{1/2} = n_0V_0$$

so that

$$n = \frac{n_0}{(1 - u^2/c^2)^{1/2}}.$$

Similarly, in Σ'

$$n' = \frac{n_0}{(1 - u'^2/c^2)^{1/2}}$$

so that

$$n' = n\frac{(1 - u^2/c^2)^{1/2}}{(1 - u'^2/c^2)^{1/2}}. \tag{10.84}$$

From equation (E9) of Appendix E

$$\frac{(1 - u^2/c^2)^{1/2}}{(1 - u'^2/c^2)^{1/2}} = \gamma(1 - vu_x/c^2). \tag{10.85}$$

Substituting in equation (10.84) we obtain

$$n' = \gamma n\left(1 - \frac{vu_x}{c^2}\right). \tag{10.86}$$

Multiplying by q gives

$$qn' = \gamma\left(qn - \frac{vqnu_x}{c^2}\right).$$

Substituting from equations (10.82), (10.80) and (10.81), we have

$$\rho' = \gamma\left(\rho - \frac{vJ_x}{c^2}\right).$$

Now, from the x' component of equation (10.83),

$$J'_x = qu'_x n'.$$

But from the velocity transformations, equation (E6) we have

$$u'_x = \frac{u_x - v}{1 - vu_x/c^2}.$$

Substituting for u'_x, and for n' from equation (10.86), we find

$$J'_x = \frac{q(u_x - v)}{(1 - vu_x/c^2)} n\gamma \left(1 - \frac{vu_x}{c^2}\right)$$

and, using equations (10.80) and (10.81),

$$J'_x = \gamma(J_x - v\rho).$$

Similarly, from equation (10.83)

$$J'_y = qu'_y n'.$$

From the velocity transformations, equation (E7),

$$u'_y = \frac{u_y}{\gamma(1 - vu_x/c^2)}.$$

Substituting for u'_y and n'

$$J'_y = \frac{qu_y}{\gamma(1 - vu_x/c^2)} n\gamma \left(1 - \frac{vu_x}{c^2}\right) = J_y.$$

Similarly,

$$J'_z = J_z.$$

Collecting the transformations, we have

$$J'_x = \gamma(J_x - v\rho) \qquad J_x = \gamma(J'_x + v\rho') \tag{10.87}$$

$$J'_y = J_y \qquad J_y = J'_y \tag{10.88}$$

$$J'_z = J_z \qquad J_z = J'_z \tag{10.89}$$

$$\rho' = \gamma\left(\rho - \frac{v}{c^2} J_x\right) \qquad \rho = \gamma\left(\rho' + \frac{v}{c^2} J'_x\right). \tag{10.90}$$

It is left as an exercise for the reader to extend the treatment to systems containing both positive and negative charges and to show that equations (10.87)–(10.90) are valid in the general case.

10.10.5. *Transformations for the fields* **D** *and* **H**

Consider the y component of the equation

$$\nabla \times \mathbf{H} = \mathbf{J} + \frac{\partial \mathbf{D}}{\partial t} \tag{10.58}$$

which is

$$\frac{\partial H_x}{\partial z} - \frac{\partial H_z}{\partial x} = J_y + \frac{\partial D_y}{\partial t}.$$

Substituting for $\partial/\partial z$, $\partial/\partial x$ and $\partial/\partial t$ from equations (10.69), (10.67) and (10.70) and for J_y from equation (10.88),

$$\frac{\partial H_x}{\partial z'} - \gamma\left(\frac{\partial}{\partial x'} - \frac{v}{c^2}\frac{\partial}{\partial t'}\right) H_z = J'_y + \gamma\left(\frac{\partial}{\partial t'} - v\frac{\partial}{\partial x'}\right) D_y$$

or

$$\frac{\partial H_x}{\partial z'} - \frac{\partial}{\partial x'}\gamma(H_z - vD_y) = J'_y + \frac{\partial}{\partial t'}\gamma\left(D_y - \frac{v}{c^2}H_z\right). \tag{10.91}$$

If Maxwell's equations are valid in Σ' then one must have in Σ'

$$\frac{\partial H'_x}{\partial z'} - \frac{\partial H'_z}{\partial x'} = J'_y + \frac{\partial D'_y}{\partial t'}. \tag{10.92}$$

Equations (10.91) and (10.92) have the same mathematical form, and if one puts

$$H'_x = H_x; \qquad H'_z = \gamma(H_z - vD_y)$$

$$D'_y = \gamma\left(D_y - \frac{v}{c^2}H_z\right)$$

then they are exactly the same.

It is left as an exercise for the reader to show by similar methods that the x and z components of equation (10.58) transform into the x' and z' components of equation (10.54) and that equation (10.55) transforms into equation (10.51), if $\mathbf{J}$ and ρ are transformed using equations (10.87)–(10.90) and if $\mathbf{H}$ and $\mathbf{D}$ satisfy the transformations:

$$D'_x = D_x \qquad\qquad D_x = D'_x \tag{10.93}$$

$$D'_y = \gamma\left(D_y - \frac{v}{c^2}H_z\right) \qquad\qquad D_y = \gamma\left(D'_y + \frac{v}{c^2}H'_z\right) \tag{10.94}$$

$$D'_z = \gamma\left(D_z + \frac{v}{c^2}H_y\right) \qquad\qquad D_z = \gamma\left(D'_z - \frac{v}{c^2}H'_y\right) \tag{10.95}$$

$$H'_x = H_x \qquad\qquad H_x = H'_x \tag{10.96}$$

$$H'_y = \gamma(H_y + vD_z) \qquad\qquad H_y = \gamma(H'_y - vD'_z) \tag{10.97}$$

$$H'_z = \gamma(H_z - vD_y) \qquad\qquad H_z = \gamma(H'_z + vD'_y) \tag{10.98}$$

It has been shown that if Maxwell's equations are valid in Σ then, if the coordinates and time are transformed using the Lorentz transformations, and if one takes the principle of constant charge as axiomatic, the transformed equations have the same mathematical form as Maxwell's equations would have if they were valid in Σ'. This is true whether or not there are material bodies present at the point. Thus Maxwell's equations satisfy the principle of relativity when the coordinates and time are transformed according to the Lorentz transformations.

Only the briefest insight into the relativistic invariance of Maxwell's equations has been given here. For more comprehensive accounts of relativistic electromagnetism, including the transformations for the potentials ϕ and $\mathbf{A}$ and the electrodynamics of moving media, the reader is referred to Rosser [12] and [13].

The covariance of Maxwell's equations and the equations for the potentials is best illustrated using four-vectors and tensors. An introductory account is given by Rosser [14].

References

1. Rosser, W. G. V., *Introductory Special Relativity* (Taylor and Francis, London, 1991).
2. Einstein, A., *Ann. Phys. Lpz,* Vol. 17, p. 891 (1905).
3. Shankland, R. S., *Amer. Journ. Phys.*, Vol. 31, p. 47 (1963).
4. Lorentz, H. A., *Astrophys. Journ.*, Vol. 68, p. 350 (1928).
5. Alväger, T., Farley, F. J. M., Kjellman, J. and Wallin, I., *Phys. Lett.*, Vol. 12, p. 260 (1964).
6. King, J. G., *Phys. Rev. Lett.*, Vol. 5, p. 562 (1960).
7. Rosser, W. G. V., *Introduction to the Theory of Relativity* (Butterworths, London, 1964), p. 87.
8. Rosser, W. G. V., see ref. 7, pp. 112–122.
9. Rosser, W. G. V., *Contemporary Physics*, Vol. 1, p. 453.
10 Rosser, W. G. V., see ref. 7, pp. 286–293.
11 Rosser, W. G. V., see ref. 7, pp. 307–310.
12 Rosser, W. G. V., see ref. 7, pp. 330–345.
13 Rosser, W. G. V., *Classical Electromagnetism Via Relativity* (Butterworths, London, 1968), pp. 175–203.
14 Rosser, W. G. V., see ref. 7, Chapter 10.

APPENDIX A

Mathematical methods

A1. A summary of the formulae of vector analysis

A1.1. *Scalar and vector products*

In this appendix, the symbols **A**, **B** and **C** will be used to denote vector quantities and the symbols ϕ and ψ will be used to denote scalar quantities. The magnitude of the vector **A** will be denoted by A or $|\mathbf{A}|$.

The scalar (or dot or inner) product of two vectors **A** and **B** will be written as $\mathbf{A} \cdot \mathbf{B}$. The magnitude of the scalar product $\mathbf{A} \cdot \mathbf{B}$ is given by

$$\mathbf{A} \cdot \mathbf{B} = AB \cos(\mathbf{A}, \mathbf{B}) = A_x B_x + A_y B_y + A_z B_z \tag{A1.1}$$

where A_x, A_y and A_z are the components of the vector **A** in a cartesian coordinate system, B_x, B_y and B_z are the components of the vector **B** and $\cos(\mathbf{A}, \mathbf{B})$ is the cosine of the angle between the vectors **A** and **B**. The scalar product of two vectors is a scalar quantity. The order of the two vectors in a scalar product does not matter, since, according to equation (A1.1), $\mathbf{A} \cdot \mathbf{B} = \mathbf{B} \cdot \mathbf{A}$. The scalar product obeys the distributive law of addition. In general we have

$$\mathbf{A} \cdot (\mathbf{B} + \mathbf{C} + \ldots) = \mathbf{A} \cdot \mathbf{B} + \mathbf{A} \cdot \mathbf{C} + \ldots \tag{A1.2}$$

The vector (or cross) product **C** of two vectors **A** and **B** will be denoted by $\mathbf{C} = \mathbf{A} \times \mathbf{B}$. The magnitude of the vector product $\mathbf{A} \times \mathbf{B}$ is $AB \sin(\mathbf{A}, \mathbf{B})$ where $\sin(\mathbf{A}, \mathbf{B})$ is the sine of the angle between the vectors **A** and **B**. The vector product $\mathbf{A} \times \mathbf{B}$ is a vector in a direction perpendicular to the plane containing **A** and **B** and is in the direction a right-handed corkscrew would advance if it were rotated from **A** to **B** through the smaller of the two angles between the two vectors **A** and **B**. If $\hat{\mathbf{i}}$, $\hat{\mathbf{j}}$ and $\hat{\mathbf{k}}$ are unit vectors in the positive x, y and z directions of a cartesian coordinate system, then

$$\mathbf{A} \times \mathbf{B} = (A_y B_z - A_z B_y)\hat{\mathbf{i}} + (A_z B_x - A_x B_z)\hat{\mathbf{j}} + (A_x B_y - A_y B_x)\hat{\mathbf{k}}. \tag{A1.3}$$

It follows from equation (A1.3) that $\mathbf{A} \times \mathbf{B} = -\mathbf{B} \times \mathbf{A}$. Hence the order of the vectors in a vector product is important. The vector product obeys the distributive law of addition. In general we have

$$\mathbf{A} \times (\mathbf{B} + \mathbf{C} + \ldots) = \mathbf{A} \times \mathbf{B} + \mathbf{A} \times \mathbf{C} + \ldots \qquad \text{(A1.4)}$$

The triple scalar product is defined as $(\mathbf{A} \times \mathbf{B}) \cdot \mathbf{C}$. It can be shown that

$$(\mathbf{A} \times \mathbf{B}) \cdot \mathbf{C} = (\mathbf{B} \times \mathbf{C}) \cdot \mathbf{A} = (\mathbf{C} \times \mathbf{A}) \cdot \mathbf{B}. \qquad \text{(A1.5)}$$

The triple vector product is defined as $\mathbf{A} \times (\mathbf{B} \times \mathbf{C})$. It can be shown that

$$\mathbf{A} \times (\mathbf{B} \times \mathbf{C}) = \mathbf{B}(\mathbf{A} \cdot \mathbf{C}) - \mathbf{C}(\mathbf{A} \cdot \mathbf{B}). \qquad \text{(A1.6)}$$

The formulae of vector analysis are often expressed in terms of the vector operator $\boldsymbol{\nabla}$, which is called del or nabla or the gradient operator. In cartesian coordinates

$$\boldsymbol{\nabla} = \hat{\mathbf{i}}\frac{\partial}{\partial x} + \hat{\mathbf{j}}\frac{\partial}{\partial y} + \hat{\mathbf{k}}\frac{\partial}{\partial z}. \qquad \text{(A1.7)}$$

A1.2. *The gradient of a scalar*

The gradient of a scalar function ϕ of position can be defined as $\boldsymbol{\nabla}\phi$, where $\boldsymbol{\nabla}$ is given by equation (A1.7). The gradient of ϕ is sometimes written as grad ϕ. In cartesian coordinates

$$\boldsymbol{\nabla}\phi = \left(\hat{\mathbf{i}}\frac{\partial}{\partial x} + \hat{\mathbf{j}}\frac{\partial}{\partial y} + \hat{\mathbf{k}}\frac{\partial}{\partial z}\right)\phi = \frac{\partial\phi}{\partial x}\hat{\mathbf{i}} + \frac{\partial\phi}{\partial y}\hat{\mathbf{j}} + \frac{\partial\phi}{\partial z}\hat{\mathbf{k}}. \qquad \text{(A1.8)}$$

The gradient of a scalar is a vector. The total differential $\mathrm{d}\phi$ of ϕ can be written in the form of a scalar product as follows

$$\mathrm{d}\phi = \frac{\partial\phi}{\partial x}\mathrm{d}x + \frac{\partial\phi}{\partial y}\mathrm{d}y + \frac{\partial\phi}{\partial z}\mathrm{d}z = \boldsymbol{\nabla}\phi \cdot \mathbf{dl}. \qquad \text{(A1.9)}$$

where

$$\mathbf{dl} = \mathrm{d}x\hat{\mathbf{i}} + \mathrm{d}y\hat{\mathbf{j}} + \mathrm{d}z\hat{\mathbf{k}}.$$

Since $\mathrm{d}\phi$ is the total change in ϕ in an infinitesimal displacement $\mathbf{dl}$, we have

$$\int_a^b \boldsymbol{\nabla}\phi \cdot \mathbf{dl} = \int_a^b \mathrm{d}\phi = \phi(b) - \phi(a) \qquad \text{(A1.10)}$$

where $\phi(a)$ and $\phi(b)$ are the values of ϕ at the initial and final positions. For any closed loop, we have

$$\oint \boldsymbol{\nabla}\phi \cdot \mathbf{dl} = \oint \mathrm{d}\phi = 0. \qquad \text{(A1.11)}$$

A1.3. *The divergence of a vector*

The divergence of a vector **A** can be defined as the scalar product $\nabla \cdot \mathbf{A}$. It is sometimes written as div**A**. In cartesian coordinates

$$\nabla \cdot \mathbf{A} = \frac{\partial A_x}{\partial x} + \frac{\partial A_y}{\partial y} + \frac{\partial A_z}{\partial z}. \tag{A1.12}$$

The divergence of a vector is a scalar quantity.

A1.4. *The curl of a vector*

The curl of a vector **A** can be defined by the vector product $\nabla \times \mathbf{A}$. It is sometimes written as curl **A**. The curl of a vector is a vector quantity. In cartesian coordinates

$$\nabla \times \mathbf{A} = \left(\frac{\partial A_z}{\partial y} - \frac{\partial A_y}{\partial z}\right)\hat{\mathbf{i}} + \left(\frac{\partial A_x}{\partial z} - \frac{\partial A_z}{\partial x}\right)\hat{\mathbf{j}} + \left(\frac{\partial A_y}{\partial x} - \frac{\partial A_x}{\partial y}\right)\hat{\mathbf{k}}. \tag{A1.13}$$

A1.5. *The Laplacian operator*

The operator $\nabla \cdot \nabla$ is generally called the Laplacian operator or the Laplacian and denoted ∇^2, so that

$$\nabla^2 \equiv \nabla \cdot \nabla. \tag{A1.14}$$

In cartesian coordinates, if the Laplacian operates on a scalar function of position, we have

$$\begin{aligned}\nabla^2\phi = \nabla \cdot \nabla\phi &= \nabla \cdot \left(\frac{\partial \phi}{\partial x}\hat{\mathbf{i}} + \frac{\partial \phi}{\partial y}\hat{\mathbf{j}} + \frac{\partial \phi}{\partial z}\hat{\mathbf{k}}\right) \\ &= \frac{\partial^2 \phi}{\partial x^2} + \frac{\partial^2 \phi}{\partial y^2} + \frac{\partial^2 \phi}{\partial z^2}.\end{aligned} \tag{A1.15}$$

A1.6. *Some useful relations*

The reader can show using cartesian coordinates that, if ϕ and ψ are scalar quantities and **A**, **B** and **C** are vector quantities

$$\nabla(\phi + \psi) = \nabla\phi + \nabla\psi \tag{A1.16}$$

$$\nabla(\phi\psi) = \phi\nabla\psi + \psi\nabla\phi \tag{A1.17}$$

$$\nabla(\mathbf{A} \cdot \mathbf{B}) = \mathbf{A} \times (\nabla \times \mathbf{B}) + \mathbf{B} \times (\nabla \times \mathbf{A}) + (\mathbf{A} \cdot \nabla)\mathbf{B} + (\mathbf{B} \cdot \nabla)\mathbf{A} \tag{A1.18}$$

$$\nabla \cdot (\mathbf{A} + \mathbf{B}) = \nabla \cdot \mathbf{A} + \nabla \cdot \mathbf{B} \tag{A1.19}$$

$$\nabla \cdot (\phi\mathbf{A}) = \phi\nabla \cdot \mathbf{A} + \mathbf{A} \cdot (\nabla\phi) \tag{A1.20}$$

$$\nabla \cdot (\mathbf{A} \times \mathbf{B}) = \mathbf{B} \cdot (\nabla \times \mathbf{A}) - \mathbf{A} \cdot (\nabla \times \mathbf{B}) \tag{A1.21}$$

$$\nabla \times (\mathbf{A} + \mathbf{B}) = \nabla \times \mathbf{A} + \nabla \times \mathbf{B} \tag{A1.22}$$

$$\nabla \times (\phi \mathbf{A}) = \phi \nabla \times \mathbf{A} + (\nabla \phi) \times \mathbf{A} \tag{A1.23}$$

$$\nabla \times (\mathbf{A} \times \mathbf{B}) = \mathbf{A}(\nabla \cdot \mathbf{B}) - \mathbf{B}(\nabla \cdot \mathbf{A}) + (\mathbf{B} \cdot \nabla)\mathbf{A} - (\mathbf{A} \cdot \nabla)\mathbf{B}. \tag{A1.24}$$

Sometimes the operator ∇ is applied more than once. The reader can show using cartesian coordinates that

$$\nabla \cdot (\nabla \times \mathbf{A}) = 0 \tag{A1.25}$$

$$\nabla \times (\nabla \phi) = 0. \tag{A1.26}$$

A very important expression in electromagnetism is the curl of the curl of a vector, namely $\nabla \times (\nabla \times \mathbf{A})$, which is used, for example, in Section 1.9.1 of Chapter 1 to go from Maxwell's equations to the wave equations. The reader can show that in cartesian coordinates

$$\nabla \times (\nabla \times \mathbf{A}) = \nabla(\nabla \cdot \mathbf{A}) - \nabla^2 \mathbf{A} \tag{A1.27}$$

where in cartesian coordinates only $\nabla^2 \mathbf{A}$ can be expressed in the form

$$\nabla^2 \mathbf{A} = \nabla^2 A_x \hat{\mathbf{i}} + \nabla^2 A_y \hat{\mathbf{j}} + \nabla^2 A_z \hat{\mathbf{k}} \tag{A1.28}$$

where the Laplacians of the scalar cartesian components A_x, A_y and A_z are each given by equation (A1.15). There are difficulties when applying the operator ∇^2 to the vector $\mathbf{A}$, when $\mathbf{A}$ is expressed in curvilinear coordinates such as spherical polars. The reader should remember that the unit vectors $\hat{\mathbf{r}}$, $\hat{\boldsymbol{\theta}}$ and $\hat{\boldsymbol{\phi}}$ in spherical polars change direction if we vary r, θ and ϕ so that $\partial \hat{\mathbf{r}}/\mathrm{d}\theta$ etc are not zero. It will be sufficient for our purposes when we use curvilinear coordinates such as spherical polars, to rewrite equation (A1.27) in the form

$$\nabla^2 \mathbf{A} = \nabla(\nabla \cdot \mathbf{A}) - \nabla \times (\nabla \times \mathbf{A}) \tag{A1.29}$$

and to assume that $\nabla^2 \mathbf{A}$ is a quantity defined by and to be determined using the right hand side of equation (A1.29).

A1.7. *Gauss' mathematical theorem*

According to Gauss' mathematical theorem, the volume integral of the divergence of a vector $\mathbf{A}$ over any finite volume is equal to the surface integral of $\mathbf{A}$ evaluated over the surface of that volume, that is

$$\int (\nabla \cdot \mathbf{A}) \mathrm{d}V = \int \mathbf{A} \cdot \mathrm{d}\mathbf{S}. \tag{A1.30}$$

The element of area $\mathrm{d}\mathbf{S}$ is equal to the magnitude $\mathrm{d}S$ of the area of the element times a unit vector $\hat{\mathbf{n}}$ in the direction of the outward normal to the surface at the position of the element of area, so that $\mathrm{d}\mathbf{S} = \mathrm{d}S\hat{\mathbf{n}}$.

Integrating equation (A1.20) over a finite volume and applying Gauss' mathematical theorem we obtain

$$\int \nabla \cdot (\phi \mathbf{A}) \mathrm{d}V = \int \phi \mathbf{A} \cdot \mathrm{d}\mathbf{S} = \int \phi (\nabla \cdot \mathbf{A}) \mathrm{d}V + \int \mathbf{A} \cdot \nabla \phi \, \mathrm{d}V. \qquad \text{(A1.31)}$$

If **A** is a constant vector $\nabla \cdot \mathbf{A} = 0$, and equation (A1.31) reduces to

$$\mathbf{A} \cdot \int \phi \, \mathrm{d}\mathbf{S} = \mathbf{A} \cdot \int \nabla \phi \, \mathrm{d}V.$$

Hence,

$$\int \phi \, \mathrm{d}\mathbf{S} = \int \nabla \phi \, \mathrm{d}V. \qquad \text{(A1.32)}$$

If we integrate equation (A1.21) over a finite volume, and assume that **A** is a constant vector, so that $\nabla \times \mathbf{A}$ is zero, we find that

$$\int \nabla \cdot (\mathbf{A} \times \mathbf{B}) \mathrm{d}V = -\mathbf{A} \cdot \int \nabla \times \mathbf{B} \, \mathrm{d}V.$$

If we apply Gauss' integral theorem, equation (A1.30) to the left hand side and then apply equation (A1.5), we find that the left hand side becomes

$$\int \nabla \cdot (\mathbf{A} \times \mathbf{B}) \mathrm{d}V = \int (\mathbf{A} \times \mathbf{B}) \cdot \mathrm{d}\mathbf{S} = \mathbf{A} \cdot \int \mathbf{B} \times \mathrm{d}\mathbf{S}.$$

Hence

$$\int \mathbf{B} \times \mathrm{d}\mathbf{S} = -\int \nabla \times \mathbf{B} \, \mathrm{d}V. \qquad \text{(A1.33)}$$

A1.8. *Stokes' theorem*

According to Stokes' theorem, the surface integral of $\nabla \times \mathbf{A}$ over any finite area is equal to the line integral of **A** around the boundary of that area, so that

$$\int (\nabla \times \mathbf{A}) \cdot \mathrm{d}\mathbf{S} = \oint \mathbf{A} \cdot \mathrm{d}\mathbf{l}. \qquad \text{(A1.34)}$$

The direction of the vector **dS** in equation (A1.34) is the direction a right-handed corkscrew would advance if it were rotated in the direction of **dl**.

A1.9. *Cylindrical coordinates*

In cylindrical coordinates r, ϕ, z if $\hat{\mathbf{r}}$, $\hat{\boldsymbol{\phi}}$ and $\hat{\mathbf{k}}$ are unit vectors in the directions of increasing r, ϕ and z, respectively and if A_r, A_ϕ and A_z are the components of $\mathbf{A}$ in the directions of $\hat{\mathbf{r}}$, $\hat{\boldsymbol{\phi}}$ and $\hat{\mathbf{k}}$ respectively, then

$$\boldsymbol{\nabla} = \hat{\mathbf{r}}\frac{\partial}{\partial r} + \hat{\boldsymbol{\phi}}\frac{1}{r}\frac{\partial}{\partial \phi} + \hat{\mathbf{k}}\frac{\partial}{\partial z} \tag{A1.35}$$

$$\boldsymbol{\nabla} \cdot \mathbf{A} = \frac{1}{r}\frac{\partial}{\partial r}(rA_r) + \frac{1}{r}\frac{\partial A_\phi}{\partial \phi} + \frac{\partial A_z}{\partial z} \tag{A1.36}$$

$$\boldsymbol{\nabla} \times \mathbf{A} = \left[\frac{1}{r}\frac{\partial A_z}{\partial \phi} - \frac{\partial A_\phi}{\partial z}\right]\hat{\mathbf{r}} + \left[\frac{\partial A_r}{\partial z} - \frac{\partial A_z}{\partial r}\right]\hat{\boldsymbol{\phi}} + \left[\frac{1}{r}\frac{\partial}{\partial r}(rA_\phi) - \frac{1}{r}\frac{\partial A_r}{\partial \phi}\right]\hat{\mathbf{k}}. \tag{A1.37}$$

When the Laplacian operator $\boldsymbol{\nabla} \cdot \boldsymbol{\nabla}$ is applied to a scalar function ψ, using equations (A1.35) and (A1.36) to evaluate $\boldsymbol{\nabla} \cdot \boldsymbol{\nabla}$ we find that in cylindrical coordinates

$$\boldsymbol{\nabla}^2\psi = \frac{1}{r}\frac{\partial}{\partial r}\left(r\frac{\partial \psi}{\partial r}\right) + \frac{1}{r^2}\frac{\partial^2 \psi}{\partial \phi^2} + \frac{\partial^2 \psi}{\partial z^2}. \tag{A1.38}$$

A1.10. *Spherical polar coordinates*

In spherical polar coordinates r, θ, ϕ, if $\hat{\mathbf{r}}$, $\hat{\boldsymbol{\theta}}$ and $\hat{\boldsymbol{\phi}}$ are unit vectors in the directions of increasing r, θ, and ϕ respectively and if A_r, A_θ, and A_ϕ are the components of $\mathbf{A}$ in the directions of $\hat{\mathbf{r}}$, $\hat{\boldsymbol{\theta}}$ and $\hat{\boldsymbol{\phi}}$ respectively then

$$\boldsymbol{\nabla} = \hat{\mathbf{r}}\frac{\partial}{\partial r} + \hat{\boldsymbol{\theta}}\frac{1}{r}\frac{\partial}{\partial \theta} + \hat{\boldsymbol{\phi}}\frac{1}{r\sin\theta}\frac{\partial}{\partial \phi} \tag{A1.39}$$

$$\boldsymbol{\nabla} \cdot \mathbf{A} = \frac{1}{r^2}\frac{\partial}{\partial r}(r^2A_r) + \frac{1}{r\sin\theta}\frac{\partial}{\partial \theta}(A_\theta \sin\theta) + \frac{1}{r\sin\theta}\frac{\partial A_\phi}{\partial \phi} \tag{A1.40}$$

$$\boldsymbol{\nabla} \times \mathbf{A} = \frac{1}{r\sin\theta}\left[\frac{\partial}{\partial \theta}(A_\phi \sin\theta) - \frac{\partial A_\theta}{\partial \phi}\right]\hat{\mathbf{r}} + \left[\frac{1}{r\sin\theta}\frac{\partial A_r}{\partial \phi} - \frac{1}{r}\frac{\partial}{\partial r}(rA_\phi)\right]\hat{\boldsymbol{\theta}} + \left[\frac{1}{r}\frac{\partial}{\partial r}(rA_\theta) - \frac{1}{r}\frac{\partial A_r}{\partial \theta}\right]\hat{\boldsymbol{\phi}}. \tag{A1.41}$$

The Laplacian of a scalar function ψ is

$$\boldsymbol{\nabla}^2\psi = \boldsymbol{\nabla} \cdot \boldsymbol{\nabla}\psi = \frac{1}{r^2}\frac{\partial}{\partial r}\left(r^2\frac{\partial \psi}{\partial r}\right) + \frac{1}{r^2\sin\theta}\frac{\partial}{\partial \theta}\left(\sin\theta\frac{\partial \psi}{\partial \theta}\right) + \frac{1}{r^2\sin^2\theta}\frac{\partial^2 \psi}{\partial \phi^2}. \tag{A1.42}$$

General reference

Spiegal, M. R., *Theory and Problems of Vector Analysis*, Shaum Publishing Co., New York, 1959.

A2. The partial derivatives of macroscopic field variables

Consider the microscopic variable $f(x, y, z, t)$. According to equation (1.147) of Chapter 1, the value of the corresponding macroscopic variable $F(x, y, z, t)$ at the field point P in Figure 1.12 is

$$F(\mathbf{r}, t) = \int f(\mathbf{r} + \mathbf{s}, t) w(\mathbf{s})\, \mathrm{d}^3\mathbf{s}.$$

If we move the field point to the position $(x + \mathrm{d}x, y, z)$, taking the weighting factor $w(\mathbf{s})$ with it, then the value of the macroscopic variable F is

$$F(x + \mathrm{d}x, y, z, t) = \int f(x + \mathrm{d}x + s_x, y + s_y, z + s_z, t) w(\mathbf{s})\, \mathrm{d}^3\mathbf{s}. \quad \text{(A2.1)}$$

It follows from Taylor's expansion theorem that

$$f(x + \mathrm{d}x + s_x, y + s_y, z + s_z, t) = f(\mathbf{r} + \mathbf{s}, t) + f'\, \mathrm{d}x + \ldots \quad \text{(A2.2)}$$

where

$$f' = \frac{\partial f(x + s_x, y + s_y, z + s_z, t)}{\partial(x + s_x)} = \frac{\partial f}{\partial x}$$

and where $\partial f/\partial x$ is evaluated at the point $(\mathbf{r} + \mathbf{s})$. Substituting from equation (A2.2) into equation (A2.1) we find that

$$\begin{aligned} F(x + \mathrm{d}x, y, z, t) &= F(x, y, z, t) + \mathrm{d}x \int \frac{\partial f(\mathbf{r} + \mathbf{s}, t)}{\partial x} w(\mathbf{s})\, \mathrm{d}^3\mathbf{s} \\ &= F(x, y, z, t) + \left\langle \frac{\partial f}{\partial x} \right\rangle \mathrm{d}x. \end{aligned} \quad \text{(A2.3)}$$

Since

$$\frac{\partial F}{\partial x} = \frac{F(x + \mathrm{d}x, y, z, t) - F(x, y, z, t)}{\mathrm{d}x}$$

using equation (A2.3) we find that

$$\frac{\partial F}{\partial x} = \left\langle \frac{\partial f}{\partial x} \right\rangle. \quad \text{(A2.4)}$$

A3. Some useful mathematical relations

Consider

$$\int_1^2 \frac{\mathrm{d}}{\mathrm{d}r}\left(\frac{\mathbf{r}}{r^2}\right)\mathrm{d}r = \int_1^2 \frac{\mathrm{d}}{\mathrm{d}r}\left(\frac{x\hat{\mathbf{i}} + y\hat{\mathbf{j}} + z\hat{\mathbf{k}}}{r^2}\right)\mathrm{d}r$$

$$= \left[\left(\frac{x}{r^2}\right)\hat{\mathbf{i}} + \left(\frac{y}{r^2}\right)\hat{\mathbf{j}} + \left(\frac{z}{r^2}\right)\hat{\mathbf{k}}\right]_1^2 .$$

If we integrate around the closed loop in Figure A1, the values of (x/r^2) etc are the same at both limits, so that

$$\oint \frac{\mathrm{d}}{\mathrm{d}r}\left(\frac{\mathbf{r}}{r^2}\right)\mathrm{d}r = 0. \qquad \text{(A3.1)}$$

Now

$$\frac{\mathrm{d}}{\mathrm{d}r}\left(\frac{\mathbf{r}}{r^2}\right)\mathrm{d}r = \frac{\mathrm{d}\mathbf{r}}{r^2} - \frac{2\mathbf{r}\;\mathrm{d}r}{r^3} . \qquad \text{(A3.2)}$$

It follows from Figure A3.1 that

$$\mathbf{dr} = \mathbf{dl} \qquad \text{(A3.3)}$$

$$\mathrm{d}r = \cos\phi \; \mathrm{d}l \qquad \text{(A3.4)}$$

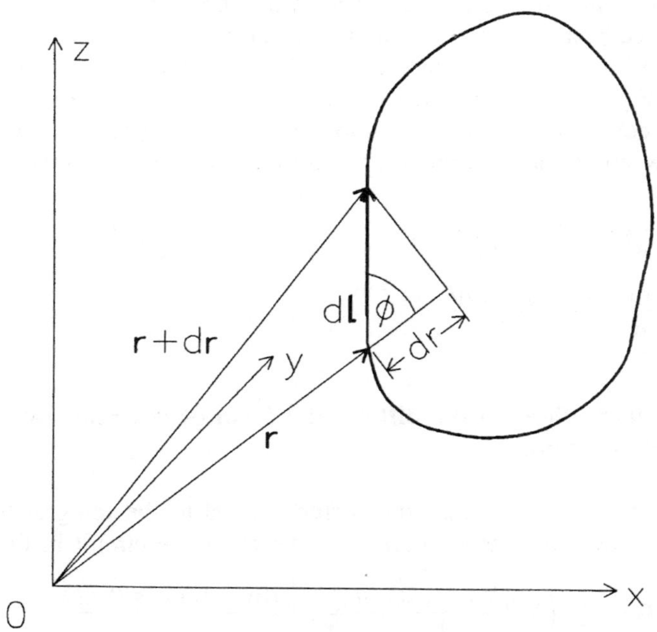

Figure A1. Derivation of some mathematical relations.

where ϕ is the angle between $\mathbf{r}$ and $\mathbf{dl}$. Integrating equation (A3.2) around the closed curve in Figure A3.1 and using equations (A3.1), (A3.3) and (A3.4) we find that

$$\oint \frac{\mathrm{d}}{\mathrm{d}r}\left(\frac{\mathbf{r}}{r^2}\right)\mathrm{d}r = \oint \frac{\mathbf{dl}}{r^2} - 2\oint \frac{\mathbf{r}\cos\phi\,\mathrm{d}l}{r^3} = 0$$

so that

$$\oint \frac{\mathbf{dl}}{r^2} = 2\int \frac{\mathbf{r}\cos\phi\,\mathrm{d}l}{r^3}. \tag{A3.5}$$

Similarly

$$\oint \frac{\mathrm{d}}{\mathrm{d}r}\left(\frac{\mathbf{r}}{r}\right)\mathrm{d}r = 0. \tag{A3.6}$$

Using equations (A3.3) and A3.4), we have

$$\frac{\mathrm{d}}{\mathrm{d}r}\left(\frac{\mathbf{r}}{r}\right)\mathrm{d}r = \frac{\mathbf{dr}}{r} - \frac{\mathbf{r}}{r^2}\mathrm{d}r = \frac{\mathbf{dl}}{r} - \frac{\mathbf{r}\cos\phi\,\mathrm{d}l}{r^2}.$$

Substituting in equation (A3.6), we find that

$$\oint \frac{\mathbf{dl}}{r} = \oint \frac{\mathbf{r}\cos\phi\,\mathrm{d}l}{r^2}. \tag{A3.7}$$

When the electric and magnetic fields due to electric circuits are discussed in Chapters 5 and 6 it is the vector $\mathbf{R}$, which goes from the current element at the source point at $\mathbf{r}_s$ to the field point at $\mathbf{r}$, that is used. To apply equations (A3.5) and (A3.7) to these circuits, assume that the field point is at the origin in Figure A3.1 and then replace $\mathbf{r}$ by $-\mathbf{R}$, r by R and ϕ by $(\pi - \theta)$ such that $\cos\phi = -\cos\theta$ in equations (A.3.5) and (A3.7), where θ is the angle between $\mathbf{R}$ and $\mathbf{dl}$. Equations (A3.5) and (A3.7) then give

$$\oint \frac{\mathbf{dl}}{R^2} = 2\oint \frac{\mathbf{R}\cos\theta\,\mathrm{d}l}{R^3}, \tag{A3.8}$$

$$\oint \frac{\mathbf{dl}}{R} = \oint \frac{\mathbf{R}\cos\theta\,\mathrm{d}l}{R^2}. \tag{A3.9}$$

A4. The corrections to the differential form of the Biot–Savart law for steady currents

According to equation (6.25) the correction $\Delta\mathbf{B}$ to the integral form of the Biot–Savart law for a steady current in the complete circuit in Figure 6.2 is

$$\Delta\mathbf{B} = \frac{\mu_0}{4\pi c}\left(\oint \frac{qN_0(\mathbf{a}\times\mathbf{R})\,\mathrm{d}l}{R^2} + 2\oint \frac{I(\mathbf{u}\times\mathbf{R})\cos\theta\,\mathrm{d}l}{R^3}\right) \tag{A4.1}$$

where we have put $u\mathbf{dl}$ equal to $\mathbf{u}\mathrm{d}l$ in equation (6.25).

We shall assume that in a given current element, there are N_0 moving positive charges per metre length each of charge q and all moving with the same velocity **u** and the same acceleration **a**. If the circuit is made up of different conductors in series, then N_0, **u** and **a** may vary subject to the condition that the current $I = qN_0u$ is the same in all parts of the circuit. The acceleration **a** also varies in regions where the circuit is curved due to the contribution of the centripetal acceleration of the moving charges.

For a constant current $I = qN_0u$, we have

$$\frac{\mathrm{d}}{\mathrm{d}R}\left(qN_0u\left\{\frac{\mathbf{u}\times\mathbf{R}}{R^2}\right\}\right)\mathrm{d}R = qN_0u\frac{\mathrm{d}}{\mathrm{d}R}\left(\frac{\mathbf{u}\times\mathbf{R}}{R^2}\right)\mathrm{d}R$$

$$=\frac{qN_0u}{R^2}\left(\mathbf{u}\times\frac{\mathrm{d}\mathbf{R}}{\mathrm{d}R}+\frac{\mathrm{d}\mathbf{u}}{\mathrm{d}R}\times\mathbf{R}\right)\mathrm{d}R$$

$$-\frac{2qN_0u(\mathbf{u}\times\mathbf{R})}{R^3}\,\mathrm{d}R. \tag{A4.2}$$

It follows from Figure 6.2 that

$$\mathrm{d}R = -\cos\theta\ \mathrm{d}l \tag{A4.3}$$

$$\mathrm{d}\mathbf{R} = -\mathrm{d}\mathbf{l}. \tag{A4.4}$$

If a charge q moves a distance **dl** in a time dt, then using equation (A4.4) we have

$$\mathbf{u} = \frac{\mathrm{d}\mathbf{l}}{\mathrm{d}t} = -\frac{\mathrm{d}\mathbf{R}}{\mathrm{d}t}. \tag{A4.5}$$

Using equation (A4.3),

$$u = \frac{\mathrm{d}l}{\mathrm{d}t} = -\frac{1}{\cos\theta}\left(\frac{\mathrm{d}R}{\mathrm{d}t}\right).$$

Rearranging,

$$\frac{\mathrm{d}R}{\mathrm{d}t} = -u\cos\theta. \tag{A4.6}$$

Using equation (A4.6) we have

$$\frac{\mathrm{d}\mathbf{R}}{\mathrm{d}R} = \frac{\mathrm{d}\mathbf{R}}{\mathrm{d}t}\frac{\mathrm{d}t}{\mathrm{d}R} = -\frac{\mathbf{u}}{u\cos\theta}. \tag{A4.7}$$

Since $\mathbf{u}\times\mathbf{u}$ is zero, it follows from equation (A4.7) that

$$\mathbf{u}\times\frac{\mathrm{d}\mathbf{R}}{\mathrm{d}R} = 0. \tag{A4.8}$$

Using equation (A4.6), we find that

$$\frac{\mathrm{d}\mathbf{u}}{\mathrm{d}R} = \frac{\mathrm{d}\mathbf{u}}{\mathrm{d}t}\frac{\mathrm{d}t}{\mathrm{d}R} = -\frac{\mathbf{a}}{u\cos\theta}. \tag{A4.9}$$

Putting $\mathbf{u} \times \mathrm{d}\mathbf{R}/\mathrm{d}R$ equal to zero in equation (A4.2) and substituting for $\mathrm{d}\mathbf{u}/\mathrm{d}R$ and $\mathrm{d}R$ from equations (A4.9) and (A4.3) respectively we find, that with $I = qN_0u$

$$\frac{\mathrm{d}}{\mathrm{d}R}\left(\frac{I(\mathbf{u} \times \mathbf{R})}{R^2}\right)\mathrm{d}R = qN_0\left(\frac{\mathbf{a} \times \mathbf{R}}{R^2}\right)\mathrm{d}l + \frac{2I(\mathbf{u} \times \mathbf{R})\cos\theta\,\mathrm{d}l}{R^3}.$$

Integrating around the complete circuit in Figure 6.2, we find that, since the integral of the left hand side is zero,

$$0 = \oint \frac{qN_0(\mathbf{a} \times \mathbf{R})\,\mathrm{d}l}{R^2} + 2I\oint \frac{(\mathbf{u} \times \mathbf{R})\cos\theta\,\mathrm{d}l}{R^3}. \qquad \text{(A4.10)}$$

Substituting in equation (A4.1), we find that

$$\Delta\mathbf{B} = 0. \qquad \text{(A4.11)}$$

APPENDIX B

Conduction current flow in stationary conductors

B1. Example of the mode of action of a source of emf

In electrostatics, the electric field inside a stationary isolated conductor is always zero, whereas, according to equation (1.42), when a conduction current flows in a stationary conductor, that forms part of a complete electrical circuit, there is an electric field inside the conductor. To illustrate how a source of emf can maintain a steady electric field inside a conductor, that forms part of a complete electrical circuit, consider the idealized Van de Graaff generator shown in Figure B1. When the Van de Graaff is operating on open

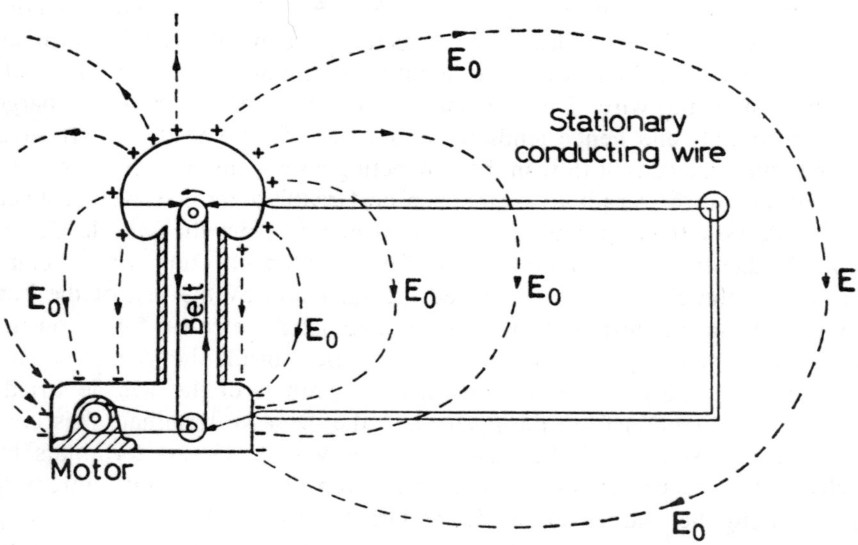

Figure B1. The Van de Graaff generator gives rise to an external electric field $\mathbf{E}_0$, which in turn gives rise to a conduction current in the stationary conducting wire. When the conduction current is steady, surface charge distributions help to guide the current flow in a direction parallel to the wire. When the current is steady, the resultant electric field inside the wire is parallel to the wire.

circuit, there are electric charge distributions of opposite signs on the positive and negative terminals, which give rise to an electric field, which will be denoted by $\mathbf{E}_0$. This electric field extends into the space outside the Van de Graaff generator, as shown by the dotted lines labelled $\mathbf{E}_0$ in Figure B1. For purposes of discussion, we shall assume that a long conducting wire is brought up, almost instantaneously, to join the terminals of the Van de Graaff. In practice, there will be transient electric currents in the wire as it is brought up to join the terminals of the Van de Graaff. We shall assume that eventually a state of dynamic equilibrium is reached when the Van de Graaff maintains a steady conduction current I in all parts of the connecting wire. Initially, the only electric field is $\mathbf{E}_0$ which is due to the charges on the terminals of the Van de Graaff. The electric field $\mathbf{E}_0$ will act on the conduction electrons inside the connecting wire, and initially there will be a conduction current flow given by $\mathbf{J} = \sigma\mathbf{E}_0$ in a direction parallel to the direction of the electric field $\mathbf{E}_0$ due to the charges on the Van de Graaff terminals, which in general is not parallel to the connecting wire. This initial direction of current flow will lead to the build up of charge distributions on the surface of the connecting wire which, when the current I in the connecting wire is constant, give a contribution $\mathbf{E}_s$ to the total electric field $\mathbf{E} = (\mathbf{E}_0 + \mathbf{E}_s)$ which, firstly prevents current flow in a direction perpendicular to the connecting wire and secondly gives the appropriate value of total electric field $\mathbf{E}$ to give the same value of conduction current in all parts of the external circuit, whatever the topology of the connecting wires, even if they are tied into knots.

Under the influence of the total electric field $\mathbf{E} = (\mathbf{E}_0 + \mathbf{E}_s)$ inside the connecting wire, electrons flow from the connecting wire into the positive terminal of the Van de Graaff and from the negative terminal of the Van de Graaff into the connecting wire. This current flow tends to reduce the total charge on the terminals and hence tends to reduce $\mathbf{E}_0$. If the Van de Graaff were not operating, the current flow in the connecting wire would reduce the charge distributions on the terminals of the Van de Graaff to zero. However, when the Van de Graaff is operating, the loss of charge from the Van de Graaff terminals, due to the conduction current flow in the connecting wire, is compensated by the electric charge carried mechanically by the belt of the Van de Graaff from one terminal to the other against the electric forces on the charges on the moving belt due to the electric field inside the Van de Graaff, which is due mainly to the charges on the terminals of the Van de Graaff and is in the direction from the positive to the negative terminal. This continuous replenishment of the charge on the Van de Graaff terminals by mechanical means prevents the conduction current flow in the connecting wire from reducing the charges on the Van de Graaff terminals to zero and making the connecting wire an equipotential with zero electric field inside the connecting wire. After the transient state is over, a state of dynamic equilibrium is reached when the charge carried by the belt of the Van de Graaff per second from one terminal to the other is equal to the conduction current flowing in the connecting wire. When the current flow is steady, the surface charge distributions giving rise to $\mathbf{E}_s$ do not change with time and they can be treated as electrostatic charge distributions, in which case $\oint \mathbf{E}_s \cdot d\mathbf{l}$ is zero.

The existence of electric fields due to the surface charge distributions associated with steady conduction current flow in a conductor was demonstrated by the photographs due to Jefimenko [1]. A typical example is shown in Figure B2. Jefimenko used two dimensional printed circuits to demonstrate the electric fields associated with current carrying conductors of different geometrical configurations connected to a Van de Graaff. The example shown in Figure B2 is similar to the experimental situation shown in Figure B1. The photograph reproduced in Figure B2 confirms that, when the current is steady, the electric field inside the conductor is parallel to the conductor. The photograph also shows that there are electric fields in the space outside the conductors. Since the different parts of the circuit in Figure B2 are at different electrostatic potentials, it is only to be expected that there are electric fields associated with these potential differences, and that these electric fields extend into the space outside the conductors in Figure B2.

According to the standard boundary conditions, derived from Maxwell's equations, the tangential component of $\mathbf{E}$ is continuous across the surfaces of the conductors in Figure B2. There is a discontinuity in the normal com-

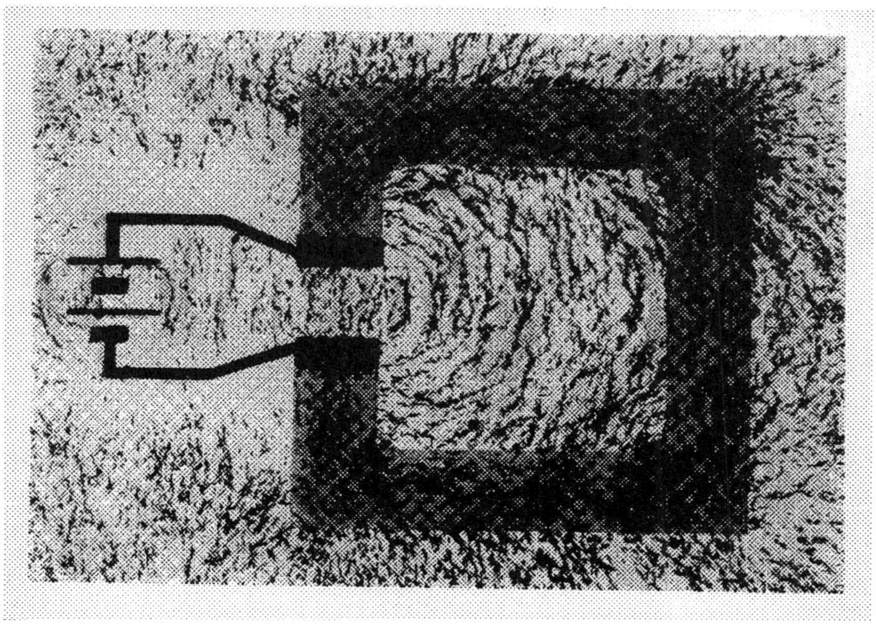

Figure B2. An example of the electric fields associated with steady electric currents due to Jefimenko [1]. A Van de Graaff was used as the source of e.m.f., and grass seeds used to show the electric field. Notice that 'inside' the conductor the electric field is parallel to the conductor. The electric field extends into the space outside the conductor. (Reproduced by permission of the American Journal of Physics.)

ponent of **D** at the surface of the conductor, which is equal to σ the charge per unit area on the surface of the conductors. Since there is no conduction current flow in the direction of the normal to the conductors after the transient state is over, the normal component of **E** is zero just inside the conductor. Due to the surface charge distributions there is a normal component of **E** equal to σ/ε_0 outside the conductor. This shows that the electric field just outside the surface of the conductor should have both a normal and a tangential component in agreement with Figure B2.

The electric fields in the space outside the conductors are generally neglected in circuit theory, since the current flow associated with these electric fields is negligible due to the extremely low value of the electrical conductivity of air, which is generally only 10^{-20} times that of a typical metallic conductor. If the circuit in Figure B2 were placed inside a conducting fluid such as mercury, there would of course be a significant current flow in the space outside the original metallic conductors.

In this section, the role of electric fields in conduction current flow was stressed, whereas in elementary courses on circuit theory it is the concept of potential differences that is invariably used. In terms of potential differences, one would say that the Van de Graaff generator in Figure B1 raises charges mechanically from the potential of one terminal to the potential of the other terminal, thus maintaining a potential difference across the terminals of the Van de Graaff. In elementary circuit theory, it is said that this external potential difference drives electric current around the external circuit and that the potential differences ϕ across the various parts of the external circuit are consistent with Ohm's law in the form $\phi = IR$. We have probed deeper to see how these potential differences in the distant parts of the circuit arise. They arise from the surface charge distributions that give the appropriate value for the total electric field inside the conductors that acts on the conduction electrons to give the same steady current in all parts of the circuit.

B2. Location of the charge distributions associated with conduction current flow

According to Maxwell's equations, for a field point inside a material medium

$$\nabla \cdot \mathbf{D} = \nabla \cdot (\varepsilon_r \varepsilon_0 \mathbf{E}) = \rho.$$

Using equation (A1.20) of Appendix A1.6 to expand $\nabla \cdot (\varepsilon_r \varepsilon_0 \mathbf{E})$ we obtain

$$\rho = \varepsilon_0 \mathbf{E} \cdot \nabla \varepsilon_r + \varepsilon_r \varepsilon_0 \nabla \cdot \mathbf{E}. \tag{B1}$$

According to equation (1.50), when the conduction current in the circuit is steady, we have

$$\nabla \cdot \mathbf{J} = 0. \tag{B2}$$

Since $\mathbf{J} = \sigma\mathbf{E}$, equation (B2) gives

$$\nabla \cdot \sigma\mathbf{E} = 0. \tag{B3}$$

Using Equation (A1.20) of Appendix A1.6 to expand $\nabla \cdot (\sigma\mathbf{E})$, equation (B3) becomes

$$\nabla \cdot (\sigma\mathbf{E}) = \sigma\nabla \cdot \mathbf{E} + \mathbf{E} \cdot \nabla\sigma = 0.$$

Hence,

$$\nabla \cdot \mathbf{E} = -\frac{\mathbf{E} \cdot \nabla\sigma}{\sigma}.$$

Substituting for $\nabla \cdot \mathbf{E}$ in equation (B1), we obtain

$$\rho = \varepsilon_0\mathbf{E} \cdot \nabla\varepsilon_r - \left(\frac{\varepsilon_r\varepsilon_0}{\sigma}\right)\mathbf{E} \cdot \nabla\sigma.$$

Substituting for $\mathbf{E}$ using $\mathbf{E} = \mathbf{J}/\sigma$ we have

$$\rho = \varepsilon_0\frac{\mathbf{J}}{\sigma} \cdot \nabla\varepsilon_r - \frac{\varepsilon_r\varepsilon_0}{\sigma^2}\mathbf{J} \cdot \nabla\sigma = \varepsilon_0\mathbf{J} \cdot \nabla\left(\frac{\varepsilon_r}{\sigma}\right). \tag{B4}$$

According to equation (B4), when the conduction current in the circuit is steady, we can have charge distributions wherever there are spatial variations of either the relative permittivity ε_r or the electrical conductivity σ. In electrical circuits this is generally where there are spatial variations of σ which is generally at the surfaces of the conductors and at the boundaries where conductors of different electrical conductivities are joined. Since both ε_r and σ are constant inside a homogenous conductor, according to equation (B4) there are no volume charge distributions inside a stationary homogenous conductor.

Problem: Assume that initially, at the time $t = 0$, there is a charge density ρ_0 inside a homogenous stationary conductor of conductivity σ and relative permittivity ε_r. Show that the charge density decreases with time according to the equation

$$\rho = \rho_0 \exp\left[-\frac{\sigma t}{\varepsilon_r\varepsilon_0}\right] \tag{B5}$$

Hint: Take the divergence of $\mathbf{J} = \sigma\mathbf{E}$, then substitute in the continuity equation (1.49) and use $\nabla \cdot \mathbf{E} = \rho/\varepsilon_r\varepsilon_0$.

Comment: If we assume that for copper $\varepsilon_r = 1$ and $\sigma = 5.7 \times 10^7$ S m^{-1} we find that the relaxation time $\varepsilon_r\varepsilon_0/\sigma$ for copper is $\sim 10^{-19}$s. This suggests that any volume charge distributions in copper should disappear very quickly.

B3. Magnitudes of the surface and boundary charge distributions associated with conduction current flow

As an example, consider the junction of a copper wire of conductivity $\sigma_1 = 5.7 \times 10^7$ S m^{-1} and a nickel wire of conductivity $\sigma_2 = 1.28 \times 10^7$ S m^{-1}, as shown in Figure B3. The wires have the same cross sectional area A and carry the same current I. Let E_1 and E_2 be the electric fields inside the copper and the nickel wires respectively. Since $I = J_1A = J_2A$, where $J_1 = \sigma_1E_1$ and $J_2 = \sigma_2E_2$, we have

$$I = \sigma_1E_1A = \sigma_2E_2A. \tag{B6}$$

Since the conductivity $\sigma_1 = 5.7 \times 10^7$ S m^{-1} of copper is bigger than the conductivity $\sigma_2 = 1.28 \times 10^7$ S m^{-1} of nickel, it follows from equation (B6), that when a steady current flows in Figure B3, the electric field E_2 inside the nickel wire must be bigger than the electric field E_1 inside the copper wire, as sketched in Figure B3. To give the larger electric field inside the nickel wire, there must be a net positive charge distribution on the boundary between the two wires. From the standard boundary conditions, derived from Maxwell's equations, the charge per unit area on the boundary is equal to $(D_2 - D_1)$. Hence the total charge Q on the boundary is

$$Q = A(D_2 - D_1) = A(\varepsilon_2\varepsilon_0E_2 - \varepsilon_1\varepsilon_0E_1) \tag{B7}$$

where ε_1 and ε_2 are the relative permittivities of copper and nickel respectively. Since from equation (B6), $E_1A = I/\sigma_1$ and $E_2A = I/\sigma_2$, equation (B7) becomes

$$Q = \varepsilon_0 I\left(\frac{\varepsilon_2}{\sigma_2} - \frac{\varepsilon_1}{\sigma_1}\right).$$

For a rough order of magnitude calculation, assume that $\varepsilon_2 = \varepsilon_1 = 1$. With $\sigma_1 = 5.7 \times 10^7$ S m^{-1} and $\sigma_2 = 1.28 \times 10^7$ S m^{-1}, we obtain

$$Q = 8.85 \times 10^{-12} I\left[\frac{1}{1.28 \times 10^7} - \frac{1}{5.7 \times 10^7}\right] = 5.36 \times 10^{-19} I. \tag{B8}$$

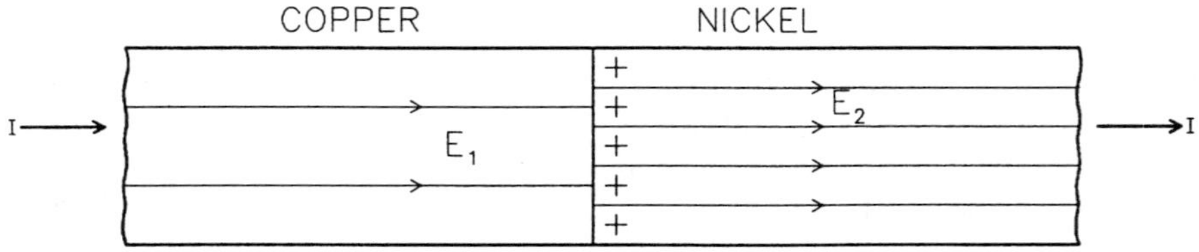

Figure B3. The junction of copper and nickel conductors. Since $\mathbf{J} = \sigma\mathbf{E}$, if $\sigma_1 > \sigma_2$ then $E_1 < E_2$ as shown. The change in the electric field arises from a charge distribution on the boundary between the condutors.

For a current of one ampere, Q is equal to 5.36×10^{-19} C, which is equal to the positive charge left if we remove an average of 3.3 electrons from the boundary. For I = 1mA, we need only remove an average of 0.0033 electrons. We can see that the average total charge, needed on the boundary between the copper and the nickel conductors in Figure B3 to give the necessary changes in the electric field inside both conductors in Figure B3, to give the same total current in both conductors, is exceedingly small.

When a wire, that is carrying a steady conduction current, is bent, the electric field lines inside the wire must also turn around the corner so that the steady conduction current continues to flow in the wire. Consider the metal wire of cross sectional area A, which has a bend at the point M in Figure B4. The steady current in the wire is I. Inside the conductor, away from the bend at M, the electric field **E** inside the wire is parallel to the wire, as shown in Figure B4. The total flux of **E** in the section LM of the wire is equal to EA. Away from the bend at M, the flux of **E** in the section MN of the wire is also equal to EA. To change the direction of **E** inside the wire at the bend at M in Figure B4, we must have extra charge distributions near the bend. To obtain a rough estimate of the extra charges needed to change the direction of **E** inside the

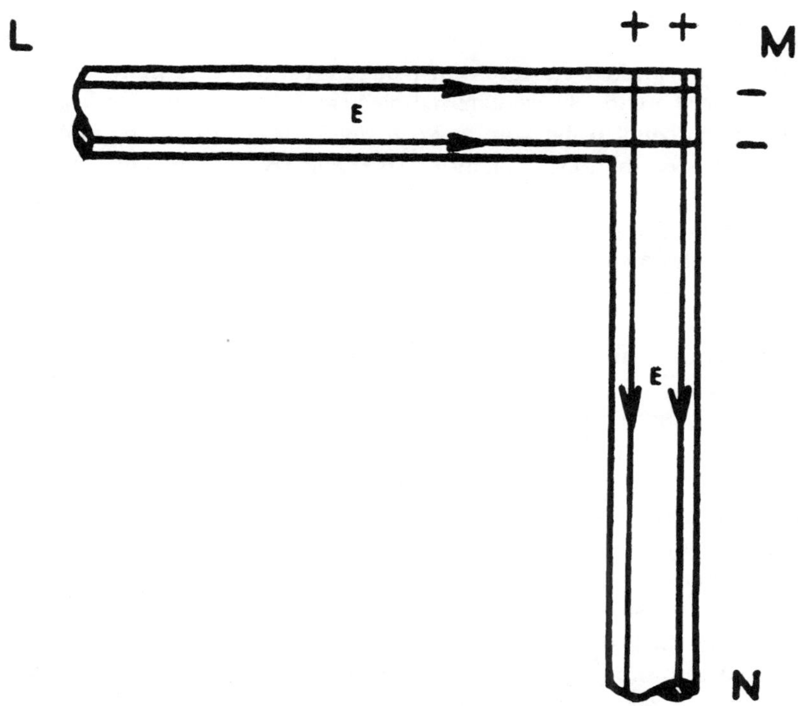

Figure B4. The electric field giving rise to a conduction current flow in the wire LMN, which corresponds to the circled section on the top right hand side corner of the conductor in Figure B1.

wire, we shall assume that the electric flux EA in the section LM of the wire terminates on a negative charge distribution near the corner at M and that the electric flux EA in the section MN of the wire starts from a positive charge distribution near the corner M, as shown in Figure B4. (The surface charge distributions away from the corner M are not shown.) Since from Gauss' flux law of electrostatics, the total electric flux from a charge of magnitude q is q/ε_0, the magnitude Q_- of the charge needed to terminate the flux EA inside the section LM of the wire is $-\varepsilon_0 EA$. Since $E = J/\sigma$ where $J = I/A$, we have

$$Q_- = -\varepsilon_0 I/\sigma. \tag{B9}$$

Similarly, the positive charge that would give an electric flux EA inside the section MN of the wire is

$$Q_+ = \varepsilon_0 I/\sigma. \tag{B10}$$

For a copper wire, $\sigma = 5.7 \times 10^7$ S m^{-1}. Since $\varepsilon_0 = 8.85 \times 10^{-12}$ F m^{-1},

$$\mathrm{Q} = \pm 1.5 \times 10^{-19} I. \tag{B11}$$

This result shows that we only need a separation of positive and negative charge of about $1.5 \times 10^{-19} I$ to change the direction of the electric field inside the wire at the bend at M in Figure B4. For $I = 1$A, this would be approximately equal to the charge on one electron. For $I < 1$A, the average charge needed would be even less. It can be seen from Figure B2 that there are electric fields outside the wire. These external electric fields are also modified in the vicinity of the bend in the wire at M in Figure B4, and this requires appropriate charge distributions. Equations (B9) and (B10) only give the charges needed to modify the electric field that is inside the conductor in Figure B4. A quantitative example of the determination of the external electric field associated with conduction current flow is given by Marcus [2].

B4. Models of conduction current flow in a stationary conductor

One type of electric current flow that is sometimes valid, for example for the beam current in a cathode ray oscilloscope, is that of individual electrons going all the way from the cathode to the anode in a vacuum. If the potential difference between the anode and the cathode is 10 kV, the electrons reach the anode with kinetic energies of 10 keV and velocities of about $0.2c$. This model is clearly not applicable to conduction current flow in a metal since, due to collisions, the mean drift velocities of the conduction electrons are only of the order of 10^{-4} m s^{-1}. Another type of current flow is a convection current due to a moving charge distribution. If the resultant charge density is ρ and the charge distribution moves with velocity $\mathbf{u}$, then the convection current density $\mathbf{J}$ is equal to $\rho\mathbf{u}$. This model does not apply to conduction current flow.

An analogy for conduction current flow, sometimes uses in elementary courses is that of the flow of a gas (or liquid) in a pipe. When gas enters one end of a pipe full of the gas, the new gas pushes some gas out from the other end of the pipe. The mechanical forces are transmitted through the gas with the speed of sound. In an ideal gas, the velocity of sound is equal to $(\gamma p/\rho_m)^{1/2}$ where p is the pressure of the gas, ρ_m is the mass density and γ is the ratio of the heat capacity at constant pressure to the heat capacity at constant volume. According to the kinetic theory of gases, $p = \frac{1}{3}\rho_m v^2$, where v is the root mean square speed of the gas molecules. Hence the speed of sound in an ideal gas is $(\gamma/3)^{1/2}v$. For a monatomic gas $\gamma = 5/3$, in which case the speed of sound is $0.75v$, illustrating that the speed of sound in a gas does not exceed the root mean square speed of the molecules of the gas. In a metallic conductor, the speeds of the conduction electrons are of the order of $c/200$, corresponding to kinetic energies of about 7 eV. If the fluid flow model were applicable to conduction current flow, one would not expect information about changes in the current in one part of a circuit to be transmitted along the connecting wires with a speed exceeding the velocities of the conduction electrons, which are typically about $c/200$. In practice, electrical signals can travel along metallic conductors with speeds close to the speed of light. This shows that the fluid flow analogy is not applicable to conduction current flow.

In the fluid flow model for the propagation of sound in a gas, it is assumed that the information is transferred by the collisions of the gas molecules. It is assumed that gas molecules only interact when they collide at separations of the order of about 10^{-10} m. This is a reasonable approximation for an ideal gas since the van der Waals forces between neutral molecules vary as approximately $1/r^7$, where r is the separation of the molecules. These forces are only significant over short distances of the order of atomic dimensions. On the other hand, the forces between electric charges are given by Coulomb's law and vary as $1/r^2$. Electrons can interact over distances much greater than 10^{-10} m and can affect more than their immediate neighbours. We have found it convenient to say that a charge gives rise to an electric field that can act on all other charges. If an electric charge is moved, its electric field is changed. According to the retarded potentials, which are developed in Section 2.3 of Chapter 2, this information is propagated in empty space, from the moving charge with the speed of light c. Thus, in principle, it is possible for information about changes in the positions and velocities of charges to be transmitted from one end of a wire to the other, via changes in the electromagnetic field, with a speed c. In practice, there is a stringent limiting factor on the speed of propagation of electrical signals, namely the speed at which the appropriate changes in the surface and boundary charge distributions, that give the changes in conduction current flow, can take place.

As an example, consider a square wave signal generator that produces square wave voltage pulses. The terminals of the signal generator are connected by long wires several kilometres long to a resistor of high resistance R, as shown in Figure B5. We shall assume that the period of the square wave voltage pulses is much longer than the time it takes for the conduction current to reach the

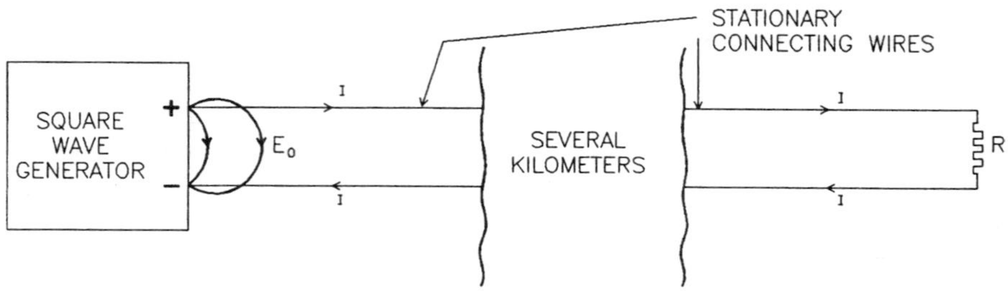

Figure B5. A square wave generator gives rise to a current flow in parallel conductors several kilometres long which are joined by the resistor *R*.

same value in all parts of the circuit. We shall assume, for purposes of discussion, that when the output from the square wave generator goes positive, the potential difference across the generator terminals builds up almost instantaneously. One the scale of several kilometres, the initial electric field $\mathbf{E}_0$ due to the potential difference across the terminals of the signal generator is approximately a dipole field and goes down as $1/r^3$, where r is the distance from the generator. After the output from the square wave generator goes positive, the changes in the electric field $\mathbf{E}_0$ are propagated away from the generator with a speed c in empty space, and, after a time delay of r/c, the changes in $\mathbf{E}_0$ can affect the conduction electrons in the connecting wires at a distance r from the signal generator. However, since $\mathbf{E}_0$ decreases as $1/r^3$ the effect of changes in $\mathbf{E}_0$ at distances of several kilometres from the signal generator, immediately after the time delay r/c, are generally negligible. The main influence of changes in the electric field $\mathbf{E}_0$, due to the potential difference across the signal generator is close to the signal generator, where changes in $\mathbf{E}_0$ give changes in the surface and boundary charge distributions on the connecting wires in the vicinity of the signal generator. The changes in the electric fields due to these changes in the charge distributions in the vicinity of the signal generator give changes in the surface and boundary charges further along the wires. In this way the appropriate surface and boundary charge distributions are built up progressively along the wires at a speed depending on how quickly the necessary charge distributions can be built up. This depends on the resistance, inductance and capacitance per metre length of the wires. For wires of finite resistance this speed is less than c, the speed of light, but is generally much greater than the speeds of individual conduction electrons. The above discussion of the build up of a steady DC current in the circuit shown in Figure B5 is similar in many respects to a transmission line problem, such as the low frequency limit of the Lecher two wire transmission line.

The neglect of transient effects, that start at a distance r from the square wave signal generator in Figure B5, at a time r/c after the output of the signal generator goes positive, is similar to the assumption made in transmission line theory that we can represent an ideal transmission line by a series of inductors and capacitors with no mutual inductance effects between the spatially separated inductors.

B5. Energy propagation in DC circuits

In general we cannot say where the energy of a system is localised. We can obtain correct answers in classical electromagnetism by attributing energy to the electromagnetic field. The Poynting vector $\mathbf{N} = \mathbf{E} \times \mathbf{H}$ is often used to calculate the rate of flow of electromagnetic energy per square metre in the electromagnetic field. At first sight the use of the Poynting vector might appear inappropriate for the steady conditions of DC current flow. It was, however shown in Section B4 that it was the resultant electric field inside the conductor due to both the source of emf and the surface and boundary charge distributions, and not the collisions of the conduction electrons, that accelerates the conduction electrons leading to Joule heating in the conductors. This suggests that the use of field concepts is appropriate. The role of the Poynting vector in DC circuits is discussed in detail in Sections 8.5 and 8.6 of Chapter 8.

References

1. Jefimenko, O., *Amer. Journ. Phys.*, Vol. 30, p. 19 (1962).
2. Marcus, A., *Amer. Journ. Phys.*, Vol. 9, p. 225 (1941).

APPENDIX C

The electric and magnetic fields due to an accelerating classical point charge

C1. Introduction

Consider the accelerating classical point charge of magnitude q, shown in Figure 3.2 of Chapter 3. The fields **E** and **B** will be determined at the field point P at the point x, y, z in Figure 3.2 at the time of observation t, using the Liénard-Wiechert potentials. The information collecting sphere passes the accelerating charge q at the retarded time $t^* = (t - [r]/c)$, where

$$[r] = c(t - t^*) \tag{C1}$$

is the distance from the retarded position of the charge at x^*, y^*, z^* to the field point P at x, y, z in Figure 3.2. We have

$$[r] = [(x - x^*)^2 + (y - y^*)^2 + (z - z^*)^2]^{1/2}. \tag{C2}$$

The fields **E** and **B** are given by

$$\mathbf{E} = -\nabla\phi - \frac{\partial \mathbf{A}}{\partial t} \tag{C3}$$

$$\mathbf{B} = \nabla \times \mathbf{A}. \tag{C4}$$

The differential operators in the operator ∇ are with respect to changes in x, y, z, the coordinates of the field point at the fixed time of observation t. The partial derivative $(\partial/\partial t)$ is with respect to the time of observation t at the fixed field point at x, y, z. However the velocity [**u**] and the acceleration [**a**] of the charge are given as functions of the retarded time t^* at the retarded position of the charge at x^*, y^*, z^*. As some of the formulae in this Appendix are long and complicated, from now on we shall not always put quantities such as $[r]$, [**u**] and [**a**], which are obviously determined at the retarded time t^*, inside square brackets.

Differentiating equation (C2) with respect to t^*, keeping x, y and z fixed, we have

$$\left(\frac{\partial r}{\partial t^*}\right)_{x,y,z} = -\frac{1}{r}\left\{(x - x^*)\left(\frac{\partial x^*}{\partial t^*}\right) + (y - y^*)\left(\frac{\partial y^*}{\partial t^*}\right) + (z - z^*)\left(\frac{\partial z^*}{\partial t^*}\right)\right\}.$$

Since $\partial x^*/\partial t^* = [u_x]$ etc, using equation (A1.1) of Appendix A1.1, we have

$$\left(\frac{\partial r}{\partial t^*}\right)_{x,y,z} = -\frac{\mathbf{r}\cdot\mathbf{u}}{r}. \tag{C5}$$

This relation can be seen geometrically. The rate of decrease of r with t^*, for fixed x, y, z, is equal to the component of the velocity of the charge at its retarded position in the direction of the field point.

For any function ϕ of x, y, z and t^*

$$\mathrm{d}\phi = \left(\frac{\partial\phi}{\partial x}\right)_{y,z,t^*}\mathrm{d}x + \left(\frac{\partial\phi}{\partial y}\right)_{x,z,t^*}\mathrm{d}y + \left(\frac{\partial\phi}{\partial z}\right)_{x,y,t^*}\mathrm{d}z + \left(\frac{\partial\phi}{\partial t^*}\right)_{x,y,z}\mathrm{d}t^*. \tag{C6}$$

If x, y and z are fixed, then $\mathrm{d}x = \mathrm{d}y = \mathrm{d}z = 0$. Dividing by $\mathrm{d}t$ we have for fixed x, y and z

$$\left(\frac{\partial\phi}{\partial t}\right)_{x,y,z} = \left(\frac{\partial\phi}{\partial t^*}\right)_{x,y,z}\left(\frac{\partial t^*}{\partial t}\right)_{x,y,z}. \tag{C7}$$

Putting $\phi = r$ in equation (C7) and using equation (C5), we find that

$$\left(\frac{\partial r}{\partial t}\right)_{x,y,z} = \left(\frac{\partial r}{\partial t^*}\right)_{x,y,z}\left(\frac{\partial t^*}{\partial t}\right)_{x,y,z} = -\frac{\mathbf{r}\cdot\mathbf{u}}{r}\left(\frac{\partial t^*}{\partial t}\right)_{x,y,z}. \tag{C8}$$

Differentiating equation (C1) with respect to t, keeping x, y and z fixed we have

$$\left(\frac{\partial r}{\partial t}\right)_{x,y,z} = c\left\{1 - \left(\frac{\partial t^*}{\partial t}\right)_{x,y,z}\right\}. \tag{C9}$$

Equating the right hand sides of equations (C8) and (C9) and rearranging, for fixed x, y, and z we have

$$\left(\frac{\partial t^*}{\partial t}\right)_{x,y,z} = \frac{1}{[1 - \mathbf{r}\cdot\mathbf{u}/rc]} = \frac{r}{s} \tag{C10}$$

where,

$$s = \left[r - \frac{\mathbf{r}\cdot\mathbf{u}}{c}\right]. \tag{C11}$$

Equation (C7) can now be rewritten in the form

$$\left(\frac{\partial\phi}{\partial t}\right)_{x,y,z} = \frac{r}{s}\left(\frac{\partial\phi}{\partial t^*}\right)_{x,y,z} = \frac{r}{[r - \mathbf{r}\cdot\mathbf{u}/c]}\left(\frac{\partial\phi}{\partial t^*}\right)_{x,y,z}. \tag{C12}$$

Putting $\mathrm{d}y = \mathrm{d}z = 0$ in equation (C6) at a fixed t and dividing by $\mathrm{d}x$, we obtain

$$\left(\frac{\partial\phi}{\partial x}\right)_{y,z,t} = \left(\frac{\partial\phi}{\partial x}\right)_{y,z,t^*} + \left(\frac{\partial\phi}{\partial t^*}\right)_{x,y,z}\left(\frac{\partial t^*}{\partial x}\right)_{y,z,t}. \tag{C13}$$

The gradient of ϕ due to changes in the coordinates x, y, z of the field point at the fixed time of observation t is

$$\nabla\phi = \hat{\mathbf{i}}\left(\frac{\partial\phi}{\partial x}\right)_{y,z,t} + \hat{\mathbf{j}}\left(\frac{\partial\phi}{\partial y}\right)_{x,z,t} + \hat{\mathbf{k}}\left(\frac{\partial\phi}{\partial z}\right)_{x,y,t}. \tag{C14}$$

A new operator ∇^* will be defined by the relation

$$\nabla^*\phi = \hat{\mathbf{i}}\left(\frac{\partial\phi}{\partial x}\right)_{y,z,t^*} + \hat{\mathbf{j}}\left(\frac{\partial\phi}{\partial y}\right)_{x,z,t^*} + \hat{\mathbf{k}}\left(\frac{\partial\phi}{\partial z}\right)_{x,y,t^*}. \tag{C15}$$

The partial differentiations in equation (C15) are with respect to the coordinates x, y and z of the field point at the fixed retarded time t^*. From equation (C13) and similar expressions for $(\partial\phi/\partial y)_{x,z,t}$ and $(\partial\phi/\partial z)_{x,y,t}$, it follows that

$$\nabla\phi = \nabla^*\phi + \left(\frac{\partial\phi}{\partial t^*}\right)_{x,y,z} \nabla t^*. \tag{C16}$$

We now have to determine ∇t^*. It follows from equations (C1) and (C14) that, at fixed t,

$$\nabla r = -c\nabla t^*. \tag{C17}$$

Putting $\phi = r$ in equation (C16) and using equation (C5), we have

$$\nabla r = \nabla^* r - \left(\frac{\mathbf{u}\cdot\mathbf{r}}{r}\right)\nabla t^*. \tag{C18}$$

Differentiating equation (C2) partially with respect to x keeping y, z and t^* fixed, and remembering that if t^* is fixed then x^*, y^* and z^* are fixed we

$$\left(\frac{\partial r}{\partial x}\right)_{y,z,t^*} = \frac{(x - x^*)}{r}.$$

Hence,

$$\nabla^* r = \frac{\mathbf{r}}{r}. \tag{C19}$$

Equation (C18) now becomes

$$\nabla r = \frac{\mathbf{r}}{r} - \left(\frac{\mathbf{u}\cdot\mathbf{r}}{r}\right)\nabla t^*. \tag{C20}$$

Equating the right hand side of equations (C17) and (C20), and rearranging we get

$$\nabla t^* = -\frac{\mathbf{r}}{sc}.$$

Substituting in equation (C16), we finally obtain

$$\nabla = \nabla^* - \frac{\mathbf{r}}{sc}\left(\frac{\partial}{\partial t^*}\right)_{x,y,z} \tag{C21}$$

where ∇ and ∇^* are defined by equations (C14) and (C15) respectively. From equation (C12), for fixed x, y and z we have

$$\left(\frac{\partial}{\partial t}\right)_{x,y,z} = \frac{r}{s}\left(\frac{\partial}{\partial t^*}\right)_{x,y,z}. \tag{C22}$$

C2. Calculation of the electric field

The electric field $\mathbf{E}$ at the field point P in Figure 3.2 is given by

$$\mathbf{E} = -\nabla\phi - \frac{\partial \mathbf{A}}{\partial t} \tag{C3}$$

where according to the Liénard-Wiechert potentials, namely equations (3.4) and (3.6) of Chapter 3

$$\phi = \frac{q}{4\pi\varepsilon_0 s} \tag{C23}$$

$$\mathbf{A} = \frac{q[\mathbf{u}]}{4\pi\varepsilon_0 c^2 s} \tag{C24}$$

where s is given by equation (C11). Substituting for ϕ and $\mathbf{A}$ in equation (C3) we have

$$\frac{4\pi\varepsilon_0}{q}\mathbf{E} = -\nabla\left(\frac{1}{s}\right) - \frac{\partial}{\partial t}\left(\frac{[\mathbf{u}]}{c^2 s}\right)_{x,y,z}. \tag{C25}$$

Using equation (C21), we have

$$-\nabla\left(\frac{1}{s}\right) = \frac{1}{s^2}\nabla s = \frac{1}{s^3}\left[s\nabla^* s - \frac{\mathbf{r}}{c}\left(\frac{\partial s}{\partial t^*}\right)_{x,y,z}\right]. \tag{C26}$$

Applying equation (C22) to $(\mathbf{u}/s)$, since $(\partial\mathbf{u}/\partial t^*) = [\mathbf{a}]$ we have

$$-\frac{1}{c^2}\frac{\partial}{\partial t}\left(\frac{\mathbf{u}}{s}\right) = -\frac{r}{c^2 s}\frac{\partial}{\partial t^*}\left(\frac{[\mathbf{u}]}{s}\right) = -\frac{r}{c^2 s^3}\left[s\mathbf{a} - \mathbf{u}\left(\frac{\partial s}{\partial t^*}\right)_{x,y,z}\right]. \tag{C27}$$

Substituting from equations (C26) and (C27) into equation (C25)

$$\frac{4\pi\varepsilon_0}{q}\mathbf{E} = \frac{1}{s^3}\left[s\nabla^* s - \frac{1}{c}\left\{\mathbf{r} - \frac{r\mathbf{u}}{c}\right\}\left(\frac{\partial s}{\partial t^*}\right)_{x,y,z} - \frac{rs\mathbf{a}}{c^2}\right]. \tag{C28}$$

We now need $\nabla^* s$ and $(\partial s/\partial t^*)_{x,y,z}$. Using equation (C11)

$$\nabla^* s = \nabla^* r - \nabla^* \left(\frac{\mathbf{r}\cdot\mathbf{u}}{c}\right). \tag{C29}$$

If t^* is constant, then x^*, y^*, z^* and $[\mathbf{u}]$ are fixed so that if $\partial[u_x]/\partial x$ etc are evaluated at a fixed retarded time they are zero. Hence

$$\nabla^*(\mathbf{r}\cdot\mathbf{u}) = \hat{\mathbf{i}}\frac{\partial}{\partial x}(xu_x) + \hat{\mathbf{j}}\frac{\partial}{\partial y}(yu_y) + \hat{\mathbf{k}}\frac{\partial}{\partial z}(zu_z)$$

$$= u_x\hat{\mathbf{i}} + u_y\hat{\mathbf{j}} + u_z\hat{\mathbf{k}} = \mathbf{u}.$$

According to equation (C19), $\nabla^* r = \mathbf{r}/r$. Hence equation (C29) becomes

$$\nabla^* s = \frac{\mathbf{r}}{r} - \frac{\mathbf{u}}{c} = \frac{1}{r}\left[\mathbf{r} - \frac{r\mathbf{u}}{c}\right] \tag{C30}$$

giving,

$$s\nabla^* s = \left[1 - \frac{\mathbf{r}\cdot\mathbf{u}}{rc}\right]\left[\mathbf{r} - \frac{r\mathbf{u}}{c}\right]. \tag{C31}$$

Differentiating equation (C11) with respect to t^*, we have

$$\left(\frac{\partial s}{\partial t^*}\right)_{x,y,z} = \left(\frac{\partial r}{\partial t^*}\right)_{x,y,z} - \left(\frac{\mathbf{r}}{c}\right)\cdot\left(\frac{\partial \mathbf{u}}{\partial t^*}\right)_{x,y,z} - \left(\frac{\mathbf{u}}{c}\right)\cdot\left(\frac{\partial \mathbf{r}}{\partial t^*}\right)_{x,y,z}.$$

Since $\mathbf{r}$ is a vector from the retarded position of the charge to the field point, for fixed x, y and z we have

$$\left(\frac{\partial \mathbf{r}}{\partial t^*}\right)_{x,y,z} = -\mathbf{u}; \quad \left(\frac{\partial \mathbf{u}}{\partial t^*}\right)_{x,y,z} = \mathbf{a}.$$

According to equation (C5), $(\partial r/\partial t^*)_{x,y,z} = -\mathbf{r}\cdot\mathbf{u}/r$. Hence

$$\left(\frac{\partial s}{\partial t^*}\right)_{x,y,z} = -\frac{\mathbf{r}\cdot\mathbf{u}}{r} - \frac{\mathbf{r}\cdot\mathbf{a}}{c} + \frac{u^2}{c}. \tag{C32}$$

Substituting from equations (C30), (C31), (C32) and (C11) into equation (C28), we have

$$\frac{4\pi\varepsilon_0}{q}\mathbf{E} = \frac{1}{s^3}\left[\left(1 - \frac{\mathbf{r}\cdot\mathbf{u}}{rc}\right)\left(\mathbf{r} - \frac{r\mathbf{u}}{c}\right)\right.$$

$$\left. - \frac{1}{c}\left(\mathbf{r} - \frac{r\mathbf{u}}{c}\right)\left(-\frac{\mathbf{r}\cdot\mathbf{u}}{r} - \frac{\mathbf{r}\cdot\mathbf{a}}{c} + \frac{u^2}{c}\right) - r\left(r - \frac{\mathbf{r}\cdot\mathbf{u}}{c}\right)\frac{\mathbf{a}}{c^2}\right].$$

The terms independent of [**a**] give

$$\mathbf{E}_V = \frac{q}{4\pi\varepsilon_0 s^3}\left[\mathbf{r} - \frac{r\mathbf{u}}{c}\right]\left[1 - \frac{u^2}{c^2}\right]. \tag{C33}$$

The terms containing [**a**] can be rewritten in the form

$$\mathbf{E}_A = \frac{q}{4\pi\varepsilon_0 s^3}\left[\left(\mathbf{r} - \frac{r\mathbf{u}}{c}\right)\frac{\mathbf{r}\cdot\mathbf{a}}{c^2} - \left(r^2 - \frac{\mathbf{r}\cdot r\mathbf{u}}{c}\right)\frac{\mathbf{a}}{c^2}\right].$$

Using equation (A1.6) of Appendix A1.1 we find that

$$\mathbf{E}_A = \frac{q}{4\pi\varepsilon_0 s^3 c^2}[\mathbf{r}]\times\left\{\left[\mathbf{r} - \frac{r\mathbf{u}}{c}\right]\times[\mathbf{a}]\right\}. \tag{C34}$$

The total electric field is

$$\mathbf{E} = \mathbf{E}_V + \mathbf{E}_A \tag{C35}$$

where $\mathbf{E}_V$ and $\mathbf{E}_A$ are given by equations (C33) and (C35) respectively. The quantities q, [**u**] and [**a**] are the magnitude of the charge, its velocity and its acceleration respectively at the retarded position of the charge in Figure 3.2 of Chapter 3, [**r**] is a vector from the retarded position of the charge to the field point, and s is given by equation (C11).

C3. Calculation of the magnetic field

Using equations (C4), (C21) and (C24), we have

$$\mathbf{B} = \nabla\times\mathbf{A} = \frac{q}{4\pi\varepsilon_0 c^2}\left[\nabla^* - \frac{\mathbf{r}}{sc}\left(\frac{\partial}{\partial t^*}\right)_{x,y,z}\right]\times\left[\frac{\mathbf{u}}{s}\right]$$

$$\frac{4\pi\varepsilon_0 c^2}{q}\mathbf{B} = \nabla^*\times\left[\frac{\mathbf{u}}{s}\right] - \frac{[\mathbf{r}]}{sc}\times\frac{\partial}{\partial t^*}\left[\frac{\mathbf{u}}{s}\right]. \tag{C36}$$

Using equations (A1.23) of Appendix A.6 to expand $\nabla^* \times [\mathbf{u}/s]$, we have

$$\nabla^*\times\left[\frac{\mathbf{u}}{s}\right] = \frac{1}{s}\nabla^*\times\mathbf{u} - \mathbf{u}\times\nabla^*\left(\frac{1}{s}\right). \tag{C37}$$

The velocity of the charge does not vary if the coordinates x, y, z of the field point are varied at fixed t^*, so that $\nabla^* \times \mathbf{u}$ is zero. Using equations (C30) and (C11), we find that

$$\nabla^*\left(\frac{1}{s}\right) = -\frac{1}{s^2}\nabla^* s = -\frac{1}{rs^2}\left[\mathbf{r} - \frac{r\mathbf{u}}{c}\right].$$

Substituting in equation (C37), and remembering that $\mathbf{u} \times \mathbf{u}$ is zero, we have

$$\nabla^* \times \left[\frac{\mathbf{u}}{s}\right] = \frac{\mathbf{u}}{rs^2} \times \left[\mathbf{r} - \frac{r\mathbf{u}}{c}\right] = \frac{\mathbf{u} \times \mathbf{r}}{rs^2}. \tag{C38}$$

Now,

$$\frac{\partial}{\partial t^*}\left(\frac{\mathbf{u}}{s}\right)_{x,y,z} = \mathbf{u}\frac{\partial}{\partial t^*}\left(\frac{1}{s}\right)_{x,y,z} + \frac{1}{s}\left(\frac{\partial \mathbf{u}}{\partial t^*}\right)_{x,y,z} = -\frac{\mathbf{u}}{s^2}\left(\frac{\partial s}{\partial t^*}\right)_{x,y,z} + \frac{\mathbf{a}}{s}.$$

Substituting for $(\partial s/\partial t^*)$ from equation (C32), we obtain

$$\frac{\partial}{\partial t^*}\left(\frac{\mathbf{u}}{s}\right)_{x,y,z} = \frac{\mathbf{u}}{s^2}\left[\frac{\mathbf{r}\cdot\mathbf{u}}{r} + \frac{\mathbf{r}\cdot\mathbf{a}}{c} - \frac{u^2}{c}\right] + \frac{\mathbf{a}}{s}. \tag{C39}$$

Substituting from equations (C38) and (C39) into equation (C36)

$$\frac{4\pi\varepsilon_0 c^2}{q}\mathbf{B} = \frac{\mathbf{u}\times\mathbf{r}}{rs^2} - \frac{\mathbf{r}}{sc}\times\frac{\mathbf{u}}{s^2}\left[\frac{\mathbf{r}\cdot\mathbf{u}}{r} + \frac{\mathbf{r}\cdot\mathbf{a}}{c} - \frac{u^2}{c}\right] - \frac{\mathbf{r}}{sc}\times\frac{\mathbf{a}}{s}. \tag{C40}$$

Picking out the terms independent of $[\mathbf{a}]$, we have

$$\mathbf{B}_V = \frac{q\mathbf{u}\times\mathbf{r}}{4\pi\varepsilon_0 c^2 s^3}\left[\frac{s}{r} + \frac{\mathbf{r}\cdot\mathbf{u}}{rc} - \frac{u^2}{c^2}\right].$$

Since,

$$s = \left[r - \frac{\mathbf{r}\cdot\mathbf{u}}{c}\right] \tag{C11}$$

$$\mathbf{B}_V = \frac{q[\mathbf{u}]\times[\mathbf{r}]}{4\pi\varepsilon_0 c^2 s^3}\left[1 - \frac{u^2}{c^2}\right]. \tag{C41}$$

Picking out the terms containing $[\mathbf{a}]$, we have

$$\mathbf{B}_A = \frac{q}{4\pi\varepsilon_0 c^3 s^3}\,\mathbf{r}\times\left[-\mathbf{u}\left(\frac{\mathbf{r}\cdot\mathbf{a}}{c}\right) - s\mathbf{a}\right]. \tag{C42}$$

Since $\mathbf{r}\times\mathbf{r} = 0$ and $rs = [r^2 - r\,\mathbf{r}\cdot\mathbf{u}/c] = \mathbf{r}\cdot[\mathbf{r} - r\mathbf{u}/c]$, equation (C42) can be rewritten in the form

$$\mathbf{B}_A = \frac{q}{4\pi\varepsilon_0 c^3 s^3}\left[\frac{\mathbf{r}}{r}\right]\times\left[\left(\mathbf{r} - \frac{r\mathbf{u}}{c}\right)(\mathbf{r}\cdot\mathbf{a}) - \mathbf{a}\left\{\mathbf{r}\cdot\left(\mathbf{r} - \frac{r\mathbf{u}}{c}\right)\right\}\right].$$

Applying equation (A1.6) of Appendix A1.1 to the terms inside the large square brackets, we finally obtain

$$\mathbf{B}_A = \frac{q}{4\pi\varepsilon_0 c^3 s^3}\left[\frac{\mathbf{r}}{r}\right]\times\left\{[\mathbf{r}]\times\left(\left[\mathbf{r} - \frac{r\mathbf{u}}{c}\right]\times[\mathbf{a}]\right)\right\}. \tag{C43}$$

The total magnetic field is given by

$$\mathbf{B} = \mathbf{B}_V + \mathbf{B}_A \tag{C44}$$

where $\mathbf{B}_V$ and $\mathbf{B}_A$ are given by equations (C41) and (C43) respectively. The quantities q, $[\mathbf{u}]$ and $[\mathbf{a}]$ are the values of the charge, velocity and acceleration of the charge at its retarded position, $[\mathbf{r}]$ is a vector from the retarded position of the charge to the field point and s is given by equation (C11).

It can be seen from equations (C33), (C34), (C41) and (C42) that

$$\mathbf{B} = \frac{[\mathbf{r}] \times \mathbf{E}}{[rc]} \tag{C45}$$

The resultant magnetic field $\mathbf{B}$ is perpendicular to both the resultant electric field $\mathbf{E}$ and the vector $[\mathbf{r}]$ from the retarded position of the charge to the field point.

APPENDIX D

Discussion of the equation $\nabla \times \mathbf{B} = \mu_0\varepsilon_0\dot{\mathbf{E}} + \mu_0\mathbf{J}$ using the field approach

Some readers may prefer to see the need to add the extra term $\mu_0\mathbf{J}$ to the right hand side of the equation

$$\nabla \times \mathbf{B} = \frac{1}{c^2}\frac{\partial \mathbf{E}}{\partial t}$$

when there is a moving charge distribution at the field point, interpreted in terms of the behaviour of the field vectors **E** and **B** at the field point, rather than using the integral methods given in Sections 4.8.1, 4.8.2 and 4.8.3. A fuller account is given by Rosser [1].

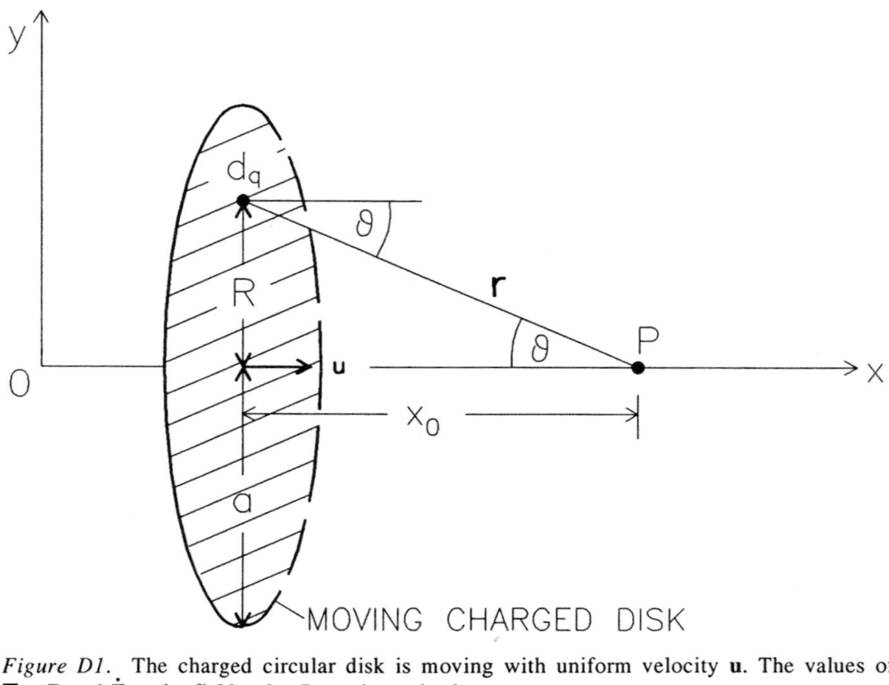

Figure D1. The charged circular disk is moving with uniform velocity **u**. The values of $\nabla \times \mathbf{B}$ and $\dot{\mathbf{E}}$ at the field point P are determined.

Consider a thin, circular disk shaped charge distribution, that is moving with a uniform velocity $u \ll c$ along the x axis of the coordinate system shown in Figure D1. The axis of the moving disk coincides with the x axis. The radius of the disk is a and its thickness is L, where $L \ll a$. The disk has a uniform charge density ρ. We shall sometimes treat the disk as a surface charge distribution of surface charge density $\sigma = \rho L$. When $u \ll c$, the expressions for the **E** and **B** fields due to a classical point charge moving with a uniform velocity $u \ll c$, reduce to equations (4.68) and (4.67) respectively.

Consider first the case before the disk reaches the field point P, which in this example is in empty space. Divide the disk into infinitesimal volume elements. Each of these can be treated as a classical point charge moving with uniform velocity and equation (4.54) can be applied to the electric and magnetic fields due to each infinitesimal volume element. Adding the contributions of all the various volume elements we find that, when the field point P is in empty space

$$\nabla \times \mathbf{B} = \frac{1}{c^2}\frac{\partial \mathbf{E}}{\partial t}. \tag{D1}$$

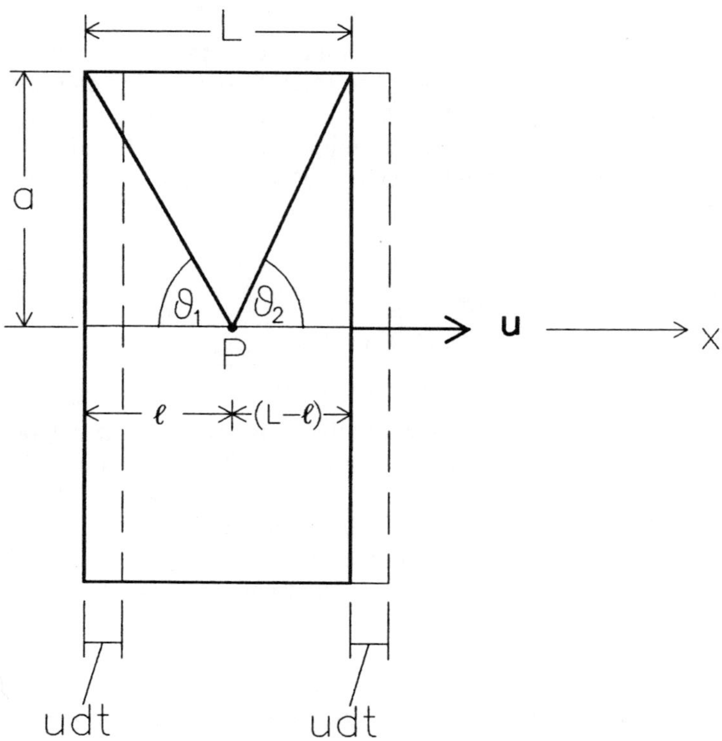

Figure D2. The charged circular disk is passing the field point P with uniform velocity **u**.

Equation (D1) is valid before the charged disk reaches P and after it has completely passed P. Hence we need only consider in detail the case when the charged disk is actually passing the field point P, as shown in Figure D2.

We shall start by determining the value of $\nabla \times \mathbf{B}$ on the x axis in Figure D1. By symmetry, the only component of $\nabla \times \mathbf{B}$ at the field point P, which is on the x axis in Figure D1, is in the $+x$ direction. Let the field point P be at a distance x_0 from the disk, as shown in Figure D1. Consider an infinitesimal area d$\mathbf{S}$ of the moving disk, which has a total charge $\mathrm{d}q = \sigma\,\mathrm{d}S = \rho L\,\mathrm{d}S$ and is at a distance R from the axis of the disk, as shown in Figure D1. The distance from the element of charge dq to the field point P in Figure D1 is $\mathbf{r}$, where $r = (R^2 + x_0^2)^{1/2}$. The angle between $\mathbf{r}$ and $\mathbf{u}$ is θ. The expression for the curl of the magnetic field due to a classical point charge moving with uniform velocity $\mathbf{u}$ is given by equation (4.52). Applying equation (4.52) to the element of charge $\mathrm{d}q = \sigma\,\mathrm{d}S$ in Figure D1, which is at a distance R from the x axis, and neglecting terms in β^2 we find that

$$
\begin{aligned}
(\nabla \times \mathbf{B})_x &= (\nabla \times \mathbf{B})_r \cos\theta - (\nabla \times \mathbf{B})_\theta \sin\theta \\
&= \frac{u\,\mathrm{d}q}{4\pi\varepsilon_0 c^2}\left[\frac{2\cos^2\theta}{r^3} - \frac{\sin^2\theta}{r^3}\right] = \frac{u(2x_0^2 - R^2)\,\mathrm{d}q}{4\pi\varepsilon_0 c^2 (R^2 + x_0^2)^{5/2}}. \qquad \text{(D2)}
\end{aligned}
$$

Now consider a circular hoop shaped element of area of the disk of internal radius R and outer radius $(R + \mathrm{d}R)$. The total charge on this element of area is $2\pi R\sigma\,\mathrm{d}R$, where $\sigma = \rho L$. Putting $\mathrm{d}q = 2\pi R\sigma\,\mathrm{d}R$, in equation (D2) to obtain the contribution of the hoop to the x component of $\nabla \times \mathbf{B}$ at the field point P in Figure D1, and then integrating from $R = 0$ to $R = a$ and using $\sigma = \rho L$ we find that

$$
\begin{aligned}
(\nabla \times \mathbf{B})_x &= \frac{\sigma u}{2\varepsilon_0 c^2}\int_0^a \frac{(2x_0^2 R - R^3)}{(R^2 + x_0^2)^{5/2}}\,\mathrm{d}R \\
&= \frac{\sigma u}{2\varepsilon_0 c^2}\left[\frac{R^2}{(R^2 + x_0^2)^{3/2}}\right]_0^a = \frac{\rho L u a^2}{2\varepsilon_0 c^2 (a^2 + x_0^2)^{3/2}}. \qquad \text{(D3)}
\end{aligned}
$$

Equation (D3) can be interpreted qualitatively as follows. The magnetic field $\mathbf{B}$ due to the moving disk is zero on the axis of the disk, but $\mathbf{B}$ is finite just off the axis and is in a direction given by the right-handed corkscrew rule, which is in the same direction on either side of the moving disk. Hence $(\nabla \times \mathbf{B})_x$ is finite and in the $+x$ direction on both sides of the moving disk and according to equation (D3) has the same value for $\pm x_0$. When $x_0 = 0$,

$$
(\nabla \times \mathbf{B})_x = \frac{\rho L u}{2\varepsilon_0 a c^2}. \qquad \text{(D4)}
$$

Assume, for the moment, that the moving disk has zero thickness. As the moving charged disk in Figure D1 approaches the field point P from the left, the electric field at P is in the $+x$ direction and is increasing in magnitude so that $\dot{\mathbf{E}}$ is positive and equation (D1) is valid. However, when the disk passes the field point P the electric field reverses direction, changing from $\sigma/2\varepsilon_0$ in the $+x$ direction to $-\sigma/2\varepsilon_0$ in the $-x$ direction. However the direction of the magnetic field $\mathbf{B}$ is unchanged and the value of $(\nabla \times \mathbf{B})_x$ is the same and given by equation (D4), showing that there is no discontinuity in $(\nabla \times \mathbf{B})_x$ associated with the discontinuity of $-\sigma/\varepsilon_0$ in the electric field $\mathbf{E}$, when the disk of zero thickness passes the field point P in Figure D1. This shows that equation (D1) must be modified when the moving charged disk of zero thickness is passing the field point. To determine the modification of equation (D1) that is required to give consistency, we shall now assume that the moving disk is of finite thickness L, where $L \ll a$, as shown in Figure D2.

As the moving charged disk of finite thickness approaches the field point P from the left in Figure D2, the electric field at P is in the $+x$ direction and is increasing in magnitude so that $\dot{\mathbf{E}}$ is positive and equation (D1) is valid. When the disk starts passing the field point P in Figure D2, the contribution of the charge to the right of the field point P to the electric field at the field point at P is in the $-x$ direction and increases in magnitude as more and more charge passes the field point P, so that $\dot{E}_x/c^2$ is now negative and is not equal to $(\nabla \times \mathbf{B})_x$, which is still positive and given by equation (D4).

The change in the electric field at the field point P in a time $\mathrm{d}t$, when the disk is passing the field point as shown in Figure D2, is the same as if we remove a slice of thickness $\mathrm{d}L = u\,\mathrm{d}t$ from the left hand side and add it to the right hand side of the moving disk at a fixed time. Removing the slice from the left hand side decreases the electric field at the point P, so that the change $\mathrm{d}E_1$ in the electric field in the $+x$ direction, which is given by the standard formula for the electric field on the axis of a charged disk, which is derived in most elementary text books, is

$$\mathrm{d}E_1 = -\frac{\rho\,\mathrm{d}L}{2\varepsilon_0}(1 - \cos\theta_1) \tag{D5}$$

where $\rho\,\mathrm{d}L$ is the charge per unit area on the slice of thickness $\mathrm{d}L$. Adding a slice of thickness $\mathrm{d}L = u\,\mathrm{d}t$ to the right hand side of the moving disk in Figure D2 also gives an electric field in the $-x$ direction at field point P. The change $\mathrm{d}E_2$ in the electric field in the $+x$ direction is

$$\mathrm{d}E_2 = -\frac{\rho\,\mathrm{d}L}{2\varepsilon_0}(1 - \cos\theta_2). \tag{D6}$$

Adding equations (D5) and (D6), we have

$$\mathrm{d}E_x = \mathrm{d}E_1 + \mathrm{d}E_2 = -\frac{\rho\,\mathrm{d}L}{2\varepsilon_0}(2 - \cos\theta_1 - \cos\theta_2)$$

where $\mathrm{d}E_x$ is the total increase in the electric field at P in the $+x$ direction. Putting $\mathrm{d}L = u\,\mathrm{d}t$, then dividing by $c^2\,\mathrm{d}t$ and rearranging we have

$$\frac{1}{c^2}\frac{\partial E_x}{\partial t} = \frac{\rho u}{2\varepsilon_0 c^2}(\cos\theta_1 + \cos\theta_2 - 2). \tag{D7}$$

If the field point P is at a distance l from the left hand side of the moving disk in Figure D2, since $L \ll a$ we have

$$\cos\theta_1 = \frac{l}{(a^2 + l^2)^{1/2}} \approx \frac{l}{a}; \quad \cos\theta_2 = \frac{(L - l)}{(a^2 + (L - l)^2)^{1/2}} \approx \frac{(L - l)}{a}.$$

Hence when $L \ll a$

$$\cos\theta_1 + \cos\theta_2 \approx \frac{L}{a}.$$

Substituting in equation (D7), we have for the field point P

$$\frac{1}{c^2}\frac{\partial E_x}{\partial t} = \frac{\rho u}{2\varepsilon_0 c^2}\left(\frac{L}{a} - 2\right) = \frac{\rho u L}{2\varepsilon_0 a c^2} - \frac{\rho u}{\varepsilon_0 c^2}. \tag{D8}$$

According to Equation (D4) when $x_0 = 0$ we have for the field point P

$$(\nabla \times \mathbf{B})_x = \frac{\rho u L}{2\varepsilon_0 a c^2}. \tag{D4}$$

Comparing equations (D8) and (D4) we see that for consistency, we must have

$$(\nabla \times \mathbf{B})_x = \frac{1}{c^2}\frac{\partial E_x}{\partial t} + \frac{\rho u}{\varepsilon_0 c^2} = \mu_0\left(\varepsilon_0\frac{\partial E_x}{\partial t} + J\right). \tag{D9}$$

This is the x component of equation (4.94).

It is left as an exercise for the reader to consider a field point inside a moving continuous charge distribution. Divide the charge distribution into two parts. Apply equation (D9) to a moving disk shaped volume at the field point and equation (D1) to the rest of the moving charge distribution, which has a circular disk shaped hole at the field point. Adding the two contributions leads to equation (4.94) The moving charge distribution can then be treated as a classical point charge so that equation (D9) leads to the Maxwell-Lorentz equation (4.95).

Reference

1. Rosser, W. G. V., *Amer. Journ. Phys.*, Vol. 43, p. 502 (1975).

APPENDIX E

The transformations of special relativity

Consider the conditions shown in Figures 10.1(a) and 10.1(b). The inertial frame Σ' is moving with uniform velocity v relative to the inertial frame Σ along their common x axis. The origins of Σ and Σ' coincide at the times $t = t' = 0$ when the y and y' axes and the z and z' axes coincide as shown in Figure 10.1(a). Assuming the validity of the principle of relativity and the principle of the constancy of the speed of light, we showed in Section 10.4 of Chapter 10 that, if an event is measured to be at the position (x, y, z) at the time t in Σ and the same event is measured to be at the position (x', y', z') at the time t' in Σ' then, according to the Lorentz transformations,

$$x' = \gamma(x - vt); \qquad x = \gamma(x' + vt') \tag{E1}$$

$$y' = y; \qquad y = y' \tag{E2}$$

$$z' = z; \qquad z = z' \tag{E3}$$

$$t' = \gamma(t - vx/c^2); \qquad t = \gamma(t' + vx'/c^2) \tag{E4}$$

where

$$\gamma = \frac{1}{(1 - v^2/c^2)^{1/2}}. \tag{E5}$$

The inverse transformations can always be obtained by interchanging primed and unprimed quantities and replacing v by $-v$. We shall assume throughout this appendix that all the transformations refer to an event that is measured to be at (x, y, z, t) in Σ and at (x', y', z', t') in Σ'. If a particle has a velocity $\mathbf{u}$ having components (u_x, u_y, u_z) in Σ and a velocity $\mathbf{u}'$ having components (u'_x, u'_y, u'_z) at the same event at (x', y', z', t') in Σ' then

$$u'_x = \frac{\mathrm{d}x'}{\mathrm{d}t'} = \frac{u_x - v}{1 - vu_x/c^2} \tag{E6}$$

$$u'_y = \frac{\mathrm{d}y'}{\mathrm{d}t'} = \frac{u_y}{\gamma(1 - vu_x/c^2)} \tag{E7}$$

$$u'_z = \frac{dz'}{dt'} = \frac{u_z}{\gamma(1 - vu_x/c^2)}. \tag{E8}$$

It can be shown by direct substitution that

$$(1 - u'^2/c^2)^{1/2} = \frac{(1 - v^2/c^2)^{1/2}(1 - u^2/c^2)^{1/2}}{(1 - vu_x/c^2)}. \tag{E9}$$

If a particle has rest mass m_0, its relativistic momentum $\mathbf{p}$ and total energy E are

$$\mathbf{p} = \frac{m_0\mathbf{u}}{(1 - u^2/c^2)^{1/2}}, \qquad E = \frac{m_0c^2}{(1 - u^2/c^2)^{1/2}}.$$

It can be shown that the transformations for the momentum $\mathbf{p}$ and the total energy E are

$$p'_x = \gamma(p_x - vE/c^2) \tag{E10}$$

$$p'_y = p_y \tag{E11}$$

$$p'_z = p_z \tag{E12}$$

$$E' = \gamma(E - vp_x). \tag{E13}$$

The transformations for force are

$$f'_x = \frac{dp'_x}{dt'} = f_x - \frac{vu_y}{c^2(1 - vu_x/c^2)}f_y - \frac{vu_z}{c^2(1 - vu_x/c^2)}f_z. \tag{E14}$$

$$f'_y = \frac{dp'_y}{dt'} = \frac{f_y}{\gamma(1 - vu_x/c^2)} \tag{E15}$$

$$f'_z = \frac{dp'_z}{dt'} = \frac{f_z}{\gamma(1 - vu_x/c^2)}. \tag{E16}$$

Full accounts of the derivations and applications of these transformations are given by Rosser [1].

Reference

1. Rosser, W. G. V., *Introductory Special Relativity* (Taylor and Francis, London, 1991).

Index